New Trends in Nonlinear Dynamics and Pattern-Forming Phenomena

The Geometry of Nonequilibrium

NATO ASI Series

Advanced Science Institutes Series

A series presenting the results of activities sponsored by the NATO Science Committee, which aims at the dissemination of advanced scientific and technological knowledge, with a view to strengthening links between scientific communities.

The series is published by an international board of publishers in conjunction with the NATO Scientific Affairs Division

A	**Life Sciences**	Plenum Publishing Corporation
B	**Physics**	New York and London
C	**Mathematical and Physical Sciences**	Kluwer Academic Publishers Dordrecht, Boston, and London
D	**Behavioral and Social Sciences**	
E	**Applied Sciences**	
F	**Computer and Systems Sciences**	Springer-Verlag
G	**Ecological Sciences**	Berlin, Heidelberg, New York, London,
H	**Cell Biology**	Paris, and Tokyo

Recent Volumes in this Series

Volume 236—Microscopic Simulations of Complex Flows
edited by Michel Mareschal

Volume 237—New Trends in Nonlinear Dynamics and Pattern-Forming Phenomena: The Geometry of Nonequilibrium
edited by Pierre Coullet and Patrick Huerre

Volume 238—Physics, Geometry, and Topology
edited by H. C. Lee

Volume 239—Kinetics of Ordering and Growth at Surfaces
edited by Max G. Lagally

Volume 240—Global Climate and Ecosystem Change
edited by Gordon J. MacDonald and Luigi Sertorio

Volume 241—Applied Laser Spectroscopy
edited by Wolfgang Demtröder and Massimo Inguscio

Volume 242—Light, Lasers, and Synchrotron Radiation: A Health Risk Assessment
edited by M. Grandolfo, A. Rindi, and D. H. Sliney

Volume 243—Davydov's Soliton Revisited: Self-Trapping of Vibrational Energy in Protein
edited by Peter L. Christiansen and Alwyn C. Scott

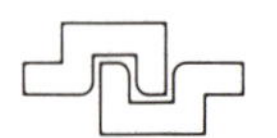

Series B: Physics

New Trends in Nonlinear Dynamics and Pattern-Forming Phenomena

The Geometry of Nonequilibrium

Edited by

Pierre Coullet

Université de Nice
Nice, France

and

Patrick Huerre

Ecole Polytechnique
Palaiseau, France

Plenum Press
New York and London
Published in cooperation with NATO Scientific Affairs Division

Proceedings of a NATO Advanced Research Workshop on
New Trends in Nonlinear Dynamics and Pattern-Forming
Phenomena: The Geometry of Nonequilibrium,
held August 2-12, 1988,
in Cargèse, France

Library of Congress Cataloging-in-Publication Data

NATO Advanced Research Workshop on New Trends in Nonlinear Dynamics
and Pattern-Forming Phenomena (1988 : Cargèse, France)
New trends in nonlinear dynamics and pattern-forming phenomena :
the geometry of nonequilibrium / edited by Pierre Coullet and
Patrick Huerre.
p. cm. -- (NATO ASI series. B, Physics ; v. 237)
"Proceedings of a NATO Advanced Research Workshop on New Trends in
Nonlinear Dynamics and Pattern-Forming Phenomena ... held August
2-12, 1988, in Cargèse, France"--T.p. verso.
"Published in cooperation with NATO Scientific Affairs Division."
Includes bibliographical references and index.
ISBN 0-306-43692-2
1. Nonlinear theories--Congresses. 2. Dynamics--Congresses.
3. Mathematical physics--Congresses. I. Coullet, Pierre.
II. Huerre, Patrick. III. North Atlantic Treaty Organization.
Scientific Affairs Division. IV. Title. V. Series.
QC20.7.N6N27 1988
530.1'55--dc20 90-48713
CIP

Printed in the United States of America

PREFACE

The basic aim of the NATO Advanced Research Workshop on "New Trends in Nonlinear Dynamics and Pattern-Forming Phenomena: The Geometry of Nonequilibrium" was to bring together researchers from various areas of physics to review and explore new ideas regarding the organisation of systems driven far from equilibrium. Such systems are characterized by a close relationship between broken spatial and temporal symmetries. The main topics of interest included pattern formation in chemical systems, materials and convection, traveling waves in binary fluids and liquid crystals, defects and their role in the disorganisation of structures, spatio-temporal intermittency, instabilities and large-scale vortices in open flows, the mathematics of non-equilibrium systems, turbulence, and last but not least growth phenomena. Written contributions from participants have been grouped into chapters addressing these different areas. For additional clarity, the first chapter on pattern formation has been subdivided into sections.

One of the main concerns was to focus on the unifying features between these diverse topics. The various scientific communities represented were encouraged to discuss and compare their approach so as to mutually benefit their respective fields. We hope that, to a large degree, these goals have been met and we thank all the participants for their efforts.

The workshop was held in Cargèse (Corsica, France) at the Institut d'Etudes Scientifiques from August 2nd to August 12th, 1988. We greatly thank Yves Pomeau and Daniel Walgraef who, as members of the organising committee, gave us valuable advice and encouragements. We are extremely grateful to Joceline Lega for her crucial role in the organization of the workshop and to Katy Arthaud for her technical and secretarial assistance. The support of Marie-France Hanseler at the Institut d'Etudes Scientifiques is gratefully acknowledged.

The meeting was sponsored by the NATO International Scientific Exchange Programme. Additional funds were provided by Commissariat à l'Energie Atomique, Centre National de la Recherche Scientifique, Direction des Recherches, Etudes et Techniques, and by the European Community.

Pierre Coullet
Patrick Huerre

CONTENTS

PATTERN FORMATION

Chemical Waves

Chemical Waves in Oscillatory Media 1
C. Vidal

Pattern Dynamics by Interaction of Chemical Waves and Hydrodynamic Flows. 11
S.C. Müller and H. Miike

Sustained Non-Equilibrium Patterns in a One-Dimensional Reaction-Diffusion Chemical System. 21
J. Elezgaray and A. Arneodo

Materials Instabilities

Pattern Selection and Symmetry Competition in Materials Instabilities 25
D. Walgraef

Numerical Simulation of Persistent Slip Band Formation. 33
C. Schiller and D. Walgraef

Rayleigh-Benard Convection

Transitions to More Complex Patterns in Thermal Convection. 37
F.H. Busse and R.M. Clever

Rayleigh-Benard Convection at Finite Rayleigh Number in Large Aspect Ratio Boxes. 47
T. Passot, M. Souli and A.C. Newell

Transition Between Different Symmetries in Rayleigh-Benard Convection. 51
S. Ciliberto, E. Pampaloni and C. Pérez-Garcia

Traveling Waves

Nonlinear Analysis of Traveling Waves in Binary Convection. 55
D. Bensimon and A. Pumir

Temporal Modulation of a Subcritical Bifurcation to Travelling Waves. 61
H. Riecke, J.D. Crawford and E. Knobloch

Experiments with Travelling Waves in Electrohydrodynamic Convection. 65
M. de la Torre Juarez and I. Rehberg

The Degenerate Amplitude Equation Near the Codimension-2 Point in Binary Fluid Convection. 69
W. Zimmermann and W. Schöpf

Travelling Waves in Axisymetric Convection. 73
L. Tuckerman and D. Barkley

About Confined States, Noise-Sustained Structures and Slugs 77
H.R. Brand

Localized Structures Generated by Subcritical Instabilities: Counterpropagating Waves. 85
O. Thual and S. Fauve

Other Areas of Interest

From the Chaos to Quasiregular Patterns 89
A.A. Chernikov, R.Z. Sagdeev and G.M. Zaslavsky

Interfacial Instabilities and Waves in the Presence of a Transverse Electric Current 95
R. Moreau, S.L. Pigny and S.A. Maslowe

Modulated Taylor Vortex Flow. 103
M. Lücke, D. Roth and H. Kuhlmann

Recent Results on the Non-Linear Dynamics of Curved Premixed Flames 107
G. Joulin

DEFECTS AND SPATIO-TEMPORAL COMPLEXITY

Patterns and Defects in Liquid Crystals 111
E. Bodenschatz, M. Kaiser, L. Kramer, W. Pesch, A. Weber and W. Zimmermann

Defects and Transition to Disorder in Space-Time Patterns of Non-Linear Waves. 125
A. Joets and R. Ribotta

Defect-Mediated Turbulence in Spatio-Temporal Patterns. 137
J. Lega

Mean Flows and the Onset of Time-Dependence in Convection 145
A. Pocheau

Experiment on Pattern Evolution in the 2-D Mixing Layer 159
F.K. Browand and S. Prost-Domasky

A Two Dimensional Model of Pattern Evolution in Mixing Layers. 171
R. Yang, P. Huerre and P. Coullet

Some Statistical Properties of Defects in Complex Ginzburg-Landau Equations 181
J-L. Meunier

Vortex Dynamics in a Coupled Map Lattice. 185
T. Bohr, A.W. Pedersen, M.H. Jensen and D.A. Rand

SPATIO-TEMPORAL INTERMITTENCY

Spatio Temporal Intermittency in Rayleigh-Benard Convection in an Annulus. 193
S. Ciliberto

Transition to Turbulence Via Spatiotemporal Intermittency: Modeling and Critical Properties. 199
H. Chaté and P. Manneville

Phase Turbulence, Spatiotemporal Intermittency and Coherent Structures 215
H. Chaté and B. Nicolaenko

Properties of Quasi One-Dimensional Rayleigh Benard Convection. 227
M. Dubois, P. Bergé and A. Petrov

Probabilistic Cellular Automaton Models for a Fluid Experiment. 237
R. Livi and S. Ruffo

Space-Time Chaos and Coherent Structures in the Coupled Map Lattices. 241
L.A. Bunimovich

SPATIALLY-DEVELOPING NON-EQUILIBRIUM SYSTEMS

Topological Defects in Vortex Streets Behind Tapered Circular Cylinders at Low Reynolds Numbers 243
C.W. Van Atta and P. Piccirillo

Large-Eddy Simulation of a Three-Dimensional Mixing Layer. 251
P. Comte, M. Lesieur and Y. Fouillet

The Effect of Nonlinearity and Forcing on Global Modes. 259
J.M. Chomaz, P. Huerre and L.G. Redekopp

MATHEMATICS OF NON EQUILIBRIUM

Mathematical Justification of Steady Ginzburg-Landau Equation Starting from Navier-Stokes. 275
G. Iooss, A. Mielke and Y. Demay

Non Linear Spatial Analysis: Application to the Study of Flows Subjected to Thermocapillary Effects. 287
P. Laure, H. Ben Hadid and B. Roux

Discretizations and Jacobian in Functional Integrals. 291
E. Tirapegui

Homoclinic Bifurcations in Ordinary and Partial Differential Equations. 295
A.C. Fowler

TURBULENCE

Large-Scale Vortex Instability in Helical Convective Turbulence. 305
A.V. Tur, S.S. Moiseev, P.B. Rutkevich and V.V. Yanovsky

Generation of Large Scale Structures in Three-Dimensional Anisotropic Incompressible Flow Lacking Parity-Invariance. 313
P-L. Sulem, U. Frisch, H. Scholl and Z.S. She

A New Universal Scaling for Fully Developed Turbulence: The Distribution of Velocity Increments 315
Y. Gagne, E.J. Hopfinger and U. Frisch

Vortex Dynamics and Singularities in the 3-D Euler Equations 321
A. Pumir and E.D. Siggia

GROWTH PHENOMENA

Directional Growth in Viscous Fingering 327
V. Hakim, M. Rabaud, H. Thomé and Y. Couder

New Results in Dendritic Crystal Growth 339
J. Maurer, P. Bouissou, B. Perrin and P. Tabeling

Microstructural Transitions During Directional Solidification of a Binary Alloy. 343
B. Billia, H. Jamgotchian and R. Trivedi

Nonstationary Cell Shapes in Directional Solidification 347
S. de Cheveigné, C. Guthmann and P. Kurowski

INDEX 351

CHEMICAL WAVES IN OSCILLATORY MEDIA

Christian Vidal

Centre de Recherche Paul Pascal (CNRS)
Université de Bordeaux I
33405 Talence, Cédex (France)

INTRODUCTION

Chemical waves are a striking phenomenon that occurs in some reacting systems evolving far from equilibrium. They consist in concentration differences which propagate through the medium at the expense of the chemical free energy content of the reactants. Thus, in contrast to the other kinds of wave (e.g. acoustic or electromagnetic, etc.), they do not carry energy. Owing to this, they exhibit several peculiar properties, such as for instance: conservation of amplitude and waveform while propagating in a dispersive medium, annihilation in head-on collisions, lack of reflection and interference. These qualitative features make them so much distinguishable that one may wonder whether the word "wave" is still appropriate ; another word, like "autowave" used in the russian literature, would emphasize how specific are these chemical propagation phenomena.

The earliest report of a chemical wave is presumably that of Luther at the beginning of this century[1]. The scientific study of chemical waves really began about two decades ago. It has reached a quantitative experimental level only during the past few years, when high precision digital image-processing devices became commonly available. The subject of chemical waves and patterns having been already reviewed several times, the reader interested in its history is refered to previous articles[2]. Thisone is rather intended to report some results obtained recently , while studying oscillatory Belousov-Zhabotinsky (BZ) media. Three issues are hereby addressed: the role of convection, the dispersion relation, and the mechanism of wave birth.

SHORT OVERVIEW ON CHEMICAL WAVES

In order to make the presentation as clear as possible, the definition of a few words needs to be recalled. First of all, a chemical wave is a reaction-diffusion process, taking place in an "active" chemical medium. By active, we mean: bistable (including metastable), excitable, or oscillatory. The phenomenon may be described as a (space and time dependent) solution of the reaction-diffusion equation:

New Trends in Nonlinear Dynamics and Pattern-Forming Phenomena
Edited by P. Coullet and P. Huerre
Plenum Press, New York, 1990

$$\frac{\partial X}{\partial t} = F(X) + D\,\nabla^2 X \qquad [1]$$

X : concentration vector (state variables)
D : matrix of diffusion coefficients
F(X) : concentration variations due to the chemical reaction.

To get sustained waves, one must keep the system out of equilibrium ; otherwise a uniform distribution would eventually be reached. Nothing is easier to do in theory: it suffices, for instance, to ascribe fixed values to at least one reactant concentration in F(X). The experimental task is much more difficult to achieve. Indeed, the feed of fresh reagent usually involves an undesirable convective flow. However, recent attempts are rather promising [3], and this problem seems to be on the way of being overcome soon. In the remaining we will only deal, of course, with the transient waves observed so far.

Now, equation [1] is known to have solutions of a special kind when the "homogeneous" (i.e. space independent) equation:

$$\frac{\partial X}{\partial t} = F(X)$$

has a solution which is a periodic function of time (limit cycle). Two limit situations have been recognized in such a case, when the reacting medium is oscillatory. In the diffusion-free limit (formally D may be set equal to zero in [1]), a spatial variation in the phase or in the frequency of the oscillation results in wave-like phenomena. Due to the gradient, concentrations do not change everywhere in the medium at the same time. Accordingly, one gets the impression that something propagates from place to place, whereas nothing moves or diffuses at all. The so-called kinematic waves[4] and pseudo-waves[5] are merely an optical illusion, entirely depending upon the initial conditions. They display striking properties. Their apparent "propagation velocity" may be very large and, in fact, has no upper limit, since it is inversely proportional to the gradient amplitude. These "waves" are not stopped by impermeable barriers, simply because they do not involve any diffusion of any kind. They may be considered as a chemical analogue of those electric signs whose lights are successively switched on an off. In other words: a wonderful artefact, not an exciting scientific problem.

If a local region of the medium oscillates faster than its surroundings, then concentration gradients develop ; as time proceeds, diffusion unavoidably enters into play. When appropriate conditions are fulfilled (weak diffusion limit), equation [1] may be cast in a form describing the subsequent evolution as a diffusion of the phase of the oscillation. This phenomenon was first predicted by Ortoleva and Ross[6], and its theory revisited later by Hagan[7]. The onset of "phase diffusion waves" (this term should be prefered to that of "phase waves", also used frequently in the literature to name the above-mentioned kinematic and pseudo-waves) requires however very specific conditions. For that matter, they are expected to be of seldom occurence ; in effect, they were not observed until quite recently[8]. Compared to "trigger waves" (see below), phase diffusion waves have singular properties: smoother concentration gradients, higher velocities, finite distance range of constant velocity, which are relevant criteria for their identification.

Most of the time, chemical oscillations exhibit a strong relaxation character, the reaction dynamics taking place over multiple time scales. As they do, also, in other active media, waves sometimes appear either spontaneously, or by external perturbation. It is the diffusion of certain species across their own concentration gradient that triggers the reaction in neighboring volume elements. In an oscillatory medium, the previous approximations no longer apply. These "trigger waves" were first seen about twenty years ago in a BZ reagent. Since then, several chemical systems (active, but not necessarily oscillatory) have been discovered, that give rise to such waves: e.g. Fe^{2+}/HNO_3, NH_2OH/HNO_3, $ClO_2^-/S_2O_3^{--}$, ClO_2^-/I^-, IO_3^-/H_3AsO_3, BrO_3^-/ferroin, etc. Nevertheless, a large majority of the available experimental data and theoretical studies deals with the ferroin-catalyzed BZ system. Consequently, this particular case, by far the best known until now, will be the only one hereafter discussed.

Going back to equation [1], it must be recalled that mathematicians[9] have shown that there exist several types of asymptotically stable waveform solutions of it (in an isotropic and infinite medium, assumed to be kept at a constant distance from its equilibrium composition). Three types of planar traveling waves may be qualitatively distinguished according to the shape of their concentration profile. For sake of clarity, a piecewise linear sketch is drawn in Table 1 to emphasize the differences. On the other hand, rotating waves lead, of course, to wavetrains in any direction going through their origin. When the physical space is two-dimensional (or is considered as such, a common approximation in "thin" layer experiments), each kind of solution gives birth to a typical pattern, named after its appearance (see Table 1). All these patterns have actually been observed, at least in one chemical medium.

Table 1. Solutions of equation [1] and and associated 2D patterns

Stable asymptotic solutions	2D patterns
* Traveling waves	
front	disc
pulse	ring
train	target
* Rotating waves	spiral

When dealing with a blue BZ oxidizing wavefront propagating through a red reduced background, the typical orders of magnitude recorded under standard laboratory conditions are the following:

steepest ferroin concentration gradient $\sim 10^{-2}$ M.mm^{-1}

wavefront width $\sim 10^{-1}$ mm

wave velocity $\sim 10^{-1}$ mm.s^{-1}.

In excitable reagents, the asymptotic velocity has been shown to change with the sulfuric acid and bromate concentrations, according to a square-root relationship. In contrast, the sensitiviness of this property to ferroïn and to malonic acid concentrations is very tiny. Over the limited range of temperature explored so far (284-318 K), the velocity also varies according to an Arrhenius law whose apparent activation energy is close to 35 $kJ.M^{-1}$, [10].

CONVECTION*

For years, people generally agreed that convection did not occur, or at least could be neglected in experiments carried out in thin layers (i.e. whose depth is about 1 mm). Yet, indications were reported[10] that convection might well take place when the upper surface of the reacting layer is left "free", that means in contact with air. Convection was also conjectured several times to have something to do with the so-called "mosaic structures", observed in BZ or BZ-like reagents[14,15,16].

This year, two experiments provided direct evidence of the very occurence of convection, even in thin layers. Quantitative data on the velocity field of the hydrodynamic flow were gathered for the first time. One experiment was performed in Dortmund, in an excitable BZ solution, the layer having a fixed depth of 0.85 ± 0.05 mm, [17]. The other one, done in Bordeaux, involved an oscillatory BZ medium; the layer was sandwiched between two rigid plates set apart at three different distances: 1.5, 2.5 and 3.5 mm, [18]. Both experiments are based on a similar technique. The hydrodynamic flow is traced with very small particles (spheres of polystyrene Ø = 0.5 μm, or of TiO_2 Ø = 2 μm), whose motion is observed with a microscope. A video imaging system, connected to a digital image-processing device, allows to measure the velocity of the particles (in fact, its horizontal component) with a precision of about ± 1 $\mu m.s^{-1}$.

The german group noticed a striking difference between two situations, depending upon whether the container (Petri dish) is left covered or uncovered. In the first case, the solution stays always at (hydrodynamic) rest. In contrast to this, a translational motion is observed when operating with an uncovered dish. There, convection sets in with a rather small velocity (a few $\mu m.s^{-1}$); furthermore, if a train of trigger waves propagates through the solution, an oscillatory hydrodynamic flow appears after a while. Its direction alternates every two wavefronts, and its velocity is one order of magnitude higher, i.e. comparable to that of the waves. This finding points out unambiguously the close connections between convection and wave propagation under certain experimental conditions.

---

* Convection-reaction patterns were reported more than ten years ago by Möckel[11] in a photochemically reacting system. It was proved later by Micheau et al[12] and by Avnir et al[13] that a physical rather than a chemical instability is at the origin of the onset of these structures. More precisely, convection sets in first, due either to evaporative cooling or to a double-diffusion process. An only then, a colored chemical reveals the non uniformity in composition generated by the hydrodynamic flow. These convection-reaction structures are beyond the scope of the topic discussed in this chapter.

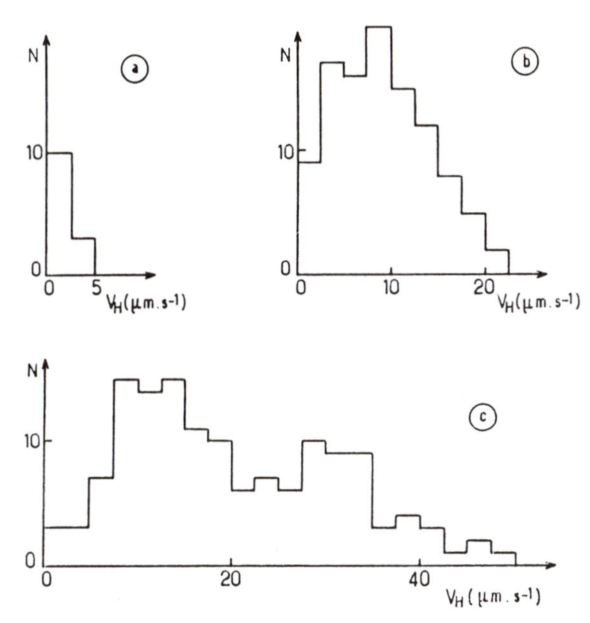

Fig. 1. Histograms of the horizontal component V_H of the particle velocities for three different layer depts (a): 1.5 mm ; (b): 2.5 mm ; (c): 3.5 mm

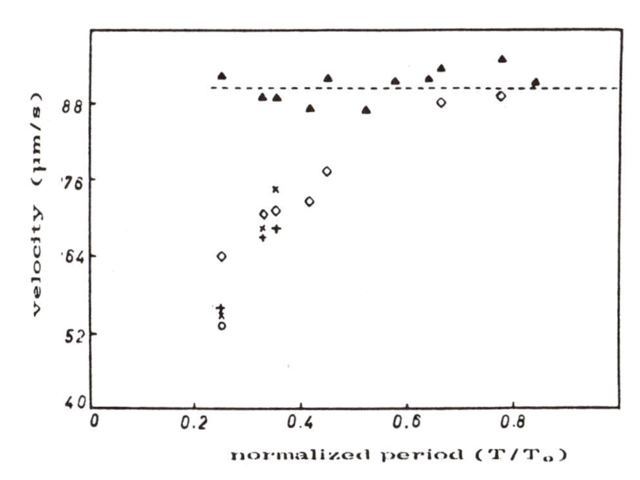

Fig. 2.

(a) Dispersion relation
- Δ first front
- ◇ second front
- + third front
- X fourth front
- o fifth front

The dashed line corresponds to the propagation velocity in the unperturbed medium. T_0 is the oscillation period.

(b) Comparison with the calculated dispersion relation (curve) for two different sets of kinetic constants.

In the Bordeaux experiment, inhomogeneities in surface tension were avoided by the "sandwich" technique. Only one control parameter was used: the layer depth. A first result is the very existence of convective motions in all cases, whereas nor evaporative cooling, nor Marangoni instability can be suspected to play a role. As shown in figure 1, the number of moving particles and their velocities strongly increase with the layer depth. A second point is the absence of connection between convection and the passage of trigger waves. More precisely, no significant change in the motion (direction and/or velocity) of a particle was noticed whenever a trigger wave passed or crossed a TiO_2 sphere. The explanation of the difference with the observations made in Dortmund presumably lies in the differences in the experimental conditions. The above-mentioned oscillatory flow is restricted to the portion of the layer close to the surface of the solution, whereas, in Bordeaux, we rather detect the motion of the "inner" part of the sandwiched layer.

Both experiments clearly emphasize the cares that must be taken to prevent the onset of convection, as long as one is interested in diffusion-reaction phenomena only. Convection may well develop in thin reacting layers, even without a free-surface condition.

DISPERSION RELATION

The theory of wave propagation in dispersive media tells us that the velocity depends upon the wavelength. This conclusion should also hold for chemical waves, despite their peculiar character. Accordingly, when dealing with a periodic pattern (such as a target or a spiral), the radial velocity of wavefronts should vary with the wavelength of the pattern. Nevertheless, until very recently, such a dependence has not been clearly reported in the literature.

An experiment especially designed to clarify this issue has been undertaken at Stanford [19]. Its principle is quite simple: by means of an external perturbation (namely a laser beam) waves were triggered in an oscillatory BZ reagent, and the velocity of the emitted wavefronts measured. The period of wavefront emission is a function of the laser power. By performing several experiments at different laser powers, one can change the wavelength of the train. The overall result is displayed in figure 2a. The black triangles correspond to the velocity of the first wavefront, that which proceeds through an unperturbed medium. Whatever the emission period (scaled by T_0, the "bulk" oscillation period), this velocity keeps a constant value within the range of experimental uncertainties. The velocity of the subsequent wavefronts is always smaller; it decreases with the number of the emitted front. As expected, the slowing down becomes more and more pronounced as the emission period diminishes. Besides, the wide range of periods over which this slowing down remains very weak (in fact , between T_0 and $T_0/2$) might explain why dispersion effects have escape notice so far.

On the basis of a two-variable model deduced from the Oregonator, Keener and Tyson have derived an explicit expression of the dispersion relation [20]. Though their calculation applies to an excitable medium, it is tempting to try to go farther, and to compare their theoretical prediction to these measurements in an oscillatory BZ reagent. To this aim, the lowest recorded velocity at each emission period (fig. 2a) is taken for the asymptotic value. One must then convert the experimental data in dimensionless variables, since the curve representing the calculated dispersion relation is drawn in a dimensionless frame of reference (fig. 2b). Using the set of scaling factors taken by Keener and Tyson thus yields the "+" signs in figure 2b. Perhaps more relevantly, one can also choose the kinetic constants so as to match together the measured velocity in the unperturbed reagent (dashed line in figure 2a) and the horizontal upper part of the dispersion curve. And so come the circles in figure 2b. Although the agreement is this time astonishingly good, one must look at it as mainly qualitative. Indeed , a deeper justification of the scaling factors needs to be provided before drawing further conclusions.

ORIGIN OF TRIGGER WAVES

Rotating waves are usually generated artificially, for instance by disrupting a small section of an expanding circular wavefront. The two free ends thus created give birth to a pair of counter-rotating waves, leading to two one-armed spirals. The very central part of each spiral is a sort of "black hole", where the concentration variations vanish. The diameter of this region is about 30 μm, and perhaps even less[21]. Fairly general models accounting for this spiral wave formation, in both oscillatory[22] and excitable[23] media, are nowadays available.

The main issue is that of the mechanism leading to the formation of an expanding wavefront, within a medium at first glance homogeneous. There exist different means for triggering, at a certain point in an active reagent, a wave which will then proceed through the solution. These external sources (e.g. a hot wire) are, of course, not puzzling at all. But the question is: how does it happen that waves also emerge "spontaneously", without any identified perturbation ? It has been first suggested that a "catalytic particle" (e.g. a dust particle, a surface defect of the container), otherwise invisible, acts as an external pacemaker. Actually, several authors reported that trigger waves no longer appear in <u>excitable</u> media, provided the solutions are thoroughly filtered and the reagent poured into an ultraclean Petri dish. The efficiency of these experimental cares in eliminating extra sources of chemical activity is, accordingly, well established. Nevertheless, they remain unsuccessful when applied to <u>oscillatory</u> media[24]: centers leading to target formation persist to appear, whatever the cares taken to reduce or avoid contamination. From this difference, one may conclude that an oscillatory reagent is much more sensitive to singularities of its boundary conditions than an excitable one. Yet another explanation might be put forward, involving local fluctuations in the oscillatory properties of the medium. A detailed stochastic analysis of this problem was derived in Brussels a few years ago [25]. The crucial point is whether a fluctuation-nucleation mechanism may, or may not give birth to an "endogeneous" center, at least from time to time. Unfortunately, a direct experimental support, worthy of confidence, to this kind of theoretical predictions seems very difficult to get, if not out of reach. No matter how large would be the statistical sample of data, at least some of the experimental conditions will always remain questionable. Further progress presumably requires to take the opposite way, namely to seed the reacting solution with particles whose physical and chemical properties are known and controled; and then, to watch the effect of this "calibrated" contamination.

Awaiting for the results of such an extended experimentation - now in progress in Bordeaux -, one can also look carefully at the very center of target patterns with a microscope. In so doing, we were recently able to observe an "homogeneous" center, that has appeared spontaneously in a thin (0.65 mm) layer of an oscillatory BZ reagent[26]. In this context, the word "homogeneous" has the following somewhat limited meaning:

i) a center free of any visible particle, down to a size of approximately 5 μm (the spatial resolution at the microscope magnification used),

ii) a center located at mid-height of the layer, that is far from the upper and lower boundaries, and having a density equal to that of the liquid.

Condition ii) precludes a surface defect or a very small gas bubble to be the observed pacemaker. On the other hand, condition i) gives only an upper limit of the size of an heterogeneity which might still be involved. Alternatively, a fluctuation would be perfectly consistent with this observation. Although the experiment is not conclusive at all, it helps us limiting the matter of controversy: unknowable catalytic particle against fluctuation-nucleation process.

Further and more refined investigation is the only way to get a convincing answer to this question.

REFERENCES

1. R. Luther, Z. Elektrochem., 12:596 (1906)
2. R. J. Field and M. Burger, in: "Oscillations and traveling waves in chemical systems", Wiley, New York (1985)
   A. Pacault and C. Vidal, J. Chim. Phys., 79:691 (1982)
   C. Vidal and P. Hanusse, Int. Rev. Phys. Chem., 5:1 (1986)
3. W. Y. Tam, W. Horsthemke, Z. Noszticius and H. L. Swinney, J. Chem. Phys., 88:3395 (1988)
4. N. Kopell, and L. N. Howard, Science, 180:1171 (1973)
5. A. Winfree, Science, 175:634 (1972)
6. P. Ortoleva and J. Ross, J. Chem. Phys., 58:5673 (1973)
7. P. S. Hagan, Adv. Appl. Math., 2:400 (1981)
8. J. M. Bodet, J. Ross and C. Vidal, J. Chem. Phys., 86:4418 (1987)
9. P. C. Fife, in: "Mathematical aspects of reacting and diffusing systems", Springer-Verlag, Heidelberg (1979)
10. P. M. Wood and J. Ross, J. Chem. Phys., 82:1924 (1985)
    I. Kuhnert and H. J. Krug, J. Phys. Chem., 91:730 (1987)
11. P. Mockel, Naturwiss., 64:224 (1977)
12. J. C. Micheau, M. Gimenez, P. Borckmans and G. Dewel, Nature, 305:43 (1983)
13. D. Avnir and M. Kagan, Nature, 307:717 (1984)
14. A. M. Zhabotinsky and A. N. Zaikin, J. Theor. Biol., 40:45 (1973)
15. K. Showalter, J. Chem. Phys., 73:3735 (1980)
16. M. Orban, J. Am. Chem. Soc., 102:4311 (1980)
17. H. Miike, S. C. Muller and B. Hess, Chem. Phys. Lett., 144:515 (1988)
18. J. Rodriguez and C. Vidal, J. Phys. Chem., in press
19. A. Pagola, J. Ross and C. Vidal, J. Phys. Chem., 92:163 (1988)
20. P. Keener and J. Tyson, Physica D, 21:307 (1986)
21. S. C. Muller, T. Plesser and B. Hess, Physica D, 24:71 and 87 (1987)

22. P. S. Hagan, SIAM J. Appl. Math., 42:762 (1982)
23. P. C. Fife, in: "Non-equilibrium dynamics in chemical systems", C. Vidal and A. Pacault eds, Springer-Verlag, Heidelberg, p. 76 (1984)
E. Meron and P. Pelce, Phys. Rev. Lett., 60:1880 (1988)
24. C. Vidal, A. Pagola, J. M. Bodet, P. Hanusse and E. Bastardie, J. Phys., 47:1999 (1986)
25. D. Walgraef, G. Dewel and P. Borckmans, J. Chem. Phys., 78:3043 (1983)
26. A. Pagola and C. Vidal, J. Phys. Chem., in press.

PATTERN DYNAMICS BY INTERACTION OF CHEMICAL WAVES AND HYDRO-DYNAMIC FLOWS

Stefan C. Müller* and Hidetoshi Miike**

* Max-Planck-Institut für Ernährungsphysiologie
Rheinlanddamm 201, D-4600 Dortmund 1, FRG
** Department of Electrical Engineering, Yamaguchi University, Ube 755, Japan

INTRODUCTION

Chemical reactions evolving under far from equilibrium conditions may exhibit self-organization in time and in space, such as oscillations in homogeneous solution or waves, that is concentration gradients travelling through space (Field and Burger, 1985; Ross et al., 1988). One of the classical systems in which such nonlinear phenomena have been investigated in much detail is the Belousov-Zhabotinskii reaction having oscillatory reaction kinetics (Zhabotinskii, 1964; Field et al., 1972, Bornmann et al., 1973). When choosing an appropriate chemical composition, this reaction can be prepared such that it remains in an excitable state when a small volume is placed into a petri dish, resulting in a liquid layer of thickness of 1 mm or less. Then, a local stimulus (e.g. a hot wire or just a dust particle) produces an excited state which propagates into the outer regions of the medium as a circular wave of excitation. Breaking a wave front leads to the formation of rotating spiral-shaped waves (Winfree, 1972; Müller et al., 1985a). Both geometric forms appear in the experiment shown in Fig. 1. Several authors have treated these structures theoretically by modelling the nonlinear coupling of the complex kinetics of this reaction and diffusion (Tyson and Keener, 1988, and references therein; Zykov, 1988).

Since the discovery of chemical waves, a variety of additional structural phenomena have been observed, especially under the condition that the experiments are performed without covering the sample dish with a glass plate. Examples are stationary structures, the so-called "mosaic" patterns, transient patterns, e.g. "droplet"-like structures, or distortions and irregular decomposition of wave fronts (Zhabotinskii and Zaikin, 1973; Orban, 1980; Showalter, 1980; Agladze et al., 1984; Müller et al., 1985b, 1986; Kuhnert et al., 1988). Some of them are illustrated in Fig. 2. So far, most of these observations have been qualitative in nature. On this basis attempts have been made to find explanations by invoking convective flow effects in the layer caused by tem-

New Trends in Nonlinear Dynamics and Pattern-Forming Phenomena
Edited by P. Coullet and P. Huerre
Plenum Press, New York, 1990

Fig. 1. Complex wave pattern in a thin layer of the Belousov-Zhabotinskii reaction. Dish diameter: 6.8 cm.

perature gradients due to evaporative cooling of the layer surface, the exothermicity of the reaction and/or inhomogeneities in chemical composition. There are some experiments in which the formation of stationary patterns has been correlated with measurements of temperature gradients and the motion of accidentally present dust particles. Frequently, qualitative comparison is made with analogous patterns in hydrodynamically unstable simple liquids (Chandrasekhar, 1961).

While the spatial characteristics of the reaction-diffusion patterns (circles and spirals) have been measured with high precision by using computerized spectrophotometric techniques (Wood and Ross, 1985; Müller et al., 1986; Pagola et

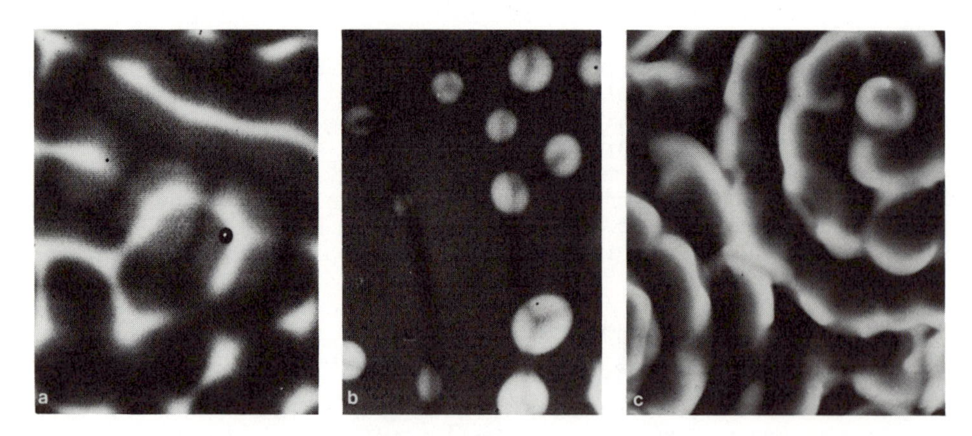

Fig. 2. Spatial patterns observed in a thin layer of the Belousov-Zhabotinskii reaction with an open liquid/gas interface. (a) Stationary mosaic pattern evolving independently of wave propagation. (b) Transient droplet-pattern forming along the boundaries of convection cells during a bulk oscillation. (c) Distortion and decomposition of chemical waves in a spiral pattern.

al., 1988), a quantitative approach to the effects of hydrodynamic flows on such patterns has been reported only recently. Miike et al. (1988a,b) presented direct evidence of hydrodynamic patterns generated in excitable media and performed detailed measurements of flow velocities. Here, we give a brief outline of how such flows are recorded quantitatively using computerized microscope video imaging techniques. As the main point in our experimental approach, a detailed correlation between chemical waves and flow patterns can be established by combining space-resolved spectrophotometric methods with space-resolved velocimetry.

MATERIALS AND METHODS

A quiescent but excitable solution of the BZ reaction was obtained by preparing a mixture of 48 mM sodium bromide, 340 mM sodium bromate, 95 mM malonic acid, and 378 mM sulfuric acid. About 5 min after mixing, the catalyst and indicator ferroin (3.5 mM) was added. All solutions were filtered with Millipore filter (0.22 μm). Subsequently polystyrene particles (diameter 0.48 μm) serving as scattering centers were mixed into the solution. With the small amount of polystyrene used, no effect on the pattern evolution was observed. A volume of the mixture was placed in a dust-free petri dish of 6.8 cm diameter. In this study the depth of the solution layer at 25 ± 1°C was usually 0.85 ± 0.05 mm. To obtain a constant period of wave propagation, a pair of spiral waves was initiated (t=0). The spiral centers were located about 2 cm away from the dish center, where measurements were carried out. Note that no CO_2 bubbles were nucleated and no waves were triggered at the dish boundaries. Patterns were investigated with and without covering the dish (air gap between layer surface and glass cover ≈ 12 mm).

The following methods based on computerized video techniques were applied: two-dimensional spectrophotometry, as previously described by Müller et al. (1986), and a two-dimensional microscope-video-imaging technique (Fig. 3). In

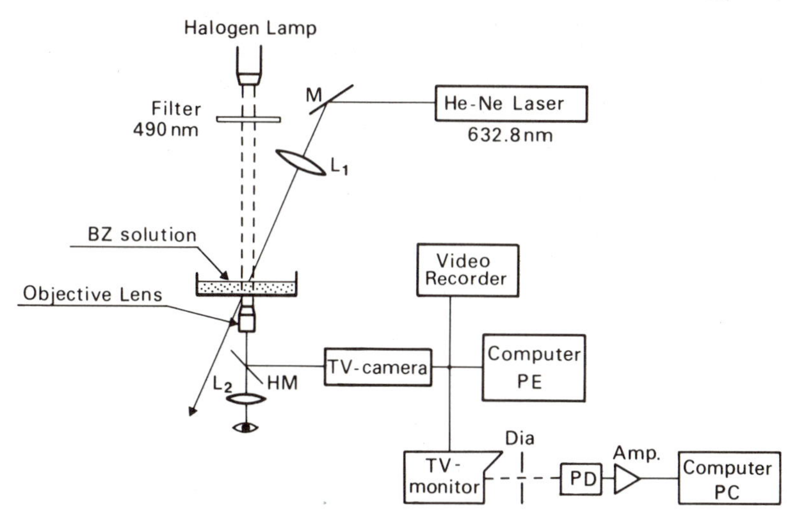

Fig. 3. Schematic diagram of the microscope video imaging system for two-dimensional velocimetry.

this apparatus a He-Ne laser focused to a small area (diameter 0.2 mm) illuminated the polystyrene particles, thus allowing the detection of local hydrodynamic flow. The microscope focal plane was usually adjusted close to the layer surface. Flow velocity was measured every 5 s by pursuing one or two particles during 1 s on a video movie displayed on a television monitor. By our also shining a homogeneous light field (490 nm) through the layer, particle motion and propagation of chemical activity could be traced simultaneously.

In addition, video movies of monochomatic transmitted light without illuminated polystyrene particles were recorded to observe the dynamic characteristic of the chemical waves. Local time traces of light absorption were obtained by placing a photodiode in front of a selected spot on the monitor (corresponding to a diameter $< 20\ \mu m$), and the local luminescence of the displayed image sequence was measured and analyzed by a computer.

RESULTS

Flow Induced by Chemical Waves

Velocity measurements were carried out at the dish center (area diameter ≈ 0.2 mm) under four different experimental conditions: (a) In a covered dish without trigger wave excitation, where no chemical pattern is observed, the motion of the suspended polystyrene particles remains very small ($v < 1.0\ \mu m/s$) and exhibits random Brownian motion. (b) If the dish is left uncovered, but no waves are triggered, there is a distinct flow with a velocity $v = 3$-$5\ \mu m/s$ (Miike et al. 1988a) which most likely is correlated with the appearance of stationary mosaic patterns (see Fig. 2a). (c) If a wave train emerging from a chemical spiral propagates through a layer in a covered dish, we observe a small but non-negligible flow parallel to wave propagation reaching a velocity $v = 4\ \mu m/s$ during the first 10 min of the experiment. In the perpendicular component we do not find any significant flow velocity. After 10 min an enhanced fluctuation with an amplitude of $10\ \mu m/s$ occurs, indicating the onset of an oscillatory change in flow direction, an effect which becomes much more pronounced in case (d) below. [We note here that recently oscillatory velocity changes of this kind with much larger amplitude and duration could be observed under the same experimental conditions, that is in wave trains in a covered sample dish (Miike, Müller, and Hess, unpublished results)]. (d) Wave trains from a spiral center propagating in an open dish readily induce a pronounced oscillatory hydrodynamic flow about 9 min after the experiment started (Fig. 4). Its maximum velocity ($\approx 40\ \mu m/s$) is comparable to the propagation speed of the waves ($\approx 50\ \mu m/s$) and its period is about twice as large as that of the passage of chemical waves through the detection area.

The oscillatory flow observed in this last case (d) leads to deformations of the wave profiles (Miike et al., 1988b). By monitoring local light intensities in a small area (diameter $< 20\ \mu m$) with a photodiode in front of the television screen (see Fig. 3), these wave deformations are reflected in periodic changes of the wave amplitude occurring after 9 min, as shown in the time trace of Fig. 5a. This measurement corroborates that in this experiment the oscillatory flow and its influence on the waves have a period twice that of the fundamental frequency of wave passage. In the power spectrum based on the second half of the time trace of Fig. 5a, the period-doubling component clearly emerges (Fig. 5b).

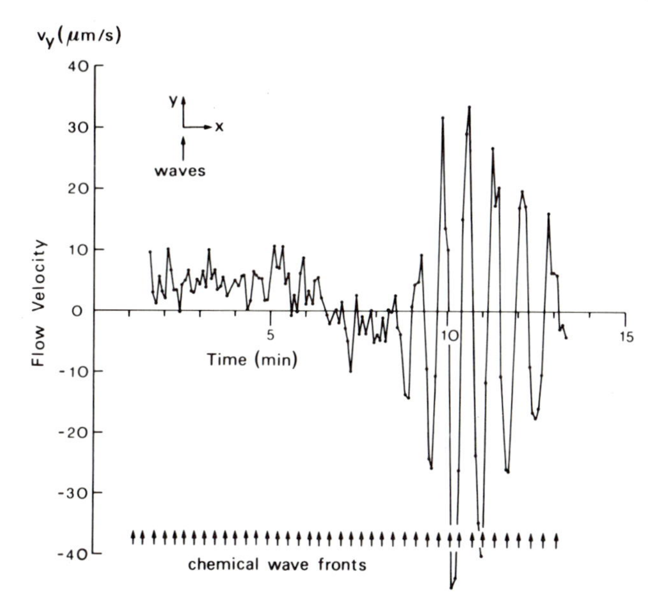

Fig. 4.
Temporal trace of flow velocity in the direction of propagating spiral waves measured at the center of an uncovered layer. Cover removed at t=2 min. Arrows indicate passage of wave fronts through the detection area.

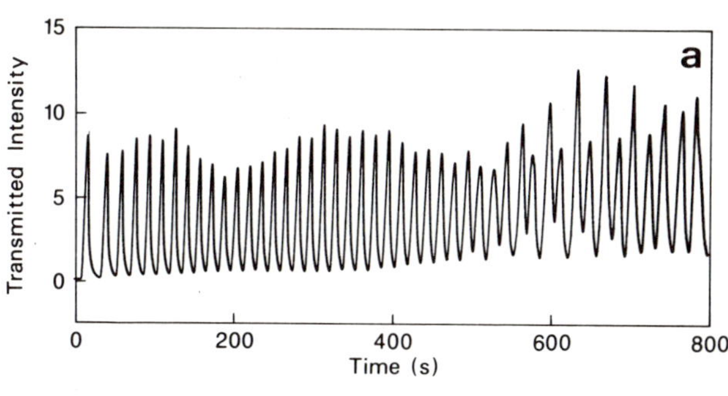

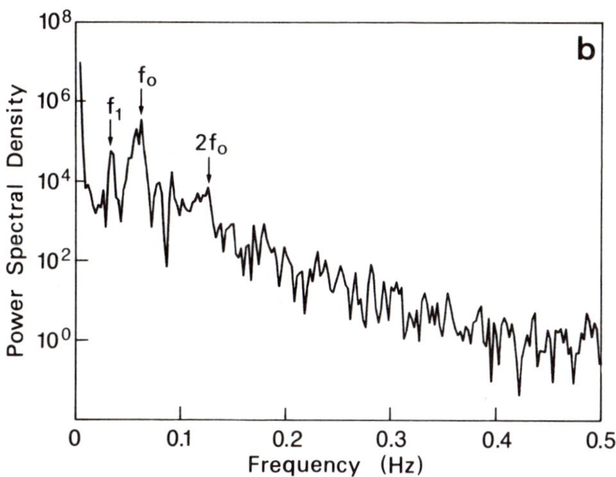

Fig. 5.
(a) Time-dependence of local transmitted light intensity and (b) its power spectrum in spiral waves with an open liquid/air interface (layer depth 0.94 mm), corresponding to the experiment of Fig. 4.

Wave Decomposition by Highly Developed Flow

Whereas the results shown in Figs. 4 and 5 represent the typical characteristics of many of the investigated flow oscillations, we also observed unusually high flow velocities accompanied by strong wave deformation and decomposition, about 5 times within 30-40 experiments performed in apparently identical sample preparations, that is with a probability of 10 to 15%. The precise conditions for this behaviour are under current investigation and clues may be found in a highly sensitive dependence on initial chemical composition, aging effects, influence of oxygen, and/or the geometric arrangement of the spiral patterns in the dish.

A time trace of extremely pronounced flow velocities measured close to the open liquid/gas interface is shown in Fig. 6. At t = 13 min an oscillatory hydrodynamic flow is induced, in this case with a period 3 times that of the passage of chemical waves. The amplitude of 50-70 μm/s is comparable to the wave propagation speed, but after about 17 min it further increases up to 300 μm/s, thus exceeding by far the speed of the waves. The period of flow oscillations is now equal to that of the wave trains.

In a similar experiment (Fig. 7) the corresponding temporal development of chemical wave front geometry is shown. At t ≈ 7 min weak and periodic deformations of the moving wave fronts appear (Fig. 7b). Later these start to vary periodically (Fig. 7c-f), as reported previously by Miike et al. (1988b). When the flow becomes more pronounced, a spatially coherent structure is recognized which overlaps periodically with the chemical fronts (Fig. 7f-c). Subsequently, for extremely pronounced flow amplitudes the oscillatory deformation turns into an irregular turbulent decomposition (Fig. 7i-k). Finally, after 25 min, the oscillatory flow stops and the wave front decomposition has led to spiral structures (Fig. 7l).

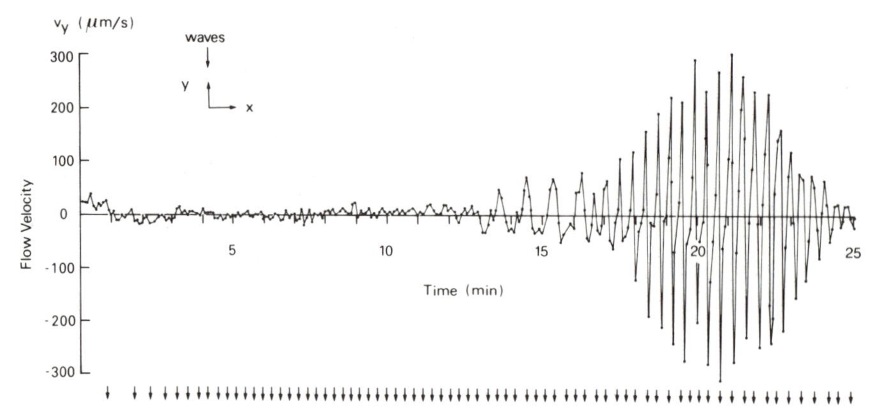

Fig. 6. Temporal trace of flow velocity measured as indicated in Fig. 4. In this experiment the velocities reach a peak value of 300 μm/s. Arrows as in Fig.4.

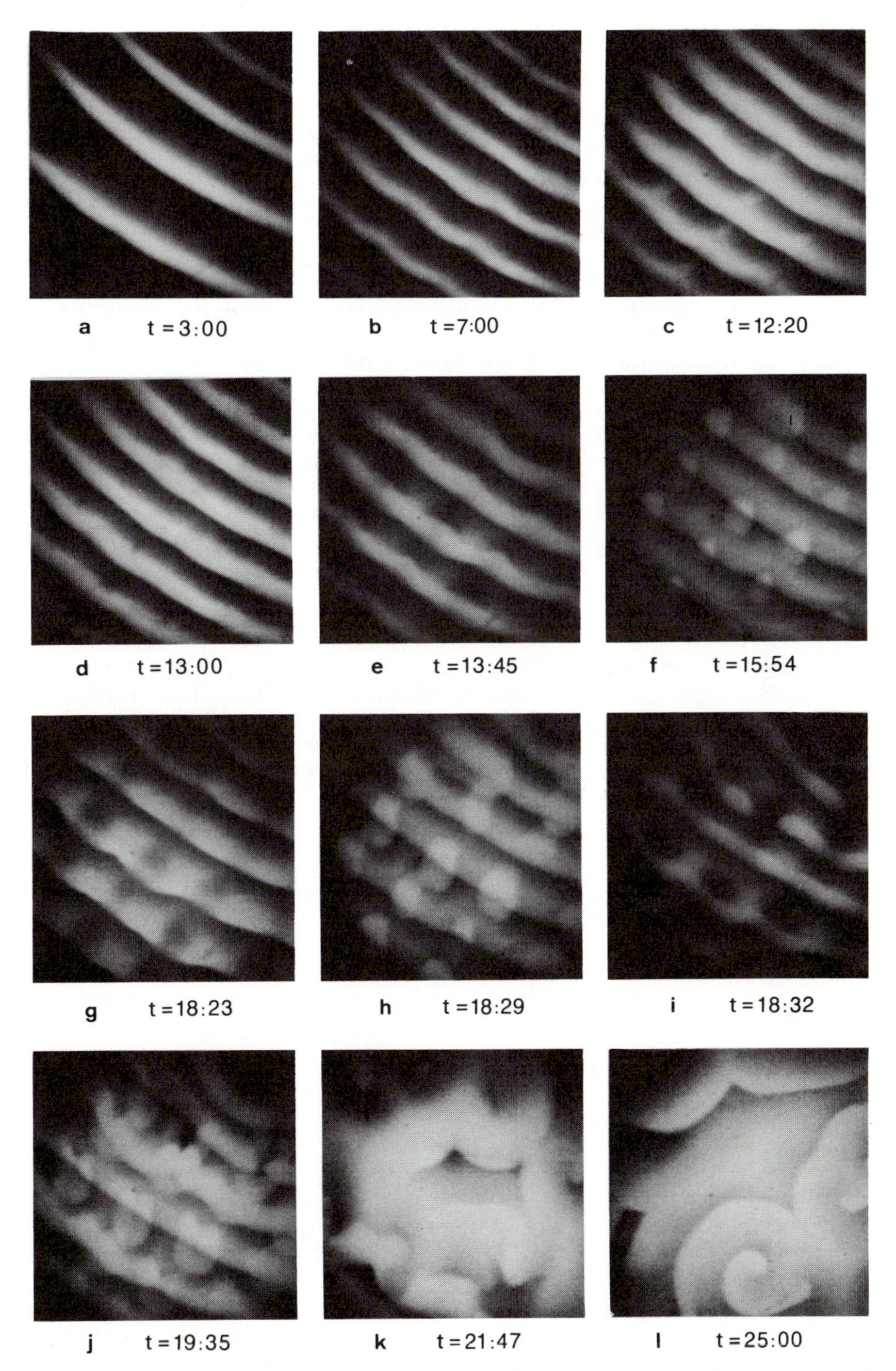

Fig. 7. Image sequence of a chemical wave train undergoing distortion and decomposition due to oscillatory flow. In this experiment flow amplitudes reached values 3 times larger than the wave propagation speed. Image section: 8.3 x 8.3 mm^2.

CONCLUDING REMARKS

Our findings establish the existence of an oscillatory hydrodynamic flow which is induced by chemical wave propagation and is responsible for oscillatory deformation and turbulent decomposition of the wave fronts. The period of flow oscillation is found to be two or three times or equal to that inherent in chemical wave propagation. These facts suggest that the flow is entrained by the periodic passage of wave trains across a given area in the sample layer and point to a nonlinear coupling between chemical waves and hydrodynamic flow. Convection caused by temperature gradients due to evaporative cooling certainly influences the details of these patterning phenomena, but it does not provide the main source. Investigations are in progress with the goal of finding similar reaction-convection couplings in covered sample dishes where evaporative effects are essentially excluded, and there are, in fact, first observations yielding oscillations under such conditions (Miike, Müller, and Hess, unpublished results). Furthermore, a more global characterization of the flow fields is needed.

Although the issue of reaction-convection coupling has been discussed on theoretical grounds (Borckmans and Dewel, 1988), e.g. in connection with photochemically induced patterns (Micheau et al., 1983; Avnir and Kagan, 1984) and spatial structures forming during biochemical oscillations (Boiteux and Hess, 1980), there exists no model yet which accounts for the dynamical behaviour presented in this article, especially the turbulent wave shapes being a consequence of flow induced by traveling concentration gradients.

ACKNOWLEDGEMENT

This work was supported by the Stiftung Volkswagenwerk, Hannover, FRG.

REFERENCES

Agladze, K.I., Krinsky, V.I., and Pertsov, A.M., 1984, Chaos in Non-stirred Belousov-Zhabotinskii Reaction is Induced by Interaction of Waves and Stationary Dissipative Structures, Nature, 308:834.

Avnir, D., and Kagan, M., 1984, Spatial Structures Generated by Chemical Reactions at Interfaces, Nature, 307:717.

Boiteux, A., and Hess, B., 1980, Spatial Dissipative Structures in Yeast Extracts, Ber. Bunsenges. Phys. Chem. 84:392.

Borckmans, P., and Dewel, G., 1988, Chemical Structures and Convection in: "From Chemical to Biological Organization", Markus, M., Müller, S.C., and Nicolis, G., eds., Springer, Heidelberg.

Bornmann, L., Busse, H., and Hess, B., 1973, Oscillatory Oxidation of Malonic Acid by Bromate, Z. Naturf. 28c:514.

Chandrasekhar, S., 1961, "Hydrodynamic and Hydromagnetic Stability", Clarendon, Oxford.

Field, R.J. and Burger, M. (eds.), 1985, "Oscillations and Traveling Waves in Chemical Systems", Wiley, New York.

Field, R.J., Körös, E., and Noyes, R.M., 1972, Oscillations in Chemical Systems, Part 2, J. Am. Chem. Soc. 94:8649.
Kuhnert, L., Pohlmann, L., and Krug, H.-J., 1988, Chemical Wave Propagation with a Chemically Induced Hydrodynamic Instability, Physica D, 29:416.
Micheau, J.C., Gimenez, M., Borckmans, P., and Dewel, G., 1983, Hydrodynamic Instabilities and Photochemical Reactions, Nature, 305:43.
Miike, H., Müller, S.C., and Hess, B., 1988a, Oscillatory Hydrodynamic Flow Induced by Chemical Waves, Chem. Phys. Lett., 144:515.
Miike, H., Müller, S.C., and Hess, B., 1988b, Oscillatory Deformation of Chemical Waves Induced by Surface Flow, Phys. Rev. Lett., 61:2111.
Müller, S.C., Plesser, Th., and Hess, B. 1985b, Surface Tension Driven Convection in Chemical and Biochemical Solution Layers, Ber. Bunsenges. Phys. Chem. 89:654.
Müller, S.C., Plesser, Th., and Hess, B., 1985a, The Structure of the Spiral Core in the Belousov-Zhabotinskii Reaction, Science 230:661.
Müller, S.C., Plesser, Th., and Hess, B., 1986, Two-dimensional Spectrophotometry and Pseudo-Color Representation of Chemical Reaction Patterns, Naturwissenschaften, 73:165.
Orban, M., 1980, Stationary and Moving Structures in Uncatalyzed Oscillatory Chemical Reactions, J. Am. Chem. Soc. 102:4311.
Pagola, A., Ross, J., and Vidal, C., 1988, Measurement of Dispersion Relation of Chemical Waves in an Oscillatory Reaction Medium, J. Phys. Chem., 92:163.
Ross, J., Müller, S.C., and Vidal, C., 1988, Chemical Waves, Science, 240:460.
Showalter, K., 1980, Pattern Formation in a Ferroin-Bromate System, J. Chem. Phys. 73:3735.
Tyson, J.J., and Keener, J.P., 1988, Singular Perturbation Theory of Traveling Waves in Excitable Media (a Review), Physica D, 32:327.
Winfree, A.T., 1972, Spiral Waves of Chemical Activity, Science, 175:634.
Wood, P.M., and Ross, J., 1985, A Quantitative Study of Chemical Wavs in the Belousov-Zhabotinskii Reaction, J. Chem. Phys., 82:1924.
Zhabotinskii, A.M., 1964, Periodic Liquid-Phase Oxidation Reactions, Dokl. Akad. Nauk SSSR, 157:392.
Zhabotinskii, A.M., and Zaikin, A.N., 1973, Autowave Processes in a Distributed Chemical System, J. Theor. Biol. 40:45.
Zykov, V.S., 1988, "Modelling of Wave Processes in Excitable Media", Manchester Univ. Press, Manchester.

SUSTAINED NON-EQUILIBRIUM PATTERNS IN A ONE-DIMENSIONAL REACTION-DIFFUSION CHEMICAL SYSTEM

J. Elezgaray and A. Arneodo

Centre de Recherche Paul Pascal
Domaine Universitaire, 33405 Talence Cedex, France

We present numerical evidence for the existence of non trivial spatio-temporal patterns in a one-dimensional reaction-diffusion chemical system, with equal diffusion coefficients. The main motivation for such a study is the fact that, besides the so-called target and spiral-wave patterns, there exists so far no unambiguous experimental evidence of nontrivial spatio-temporal patterns, resulting solely from the interaction between reaction and diffusion processes and not from convective or interfacial effects. According to a common belief, this situation arises because all the diffusion coefficients of the different chemical species are approximately equal under general experimental conditions. In fact, it can be shown that, in the limit of equal diffusion coefficients, the diffusion coupling cannot destabilize a stable homogeneous steady state. Our purpose is to show that one can overcome this difficulty when imposing concentration gradients to the system.

Stimulated by similar considerations, the experimental groups of the University of Texas (Tam et al., 1988), and the University of Bordeaux (Ouyang et al., 1988), have developed the following Open Couette-Flow Reactor (OCFR): two Continuously Stirred Tank Reactors (CSTR) are connected by a Couette-Taylor flow, with the inner cylinder rotating and the outer cylinder at rest. Chemicals injected into the CSTRs diffuse and react in the annular region between the two cylinders. At large Reynolds numbers the Taylor vortices can be considered as homogeneous in the radial and azimuthal directions. When considered at length scales larger than the vortex scale, the mass transport is diffusive (Tam et al., 1987). Consequently the OCFR can be considered as a 1D array of homogeneous cells, coupled by diffusion, with a unique (tunable) diffusion coefficient D for all chemical species. The number of pairs of vortices is rather low ($\sim$70) so that this system is neither a true extended system nor a low dimensional system. The role of the two CSTRs is to maintain non-equilibrium boundary conditions, by imposing a concentration gradient to the system (Dirichlet boundary conditions). When considering a bistable chemical reaction, e.g., the Chlorite-Iodide- Malonic acid reaction, appropriate feeding conditions can be found so that the left-end CSTR is in a (Iodide) reduced state and the right-end in an (Iodide) oxidized state. Under such conditions, a stationary pattern appears in the Couette flow (Ouyang et al., 1988): two rather homogeneous regions corresponding to each of these states respectively, are separated by a sharp transition front. This (trivial) spatial pattern is due to the existence of a switching process in the kinetics of the reaction. However, by varying the chemical composition in the end CSTRs, this front becomes unstable through a Hopf bifurcation, and a periodically oscillating front structure is obtained. Recently, more complex spatio-temporal patterns involving several switches between the reduced and oxidized states (multifront structures) have been observed in either a steady or an oscillating state (Ouyang et al., 1988).

New Trends in Nonlinear Dynamics and Pattern-Forming Phenomena
Edited by P. Coullet and P. Huerre
Plenum Press, New York, 1990

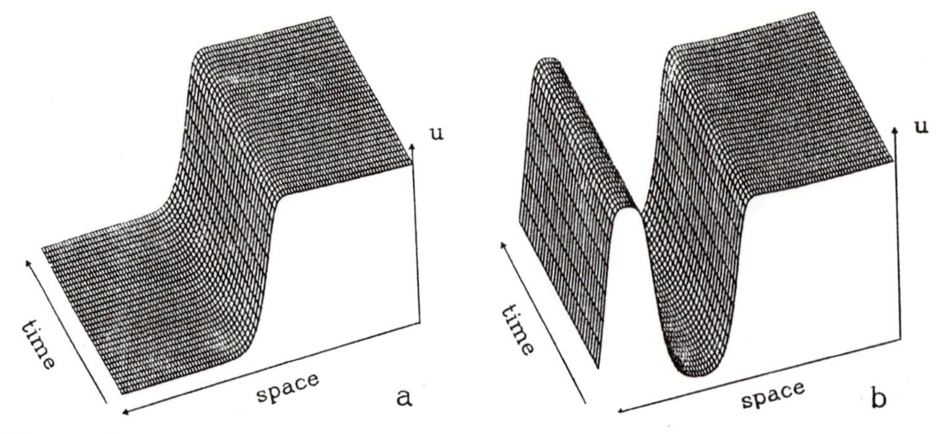

Fig. 1. Numerical stationary patterns: (a) a single-front pattern; (b) a multifront pattern.

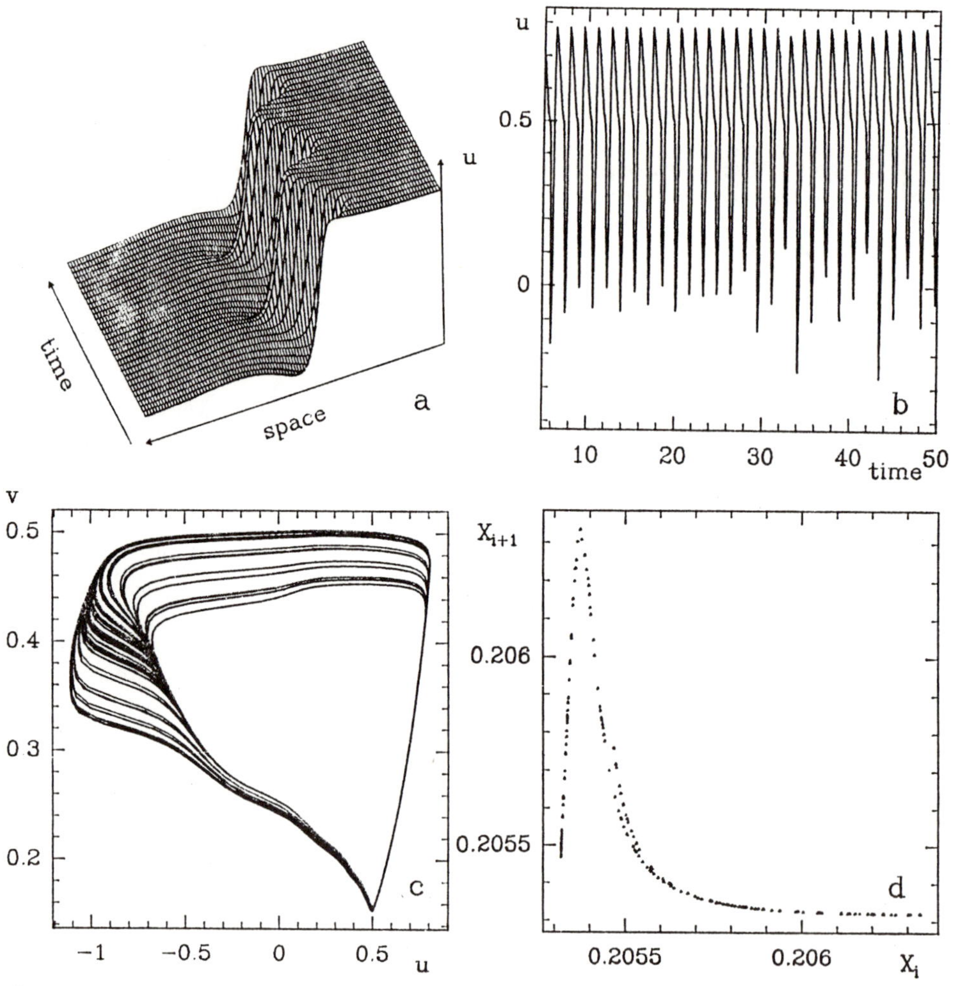

Fig. 2. Numerical oscillating patterns: (a) a periodically oscillating front; (b) time series recorded in an intermediate reacting cell visited by a chaotically oscillating front; (c) Phase portrait corresponding to the time series in (b); (d) 1D map constructed from the phase portrait in (c).

In order to understand such phenomena, we propose a simple reaction-diffusion model (Arneodo et al., 1987). More precisely, each intermediate cell is modelled by a two-variable Van der Pol-like equation, which mimics the excitable character of the chemical reaction used in the experiment. The main characteristic of this reaction term is the existence of a pleated slow manifold, consisting of three branches, among which two are attracting. These two branches account for the two families of reduced and oxidized steady states in the chemical reaction. In the absence of diffusion coupling, each cell is in a stable steady state. The equation of the slow manifold is such that the Hopf bifurcation of the reaction term is subcritical, as it is often the case in most oscillating reactions. The diffusion coefficients of the different species are set equal to D. We impose Dirichlet boundary conditions: the concentrations at the right (resp. left) are kept fixed in the upper (resp. lower) branch of the slow manifold. At high values of D, all the solutions are attracted to a single stable steady pattern, which corresponds to a spatial transition between the two attracting branches of the slow manifold (fig. 1a). At low values of D, more complicated patterns involving several fronts, i. e., several spatial swichtings between these two branches, can develop as shown in fig. 1b. The complexity of these spatial patterns is actually determined by the details of the interplay between the reaction and the diffusion processes. Moreover, when decreasing D from high values, the transition front (fig. 1a) becomes sharper and sharper, eventually undergoing a Hopf bifurcation to a diffusion-induced periodically oscillating front (fig. 2a). The simulations shown in figs. 1 and 2 have been performed using a numerical code which requires a spatial discretization of the diffusion (laplacian) term. Therefore, by varying the spatial resolution, one can bring our system close to either the OCFR or the continuous equations. It is thus interesting to note that the main dynamical properties are preserved when increasing the spatial resolution, which strongly indicates that these spatio-temporal patterns are very likely to be found in continuous 1D reaction-diffusion chemical systems.

When further scanning the parameter space, conditions can be found where the oscillating front pattern displays very complicated temporal behaviour. In fig. 2b we show aperiodic temporal oscillations recorded at a given spatial point. The corresponding phase portrait and 1D map are illustrated in fig. 2c and 2d. The well-defined unimodal character of this 1D map is a signature of the chaotic nature of these oscillations. It is somewhat puzzling to remark that the phase portraits (fig. 2c) obtained in our simulations are strikingly similar to the strange attractors observed in the BZ reaction when conducted in a CSTR (Argoul et al., 1987). Very much like in the homogeneous BZ reaction, period-doubling bifurcations are observed as precursors to chaotically oscillating fronts (so far, we could not ensure the existence of chaotically oscillating fronts in the continuous limit). Periodic-chaotic sequences have already been observed in complicated kinetic models of (homogeneous) reacting media (Argoul et al., 1987). Our point here is to put the emphasis on the diffusion-induced chaos as a new mechanism leading to chemical chaos. Another interesting issue is the existence of spatio-temporal turbulence: actually we expect these front patterns to lose their spatial coherence in the limit $D \rightarrow 0$; this should be the signature of the transition to an extended system. We are very grateful to J. Boissonade, P. DeKepper, Q. Ouyang, J. C. Roux, W. Horsthemke, H. L. Swinney, N. Y. Tam and J. Vastano for communicating their experimental results prior to publication.

REFERENCES

Argoul, F., Arneodo, A., Richetti, P., Roux, J. C. and Swinney, H.L., 1987, Chemical chaos: from hints to confirmation, Account Chem. Phys., 20:436.

Arneodo, A., and Elezgaray, J., 1987, Oscillating fronts in a one-dimensional reaction-diffusion system, in: "Spatial Inhomogeneities and Transient Behaviour in Chemical Kinetics," Manchester University Press.

Ouyang, Q., Boissonade, J., Roux, J.C. and DeKepper, P., 1988, Sustained reaction-diffusion structures in an open reactor, preprint.

Tam, W.Y., and Swinney, H.L., 1987, Mass transport in turbulent Couette-Taylor flow, Phys.Rev.A, 36:1374.

Tam, W.Y., Vastano, J.A., Swinney, H.L. and Horsthemke, W., 1988, Bifurcations to spatio-temporal chaos in a reaction-diffusion system, preprint.

PATTERN SELECTION AND SYMMETRY COMPETITION IN MATERIALS INSTABILITIES

Daniel Walgraef †

Faculté des Sciences
Université Libre de Bruxelles, CP 231
B-1050 Brussels, Belgium

INTRODUCTION

One of the most natural and still intriguing behavior of complex physico-chemical systems driven sufficiently far from thermal equilibrium is their ability to undergo symmetry-breaking instabilities leading to the spontaneous formation of coherent structures over macroscopic time and space scales (Nicolis and Prigogine, 1977). Such a behavior has been widely studied in various fields including physics, biophysics, chemistry and materials science. The question of why order appears spontaneously and which patterns are selected among a large manifold of possibilities remains a major theme of experimental and theoretical research (Swinney and Gollub, 1981, Nicolis and Baras, 1983, Hlavacek, 1985). Even though this research was at first fundamental in nature, it now appears more and more to be of technological importance.

Classical types of ordering phenomena such as the onset of convective patterns in hydrodynamical systems and the development of spatio-temporal structures in chemical systems are discussed in these proceedings. Let me nevertheless recall that in Rayleigh-Bénard, Bénard-Marangoni or liquid crystal instabilities, periodic roll patterns may be obtained (Normand et al., 1977), but also square or hexagonal planforms (Wesfreid and Zaleski, 1984). In any case, the most striking aspect of these instabilities is the symmetry breaking of the macroscopic fluid properties, namely translational and rotational symmetries in the horizontal plane. As in phase transitions this induces long wavelength fluctuations and defects such as dislocations, disclinations or even grain boundaries. which seem to play a major role in the disorganisation of hydrodynamical patterns (Ocelli et al., 1983) .

Another paradigm for self-organisation phenomena is the Belousov-Zhabotinsky reaction. In this case, due to the feeding of the chemical system with reacting species or to an initial excess of some of the products, it appears that, outside the linear domain around equilibrium, there is a thermodynamic threshold beyond which the uniform steady state becomes unstable and is replaced by new classes of regimes having completely different spatio-temporal properties, including uniform oscillations and waves having the shape of target patterns or Archimedean spirals (Winfree, 1980, Walgraef, 1988).

When one considers the huge complexity of the kinetic equations required for the description of such systems, the simple shape of these waves may be puzzling. Nevertheless these shapes may be predicted with simple but powerful concepts related to bifurcation theory and symmetry-breaking considerations. These examples show that in non equilibrium

† Senior Research Associate with the National Fund for Scientific Research (Belgium).

New Trends in Nonlinear Dynamics and Pattern-Forming Phenomena
Edited by P. Coullet and P. Huerre
Plenum Press, New York, 1990

pattern forming systems, as in phase transitions, the global properties of the structures do not depend on microscopic properties but rather on the symmetries of the problem and on the characteristics of the bifurcation.This seems also to be the case for many patterns and microstructures which form in driven or metastable materials and we expect that the methods developed for the study of instabilities and bifurcations of complex dynamical systems should be useful in the understanding of materials behavior as I will try to illustrate it briefly below.

Let me first emphasize that during the last years the whole field of materials science and related technologies has experienced a complete renewal. Effectively, by using techniques corresponding to strong non equilibrium conditions, it is now possible to escape from the constraints of equilibrium thermodynamics and to process totally new materials structures including different types of glasses, nano- and quasicrystals, superlattices, ... These techniques include ion implantation, laser beam surface melting as well as electron beam heating. For example in laser annealing, microstructures with superior resistance to friction, corrosion, ... may be frozen into place. Ultra-rapid solidification of alloys may trigger the formation of quasicrystalline structures. Point defect and void patterning in irradiated systems are related to the precipitation of solid solutions. Finally the self-organisation of dislocation populations in stressed or irradiated materials is associated with plastic instabilities and deformation localization. As materials with increased resistance to fatigue and fracture are sought for actual applications, a fundamental understanding of the collective behavior of dislocations and point defects is highly desirable. Since usual thermodynamical and mechanical concepts are not adapted to describe such situations, progress in this direction should be related to the explicit use of genuine nonequilibrium techniques, nonlinear dynamics and instability theory (Walgraef, 1987).

The peculiarities of materials instabilities, on the other hand, leads to fundamental questions in pattern selection theory. These questions are related to the effect of anisotropy, flow fields or modulated forcings on the competition between patterns with different symmetries in two and three-dimensional active media. We will see below how these problems arise and what kind of answer may be obtained, at least near instability points.

DISLOCATION PATTERNS AND DYNAMICAL INSTABILITIES

The occurence of pattern forming instabilities seems natural for defect populations in driven metals and alloys. For example, in the case of plastic deformation, the evidence of instabilities and pattern formation is overwhelming . It is now well-known for about fourty years that the elementary carriers of plastic deformation of metals are dislocations. In spite of their complex shapes and interactions it has been observed that they form well organized spatial structures. At high temperatures these regular structures mainly consist of planar networks which form the walls of cellular planforms. At lower temperatures these cells become thicker and more diffuse. Under cyclic loading the tendency to form ordered structures is even more pronounced. Especially the so-called persistent slip bands (PSB), a regular ladderlike structure embedded in a less regular vein or matrix structure has been observed in crystals oriented for single slip (Thompson et al., 1956, Essmann and Mughrabi, 1979, Tabata et al., 1983). In the case of multiple slip, PSB with different orientations may coexist leading to "labyrinth structures" (Ackermann et al., 1984, Jin and Winter, 1984).

Irradiated materials and alloys also present several types of spatial structures corresponding to dislocation or vacancy loop ordering, void lattices,...(Martin and Kubin, 1988). In some cases, the conditions for instability and pattern formation have been determined (Martin, 1984, Murphy, 1987). However it is of interest to investigate the geometry, the symmetries, and the stability ranges of the selected structures. As indicated by Krishan (1982), the nature of the selection mechanisms and how the selected patterns are influenced by the underlying lattice symmetry, or by other materials and irradiation conditions, remain largely unexplored.

Up to now, most of the analysis of microstructural ordering were restricted to one-dimensional systems. This is mainly due to the difficulties of the post-bifurcation analysis and the fact that the complexity of the dynamics does not allow, in general, the attainement of analytic solutions for the various concentrations. However we actually know that, near the instability or bifurcation points, the dynamics may be reduced to much simpler forms by taking advantage of the time and space scale separation between stable and unstable modes and by projecting the dynamics on its unstable manifold (Guckenheimer et al.,1980). The resulting slow mode dynamics leads then to amplitude equations for the patterns and allows the discussion of their selection and stability properties in higher dimensions. As a result, I would like to show here that some typical problems of pattern formation in driven materials may be adressed in this framework.

In particular, deformed or irradiated metals and alloys, structures with different symetries may be simultaneously stable beyond the primary bifurcation. For example, under irradiation, when the diffusion and interactions of point defects are isotropic the maxima of the vacancy loop density may correspond to bcc lattices or planar arrays. Hence, these structures could be in non parallel orientation, i.e. with a structure different from the structure of the host lattice.

In the case of plastic instabilities, the dislocations are the essential entities involved in the reaction-transport processes underlying the dynamics of the system. Since the motion of mobile dislocations is restricted to a limited number of slip planes, anisotropy effects are typical of this problem and related to the strain and crystalline properties of the samples. As a result different types of microstructures appear in metals oriented for single or multiple slip , under uni- or biaxial stresses , in mono- or polycrystals,... (Ackermann et al., 1984, Buchinger et al., 1984, Balakrishnan and Bottani, 1986)

Hence, since the symmetry of the defect structures is a crucial issue in irradiated or deformed materials (Cahn, 1988, Walgraef and Aifantis, 1988), a careful study of the post-bifurcation regime is needed to test the relevance of particular kinetic models to the interpretation of experimental observations.

AMPLITUDE EQUATIONS AND PATTERN SELECTION

Isotropic systems

Let me consider here the example of irradiated crystalline materials of cubic symmetry. The dynamical behavior of such materials may be described through the rate theory of radiation damage which is based on the coupled dynamics of vacancies, interstitials and vacancy loops. This model was originally developed by Bullough, Eyre and Krishan (1975), and expanded later by by Ghoniem and Kulcinski (1978) to include the dynamics of point defects in the fully dynamic rate theory. The network dislocations which are also present in the material are assumed to have a constant uniform distribution and interstitial loops are neglected. The basic mechanisms which are taken into account correspond to the diffusion, the creation and the recombination of point defects, their migration to the loops and network dislocations and the thermal emission of vacancies from loops and network dislocations. Furthermore, Part of the vacancies are instaneously trapped into vacancy loops by a cascade effect which is the dominant creation mechanism of these loops.

In this spirit, the following kinetic equation for the defect concentrations were recently proposed by Murphy (1987) :

$$\begin{aligned}
\partial_t c_i &= K - \alpha c_i c_v + D_i \nabla^2 c_i - D_i c_i (Z_{iN}\rho_N + Z_{iL}\rho_L) \\
\partial_t c_v &= K(1-\epsilon) - \alpha c_i c_v + D_v \nabla^2 c_v - D_v (Z_{vN}(c_v - \bar{c}_{vN})\rho_N + Z_{vL}(c_v - \bar{c}_{vL})\rho_L) \\
\partial_t \rho_L &= \frac{1}{|\vec{b}| r_L^0} [(\epsilon K - \rho_L (D_i Z_{iL} c_i - D_v Z_{vL}(c_v - \bar{c}_{vL}))] \qquad (1)
\end{aligned}$$

where c_v corresponds to vacancies, c_i to interstitials, ρ_L and ρ_N to vacancy loops and network dislocation densities. K is the displacement damage rate and ϵ the cascade collapse efficiency; D_i and D_v are the diffusion coefficients, α is the recombination coefficient, $\vec{b}$ the Burgers vector, r_L^0 the mean vacancy loop radius and $Z_{..}$ the bias factors which will be approximated by $Z_{iL} = Z_{iN} = 1$ and $Z_{vL} = Z_{vN} = 1 + B$. $\bar{c}_{vN}$ and $\bar{c}_{vL}$ are the thermally emitted vacancies from network dislocations and vacancy loops. The various coefficients appearing in these rate equations may be computed theoretically or related to measurable quantities (Ghoniem and Kulcinski, 1978).

The linear stability analysis of the uniform steady state shows that a pattern forming instability occurs when the dislocation bias overcomes the cascade collapse efficiency, i.e. :

$$B > \epsilon[1 + \sqrt{\frac{\rho_N}{\rho_L^0}}]^2 \tag{2}$$

for a critical wavelength given by:

$$\lambda_c = 2\pi[\frac{D_v(\bar{c}_{vL} - \bar{c}_{vN})}{(1+B)\epsilon K \rho_N}]^{1/4}$$

and we see that the wavelength decreases with increasing network dislocation density, cascade collapse efficiency and damage rate. On the other hand, its temperature dependence is more difficult to asses since D_v is an increasing function of the temperature while $(\bar{c}_{vL} - \bar{c}_{vN})$ is a decreasing function of the temperature and its global behavior may vary from material to material. As an example, consider 316 L steel irradiated at 500^0 C with a displacement damage rate of 10^{-6} dpa s^{-1} : the critical wavelength is nearly 1.24 μm for solution-annealed material with a typical dislocation density of 10^{13} m^{-2}. The wavelength is smaller for cold-worked material (Walgraef and Ghoniem, 1988), on the order of 0.39 μm for a dislocation density of 10^{15} m^{-2} .

However the orientation and the number of wavevectors underlying the possible structures are not fixed and the associated degeneracy may only be partly removed by a nonlinear analysis in the post bifurcation regime. Effectively, as usual for isotropic reaction-diffusion systems, the slow mode dynamics near this pattern-forming instabilities may be written as (Walgraef, 1988, Walgraef and Ghoniem, 1988) :

$$\tau\partial_t\sigma(\vec{x},t) = [\frac{b-b_c}{b_c} - \xi_0^2(q_c^2 + \nabla^2)^2]\sigma(\vec{x},t) + v\sigma^2(\vec{x},t) - u\sigma^3(\vec{x},t) \tag{3}$$

where σ is the order parameterlike variable while b is the bifurcation parameter and b_c its critical value. u and v are system dependent and may be explicitely calculated for each particular dynamical system. In the present case, $b = B/\epsilon$, $\xi_0^2 \propto (\rho_L^0/\rho_N)^{-1/2}$, $v = 2(\rho_L^0/\rho_N)^{-3/2}$ and $u = 2(\rho_L^0/\rho_N)^{-5/2}$.

It is well known that, for $b > b_c$, the stable solutions of this Landau-Ginzburg type of dynamics correspond to :

(1) roll or wall structures associated with spatial modulations of the order parameter in one direction. They appear via a second orderlike transition, or supercritical bifurcation, at $b = b_c$.

(2) rodlike hexagonal or triangular structures appearing via first orderlike transition, or subcritical bifurcations, defined by the following amplitude equations:

$$\sigma(\vec{x},t) = \sum_{i=1}^{3} A_i(\vec{x},t)e^{i\vec{q}_i.\vec{x}} + c.c. \qquad |\vec{q}_i| = q_c, \qquad \vec{q}_1 + \vec{q}_2 + \vec{q}_3 = 0$$

$$\tau\partial_t A_i = [\frac{b-b_c}{b_c} + \frac{4\xi_0^2}{q_c^2}(\vec{q}_i.\vec{\nabla})^2]A_i + vA_{i-1}^*A_{i+1}^* - 3u[|A_i|^2 + 2\sum_{j\neq i}|A_j|^2]A_i \tag{4}$$

The steady state is given by ($|A_i| = A = \frac{1}{30u}[v + \sqrt{v^2 + 60u\frac{b-b_c}{b_c}}]$).

(3) bcc lattices or filamental structures of cubic symmetry, also associated with a subcritical bifurcation and defined similarly to hexagonal structures but with six pairs of wavevectors. The corresponding steady state is then given by:

$$\sigma(\vec{x}) = A[\cos\frac{q_c}{\sqrt{2}}x\cos\frac{q_c}{\sqrt{2}}y + \cos\frac{q_c}{\sqrt{2}}y\cos\frac{q_c}{\sqrt{2}}z + \cos\frac{q_c}{\sqrt{2}}z\cos\frac{q_c}{\sqrt{2}}x]$$

with $A = \frac{1}{33u}[v + \sqrt{v^2 + 33u\frac{b-b_c}{b_c}}]$.

When the bifurcation parameter is increased, the 2d and 3d structures may in turn become unstable (the hexagonal structure for $\frac{b-b_c}{b_c} > \frac{4v^2}{3u}$ and the bcc structure for $\frac{b-b_c}{b_c} > 3\frac{v^2}{u}$).

Hence, between threshold and $3v^2/u$, bcc structures should be expected while above this limit the structure should consist of regularly spaced planes of maximum density, even in isotropic media.

In the case of irradiated materials described by the dynamical system (1), the order parameterlike variable is proportional to the vacancy loop density and the coefficients appearing in the amplitude equations are directly related to the physical parameters of the problem. The stability domains of bcc and planar structures and the critical wavelength may for example be computed versus network dislocation density, displacement damage rate or temperature as analyzed by Walgraef and Ghoniem (1988).

In the case of an anisotropic diffusion of interstitials as in hcp materials where the mobility of interstitials is much larger in the basal planes than between these planes, it is easy to show with the same method, that the stable patterns for vacancy loops correspond to planar arrays with planes of maximum density parallel to the planes of high interstitial mobility in agreement with experimental observations (Evans, 1987). It is also interesting to note that recent experimental observations by Jäger et al. (1988) indicate that spatial microstructure modulation is a general phenomenon under ion-irradiation conditions where cascade are produced. While the discussion outlined here is in general agreement with these experimental observations, it should be viewed as a step towards a generalized theoretical explanation of the nature of microstructure ordering under irradiation.

Anisotropic Systems

A reaction-transport model describing the collective behavior of dislocations populations during fatigue experiments and presenting patterning instabilities leading to the formation of PSB structures has recently been proposed by Walgraef and Aifantis (1985, 1986). This model was originally derived for monocrystals oriented for single slip, the primary slip direction being parallel to the x axis, and submitted to cyclic loading of frequency ν. In this case, after the hardening period the forest of immobile dislocations is already well developped and two types of dislocations were considered : "trapped" or nearly immobile ones, of density ϱ_i , and "free" or mobile ones gliding on the primary slip plane and of density ϱ_m . For each family of dislocations a balance equation is set up and by taking into account their basic reation-transport processes, the following dynamical system is obtained:

$$\partial_t\rho_i = \nabla_i(D_{ij}^{(0)} - D_{ijk}^{(1)}\nabla_k^2)\nabla_j\rho_i + f(\rho_i) - b\rho_i + \sum_n c_n\rho_i^n\rho_m$$

$$\partial_t\rho_m = D_M\nabla_x^2\rho_m + b\rho_i - \sum_n c_n\rho_i^n\rho_m \tag{5}$$

where $f(\rho_i)$ represents creation and annihilation of trapped dislocations, $b\rho_i$ the stress induced freeing of trapped dislocations and $\sum_n c_n\rho_i^n\rho_m$ the pinning of mobile dislocations by the forest.

Since the motion of mobile dislocations in the primary slip direction is much faster than any other, the uniform steady state (ρ_i^0, ρ_m^0) becomes unstable, even for $D^{(0)}$ positive, versus density modulations along the x direction at a bifurcation point given by :

$$\beta = \beta_c = (\sqrt{a} + \sqrt{\gamma D_{xx}^{(0)}/D_M})^2, \quad \vec{q} = q_c \vec{1}_x, \quad q_c = \frac{2\pi}{\lambda_c} = (\frac{a\gamma}{D_{xx}^{(0)} D_M})^{1/4} \tag{6}$$

where $a = -f^{(1)}(\rho_i^0)$, $\gamma = \sum_n c_n \rho_i^{0n}$ and $\beta \propto b_c$.

Hence, beyond this pattern forming instability, ladder-like structures are expected to develop with a wavelength λ_c which is a material property depending on the dislocation mobilities, and on their multiplication and pinning rates. On expressing the links between the parameters of the model and experimental quantities, it may be shown (Schiller and Walgraef, 1987) that the wavelength satisfies usual phenomenological relations such as $q_c \propto \sqrt{\rho_i^0} \propto \sqrt{\tau_s}$, where τ_s is the resolved shear stress.

When $D_{ij}^{(0)} < 0$, a diffusional instability may occur leading first to the formation of cellular structures which may be associated with the vein structure of the matrix. By increasing the stress intensity, the freeing of trapped dislocations becomes more and more efficient and, as a result of the highly anisotropic motion of free dislocations, this cellular structure is destabilized versus ladder-like structures with wavevectors parallel to the primary slip direction (Walgraef and Aifantis, 1986, Walgraef and Schiller, 1987).

The experimental observations of dislocation patterning in fatigued crystals oriented for double slip show interesting additional features (Jin and Winter, 1984). Effectively, according to the interactions between the dislocations associated to each slip system, different types of structure emerge. For example, if the interaction is strong, leading to locking effects of the Lomer-Cottrel type, PSB and ladder-like structures associated to the two primary slip systems develop in separated domains. On the other hand, if the interaction produces a third type of dislocation with a different orientation for its Burgers vector, cellular structures may emerge.

In this case, the slow mode dynamics associated with the reaction-transport model is, in the three-dimensional space ($0x$ and $0y$ being the primary slip directions) :

$$\tau_0 \partial_t \sigma = [\epsilon - d(q_c^2 + \nabla^2)^2 - \kappa \nabla_x^2 \nabla_y^2 + d_z \nabla_z^2]\sigma - v\sigma^2 - u\sigma^3 \tag{8}$$

Hence, steady state solutions corresponding to modulations of uniform amplitude are:

$$\sigma_x = 2R_0 \cos(q_c x + \phi_0), \quad \sigma_y = 2R_0 \cos(q_c y + \psi_0), \quad R_0 = \sqrt{\frac{\epsilon}{3u}}, \quad \phi_0, \psi_0 = \text{constant} \tag{9}$$

Since the two slip directions are equivalent as far as the pattern forming instability is concerned, one could also expect the formation of cellular patterns of the type:

$$\sigma_0 = A_x exp iq_c x + A_x^* exp - iq_c x + A_y exp iq_c y + A_y^* exp - iq_c y, \tag{10}$$

and of amplitude given by :

$$\begin{aligned} \tau_0 \partial_t A_x &= \epsilon A_x + (\xi_\parallel^2 \nabla_x^2 + \xi_\perp^2 \nabla_y^2 + \xi_z^2 \nabla_z^2) A_x - 3u A_x (|A_x|^2 + 2|A_y|^2) \\ \tau_0 \partial_t A_y &= \epsilon A_y + (\xi_\parallel^2 \nabla_y^2 + \xi_\perp^2 \nabla_x^2 + \xi_z^2 \nabla_z^2) A_y - 3u A_y (|A_y|^2 + 2|A_x|^2) \end{aligned} \tag{11}$$

The linear stability analysis of the uniform amplitude steady state, $|A_x^0| = |A_y^0| = \sqrt{\epsilon/9u}$, shows that the square planforms are unstable. Hence the only possible structures correspond to PSB associated with one or the other slip systems, or to mixed structures corresponding to regions where the two slip systems dominate alternatively. But as a consequence of the instability of square patterns, when one wall structure nucleates in some region of the crystal it empedes the other one to develop in the same region. For example, if one considers amplitude variations in one direction only, chains of alternating kink-antikink solutions of the amplitude equation (11) exist, corresponding to a succession of domains where A_x and A_y are alternatively zero and nearly equal to $\sqrt{\epsilon/3u}$. As recently discussed by Coullet et al. (1987), these solutions are dynamically unstable with very long transients. But random pinning of these kink and antikink solutions may occur as a result of the interactions of these "defects" with inhomogeneities of the medium, or due to oscillatory oscillations between the defects themselves. Since the wall structures are initiated within the relatively regular vein structure of the matrix, the experimentally observed alternating domains with PSB associated to one or the other slip system are consistent with their description within the amplitude equation formalism.

CONCLUSION

The competition between defect motion, creation and interactions in stressed, strained or irradiated solids is able to destabilize uniform distributions and, as a result, induce the nucleation of various types of defect microstructures. This effect is of primary importance since it affects the macroscopic properties of the material. It may for example trigger the formation of micro-cracks, lead to strain oscillations and bursts, and are at the origin of the precipitation of solid solutions and of the nucleation of voids and of void lattices in irradiated metals and alloys (Walgraef, 1987, Martin and Kubin, 1988). Hence a coherent description of materials instabilities associated with the spatio-temporal organisation of defects will hopefully lead to a deeper understanding of these phenomena. Due to the strong nonequilibrium conditions under which they occur, classical mechanical or thermodynamical considerations are not sufficient and we need an important input from nonlinear dynamics and instability theory. Effectively, despite the huge complexity of the defect dynamics, even in the case of phenomenological models, valuable information may be obtained via the reduced dynamics near instability points leading to a possible description of the pattern selection and stability properties in the post-bifurcation regime. The selected structures strongly depend on the characteristics of the dominant nonlinearities near threshold but also on the underlying crystal structure via the anisotropies and, as in many other self-organization phenomena far from thermal equilibrium, on the experimental procedures. For the time being, the results obtained by the combination of bifurcation analysis, amplitude equation formalism and numerical simulations are in general qualitative agreement with experimental observations and the discussion outlined here should be viewed as a first step towards a general theoretical explanation of the macroscopic behavior of driven or degrading solids.

Acknowledgments: Stimulating discussions with E.C.Aifantis, P.Coullet, N.Ghoniem and J.E.Wesfreid are gratefully acknowledged. Parts of this work were supported by a NATO grant (082/84) for international collaboration in research .

References

F.Ackerman, L.P.Kubin, J.Lepinoux and H.Mughrabi, *Acta Metall.* **32** (1984), p. 715.

V.Balakrishnan and C.E.Bottani, "*Mechanical Properties and Behavior of Solids : Plastic Instabilities*," World Scientific, Singapore, 1986.

L.Buchinger, S.Stanzl and C.Laird, *Phil.Mag.* **50** (1984), p. 275.

R.Bullough, B.L.Eyre and K.Krishan, *J.Nucl.Mat.* **44** (1975), p. 121.

R.W.Cahn, *Nature* **329** (1987), p. 284.

P.Coullet, C.Elphick and D.Repaux, *Phys.Rev.Lett.* **58** (1987), p. 431.

U.Essmann and H.Mughrabi, *Phil.Mag.* **40** (1979), p. 731.

J.H.Evans, *Mater.Sci.Forum* **15-18** (1987), p. 869.
N.M.Ghoniem and G.L.Kulcinski, *Radiation Effects* **39** (1978), p. 47.
J.Guckenheimer, J.Moser and S.Newhouse, "*Dynamical Systems*," Birkhauser, Boston, 1980.
V.Hlavacek, "*Dynamics of Nonlinear Systems*," Gordon and Breach, New York, 1985.
N.Y. Jin and A.T.Winter, *Acta Metall.* **32** (1984), p. 1173.
W.Jäger, P.Ehrhart and W.Schilling, in "*Nonlinear Phenomena in Materials Science*," G.Martin and L.P.Kubin eds., Transtech, Aedermannsdorf (Switzerland), 1988, p. 279.
K.Krishan, *Radiat.Eff.* **66** (1982), p. 121.
G.Martin, *Phys.Rev.* **B30** (1984), p. 1424.
G.Martin and L.P.Kubin eds., "*Nonlinear Phenomena in Materials Science*," Transtech, Aedermannsdorf (Switzerland), 1988.
S.M.Murphy, *Europhys.Lett* **3** (1987), p. 1267.
G.Nicolis and I.Prigogine, "*Self-Organization in Non Equilibrium Systems*," Wiley, New York, 1977.
G.Nicolis and F.Baras, "*Chemical Instabilities, Applications in Chemistry, Engineering, Geology and Materials Science*," Reidel, Dordrecht, 1983.
C.Normand, Y.Pomeau and M.Velarde, Convective Instabilities, a Physicist Approach, *Rev.Mod.Phys.* **49** (1977), 581-624.
R.Ocelli, E.Guazzelli and J.Pantaloni, *J.Physique Lett.* **44** (1983), p. L567.
C.Schiller and D.Walgraef, *Acta Metall.* **36** (1987), p. 563.
H.L.Swinney and J.P.Gollub, "*Hydrodynamic Instabilities and the Transition to Turbulence*," Springer, Berlin, 1981.
T.Tabata, H.Fujita, M.Hiraoka and K.Onishi, *Phil.Mag.* **A47** (1983), p. 841.
N.Thompson, N.Wadsworth and N.Louat, *Philos.Mag* **1** (1956), p. 113.
D.Walgraef and E.C.Aifantis, *J.Appl.Phys.* **58** (1985), p. 688.
D.Walgraef and E.C.Aifantis, *Int.J.Engng.Sci.* **23** (1986), 1351, 1359,1364.
D.Walgraef and C.Schiller, *Physica D* **27** (1987), p. 423.
D.Walgraef, "*Patterns, Defects and Microstructures in Nonequilibrium Systems*," Martinus Nijhoff, Dordrecht, 1987.
D.Walgraef and E.C.Aifantis, *Res Mechanica* **23** (1988), p. 161.
D.Walgraef, in "*Nonlinear Phenomena in Materials Science*," G.Martin and L.P.Kubin eds., Transtech, Aedermannsdorf (Switzerland), 1988, p. 77.
D.Walgraef, "*Structures Spatiales loin de l'Equilibre*," Masson, Paris, 1988.
D.Walgraef and N.M.Ghoniem, Spatial Instabilities and Dislocation Loop Ordering in Irradiated Materials, UCLA preprint.
J.E.Wesfreid and S.Zaleski, "*Cellular Structures in Instabilities*," Lecture Notes in Physics 210, Springer, Berlin, 1984.
A.T.Winfree, "*The Geometry of Biological Time*," Springer, Berlin, 1980.

NUMERICAL SIMULATION OF PERSISTENT SLIP BAND FORMATION*

C. Schiller and D. Walgraef

Service de Chimie Physique, CP 231, Université Libre de Bruxelles
Boulevard du Triomphe, 1050 Bruxelles, Belgium
Bitnet nr.: ulbg028 bearn

During the transformation of the vein structure into the persistent slip band structure in a copper crystal under fatigue (i.e. cyclic stress) the dislocation distribution undergoes complex microscopic changes. A roughly homogeneous distribution is converted into a periodic roll structure of many hundreds of wavelengths that crosses the crystal through its whole cross section. The essential features of a dislocation arrangements are incorporated in a description of the dislocations by a space and time dependent density $\varrho(x,t)$. It has also been recognized that the dislocation population can be divided into two groups, namely mobile dislocations, mainly of the screw type, which are free to move in the slip plane, and trapped ones, consisting mainly of dislocation dipoles with Burgers vectors of opposite sign; they move much less, since they are immobilized by the dislocation forest and also because they are only sensitive to dipole stress fields. For each family, a balance equation is set up. The source and the flux terms are determined by the following basic processes:

• On time scales much larger than the fatigue process period the dislocations move by an effective *diffusion* process. This is a consequence of the back and forward motion of the mobile dislocations during each fatigue cycle and introduce for both dislocation families a flux term $\Delta\varrho$ in the balance equations.

• the *pinning* of mobile dislocations by dipoles or multipoles leads to a sink term in the balance equation for the mobile dislocation density, and to a corresponding source term in the other equation. This term will, in general, be nonlinear and its exact form depends on the microscopic mechanisms of the process.

• the *production* of dislocations by Frank-Read or Bardeen-Herring sources adds another source term $g(\varrho_i)$.

• the *freeing* of trapped dislocations by the stress means their transformation to mobile ones. This is due either to the breaking of the dipoles by the applied stress or to the destruction of the dipoles by the interaction with arriving secondary dislocations. This process introduces a linear source term $b\varrho_i$, where the parameter b, specifying the freeing rate, is zero up to a critical level of stress, after which it reaches suddenly a finite value. This value is given by the strain rate, the macroscopic quantity determining the freeing of dislocations.

*This chapter is an abstract of the paper with the same title by C.Schiller and D.Walgraef, Acta Met. **36** (1988) 563–573.

New Trends in Nonlinear Dynamics and Pattern-Forming Phenomena
Edited by P. Coullet and P. Huerre
Plenum Press, New York, 1990

With the processes described above, the equations for the mobile and immobile dislocation densities read:

$$\begin{aligned} \dot{\varrho_i} &= g(\varrho_i) - b\varrho_i + \gamma \varrho_{mo} \varrho_i^2 + D_i \Delta \varrho_i \\ \dot{\varrho_{mo}} &= b\varrho_i - \gamma \varrho_{mo} \varrho_i^2 + D_m \Delta \varrho_{mo} \end{aligned} \tag{1}$$

where $g(\varrho_i)$ is the function describing the generation of the slow dislocations by the applied stress. The freeing rate b plays the role of the so-called bifurcation parameter; its magnitude determines the type of the solutions of eq. (1). We limit the following discussion to one spatial dimension, the direction of primary slip, since PSB are essentially one-dimensional periodic structures. Taking the available data on the diffusion coefficients, dislocation densities and annihilation lengths into account, one can deduce a set of numerical parameters that is compatible with the observations.

In the experiments, the persistent slip bands usually start at the border of the crystal; this is due mainly by a different value of the local stress. This situation can be simulated by taking a value for the bifurcation parameter b which depends on the position inside the crystal. Since b depends on the stress, it is expected to be larger at the border than in the bulk. This is simulated in fig. 1, where a position dependent b was taken. The evolution shows a beginning of a pattern at the border as expected; second, it shows that slightly irregular wavelengths can appear in a PSB. One sees that thin walls of high concentration of dislocation dipoles are formed, spaced by wide valleys of low concentration; the concentration of mobile dislocations is so small that it is almost not visible in the pictures. The structure remains stable in time after it has been built up. Its wavelength is $\lambda = 1.4\,\mu$m, quite far the critical value of $\lambda = 0.6\,\mu$m, which is predicted to be the pattern wavelength only near b_c. One sees that numerical simulations are necessary for situations far away from the bifurcation point. The walls occupy about 9% of the length, the value observed being in the range 10% to 20%. Third, the pattern starts already with a very small perturbation of the steady state; it is so small that it is invisible in fig. 1. These findings show that the high local stress concentration at the surface considerably enhances the formation of PSB's. These results are confirmed by the observations of PSB formation near regions of surface damage; it is known that regions with a higher local stress give rise to PSB much before the surroundings. In fig. 1 the propagation speed is predicted by nonlinear analysis to be (for $b \approx b_c$)

$$v = 4\frac{q_c^2 D_m}{T}\sqrt{D_i(b - b_c)} \tag{2}$$

near the bifurcation. In our simulations we find a propagation of 0.4 to 0.5 wavelengths per time unit, whereas eq. (2) gives a value of 1.7. The difference is due to the distance from the bifurcation point. The present model estimate is too small compared with the experiments. Valuable information could be gained to improve eq. (1) if the formation process itself could be observed. Work is in progress to find better estimates on the front propagation velocity and its stability in one and two dimensions.

The presented model is a first step towards a description of PSB by kinetic equations for dislocation densities. The approach is very encouraging since it reproduces many features observed about PSB's:

◇ A pattern with a *regular* wall distance forms above a stress threshold. With physically reasonable values for the model parameters the true wavelength can be reproduced.

◇ The expression for the wavelength behaves as expected for different temperatures, materials and dislocation properties. It is proportional to the inverse square root of the dislocation density.

◇ The thin dislocation walls made up of dislocation dipoles spaced by large valleys containing mostly screw dislocations are reproduced.

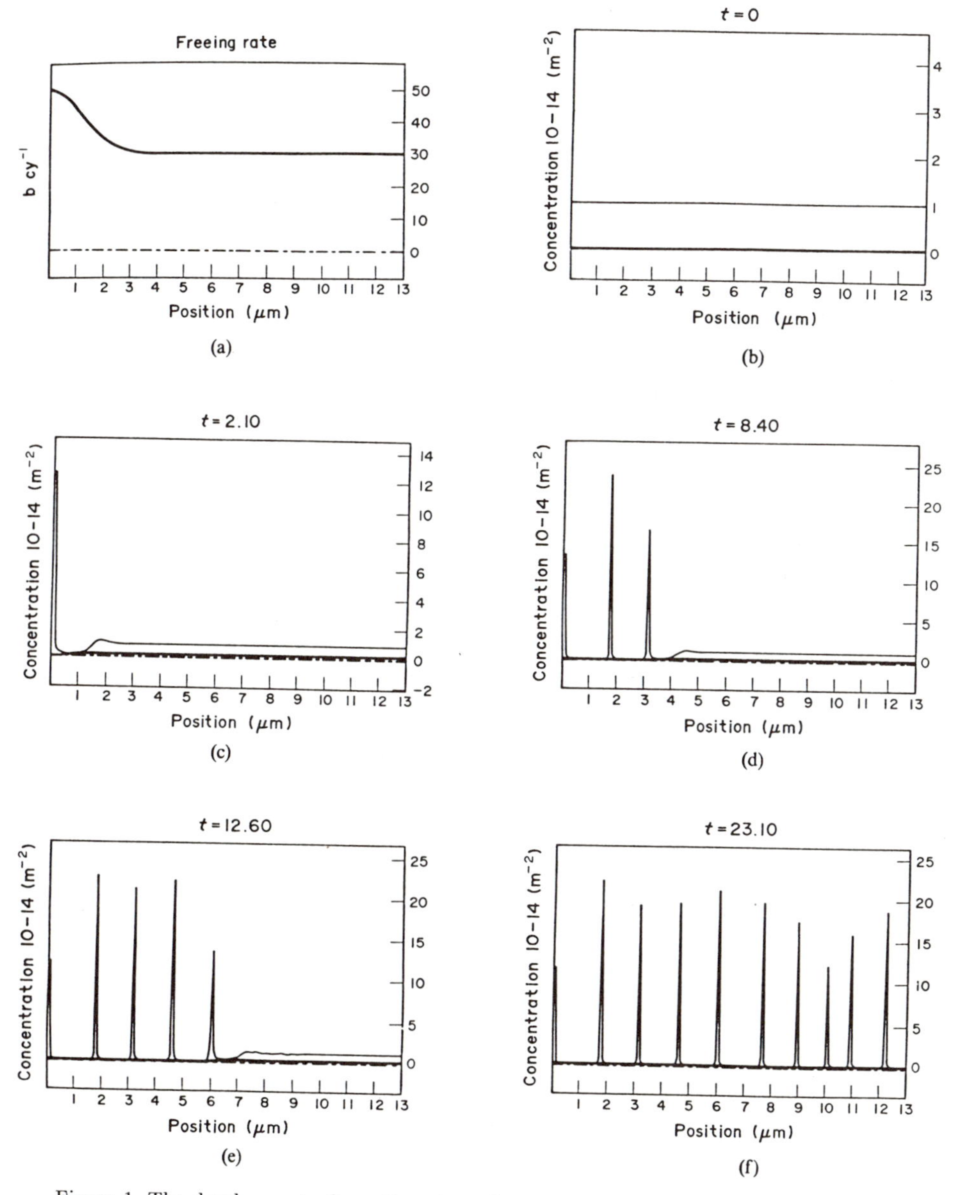

Figure 1. The development of a pattern in a situation with high local stresses at the left border. The stresses lead to a higher value for b at the border, shown in the first graph.

◇ Dislocation densities and their relative percentages are found of the correct order of magnitude. The values of the model parameters were determined very summarily; nevertheless no striking contradiction with experiments occurs.
◇ The formation process reproduces well the growth in its initiation at the specimen border and the inhibition by surface treatment; the time scales are of the correct order of magnitude.

On the other hand, several aspects must be improved:
◇ The description of interactions between dislocations and of the multiplication process is oversimplified and gives no connection to the individual microscopic processes, since they are globally represented by a cubic nonlinearity.
◇ Most parameters in the model are assumed to be independent of stress, temperature or other changing conditions. The local stress was assumed to be equal to the applied stress. The sensitivity of these results on these influences should be tested.
◇ The effect of more realistic boundary conditions including the formation of extrusions and intrusions at the border should be studied.
◇ Up to now only one dimensional aspects of the reaction-diffusion model were tested. Hence important PSB features such as their relatively constant width or their initiation from the matrix structure could not be discussed.

Most of the drawbacks will be eliminated in forthcoming refinements of the model; preciser modelling of the dislocation interactions will allow for better quantitative reproduction and possibly a true calculation of the wall distance.

In summary one can say that the model presented (eq. 1) shows that a phenomenological description of dislocation populations with the help of kinetic equations for the dislocation densities based on the competition between their movement and their multiplication and interactions, reproduces, despite its simplicity, a large collection of experimental data on Persistent Slip Bands (PSB); the main points of the approach are confirmed and it has been shown that the formation of PSB is an example of self-organisation.

Bibliography

[1] H. Mughrabi, F. Ackermann & K. Herz, *Persistent slipbands in fatigued face-centered and body-centered cubic metals,* in J.T. Fong ed., *Fatigue Mechanisms,* Proceedings of an ASTM-NBS-NSF symposium, Kansas City, Mo., May 1978, ASTM-STP 675, pp. 69–105, American Society for Testing and Materials, 1979.

[2] D. Walgraef & E.C. Aifantis, *On the formation and stability of dislocation patterns I, II, III,* Int. J. Eng. Sci. **23** (1985) pp. 1351–1358, 1359–1364, 1365–1372.

[3] G. Dee & J.S. Langer, *Propagating pattern selection,* Phys. Rev. Letters **50** (1983) pp. 383–386.

TRANSITIONS TO MORE COMPLEX PATTERNS IN THERMAL CONVECTION

F.H. Busse and R.M. Clever

Institute of Physics, University of Bayreuth, 8580 Bayreuth, FRG, and Institute of Geophysics and Planetary Physics, UCLA, Los Angeles, USA

SUMMARY

The instabilities of convection rolls in a fluid layer heated from below are reviewed and results of recent computations on three-dimensional knot convection flow and on travelling wave convection are reported. Periodic boundaries in the horizontal directions and rigid, thermally well conducting boundaries at top and bottom have been assumed. The analysis of the stability of the three-dimensional convection patterns indicates transitions to a variety of time-dependent forms of convection which in part can be related to experimental observations. Further experimental work on low Prandtl number fluids appears to be desirable for the study of convection in the form of standing waves.

1. INTRODUCTION

Thermal convection in an extended fluid layer heated from below is unique among hydrodynamic instabilities in that a maximum number of symmetries is available and in that more transitions from simple to more complex patterns of flow can be followed than in other fluid systems. Moreover, there are a number of external parameters available which either reduce the symmetry such as an imposed horizontal magnetic field or do not change the symmetry of the fluid layer, but provide a useful control parameter and introduce new degrees of freedom such as a vertical magnetic field. In this review we give a brief survey of patterns of nonlinear solutions and of their instabilities using symmetry considerations as a guide. In section 2 the instabilities of convection rolls will be considered, in section 3 three-dimensional knot-solutions will be discussed and in section 4 some properties of travelling wave convection in low Prandtl fluid will be outlined. The paper closes with some concluding remarks on open problems and work in progress.

2. INSTABILITIES OF CONVECTION ROLLS

We consider steady two-dimensional solutions of the basic equations describing convection rolls in a horizontal layer heated from below. The

New Trends in Nonlinear Dynamics and Pattern-Forming Phenomena
Edited by P. Coullet and P. Huerre
Plenum Press, New York, 1990

Table 1. Properties of Instabilities of Convection Rolls in a Layer with Rigid Boundaries. The quantity $2\pi/\Omega$ is a measure of the circulation time in rolls.

Instability	Symmetry Class	b	d	σ_i	Symbol
cross-roll	OC	$>\alpha_c$	0	0	CR
knot	OC	$<\alpha_c$	0	0	KN
dual blob	OC	$\lesssim\alpha_c$	0	$\sim 2\Omega$	DB
zig-zag	ES	$\lesssim\alpha_c$	0	0	ZZ
oscillatory	ES	$<\alpha_c$	0	$\sim\Omega$	OS
single-blob	EC	$>\alpha_c$	0	$\sim\Omega$	SB
skewed-varicose	E	$<<\alpha_c$	$<<\alpha_c$	0	SV
Eckhaus	E	0	$<<\alpha_c$	0	EC

deviation θ of the temperature from the temperature distribution of the state of pure conduction can be used as the representative dependent variable of the convection flow. Using a Cartesian system of coordinates with the z-coordinate in the direction opposite to gravity and the x-coordinate in the direction of the axis of the rolls we can write the solution for θ in the form

$$\theta = \sum_{m,n} b_{mn} \cos m\alpha y \sin n\pi(z+\tfrac{1}{2}) \tag{2.1}$$

Anticipating that the solutions of interest possess a vertical plane of symmetry, we have located the origin on the intersection between such a plane and the median plane of the layer. Assuming a Boussinesq fluid and symmetric boundary conditions we find that the convection rolls bifurcating from the static solution of the problem at the critical Rayleigh number R_c satisfy the additional symmetry property

$$\theta(y,z) = -\theta(\tfrac{\pi}{\alpha}-y,-z) \tag{2.2}$$

According to this property all coefficients b_{mn} with odd m+n vanish in the representation (2.1). Solutions of the form (2.1) can be obtained for a wide range of the R-α-P parameter space where R is the Rayleigh number and P is the Prandtl number. We refer to Busse (1967), Clever and Busse (1974, 1978), Busse and Clever (1979) and Bolton, Busse and Clever (1986).

General three-dimensional infinitesimal disturbances of the steady solution given by (2.1) can be written in the form

$$\theta = \sum_{m,n} \tilde{b}_{mn} \exp\{im\alpha y+idy+ibx+\sigma t\} \sin n\pi(z+\tfrac{1}{2}) \tag{2.3}$$

where Floquet's theorem has been used. Because of property (2.2) the general disturbances of the form (2.3) separate into two classes. Those of class E have vanishing coefficients $\tilde{b}_{mn}$ for odd n+m, while the coefficients of class O vanish for even n+m. It turns out that the real

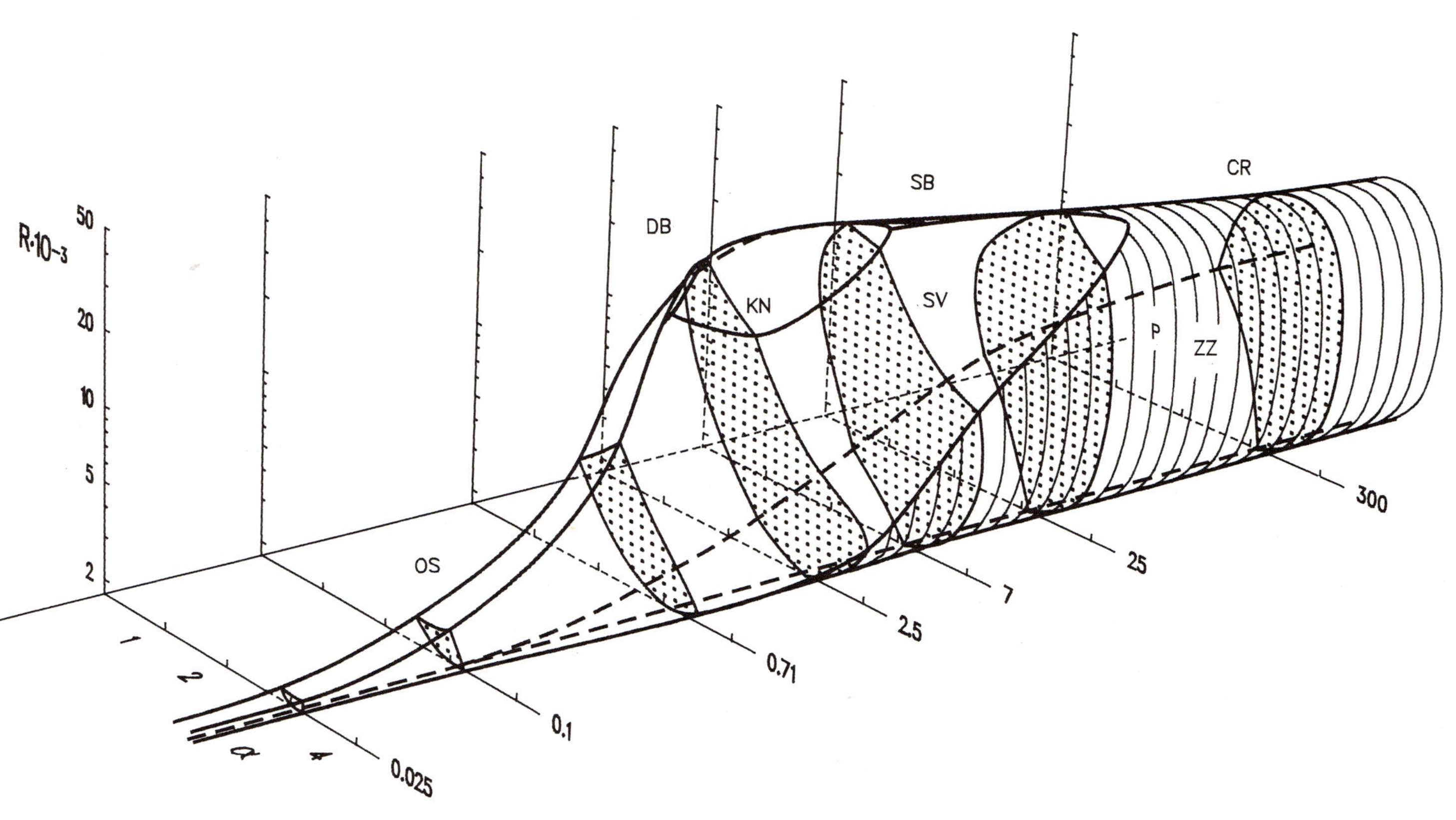

Figure 1. Region of stable roll solution in the R-α-P-parameter space. The stability region is bounded by several surfaces corresponding to the instabilities listed in Table 1.

part of the growthrate σ often reaches a maximum for $d = 0$ in which case an additional symmetry appears. The disturbances can be separated in this case into those that are symmetric in y and those that are antisymmetric. We shall denote the former subclass by C (for cosine) and the latter subclass by S (for sine). All types of disturbances that have been determined as instabilities of the steady roll solution (2.1) are listed in table 1. As is evident from this table most of the instabilities of convection rolls can be distinguished by their symmetry property if the symmetry in time is also included. But the knot- and the crossroll-instability have the same symmetry properties and can be distinguished as two separate maxima of the growthrate σ as a function of b when they appear together. The boundaries of the region of stable rolls corresponding to the instabilities listed in table 1 are shown in figure 1. For details of the analysis we refer to the above-mentioned papers in which the stability diagrams have been obtained for special values of the Prandtl number. Using those results and a few diagrams of stability boundaries for fixed values of α and R the surfaces drawn in figure 1 have been obtained by interpolation.

3. THREE-DIMENSIONAL SOLUTIONS IN THE FORM OF KNOT CONVECTION

Except for the Eckhaus instability all instabilities of table 1 lead to three-dimensional forms of convection and in all cases with $d = 0$ the resulting convection flow assumes a regular pattern at least in the ideal case of uniformily excited disturbances. In the case of the skewed varicose instability it is not clear whether it can lead to a spatially periodic pattern of convection. If such a pattern exist it is likely to be unstable.

The simplest form of three-dimensional convection bifurcating from two-dimensional rolls is bimodal convection. This form of convection predominates in high Prandtl number fluids at Rayleigh numbers above $2 \cdot 10^4$. For experimental and theoretical studies of this form of convection we refer to Busse and Whitehead (1971), Whitehead and Chen (1976) and Frick et al. (1983). In terms of symmetry properties the transition to steady knot convection induced by the knot instability is analogous to the transition to bimodal convection; the actual appearance of the two types of three-dimensional convection is quite different, however. In figure 2 a typical example of knot convection is shown. The concentration of the rising and falling fluid into thin plumes is a characteristic feature of knot convection and is responsible for its name since the plumes become visible as "knots" in the shadow-graph observations (Busse and Clever, 1979). At higher Rayleigh number spoke like structures corresponding to fluid breaking away from the thermal boundary layers become noticeable around the stems of the plumes. These spokes indicate that knot-convection is a representative example of spoke pattern convection which is the ubiquitous form of convection observed in experiments with fluids of moderate Prandtl numbers for Rayleigh numbers from a few times 10^4 to 10^6.

In the experimentally realized spoke pattern convection the spokes are strongly time dependent and the origin of this time dependence can be traced to hot or cold blobs of fluids circulating around with the convection velocity. This kind of time dependence originates from the blob instabilities listed in table 1. Since they occur in the same general parameter regime as the knot instability, it is not surprising that they reappear as instabilities of steady knot convection once the latter form of convection has been established. Conversely, the characteristic features of knot convection will appear when the transition from rolls to standing blob oscillations has occurred first.

Figure 2. Steady knot convection for $R = 4 \cdot 10^4$, $P = 2.5$, $\alpha_x = 1.1$, $\alpha_y = 1.5$. Isotherms (upper row), lines of constant vertical velocity (middle row) and streamlines of the component of the velocity field associated with vertical vorticity (lower row) are shown in the x,y-plane for $z = -0.3$ (left side) and $z = 0$ (right side) except for the upper left picture which shows $\partial(\theta - \bar{\theta})/\partial z$ at $z = -0.5$ where $\bar{\theta}$ is the horizontally averaged component of θ. The truncation parameter $N_T = 14$ has been used in the representation (3.1), i.e. coefficients with $l+m+n > N_T$ have been neglected.

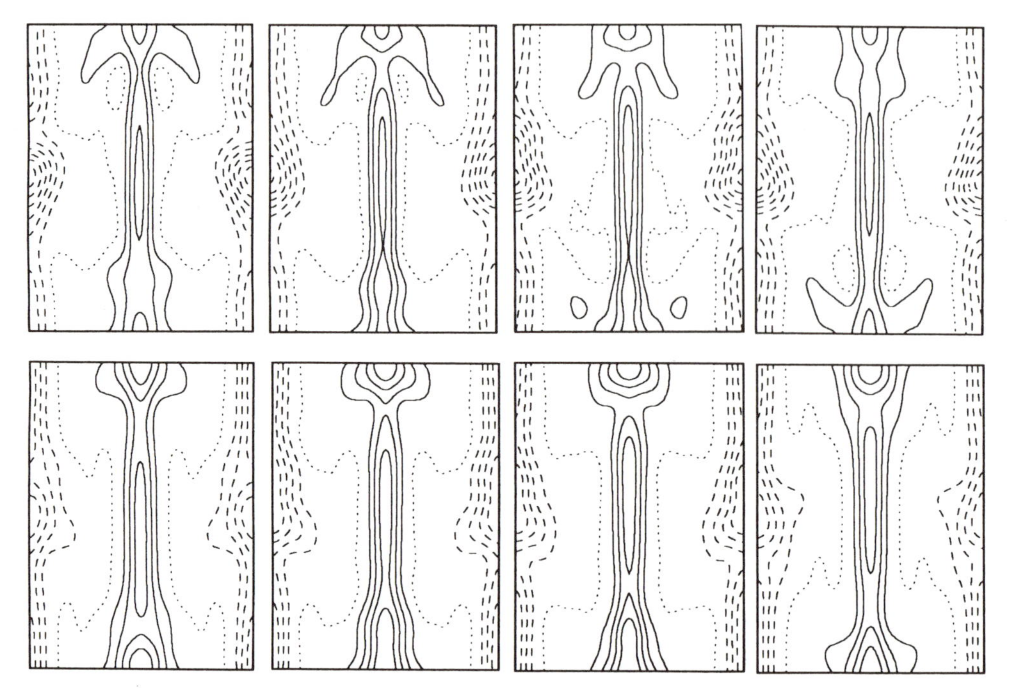

Figure 3. Convection in the form of oscillatory knot convection for $R = 3.5 \cdot 10^4$, $P = 4$, $\alpha_x = 1.5$, $\alpha_y = 2.0$. The upper and lower rows show lines of equal vertical velocity on the planes $z = -0.3$ and $z = 0$, respectively, for one half period at four equidistant times, 0.006 thermal time scales d^2/κ apart.

But this latter case can not be investigated as easily since the blob instabilities break two symmetries at once namely the invariances with respect to translation along the axis of the rolls and with respect to translation in time.

Since steady knot convection can be described by an expression for θ of the form

$$\theta = \sum_{l,m,n} b_{lmn} \cos l\alpha_x x \cos m\alpha_y y \sin n\pi(z+\tfrac{1}{2}) \tag{3.1}$$

where non-vanishing coefficients b_{lmn} are obtained only for even l+m+n, the general disturbances of the form

$$\tilde{\theta} = \sum_{l,m,n} \tilde{b}_{lmn} \exp\{il\alpha_x x + idx + im\alpha_y y + iby + \sigma t\} \sin n\pi(z+\tfrac{1}{2}) \tag{3.2}$$

separate in two classes, those with non-vanishing coefficients $\tilde{b}_{lmn}$ for even l+m+n and those with odd l+m+n. In the special case $b = d = 0$ additional symmetries with respect to the x- and y-dependences of the disturbances are realised and eight different types of instabilities can be distinguished

$$\text{ECC, ESC, ECS, ESS, OCC, OSC, OCS, OSS} \tag{3.3}$$

The notation used here is analogous to that introduced in table 1 in that the first C or S denotes symmetry or antisymmetry in x while the last letter does the same with respect to the y-dependence. According to experimental observations the most important instabilities do indeed correspond to case b = d = 0 and the computations of Clever and Busse

Figure 4. Convection in the form of symmetric travelling waves for $R = 3000$, $P = 0.025$, $\alpha_x = 2.2$, $\alpha_y = 2.9$. Lines of constant vertical velocity (left), of isotherms (middle) and of constant streamfunction ψ (right) are shown in the plane $z = 0$. The component of the velocity field associated with vertical vorticity is described by $\tilde{v} = \nabla\times\tilde{k}\psi$ where $\tilde{k}$ is the vertical unit vector.

(1988) have thus been restricted to this case. A special property of the instabilities with CS or SC in (3.3) is that they include a mean flow component which is symmetric with respect to the median plane z = 0 in the case of class E and antisymmetric in the case of class O. Such a property is not realised for the instabilities of table 1 since those occur all for finite values b or d. Instabilities corresponding to every subset listed in (3.3) seem to occur in the case of knot convection depending on the parameters α_x, α_y, and P. The mechanism of instability is similar to the dual blob instability of table 1 in all cases and only phase relationships differ somewhat for the different cases (3.3). An example of oscillatory knot convection induced by the ESC instability is shown in figure 3.

4. TRAVELLING WAVE CONVECTION

Symmetric travelling wave convection is realized through the Hopf bifurcation from steady rolls in form of the oscillatory instability. In the frame of reference moving with wavespeed c the solution describing travelling wave convection becomes steady,

$$\theta = \sum_{lmn} \{b_{lmn}\cos l\alpha_x(x-ct)+\hat{b}_{lmn}\sin l\alpha_x(x-ct)\}\begin{Bmatrix}\cos m\alpha_y y\\ \sin m\alpha_y y\end{Bmatrix}\sin n\pi(z+\tfrac{1}{2}) \quad (4.1)$$

where the upper function in the wavy bracket applies for even l and the lower function corresponds to odd l. As in the case of rolls the coefficients b_{lmn} and b_{lmn} are non-vanishing only for even m+n. A typical example of finite amplitude travelling wave convection is shown in figure 4. It is worth noting that a symmetric mean flow along the x-direction is associated with the travelling wave and that it is proportional to the square of the wave amplitude as long as the latter is sufficiently small. While the mean flow is almost negligible at the Prandtl number of air it can become quite important in low Prandtl number convection (Clever and Busse, 1989).

In the analysis of the stability of symmetric travelling wave convection symmetry properties can be used again if the analysis is restricted to disturbances fitting the same horizontal periodicity interval as the travelling wave solution. Four classes of disturbances can be distinguished when the representation

$$\tilde{\theta} = \sum_{l,m,n} [(\tilde{b}_{lmn} \cos l\alpha_x(x-ct) + \check{b}_{lmn} \sin l\alpha_x(x-ct)] \begin{Bmatrix} \sin m\alpha_y y \\ \cos m\alpha_y y \end{Bmatrix} \sin n\pi(z+\tfrac{1}{2}) e^{\sigma t} \tag{4.2}$$

for the disturbance temperature field is assumed. The summation over m,n can be extended either over non-negative integers with even m+n or those with odd m+n, and the upper function in the wavy bracket can be chosen for odd l and the lower function for even l as in the case of the symmetric travelling wave solution, or the choice can be made the other way around. In the analysis of Clever and Busse (1987, 1989) it is found that only disturbances exhibiting the symmetry opposite to that of symmetric travelling wave convection in both respects give rise to growthrates σ with positive real parts in the parameter regime that has been investigated. For Prandtl numbers of the order unity these disturbances lead to asymmetric travelling wave convection while for small Prandtl numbers of the order 0.1 or smaller a transition to standing oscillations takes place. Experimental observations for the latter phenomenon do not seem to be available yet, however.

5. CONCLUDING REMARKS

As thermal convection evolves from simple rolls towards more complex patterns, various degrees of freedom become occupied such as vertical vorticity, time dependence and mean flow components of motion all of which vanish for rolls. The new degrees of freedom are introduced by a variety of instabilities which break one or more of the remaining symmetries of the convection flow. At moderate and high Prandtl numbers the thermal boundary layers play a crucial role in the mechannisms of instability while for low Prandtl number fluids the momentum advection terms exert a dominant influence.

The horizontally periodic convection flows reviewed in this paper represent idealized solutions that can not easily be realized in laboratory experiments. These solutions retain a maximum of symmetries at each step of the evolution of nonlinear convection and thereby they permit a separation of different mechanisms of instability. In typical experimental situations symmetries can be realized only approximately and thus most of the transitions correspond to imperfect bifurcations. In some instances small asymmetries can play an important role in triggering transitions. The transition to time dependent convection in high Prandtl number fluids is an example for this effect as is discussed in Busse (1978). Nevertheless, the analysis of strictly periodic patterns must precede systematic attempts to understand transitions in more general patterns.

Of particular interest has been the generation of mean flows by convection in horizontal layers. This phenomenon has been observed by Krishnamurti and Howard (1981) and an idealized mathematical model has been presented by Howard and Krishnamurti (1986). The analysis reviewed in the present paper suggests that the oscillatory instability of knot convection often gives rise to mean flows in the relevant Prandtl number regime. The different types of oscillatory knot convection with and without mean flow require further study. Transitions to quasiperiodic and chaotic time dependence seem to occur. But the sensitivity of these phenomena with respect to the truncation parameter demands expensive numerical computations.

ACKNOWLEDGEMENTS

The research reported in this paper has been supported by the Atmospheric Sciences Section of the U.S. National Science Foundation.

REFERENCES

Bolton, E.W., Busse, F.H., and Clever, R.M., 1986, Oscillatory instabilities of convection rolls at intermediate Prandtl numbers, J. Fluid Mech., 164: 469-485.

Busse, F.H., 1967, On the stability of two-dimensional convection in a layer heated from below, J. Math. Phys., 46: 140-150.

Busse, F.H., 1978, Nonlinear Properties of Convection, Rep. Progress Physics, 41: 1929-1967.

Busse, F.H., and Clever, R.M., 1979, Instabilities of convection rolls in a fluid of moderate Prandtl number, J. Fluid Mech., 91: 319-335.

Busse, F.H., and Whitehead, J.A., 1971, Instabilities of convection rolls in a high Prandtl number fluid, J. Fluid Mech., 47: 305-320.

Clever, R.M., and Busse, F.H., 1974, Transition to time-dependent convection, J. Fluid Mech., 65: 625-645.

Clever, R.M., and Busse, F.H., 1978, Large wavelength convection rolls in low Prandtl number fluids, J. Appl. Math. Phys. (ZAMP), 29: 711-714.

Clever, R.M., and Busse, F.H., 1987, Nonlinear Oscillatory Convection, J. Fluid Mech., 176: 403-417.

Clever, R.M., and Busse, F.H., 1988, Three-dimensional knot convection in a layer heated from below, J. Fluid Mech., in press.

Clever, R.M., and Busse, F.H., 1989, Nonlinear oscillatory convection in the presence of a vertical magnetic field, J. Fluid Mech., in press.

Frick, H., Busse, F.H., and Clever, R.M., 1983, Steady three-dimensional convection at high Prandtl number, J. Fluid Mech., 127: 141-153.

Howard, L.N., and Krishnamurti, R., 1986, Large-scale flow in turbulent convection: a mathematical model, J. Fluid Mech., 170: 385-410.

Krishnamurti, R., and Howard, L.N., 1981, Large-scale flow generation in turbulent convection, Proc. Nat. Acad. Sci. USA, 78: 1981-1985.

Whitehead, J.A., and Chan, G.L., 1976, Stability of Rayleigh-Bénard convection rolls and bimodal flow at moderate Prandtl number, Dyn. Atmosph. Oceans, 1: 33-49.

RAYLEIGH-BENARD CONVECTION AT FINITE RAYLEIGH NUMBER IN LARGE ASPECT RATIO BOXES

T. Passot, M. Souli, and A.C. Newell

Dept. of Mathematics
University of Arizona
Tucson, AZ 85721

I INTRODUCTION

Our goal in these studies is to derive the phase-amplitude-mean drift equations for the Oberbeck-Boussinesq equations at finite Prandtl number. As a step along the way, we have derived the phase-amplitude equation in the infinite Prandtl number limit. Because the Oberbeck-Boussinesq equations are rather complicated, we illustrate the ideas and numerical methods using the real Swift-Hohenberg equation, which is a reasonable approximation to the full equations in the inertia-less limit. We consider finite Rayleigh numbers, and large aspect ratio boxes. The phase equation will come from a solvability condition in a multi-scale expansion. The method is analogous to the nonlinear WKB technique developped by Whitham (1973). Unlike the earlier case treated by Cross and Newell (1984), the equation considered here has no analytical solution for the nonlinear steady case so that this problem will contain all the ingredients used in the real convection problem. A phenomenological mean drift is introduced in the phase equation; this will be well justified with Oberbeck-Boussinesq equations. Numerical results for the long wavelength instabilities of the phase are presented.

II ANALYSIS OF THE MODEL

We consider here the real version of the Swift-Hohenberg model, which reads :

$$\partial_t w + (\Delta + 1)^2 w - Rw + w^3 = 0. \tag{2.1}$$

where w is a real scalar field (e.g. vertical velocity), R is the stress parameter which plays the role of the Rayleigh number, and Δ is the two-dimensional Laplacian: $\Delta = \partial_x^2 + \partial_y^2$.

This equation admits steady periodic solutions of the form: $w(x, y) = f(\theta)$ where $\theta = \vec{k}.\vec{x}$, which correspond to straight parallel rolls of fixed wavevector $\vec{k}$. By requiring f to be periodic, we get the eikonal equation and thus the range of values of $|\vec{k}|$ for which f exists. We have: $(k^2\partial_\theta^2 + 1)^2 f - Rf + f^3 = 0$. Let us write f in

New Trends in Nonlinear Dynamics and Pattern-Forming Phenomena
Edited by P. Coullet and P. Huerre
Plenum Press, New York, 1990

a Galerkin basis: $f = \Sigma_{m=-\infty}^{m=+\infty} \alpha_m A^{|m|} e^{im\theta}$. We will get approximate solutions of f by substituting this series into (2.1) and taking leading order terms in power of A, in each coefficient of $e^{im\theta}$. This procedure has to be further refined for large values of the amplitude A by reexpanding each coefficient α_m again in powers of A. We will just give the first few terms of the equation satisfied by A, taking $\alpha_1 = 1$:

$$\frac{3}{(1-9k^2)^2 - R} A^4 - 3A^2 + R - (1-k^2)^2 = 0. \tag{2.2}$$

The requirement $A^2 \geq 0$ gives the boundary of the marginal stability curve: $\sqrt{1-\sqrt{R}} < k < \sqrt{1+\sqrt{R}}$. The critical value of R above which such conditions can be satisfied is $R_c = 0$ for which $k_c = 1$ is the only possible value of k.

We will now consider deformations of this straight roll pattern. Observations of convection experiments (Gollub and Carrier 1983) show that we can assume a large scale variation of the wavenumber k. Let us look for solutions of (2.1) in term of a multi-scale expansion, introducing new slow scales: $X = \epsilon x, Y = \epsilon y, T = \epsilon^2 t$, where ϵ is the inverse aspect ratio of the convective box. T is scaled according to the horizontal diffusion time scale, which is the relevant one in the convection context. The solution will still be locally periodic. Then one has to assume that $\vec{k} = \nabla_x \theta$ is a function of X, Y, T only; we thus are led to introduce a slow phase $\Theta = \epsilon\theta$ and we have: $\nabla_X \Theta = \vec{k}(X, Y, T)$. Our unknown w will be expanded in the form:

$$w = w_0(\theta, X, Y, T) + \Sigma_{p=1}^{p=+\infty} \epsilon^p w_p(\theta, X, Y, T) \tag{2.3}$$

where $w_0(\theta, X, Y, T) = f(\theta, A, k)$. The parameters A and k are related through the eikonal equation and now depend on large scales. We have a more general framework than the usual one used for deriving the N.W.S. equation (Newell and Whitehead, 1969), since now, no orientation of the rolls is prefered and R is not constraint to be close to R_c. Our small parameter ϵ is not any more the deviation from critical but is now the inverse aspect ratio. Substituting the expansion (2.3) into (2.1) will lead to a set of linear ordinary differential equations for each $p \geq 1$. The non-trivial solvability conditions obtained at each order are the successive approximations of the equation for $\Theta_T(X, Y, T)$. We obtain at order ϵ:

$$((k^2\partial_\theta^2+1)^2 - R + 3w_0^2)w_1 = -(\partial_\theta w_0)\Theta_T - (D_1(1+k^2\partial_\theta^2) + (1+k^2\partial_\theta^2)D_1)\partial_\theta w_0, \tag{2.4}$$

where $D_1 = 2\vec{k}.\nabla + \nabla\vec{k}$. The translational invariance of the problem assures us that $\partial_\theta f$ is a null eigenvector of the linear operator. This operator is self-adjoint, the Fredholm alternative on (2.4) will thus simply be: $< F_1 | \partial_\theta f >_\theta = 0$, where F_1 is the r.h.s. of (2.4). This gives rise to an equation of the form:

$$\tau(k)\Theta_T + \alpha(k)\Theta_{XX} + \beta(k)\Theta_{XY} + \gamma(k)\Theta_{YY} = 0 \tag{2.5}$$

where τ, α, β, γ are nonlinear functions of $k = |\nabla\Theta|$. Due to isotropy, (2.5) can be rewritten in the form: $\tau\Theta_T + \nabla.(\vec{k}B(k)) = 0$. Note here that a more complete expansion of the variable θ in the form: $\theta = \frac{\Theta}{\epsilon} + \Phi(X, T)$, which may be necessary

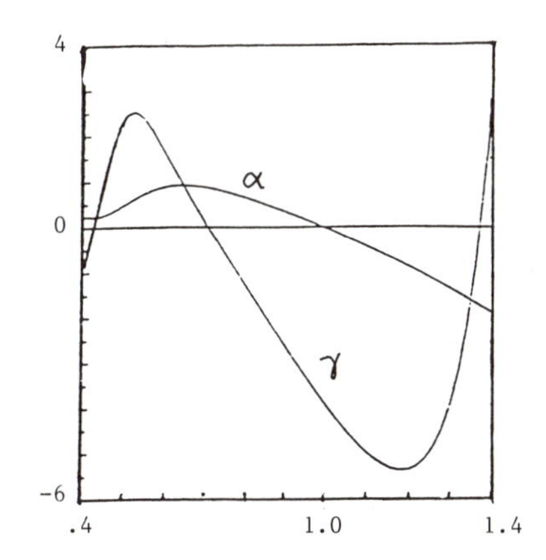

Fig. 1 Graph of α and γ for $R = 2$, displaying the stability region where they both are negative.

Fig. 2 Isovalues of the phase for a zigzag instability at $R = 2.$ for $k_0 = 0.8$ and $\mu = 0.01$.

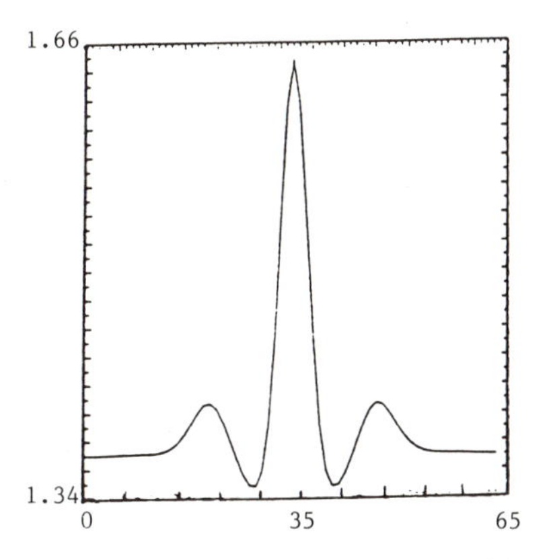

Fig. 3 This figure shows the graph of $k = \nabla\Theta$ for an Eckaus instability solving the 1D equation for k. At $T = 0$, $k = 1.38 - 0.01\cos(x)$, $R = 2$, $\mu = 0.07$. Observe a region where the gradient of the phase tends to be infinite corresponding to an elimination of a pair of rolls. The overshoots at the edges are due to the bilaplacian. We prolonge the coefficients α, β, γ for every value $k > k_{max}$ by constant values.

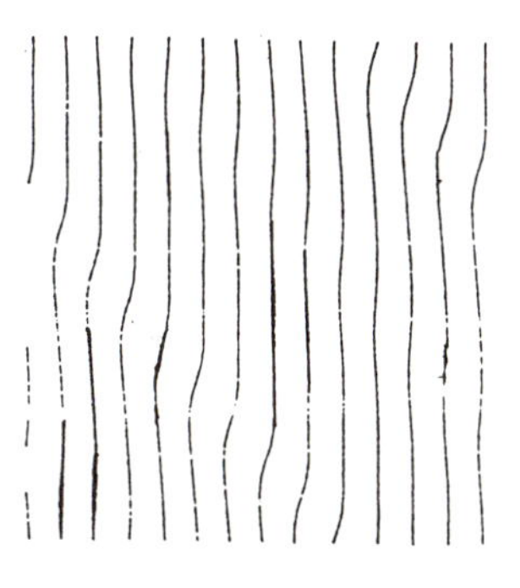

Fig. 4 Isovalues of the phase for a skew-varicose instability at $R = 2.$ for $k_0 = 1.37$, $\mu = 0.08$ and $\gamma = 20$.

when we are close to critical, gives no contribution to the phase equation at this order, since each term containing Φ is non secular. We now briefly describe the method used to compute the coefficients τ, B. For a set of values of k inside the marginal curve, f is computed with a Newton's method; in order to converge, the solution of (2.2) is taken at the first iteration. The Galerkin coefficients are then interpolated by cubic spline polynomials. This is needed to compute terms of the form $(\vec{k}\vec{\nabla})\partial_\theta f = (\vec{k}\vec{\nabla})k^2 \frac{\partial^2 f}{\partial(k^2)\partial\theta}$. The scalar product is then taken with $\partial_\theta f$. This is done assuming $k_y = 0$. B is then calculated by a Runge-Kutta method integrating: $\frac{dB}{dk} = \frac{B}{k}(\frac{\alpha}{\gamma} - 1)$ with: $\frac{dB}{dk} = \frac{\alpha}{k}$ at $k = k_B$ where $\gamma(k_B) = 0$.

III NUMERICAL RESULTS

The phase equation (2.5), determined from the solvability condition at the order ϵ, will be completed by adding a correction term which prevents small scale fluctuations to grow up. This term arises naturally when one applies the solvability condition at order ϵ^2. A phenomenological mean drift is added to the equation. One of the main contribution of the mean drift is that it changes the Eckaus instability into the skew varicose one. So the complete phase equation reads:

$$\tau(\Theta_T + \vec{k}.\vec{U}) + \nabla.(\vec{k}.B(k)) + \tau\mu\nabla^4\Theta = 0 \qquad (3.1)$$

$\vec{U}$ is the solenoidal component of $\gamma(\vec{k}\nabla.(\vec{k}A^2))$ where γ is a coupling constant. We are solving equation (3.1) for a periodic perturbation around the basic solution $\Theta = \vec{k}_0.\vec{X}$ of straight parallel rolls of wavenumber k_0, using a pseudo spectral method on a 32 by 32 grid. The temporal scheme is Adam's Bashforth coupled with Crank Nicolson for the implicit terms. The nonlinear term $\alpha(k)\Theta_{xx}$ is decomposed into $(\alpha - \alpha_{max}) * \Theta_{xx} + \alpha_{max} * \Theta_{xx}$; the first part is treated explicitly and the second implicitly. This allows for larger time steps.

From the resolution of equation (3.1), we recover the region of stable straight rolls (Fig. 1), and the long wavelength instabilities, such as the zigzag (Fig. 2) and Eckhaus (Fig. 3) instabilities without effect of mean drift as detailed by Cross and Newell (1984). We checked the boundaries of the stable region by directly solving the eigenvalue problem obtained by linearizing (2.1) around the nonlinear basic solution as done by Clever and Busse (1974). We also recover the skew varicose instabilities (Fig. 4) with a non-zero mean drift.

References

Clever, R. M., and Busse, F. H., 1974, Transition to time-dependant convection, J. Fluid Mech. 65:625.
Cross, M. C., and Newell, A. C., 1984, Convection patterns in large aspect ratio systems, Physica 10D:299.
Gollub, J. P., and Mc Carrier, A. R., 1983, Doppler imaging of the onset of turbulent convection, Phys. Rev. A26:3470.
Newell, A. C., and Whitehead, J. A., 1969, Finite bandwidth amplitude convection, J. Fluid Mech. 38:279.
Whitham, G. B., 1973, Linear and nonlinear waves. Wiley interscience, New York.

TRANSITION BETWEEN DIFFERENT SYMMETRIES IN RAYLEIGH-BENARD CONVECTION

S. Ciliberto, E. Pampaloni and C. Pérez-Garcia (†)

Istituto Nazionale di Ottica
Largo E. Fermi 6
50125 Arcetri-Firenze, Italia

Convective patterns with different symmetries can develop in Rayleigh-Bénard convection. When the transport coefficients of the fluid are temperature dependent (non-Boussinesq conditions) a hexagonal pattern is stable near threshold, but it is replaced by a pattern of rolls when the heating rate is increased still further. We present here recent experimental results on the transition between a hexagonal pattern and a pattern of rolls in convection in pure water under non-Boussinesq conditions. The convective cell is cylindrical with a liquid depth $d = 2.00$ mm and a diameter $D = 72$ mm. This gives an aspect ratio $\Gamma = D/2d = 18$. The Prandtl number of water at the mean temperature of 28^0 C is $P = 5.81$. The general features of the pattern are determined qualitatively by a shadowgraph technique. Heat-flow and optical measurements enable us to obtain quantitatively local and global characteristics of the pattern. The optical technique is based on the deflections of a laser beam that crosses the fluid layer in the vertical direction. This technique allow us to reconstruct the temperature field averaged on the vertical direction. More details about the experimental set-up can be found elsewhere.[1]

For $\Gamma = 18$ the corresponding critical Rayleigh number is $R_C = 1714$.[2] In our experiment this is equivalent to a critical temperature difference $\Delta T_C = 12.58$ K. This ΔT induces variations in the termal expansion coefficient and in the viscosity sufficient to give rise to an hexagonal pattern, but still small to be considered as perturbations in the nonlinear regime, as assumed in the theoretical calculations.[3] The critical temperature, obtained from heat-flow measurements and from the deflection technique turns out to be $\Delta T_C = 12.60 \pm 0.02$ K, in good agreement with the theoretical one.

The shadowgraph images in Fig. 1 show typical patterns when increasing $\varepsilon = (\Delta T - \Delta T_C)/\Delta T_C$. A transition from a hexagonal pattern and a pattern of rolls is clearly seen. Very near threshold the pattern consists of a regular array of hexagons surrounded with some zones in which rolls seem to destabilize the pattern. It is remarkable to observ that the pattern is very regular and stable. Fig. 1 shows that finally one set of rolls survives, although some grain boundaries with hexagons are present.

The deflection technique gives information of the temperature gradients ($\partial T/\partial x$, $\partial T/\partial y$) averaged on the vertical direction. After integration one can obtain the temperature field $T(x,y)$. In Fig. 2 two examples of the reconstruction of $T(x,y)$ in the central part of the scanning area for $\varepsilon = 0.02$

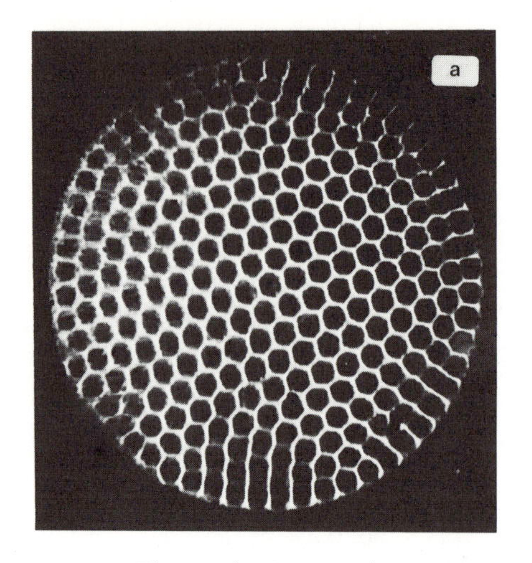

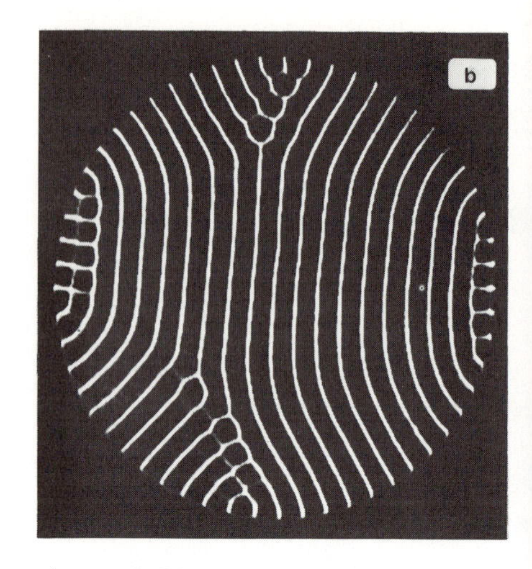

Figure 1. Convective patterns for a) $\varepsilon = 0.00$ and b) $\varepsilon = 0.14$

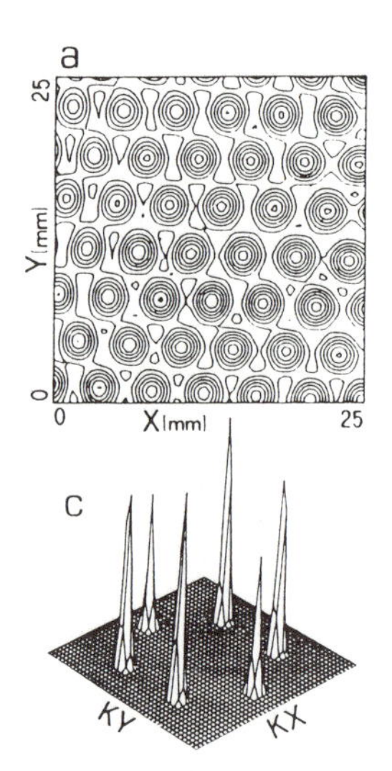

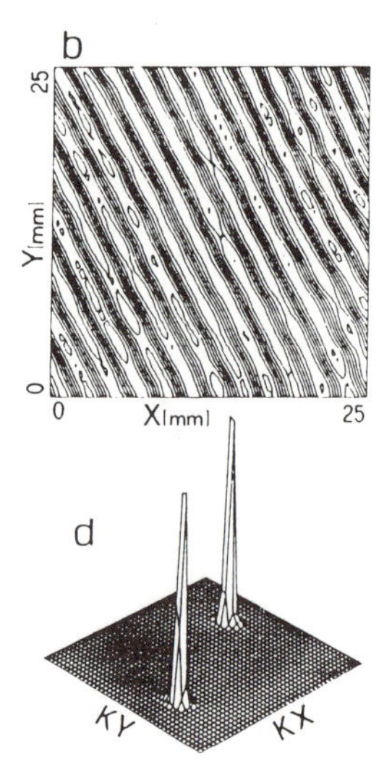

Figure 2. Temperature-field reconstruction from the horizontal gradients. a) Hexagonal pattern for $\varepsilon = 0.02$. b) Rolls for $\varepsilon = 0.14$. Fourier spectra of the temperature field in the full scanning area for the same values of ε as in a) and b): c) hexagons; d) rolls.

and $\varepsilon = 0.14$. Figs. 2 c-d illustrate the Fourier spectrum of $T(x,y)$ in the full scanning area. The peaks in these figures are very sharp and their amplitudes are comparable. This confirms that almost perfect patterns develop.

By integration of the amplitudes of these peaks one can obtain the normalized Nusselt number $N' = (N-1)\, R/R_c$,[2] (N is the Nusselt number). Measurements for different ε allow to determine the dependence of N on this parameter. The dependence is shown in Fig. 3. Experimental points lie on two straight lines, one for hexagons ($0.00 \leq \varepsilon \leq 0.04$) and one for rolls ($0.04 \leq \varepsilon \leq 0.14$). The slopes, obtained from a best fit of the data, are $\gamma_h = 0.86 \pm 0.03$ and $\gamma_r = 1.13 \pm 0.02$. They differ from the theoretical ones for $\Gamma = \infty$.[3] However, The [3]ratio between these slopes, which must be independent of Γ, is $\gamma_r / \gamma_h =$

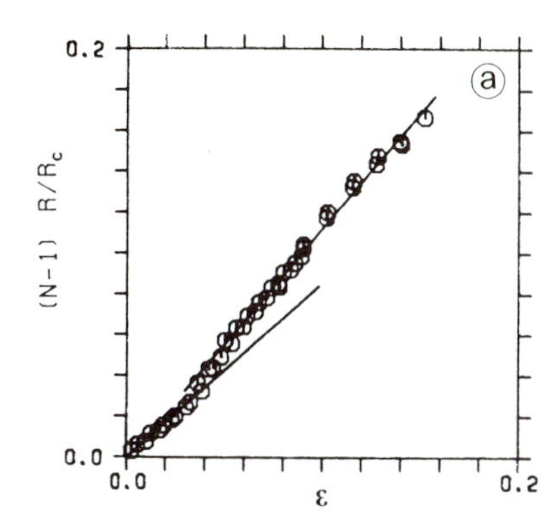

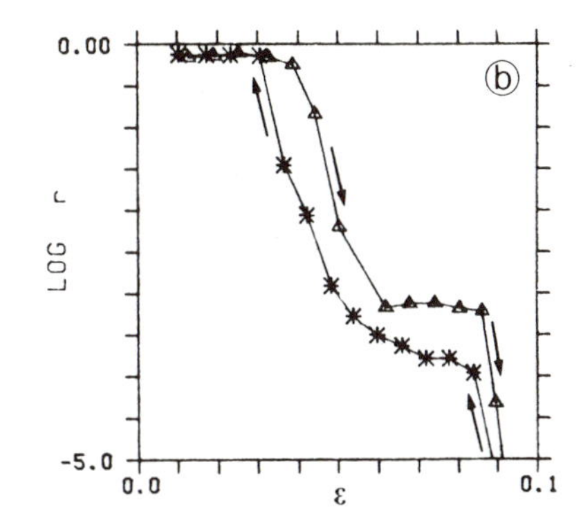

Figure 3. a) Normalized Nusselt number N'.b) Ratio between modes (see the text). The arrows indicate the different evolution of the vanishing modes.

1.30±0.05 in very good agreement with the theoretical value $\gamma_r / \gamma_h = 1.294$ obtained from ref. 3.

The square amplitudes ϕ_i of the three sets of rolls are obtained by integration of the square amplitude of the different modes on the conjugate peaks in the Fourier spectrum T(x,y)[1]. In order to characterize the transition from hexagons to rolls, we have plotted in Fig. 3b the ratio $r = (\phi_2 + \phi_3)/2\phi_1$ as a function of ε on a log-lin scale. Here ϕ_1 is the amplitude of the surviving mode and ϕ_2, ϕ_3 are the amplitudes of the two modes (at $2\pi/3$ rad with respect to the first one) which finally vanish. The high plateau $r \simeq 1$ corresponds to hexagons, while the transition to a pattern of rolls is characterized by the limit $r \to 0$. From this figure one can deduce that the transition occurs between $\varepsilon_r = (3.0\pm0.1)\times10^{-2}$ and $\varepsilon_h = 0.090\pm5\times10^{-3}$. (Here ε_r denotes the minimum value of ε for which the rolls are stable and ε_h the maximum ε for which hexagons are stable). It is smooth and hysteretic: smooth because the pattern of hexagons is not suddenly replaced by a pattern of rolls; hysteretic, because the pattern is not the same by increasing or decreasing ε. These values are lower than the theoretical ones[3] $\varepsilon_r = 5.3\times10^{-2}$ $\varepsilon_h = 0.18$ for a laterally infinite system. Shadowgraph and deflection techniques show that the rolls normal to the lateral boundaries tend to destabilize the hexagonal pattern, then reducing the slopes $_{h,r}$ and the thresholds $\varepsilon_{r,h}$.. Our observations suggest that the role of these defects cannot be fully characterized by global techniques, but these must supplemented by local measurements. Moreover, a general theory, taking into account the role of defects in such transition becomes necessary for enlightement on this problem.

ACKNOWLEDGEMENTS

We thank G. Ahlers, F.T. Arecchi, P. Bigazzi, P. Coullet, P. Hohenberg and M. Lücke for fruitful comments and L. Albavetti and M. D'Uva for very efficient technical assitance. This work has been partially supported by a EEC grant. One of us (C.P-G) also acknowledges the support of the Fundación " Conde de Barcelona " (Barcelona, Spain)

REFERENCES

(†) On leave from Dpto. Física, Univ. Autónoma de Barcelona.
([1]) S. Ciliberto, E. Pampaloni and C. Pérez-García, Phys. Rev. Lett., 61: 1198 (1988).
([2]) G. Ahlers, M.C. Cross, P.C. Hohenberg and S. Safran, J. Fluid. Mech., 110: 297 (1981).
([3]) F.H. Busse, J. Fluid. Mech., 30: 625 (1967).

NON LINEAR ANALYSIS OF TRAVELING WAVES IN BINARY CONVECTION

David Bensimon[1] and Alain Pumir[1,2]

[1] Laboratoire de Physique Statistique, Ecole Normale Supérieure
24, rue Lhomond, 75231 Paris Cedex, France

[2] SPT, CEN Saclay, 91191 Gif-sur-Yvette, France

Introduction

Much effort has been devoted recently to the phenomenon of convection in binary mixtures. A rich variety of dynamical states has been found experimentally[1-5]. Contrary to what happens for pure fluid convection the conducting state gets destabilized at finite frequency through a Hopf bifurcation[6]. In certain circumstances, a system of propagative rolls appears, with a subcritical transition when the control parameter is increased. These traveling waves have non trivial interactions that lead to novel and rather complex phenomena. Binary mixture convection is therefore an appropriate tool for studying chaotic behavior in large physical systems and the so called 'weak turbulence' regime.

The emergence of nonlinear traveling waves seems to be of primary importance in these systems; it is therefore appropriate to ask first simple questions about their structures. It has been recognized experimentally that when the temperature is increased, traveling waves of finite amplitude and finite velocity appear[1,2]. As the control parameter is increased, the velocity of propagation decreases to zero. At this point, steady convection sets in (until it is destroyed by secondary instabilities).

The linear stability is very well understood[6,7,8]. Besides the temperature stratification that leads to a destabilization of the conducting state of a pure fluid, when heated from below, a concentration stratification is generated, due to the Soret effect. It may either tend to bring light fluid in the hot region, thus making the conducting state even more unstable, and lowering the threshold of instability, or, on the contrary, tend to bring heavy fluid in the hot region, thus diminishing the mechanism of instability, and increasing the convection threshold. This depends on the separation ratio, $\psi = -(\alpha'/\alpha)S_T\, c_0(1-c_0)$, where α' is the solutal expansion, α the thermal expansion, S_T the Soret coefficient and c_0 the average concentration. When $\psi < 0$, the conducting state is destabilised through a Hopf bifurcation, corresponding to traveling waves. The transition is subcritical. This is due to the fact that the Lewis number, L is very small, and to a well understood property of the advection-diffusion of a passive scalar in a flow with recirculating zones, at high Péclet number. Namely, the concentration gradients are expelled from the zones with closed streamlines. Therefore, once the convection regime is started, the concentration stratification is destroyed, and one is left with an (almost) pure fluid, above threshold.

In fact, because of this effect, a weakly nonlinear analysis[9] analogous to what has been developed for pure fluid convection is not very useful, in most of the experimental situations. In order to make further progress, we develop in this paper a different strategy. As argued before, because of the expulsion of the concentration gradients from the rolls, a good starting

New Trends in Nonlinear Dynamics and Pattern-Forming Phenomena
Edited by P. Coullet and P. Huerre
Plenum Press, New York, 1990

point for doing perturbation theory is the convective state for a pure fluid. When $\psi = 0$, the equation for the concentration field decouples from the equations for the velocity and temperature field. By treating ψ as a small parameter, together with the convection amplitude ε, one has to solve the equation for the concentration field, γ, and then to study the effect of γ on the velocity and temperature fields. A traveling wave velocity is introduced as a solvability condition at first order, and an amplitude equation results from a second order calculation. By solving these equations, one obtains, among other things the velocity of traveling waves as a function of the the temperature gradient across the cell.

In this paper, we sketch the calculation of the properties of the traveling wave in a simplified case and present some results; a more detailed analysis can be found elsewhere[10].

Basic equations

A convenient limit for studying theoretically convection phenomena is the case where the Prandtl number, $Pr = \nu/\kappa$ is infinite, where ν is the viscosity of the fluid and κ the heat diffusion coefficient. Denoting the horizontal direction by x and the vertical direction by z, it is convenient to introduce the streamfunction φ, defined by : $\mathbf{u} = (-\partial_z\varphi, 0, \partial_x\varphi)$. The variables in the equations of motion are rescaled by using the height of the cell, d as the unit of length, the diffusion time $\tau_d = d^2/\kappa$ as the unit of time, δT as the unit of temperature and $\delta T\, S_T\, c_0(1-c_0)$ as the unit of concentration. The temperature fluctuation θ is also defined by $T = -z + \theta$. The Boussinesq equations read:

$$\nabla^4 \varphi + R\,\partial_x \theta = R\,\psi\,\partial_x \gamma \qquad (1a)$$

$$\nabla^2 \theta + \partial_x \varphi = \partial_t \theta + \mathbf{u}.\nabla\theta \qquad (1b)$$

$$L\nabla^2 \gamma = \partial_t \gamma + \mathbf{u}.\nabla\gamma \qquad (1c)$$

R is the usual Rayleigh number:

$$R = \frac{g\alpha d^3 \delta T}{\nu\kappa}$$

It should be noticed that we have slightly changed the definition of γ in Eqs (1a,b,c). In the literature, one rather uses a variable of concentration Γ related to ours by $\Gamma = \gamma + LT/(1-L)$. As it has already been stated before, when $\psi = 0$, the equation for the concentration field (1c) completely decouples from the equations for the velocity and temperature field. This is the starting point of our analysis. The boundary conditions we will consider are :

a/ free - free - permeable boundary conditions :

$$\varphi = \partial^2_z \varphi = 0 \text{ on } z = 0,1 \qquad (1d)$$

$$\gamma = 0 \text{ on } z = 0$$

$$\gamma = 1 \text{ on } z = 1$$

b/ rigid - rigid - impermeable boundary conditions on $z = 0,1$:

$$\varphi = \partial_z \varphi = 0 \qquad (1e)$$

$$\partial_z \gamma = 1 - \partial_z \theta$$

Now, we look for traveling wave solutions :

$$\zeta(x,z,t) = \zeta(x - v_0 t, z).$$

where ζ stands for φ, θ or γ. We define the distance to the critical Rayleigh number R_c by $\varepsilon^2 = (R-R_c)/R_c$, where R_c is the critical Rayleigh number for a pure fluid with the same boundary conditions. It is suggested by experimental data as well as by the linear stability results that, in the domain of existence of traveling wave solutions, $\varepsilon \approx (-\psi)^{1/2}$. The velocity of the traveling wave, v_0 is also expected to be of the same order as ε. Therefore, one can define $A = v_0/\varepsilon$ and $B = \psi/\varepsilon^2$. Expanding the fields φ, θ and γ in power series of ε, and substituting into equations (1a,b,c), one obtains, at order ε:

$$R_c^{-1} \nabla^4 \varphi_1 + \partial_x \theta_1 = 0 \qquad (2a)$$

$$\nabla^2 \theta_1 + \partial_x \varphi_1 = 0 \qquad (2b)$$

$$p_0^{-1} \nabla^2 \gamma_0 + \partial_x \gamma_0 - A^{-1} \mathbf{u}_1 . \nabla \gamma_0 = 0 \qquad (2c)$$

with $p_0 = v_0/L$. Equations (2a,b) correspond to the Rayleigh Bénard problem at infinite Prandtl number at the onset of convection. Equation (2c) is more difficult to solve. It is tractable analytically in the limit $A^{-1} \mid \mathbf{u}_1 \mid << 1$, as we will explain below, or in the limit : $A^{-1} \mid \mathbf{u}_1 \mid >> 1$ by boundary layer techniques[10], as explained in ref. 11,12. In between, it can be solved numerically. At order ε^2, one gets :

$$R_c^{-1} \nabla^4 \varphi_2 + \partial_x \theta_2 = B \partial_x \gamma_0 \qquad (3a)$$

$$\nabla^2 \theta_2 + \partial_x \varphi_2 = -A \, \partial_x \theta_1 + \mathbf{u}_1 . \nabla \theta_1 \qquad (3b)$$

$$p_0^{-1} \nabla^2 \gamma_1 + \partial_x \gamma_1 - A^{-1} \mathbf{u}_1 . \nabla \gamma_1 = A^{-1} \mathbf{u}_2 . \nabla \gamma_0 \quad (3c)$$

and, to order ε^3:

$$R_c^{-1} \nabla^4 \varphi_3 + \partial_x \theta_3 = -\partial_x \theta_1 + B \, \partial_x \gamma_1 \qquad (4a)$$

$$\nabla^2 \theta_3 + \partial_x \varphi_3 = -A \partial_x \theta_2 + (\mathbf{u}_2 . \nabla \theta_1 + \mathbf{u}_1 . \nabla \theta_2) \qquad (4b)$$

In order to solve the equations at order 2 and 3 for the velocity and for the temperature field, one has to solve a linear system, with a nontrivial kernel. One thus has to insist that the right hand side is orthogonal to the zero modes of the adjoint. These zero modes are $(\varphi_1, -\theta_1)$ and $(\partial_x \varphi_1, -\partial_x \theta_1)$. For equations (3a,b), the first orthogonality condition is automatically satisfied, whereas the second gives rise to a non trivial condition :

$$A = -B \frac{<\partial_x \gamma_0, \partial_x \varphi_1>}{<(\partial_x \theta_1)^2>} \qquad (5)$$

This equation determines the velocity of the traveling wave. Likewise, at next order (Eqs (4a,b)) the solvability conditions reads :

$$<\partial_x \theta_1, \varphi_1> + <\mathbf{u}_2 . \nabla \theta_1 + \mathbf{u}_1 . \nabla \theta_2, \theta_1> = B <\partial_x \gamma_1, \varphi_1> + A <\partial_x \theta_2, \theta_1> \qquad (6)$$

From this equation one can deduce the amplitude of the traveling wave. In the next section, we show how to solve equation (2c), (3c) in the limit of large velocity of the travelling wave (compared to the convection velocity) in the case of free - free - permeable boundary condition and determine the nature of the bifurcation.

Solution of the convection diffusion equation in the free-free-permeable case

Although this case is not relevant for experimental purposes, it is tractable analytically and provides a good understanding of the phenomenon. The solution of the Bénard problem with boundary conditions (1d) is :

$$\varphi_1 = C \sin(kx) \sin(\pi z)$$

$$\theta_1 = \frac{kC}{k^2 + \pi^2} \cos(kx) \sin(\pi z)$$

In this case, the critical Rayleigh number is $R_c = (k^2 + \pi^2)/k^2$, and $k = \pi/\sqrt{2}$. The equation to be solved for the concentration field is :

$$p_0^{-1} \nabla^2 \gamma_0 + \partial_x \gamma_0 - g \, \hat{u}_1 . \nabla \gamma_0 = 0 \qquad (7)$$

with $g = A^{-1} C = \varepsilon C / v_0 \approx |\mathbf{u}_1| / v_0$ and $\hat{\mathbf{u}}_1 = C^{-1} \mathbf{u}_1$. We look for a solution in power series of g :

$$\gamma_0 = \gamma_{00} + g\,\gamma_{01} + g^2\,\gamma_{02} + \cdots \qquad (8)$$

In order to enforce the boundary conditions (1d), the γ_{0n} must satisfy : $\gamma_{00}|_0 = 0$, $\gamma_{00}|_1 = 1$, and for $n = 1,2,\ldots$, $\gamma_{0n}|_{0,1} = 0$. The resulting set of equations can be solved to any desired order, and the solution can be written as :

$$\gamma_0 = z + A^{-1} f(g,p_0)\,\varphi_1 + \gamma_R \qquad (9a)$$

$$\gamma_R = \sum_{m,n} \{\gamma^s_{m,n} \sin(mkx) + \gamma^c_{m,n} \cos(mkx)\} \sin(n\pi z) \qquad (9b)$$

with $mn \neq 1$. For $m \neq 0$, $\gamma_{m,n}{}^c$ is $O(p_0^{-1})$, and thus negligible. The important point is that φ_1 and γ_R are orthogonal, as well as their x derivatives. In the small g limit, f can be evaluated to any desired order. In fact,

$$f^{-1}(g,p_0) = 1 + \left(\frac{k^2+\pi^2}{p_0 k}\right)^2 + \frac{k^2+\pi^2}{8} g^2 - \frac{(k^2+\pi^2)(k^2+9\pi^2)}{64} g^4 + O(g^6, g^2 p_0^{-2}) \qquad (10)$$

One finds for the traveling wave velocity :

$$v_0^2 = -\psi\, a_1 f(g, p_0) \qquad (11)$$

with $a_1 = (\pi^2 + k^2)^2/k^2$. At onset, $(g \to 0)$, the result of linear stability analysis for the frequency of the wave is recovered. It has to be noticed that the formula for the traveling wave velocity in terms of f is general : even with different boundary conditions, equations (9,11) still hold. Of course, equation (10) has to be modified.

In order to completely determine the properties of the traveling wave, one also needs to find the amplitude of convection (the constant C above). It can be obtained by using the solvability condition at next order. After some algebra, we find :

$$C\left(1 + Bf_1(g, p_0)\right) = b_1 C^3 \qquad (12)$$

with $b_1 = k^2/8(k^2+\pi^2)$ and $f_1 \approx f$. Together with the equation for the traveling wave velocity, this equation determines completely the properties of the roll pattern.

Near onset ($g \ll 1$), reporting equation (10) into (12), one obtains the equation :

$$(|B|^{-1} - 1)\, g - (a_1 b_1 - h_1) g^3 - (h_2 + h_1(h_1 - a_1 b_1))\, g^5 = 0 \qquad (13)$$

with $h_1 = (\pi^2+k^2)/8$ and $h_2 = (k^2+\pi^2)(k^2+9\pi^2)/64$. The bifurcation from the conduction state (characterized by $g = 0$) occurs at $|B| = 1$, that is for $\varepsilon = \sqrt{-\psi}$. The nature of the bifurcation is determined by the sign of the cubic term : supercritical (subcritical) when $(a_1 b_1 - h_1) > 0$ (<0). In the case of free-free-permeable boundary conditions, $(a_1 b_1 - h_1) = 0$. The bifurcation is still supercritical, the amplitude of convection growing like the quartic root of the distance to the threshold, $\varepsilon^2 + \psi$.

$$\varepsilon C = 3\left(\frac{-64\psi}{57}\right)^{1/4} (\varepsilon^2 + \psi)^{1/4} \qquad (14)$$

Such a behavior has been found in a somewhat similar context[13].

Rigid - rigid - impermeable boundary conditions

As explained before, much of the previous analysis can be also applied to the more physical case of rigid-rigid-impermeable boundary conditions. We present the results only (the details of the full analysis can be found elsewhere[10]). In this case, the bifurcation from the conducting state is inverted. (the coefficient $(a_1 b_1 - h_1)$ is negative). A numerical solution of the equations (11,12) allows a determination of the full curve of the convection amplitude vs. the reduced temperature (ε^2/ψ) (fig.1) and of the velocity of the traveling wave vs. the reduced temperature. (fig.2). It can be seen that the traveling wave solutions seem to disappear when the temperature is increased (the velocity of the waves tends to 0). Indeed, this can be understood analytically, by using boundary layer techniques, appropriate for solving the concentration equation (2c) in the case where the flow is made of recirculating

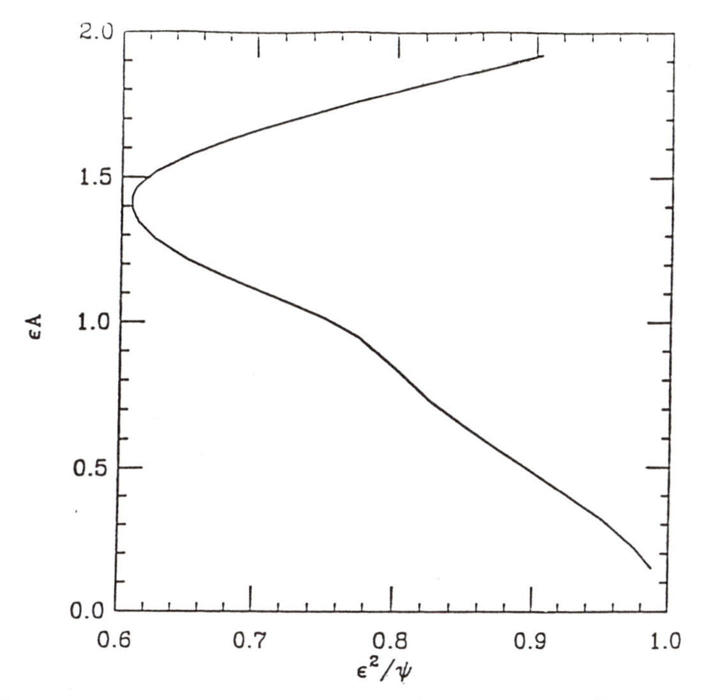

Fig. 1 . The amplitude of convection as a function of the parameter ε^2/ψ for a value of the $L = 10^{-2}$ and $\psi = 0.25$.

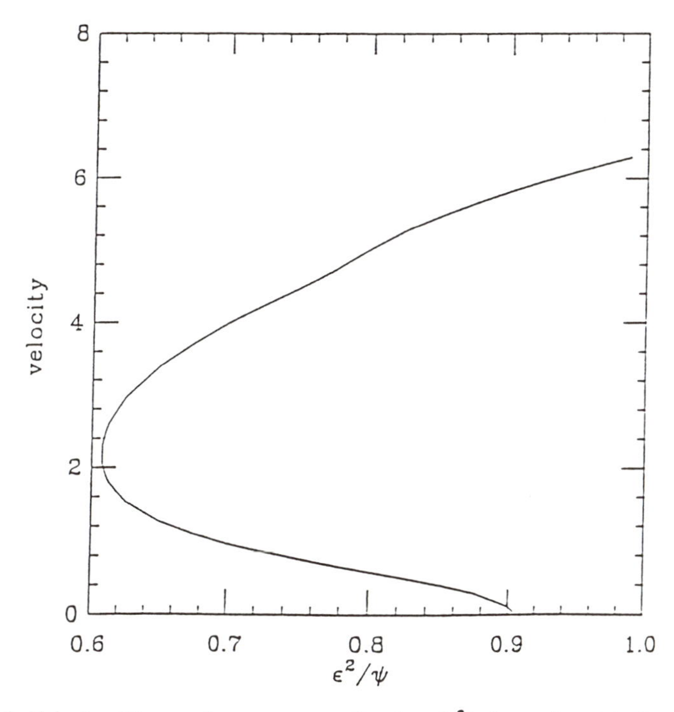

Fig. 2 . Velocity of the traveling wave v_0 as a function of ε^2/ψ for a value of the $L = 10^{-2}$ and $\psi = 0.25$.

zones at high Péclet number. The results of the boundary layer calculation agree very well with the numerical calculations[10].

Physically, the upper branch of fig.1 corresponds to the observed traveling wave regime[1-4]. In agreement with experimental observations, the amplitude of the convection in this regime is equal to the amplitude of the pure fluid convection. From fig. 2, we expect a very fast decrease of the velocity of the waves, as ε varies by no more than 10%. Again this is in good agreement with experimental observations. However, the transition from traveling waves to steady overturning convection is predicted to be critical, *i.e.*, forward (fig. 2). It is experimentally observed to be hysteretic[1]. This could be due to a stabilasation of the steady overturning state by the lateral walls. Results from an experiment in an annulus are awaited.

Although our theory seems to give at least a qualitative description of some of the experimental observations, it does not explain why several regimes can coexist in the same cell[4]. It is conceivable that some of the fast traveling waves patterns that have been computed, which are unstable in an infinite geometry, can be partially restabilized in a confined geometry. Obviously, quantitative comparisons between the present theory and experimental data are needed.

Acknowledgement

It is a pleasure to acknowledge Boris Shraiman for many discussions.

References

1) E. Moses and V. Steinberg, Phys. Rev. **A34**, 693 (1986)
2) E. Moses, J. Fineberg and V. Steinberg, Phys. Rev. **A35**, 2757 (1987)
3) R.W. Walden, P. Kolodner, A. Passner and C. Surko, Phys. Rev. Lett. **61**, 2030 (1985)
4) P. Kolodner, D. Bensimon and C. Surko, Phys. Rev. Lett. **60**, 1723 (1988)
5) T.S. Sullivan and G. Ahlers, Phys. Rev. Lett. **61**, 78 (1988)
6) D.T.J. Hurle and F. Jakeman, J. Fluid Mech. **47**, 667 (1971)
7) B.J.A. Zielinska and H.R. Brand, Phys. Rev. **A35**,4349 (1987)
8) M.C. Cross and K. Kim, Phys. Rev. **A37**, 3909 (1988)
9) E. Knobloch, Phys. Rev. **A34**, 1536 (1986)
10) D. Bensimon, A. Pumir and B. Shraiman, in preparation
11) B. Shraiman, Phys. Rev. **A36**, 261 (1987)
12) M.N. Rosenbluth, H.L. Berk, I. Doxas and W. Horton, Phys. Fluids **30**, 2636 (1987)
13) C.S. Bretherton and E.A. Spiegel, Phys. Lett. **96A**, 152 (1983)

TEMPORAL MODULATION OF A SUBCRITICAL BIFURCATION TO TRAVELLING WAVES

Hermann Riecke[+], John David Crawford[+] and Edgar Knobloch[*]

[+]Institute for Nonlinear Science, University of California
San Diego, La Jolla, Ca. 92093
[*]Department of Physics, University of California, Berkeley, Ca. 94720

The Hopf bifurcation to travelling and standing waves which arises in a variety of systems has drawn considerable attention during the past few years. Recently it has been shown that modulating a control parameter $\mathcal{R}$ in time has a strong influence on the selection properties of such a system if the modulation frequency is close to a resonance with the natural frequency of these waves (Riecke et al., 1988; Walgraef, 1988; Rehberg et al., 1988): if stable travelling waves bifurcate supercritically in the unmodulated system, the modulation lowers the threshold for standing waves below that for travelling waves. This can lead to stable standing waves in a parameter regime where travelling waves do not exist. However, in the system which has been studied experimentally the most (binary mixture convection) the unmodulated instability to travelling waves is subcritical. This naturally raises the question whether in such a system the modulation can still stabilize standing waves even though the travelling waves appear already below threshold. This is the topic of the present paper.

Following our previous work (Riecke et al., 1988, hereafter RCK; see also Walgraef, 1988) we consider a Hopf bifurcation in the presence of translation and reflection symmetry and assume the control parameter to be modulated at a frequency close to twice the Hopf frequency. To fifth order in the amplitudes we obtain

$$\partial_t\eta = a\eta + b\zeta + c\eta(|\eta|^2+|\zeta|^2) + g\eta|\zeta|^2 + q\eta|\eta|^2|\zeta|^2 + s\eta|\zeta|^4 + p\eta|\eta|^4$$
$$\partial_t\zeta = a^*\zeta + b^*\eta + c^*\zeta(|\eta|^2+|\zeta|^2) + g^*\zeta|\eta|^2 + q^*\zeta|\eta|^2|\zeta|^2 + s^*\zeta|\eta|^4 + p^*\zeta|\zeta|^4, \quad (1)$$

where all the coefficients are complex ($a=a_r+ia_i$, etc) and η and ζ are the (complex) amplitudes of the left- and right-travelling wave-components. The linear coefficients, which are assumed to be small, are related to the external parameters via $\mathcal{R}=\mathcal{R}_c+a_r\mathcal{A}+b\mathcal{C}\cos\omega_e t$ and $a_i=\omega_h+a_r\mathcal{B}-\omega_e/2$ with $\mathcal{A}$, $\mathcal{B}$ and $\mathcal{C}$ being O(1) quantities. For the case of a subcritical bifurcation of the travelling waves, c_r is positive and in order for this local analysis to be applicable one has to assume $|c_r|$ to be small.

Due to the translational symmetry (1) can be simplified by going to

$$
\begin{aligned}
\partial_t x &= F_1 \equiv a_r x + by\cos\psi + c_r x(x^2+y^2) + g_r xy^2 + p_r x^5 + q_r x^3y^2 + s_r xy^4, \\
\partial_t y &= F_2 \equiv a_r y + bx\cos\psi + c_r y(x^2+y^2) + g_r x^2y + p_r y^5 + q_r x^2y^3 + s_r x^4y, \\
\partial_t \psi &= F_3 \equiv 2a_i + n_i(x^2+y^2) + (p_i+s_i)(x^4+y^4) + 2q_i x^2y^2 - b\sin\psi\,(x^2+y^2)/xy,
\end{aligned}
\tag{2}
$$

with n=2c+g. Here use has been made of the fact that b can be assumed real by a suitable choice of phases. Since not all cubic coefficients vanish at the singular point under consideration, nonlinear coordinate transformations can still reduce (2) considerably. In addition, we are here interested only in the static solutions of (2), which correspond to either travelling waves or standing waves which are phase-locked to the external modulation. Therefore it is sufficient to look for the solutions of

$$(1+\beta_1 x'^2+\beta_2 y'^2)F_1(x,y,\psi)=0, \quad (1+\beta_2 x'^2+\beta_1 y'^2)F_2(x,y,\psi)=0, \quad F_3(x,y,\psi)=0$$

where $x=x'(1+\alpha x'^2+\gamma y'^2+\ldots)$, $y=y'(1+\gamma\, x'^2+\alpha y'^2+\ldots)$. With a suitable choice of α, β_1, β_2 and γ one obtains (after dropping the primes)

$$
\begin{aligned}
0 &= a_r x + by\cos\psi + c_r x\,(x^2+y^2) + g_r xy^2 + p_r x^5, \\
0 &= a_r y + bx\cos\psi + c_r y\,(x^2+y^2) + g_r x^2y + p_r y^5, \\
0 &= xy\left[2a_i + n_i(x^2+y^2)\right] - b\sin\psi\,(x^2+y^2).
\end{aligned}
\tag{3}
$$

The bifurcation to standing waves (x=y) is assumed supercritical and non-degenerate ($2c_r+g_r<0$). Therefore the quintic term adds only a small correction; these waves are given by essentially the same expressions as in the non-degenerate case discussed in RCK. Their stability, however, depends sensitively on c_r as will become apparent below.

Turning to the travelling waves (TW) we note that, as in the non-degenerate case, pure right- or left-travelling waves do not exist in the presence of the modulation ($b\neq0$). If we introduce the quantities $A=x^2+y^2$ and B=xy and eliminate ψ we obtain, using that $p_r A \ll g_r$,

$$
\begin{aligned}
&4a_r a_i^2 + 4A(a_r a_i n_i + a_i^2 c_r) + A^2\left[a_r(g_r^2+n_i^2) + 4a_i n_i c_r + p_r(4a_i^2-b^2)\right] + \\
&\qquad + A^3\left[c_r(g_r^2+n_i^2) + 4p_r a_i n_i\right] + p_r(g_r^2+n_i^2)A^4 = 0, \\
&a_r + c_r A + p_r\left[A^2 - B^2\right] = 0
\end{aligned}
\tag{4}
$$

with the condition that A>2B. The phase diagram for TW is determined by their saddle-node line SN_T (provided $p_r<0$) and the line M on which they merge with the phase-locked standing waves (SS). The latter is given by the limit $A\to2B$ which yields

$$b_M^2 = a_i^2 + \left[-a_r(g_r^2+n_i^2) + 2\left[c_r^2(g_r^2+n_i^2)/3p_r - a_i n_i c_r\right]\left[1\pm\sqrt{1-3p_r a_r/c_r^2}\,\right]\right]/3p_r \tag{5}$$

where the "-" branch is only relevant for $a_r<0$. The expression for SN_T is rather complicated in the general case. For $a_i=0$, however, it simplifies to

$$b_{SN}^2 = \left[a_r - c_r^2/4p_r\right](g_r^2+n_i^2)/p_r. \tag{6}$$

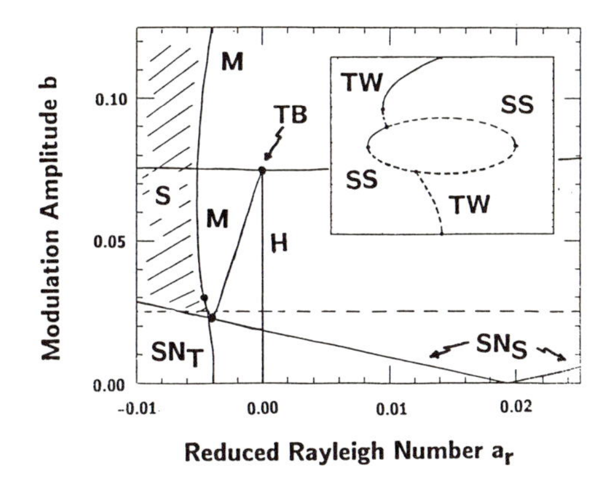

Fig. 1 *Phase diagram, inset: bifurcation diagram (cut along dashed line)* (a_i=-0.075, c=0.125+i, g=-1+i, p_r=-1)
H: Hopf bifurcation to TW, S: *steady bifurcation to* SS,
SN_T, SN_S: *saddle-node bifurcations of* TW and *of* SS, respectively.
M: TW *merge with* SS. *Standing waves are stable in the shaded region.*

Results for the case $a_i \neq 0$ are shown in fig. 1. The neutral curve for TW is given by H and that for SS by S. As discussed in RCK the direction of the bifurcation to SS at the codimension-2 point TB (a_r=0, $b^2=a_i^2$) depends on the sign of $a_i n_i$. Thus for $a_i n_i<0$ - as shown in fig. 1 - both TW and SS bifurcate subcritically close to a_r=0 and undergo a saddle-node bifurcation at SN_T and SN_S, respectively. Along M one of the TW merges with one of the SS.

The stability analysis of the TW is somewhat involved and will not be described here. For the SS that analysis is simplified due to the remaining reflection symmetry. It turns out that their stability is governed by the steady bifurcation to TW at M, which is given by (5), and a Hopf bifurcation to standing waves which are not phase-locked to the modulation (SW). Since the latter involves only standing wave components it is insensitive to the sign of c_r (cf. RCK). Thus the standing waves are in fact stable within this framework in the shaded region to the left of M and can be excited by a sufficiently large modulation amplitude. This conclusion is supported by recent experiments in binary mixtures (Rehberg et al., 1988). The question of stability with respect to spatial modulation is deferred to later work.

Acknowledgements

The authors acknowledge support by the German Science Foundation (DFG), by DARPA through contract AFOSR F49620-87-C-0117 and by the NSF through grant DMS-8814702.

REFERENCES

Rehberg, I., Rasenat, S., Fineberg, J., de la Torre-Juarez, M. and Steinberg, V., 1988, **Phys. Rev. Lett.** (submitted).
Riecke, H., Crawford, J.D. and Knobloch, E., 1988, **Phys. Rev. Lett.**, 61, 1942.
Walgraef, D., **Europhys. Lett.**, (to appear).

EXPERIMENTS WITH TRAVELLING WAVES IN ELECTROHYDRODYNAMIC CONVECTION

Manuel de la Torre Juarez and Ingo Rehberg

Physikalisches Institut, Universität Bayreuth, 8580-Bayreuth (FRG)

The experimental setup is the following: A nematic liquid crystal (purified MBBA or Merck-PhaseV) is sandwiched between two transparent electrodes rubbed to produce a prefered direction on the orientation of the molecules. The thickness (13-50μm) was adjusted with polymeric mylar sheets, and the whole cell was sealed. The cell is embedded in an isothermal (±0.1K) box. A shadowgraphic image of the pattern (Rasenat et al. 1989) is observed with a CCD-camera mounted on a polarising microscope and digitized (512x512 pixels, 256 grey scales). Applying an ac-voltage to the cell leads to convection setting in at a critical driving voltage V_c, as shown in Fig.1. Here a direct bifurcation to travelling waves (TW) is observed in the whole conductive regime and even in the dielectric regime, unlike measurements shown elsewhere (Hirakawa & Kai 1977, Joets & Ribotta 1988, Rehberg et al. 1988a,b). The solid line is obtained by linear stability analysis of the Leslie-Erickson equation (Zimmermann & Thom 1988), where MBBA parameters (25°C) are used and the electrical conductivity σ has been adjusted to fit the threshold voltage where the dielectric and the conductive instability meet. This theory predicts a steady bifurcation. A hopf bifurcation from the spatially homogenous state, which would explain the TW observed in the experiment, has not been found for a range of material parameters likely to include the realistic values for MBBA. The agreement of our experiments (Rehberg et al. 1988a) with the theoretical predictions for temporal modulation of a direct hopf bifurcation with O(2)-symmetry (Riecke et al. 1988, Walgraef 1988), however, strongly supports the idea that a hopf bifurcation is responsible for the observed TW.

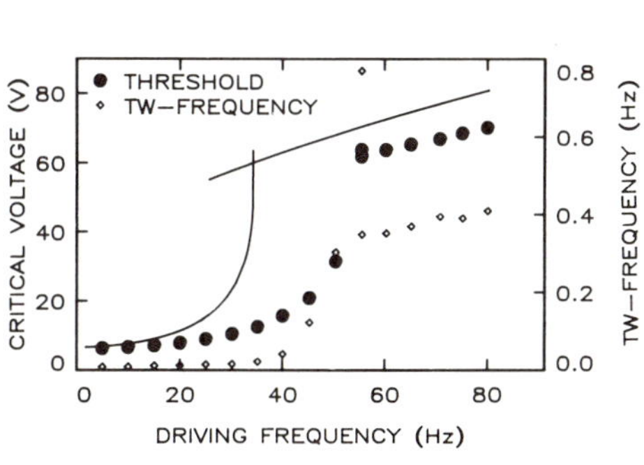

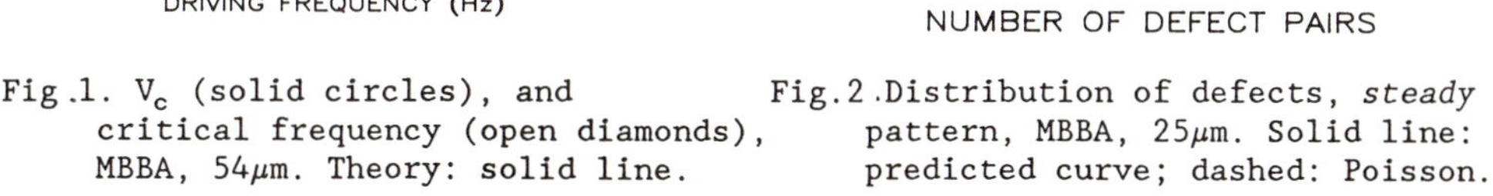

Fig.1. V_c (solid circles), and critical frequency (open diamonds), MBBA, 54μm. Theory: solid line.

Fig.2.Distribution of defects, *steady* pattern, MBBA, 25μm. Solid line: predicted curve; dashed: Poisson.

New Trends in Nonlinear Dynamics and Pattern-Forming Phenomena
Edited by P. Coullet and P. Huerre
Plenum Press, New York, 1990

In the experiment (Rehberg et al. 1988b), the transition to turbulence in TW convection is characterized by the appearance of topological defects. Their statistical distribution is in very good agreement with a prediction based on the idea that the probability to create a defect pair is only dependent on $\epsilon=V/V_c-1$, while the annihilation rate is proportional to the number of defects squared (Gil et al. 1988). These arguments should apply to TW as well as to steady patterns. Fig.2 shows the curve obtained in steady convection. The squared poisson distribution (solid line) fits the data well.

We performed measurements in a 13μ cell (Phase V, cutoff frequency 100Hz) in order to measure some of the coefficents of the amplitude equations

$$\partial_t A_1=(\mu+i\nu)A_1+\chi A_2-(1+i\beta)A_1|A_1|^2+(\gamma+i\delta)A_1|A_2|^2$$
$$\partial_t A_2=(\mu-i\nu)A_2+\chi A_1-(1-i\beta)A_2|A_2|^2+(\gamma-i\delta)A_2|A_1|^2 \quad (1)$$

which govern the dynamics of oscillatory states under the influence of a temporal forcing $V(t)=1+\epsilon+b\cos(\omega_e t)$, with $\omega_e\approx 2\omega_h$, ($\omega_h$: critical frequency) where A,B,C defined by $\mu=\epsilon/A$, $\nu=B\mu-(\omega_e/2-\omega_h)$, $\chi=b/C$ have to be determined experimentally (Riecke et al. 1988).

Fig. 3 shows a measurement of the onset of travelling waves. Light intensity scans on a line perpendicular to the roll axis are taken and subtracted from a reference line measured without convection. The temporal average of the rms-values of these differences provides information about the amplitude of convection and is plotted as a function of the driving voltage. The critical threshold for the onset of convection is thus determined to be $8.77V_{eff}$. The finite RMS-values obtained at negative ϵ are caused by intensity fluctuations and camera noise. The frequency of the TW observed above this voltage is shown to decrease monotonically with increasing ϵ. The solid line is obtained by linear regression analysis allowing the determination of the critical frequency ω_h.

The linear stability analysis of equations (1) shows that the convection threshold for $\mu<0$ is defined by the relation $\mu^2+\upsilon^2=\chi^2$ which defines an hyperbola in the ω_e-b plane. Fig.4 shows the corresponding measurement together with a fit to a hyperbola, where A,B,C and ω_h are adjustable parameters. Thus the linear coefficients of the amplitude equations (1) can be determined, although more measurements are needed to improve the accuracy of the results.

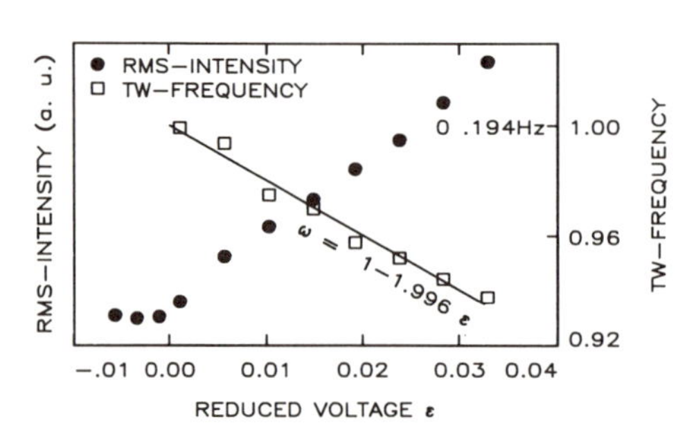

Fig. 3. The amplitude of convection increases with ϵ, and the frequency of the TW decreases. The solid line is the linear regression line. (Phase V,13μm)

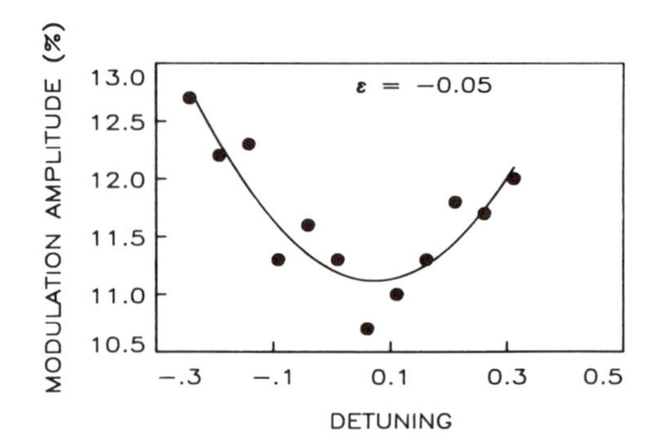

Fig. 4. The threshold of convection in the ω_e-b plane (Phase V, 13μm) The adjusted hyperbola (solid line) is described by A=0.453s, B=-0.130, C=1.00s, ω_h=0.198s^{-1}

One can then even calculate the nonlinear parameter β=B-A$\cdot\Delta\omega_{TW}/\Delta\epsilon$, thus separating the linear from the nonlinear frequency shift (unlike Chiffaudel et al. 1987). By the slope of the frequency shown in Fig.3, β is determined to be -0.183. Next we study the nonlinear resonance curve obtained by measuring the amplitude of the convection as a function of the modulation frequency for a fixed negative ϵ and a fixed modulation amplitude. The result is shown in Fig.5. Note that the transitions to standing waves are supercritical ones. The solid line is the formula given by Riecke et al. (1988), where the unknown nonlinear parameters δ and γ had been adjusted to produce a reasonable similarity to the measurement.

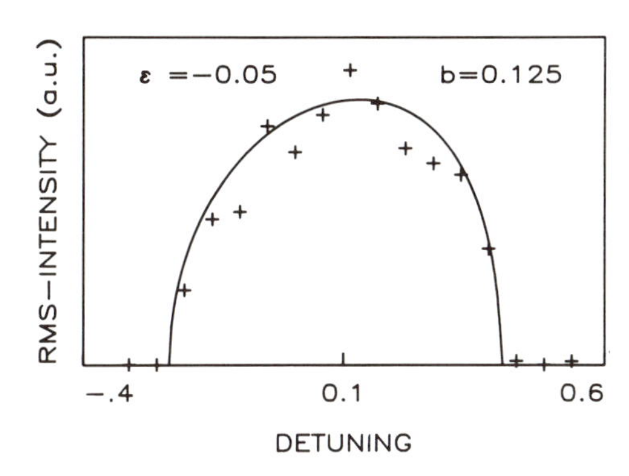

Fig. 5

The amplitude of the standing wave convection measured as a function of the modulation frequency (+). The solid line is the theoretical curve, with δ and γ adjusted. Note that convection occurs at the subcritical value ϵ=-0.05.

Acknowledgements

This work was partially supported by the Deutsche Forschungsgemeinschaft and the Stiftung Volkswagenwerk. We thank W. Zimmermann for calculating the linear stability curve of Fig.1.

References

Chiffaudel,A., Fauve,S., 1987,
Strong Resonance in Forced Oscillatory Convection, Phys.Rev. A35:4004.
Gil,L., Lega,J., & Meunier,J.L., 1988,
Statistical Properties of the Topological Turbulence, to be published.
Hirakawa,K., Kai,S., 1977 Analogy Between Hydrodynamic Instabilities in
Nematic Liquid Crystal and Classical Fluid Mol.Cryst.Liq.Cryst., 40:261.
Joets,A., & Ribotta,R., 1988, Localized, Time-Dependent State in the
Convection of a Nematic Liquid Crystal, Phys.Rev.Lett., 60: 2164.
Rasenat,S., Hartung,G., Winkler,B.L., Rehberg,I., 1989, The Shadowgraph
Method in Convection Experiments, Experiments in Fluids, in print.
Rehberg,I., Rasenat,S., Fineberg,J., de la Torre Juarez,M., Steinberg,V.,
1988 Temporal Modulation of Travelling Waves, Phys.Rev.Lett.,61:2449.
Rehberg,I., Rasenat,S., Steinberg,V.,1988 Travelling Waves and Defect-
initiated Turbulence in Electroconvecting Nematics, to be published.
Riecke,H., Crawford,J.D., Knobloch,E., 1988 Time-Modulated Oscillatory
Convection, Phys.Rev.Lett.,61: 1942.
Walgraef,D., 1988, External Forcing of Spatio-Temporal Patterns, to be publ.
Zimmermann,W., & Thom,W. 1988 Symmetry Breaking Effects in the
Electrohydrodynamic Instability in Nematics, to be published.

THE DEGENERATE AMPLITUDE EQUATION NEAR THE CODIMENSION-2 POINT IN BINARY FLUID CONVECTION

Walter Zimmermann and Wolfgang Schöpf

Physikalisches Institut
Universität Bayreuth
8580 Bayreuth, Fed. Rep. Germany

1. *Introduction* - Hydrodynamical systems like Rayleigh-Bénard convection play a major role in the investigation of self-organization in nonequilibrium systems and in various bifurcation problems. In binary fluid mixtures one can either have a transition from the heat conducting state to the convective state via a stationary bifurcation or via a Hopf bifurcation, depending on the separation ratio Ψ. The instabilities set in at the critical Rayleigh number $R_c^s(\Psi)$ for the stationary bifurcation or at $R_c^h(\Psi)$ for the Hopf bifurcation, in the latter case with a critical frequency $\omega_c(\Psi)$. At the codimension-2 (CT) point Ψ_{CT}, both instabilities have the same threshold $R_c^s = R_c^h =: R_{CT}$. The possibility of interesting dynamical phenomena near the CT-point, which are otherwise not expected at threshold, has induced considerable experimental and theoretical work [1-5 and ref. cited therein].

For realistic - rigid and impermeable - boundary conditions at the upper and lower plate of the convection cell it has been shown recently that at the CT-point the critical wavenumbers for the two instabilities, k_c^s and k_c^h, are slightly different [2-4]. This feature has often not been included in previously derived amplitude equations [1]. Small values of $(k_c^s - k_c^h)$ have been allowed in a generalised amplitude equation, second order in time, describing the dynamics near the CT-point [5]. Furthermore spatial degrees of freedom have been included in the analysis of ref.5 for the first time. These changes lead to a qualitative different behaviour near the CT-point. Especially the stability behaviour of the travelling waves (TW) or the standing waves (SW) is substantially changed. Before a secondary bifurcation or the heteroclinic orbit is reached the TW often become unstable due to the Benjamin-Feir instability. Also TW are Benjamin-Feir unstable already at threshold in some neighbourhood of the CT-point if the stationary pattern bifurcates subcritically [5].

For the calculations in ref.5 only the signs of the relevant coefficients for the CT-amplitude equation were appropriately chosen, because their numerical values were unkown. Recently the nonlinear coefficients of the "simple" amplitude equations for the stationary pattern, the TW and the SW have been calculated rigorously [4]. Using these results we are now able to give the values of the coefficients of the CT-amplitude equation and this is the aim of the present article.

2. *Simple amplitude equations* - For a codimension-1 bifurcaton the near-threshold behaviour of convection for small values of the reduced Rayleigh

New Trends in Nonlinear Dynamics and Pattern-Forming Phenomena
Edited by P. Coullet and P. Huerre
Plenum Press, New York, 1990

number $\epsilon^2 := (R - R_c)/R_c$ can be described by

$$\vec{u}(x,z,t) = \epsilon\cdot[A(X,T)\cdot e^{i(\omega_c t+k_c x)} + B(X,T)\cdot e^{i(\omega_c t-k_c x)}]\vec{U}(z) + c.c. \quad (1)$$

Here $\vec{u}$ describes for $\omega_c = 0$ the SP (then one may set B = 0) and for $\omega_c \neq 0$ the TW if A = 0 (or B = 0), and the SW if A = B. $\vec{u}$ includes the system variables like the temperature, the concentration and the velocitiy field. The amplitudes A (and B) depend on the slow variables $X = \epsilon x$ and $T = \epsilon^2 t$ (We disregard the second space-dimension). For all these cases the amplitude equation for A (and also for B) reads

$$\tau\cdot\partial_T A = (1 + ic_0)\cdot A + \xi^2(1 + ic_1)\cdot\partial_X^2 A - \alpha\cdot(1 + ib)\cdot|A|^2 A. \quad (2)$$

The relaxation time τ, the coherence length ξ, the frequency dispersion c_1 and the frequency shift c_0 are determined from the threshold behavior [3,4]. For TW and SW these linear coefficients coincide, and for the SP all imaginary parts vanish ($c_0=c_1=b=0$). The sign of α determines whether the bifurcation is subcritical ($\alpha<0$) or supercritical ($\alpha>0$), while b is the nonlinear frequency renormalization. The recent calculations show, that at the CT-point the stationary bifurcation is subcritical ($\alpha_{SP}<0$) and that the TW and the SW bifurcate supercritical ($\alpha_{TW}>0$ and $\alpha_{SW}>0$). Furthermore the TW are stable with respect to the SW [4].

3. *Codimension-2 point and degenerate amplitude equation* - In the vicinity of the CT-point (R_{CT},Ψ_{CT}) the linear dynamics can be approximated by a second order polynomial $\sigma^2-e\sigma-d = 0$, where e and d are determined from the linear stability calculation. The condition e=0 (d=0) gives the neutral curve $R_o^h(k)$ ($R_o^s(k)$). An expansion of e and d with respect to small deviations from the critical values R_c^h, k_c^h and R_c^s, k_c^s up to the lowest relevant order leads to $e = \beta[R-R_c^h - \xi_h^2 R_c^h(k-k_c^h)^2]$, $d = \delta[R-R_c^s - \xi_s^2 R_c^s(k-k_c^s)^2]$, with positive constants β and δ.

The degenerate amplitude equation reflecting this linear behavior, is [5]

$$\partial_T^2 D - \eta\{r + (\partial_X - ip)^2\}\partial_T D + \eta(f_2 + f_3)|D|^2\partial_T D + \eta f_3 D^2\partial_X D^* \\ - \{r + s + a\partial_X^2 - f_1|D|^2\}D = 0. \quad (3)$$

The amplitudes of the physical quantities are of order $\eta|D|$ and the fast variation with wavenumber k_c^s is separated out. We have introduced $(R-R_c^h) = \eta^2(\delta/\beta^2)r$, $(R_c^h-R_c^s) = \eta^2(\delta/\beta^2)s$, $(k_c^h-k_c^s) = \eta(\delta/\xi_h^2 R_c^h\beta^2)^{1/2}p$ with the formal expansion parameter η and $a = \xi_s^2/\xi_h^2$. For the time and the space coordinates we now use the following scaling: $T = \eta\cdot\delta/\beta\cdot t$, $X = \eta\cdot(\delta/\xi_h^2 R_c^h\beta^2)^{1/2}\cdot x$. With an appropriate choice of p one fixes η to a small value so that r and s are left as two control parameters in eq.(3). Terms of order η like $i\eta\gamma_1\partial_X D$, $i\eta\gamma_2\partial_X^3 D$, $i\eta\gamma_3|D|^2\partial_X D$ and $i\eta\gamma_4 D^2\partial_X D^*$ are left out in eq.(3) for the same argument as in ref.5. The coefficients f_i and especially their numerical values can in principle be determined by a weakly nonlinear analysis from the full dynamic equations and their signs determine whether the bifurcations are subcritical or supercritical.

4. *Determination of the coefficients* - Eq.(3) has stationary, TW and SW solutions and the associated simple amplitude equations can be derived from eq.(3) as special limits. This procedure provides the following relations between the coefficients of eq.(2) and eq.(3):

$$\beta = 2/(\tau_h R_{CT}) \;, \qquad \delta = 2\xi_h^2(k_c^h - k_c^s)^2/(\tau_h \tau_s R_{CT})$$

$$f_1 = R_{CT}\alpha_{SP}\beta^2/\delta \;, \quad f_2 = R_{CT}\alpha_{TW}\beta^2/\delta \;, \quad f_3 = R_{CT}(\alpha_{SW} - \alpha_{TW})\beta^2/2\delta \tag{4}$$

$$f_1/f_2 = -b_{TW}\cdot\beta\omega_c/\delta \;, \qquad f_1/(f_2+2f_3) = -b_{SW}\cdot\beta\omega_c/3\delta$$

Because of this multiplicity for the coefficients f_i we are now able to derive some consistency conditions between the coefficients of the different amplitude equations for the stationary pattern, the TW and the SW, which must be valid at the CT-point:

$$\alpha_{TW}b_{TW} = -\alpha_{SP}\delta/\omega_c\beta \;, \; \alpha_{SW}b_{SW} = -3\alpha_{SP}\delta/\omega_c\beta \quad [\Rightarrow \alpha_{SW}b_{SW} = 3\cdot\alpha_{TW}b_{TW}] \tag{5}$$

Our numerical results obtained earlier [4] are consistent with eq.(5).

In binary fluid mixtures we have two additional dimensional material parameters, the Prandtlnumber P and the Lewisnumber L. In table I the numerical values of various coefficients are given for different combinations of L and P. These combinations are often used in the literature and are assumed to be valid for water-alcohol mixtures (L=0.02, P=17) and for He^3/He^4-mixtures (L=0.03, P=0.6 or L=0.04, P=0.75).

Table I

L	P	R_{CT}	k_c^s	k_c^h	ω_{CT}	a	$\beta\cdot 10^3$	$\delta\cdot 10^3$	f_1	f_2	f_3
0.02	17	1726.5	3.127	3.105	0.016	0.999	11.2	2.19	-667	10.4	833
0.03	0.6	1758.3	3.143	3.075	0.046	0.984	6.13	1.81	-290	4.91	338
0.04	0.75	1769.2	3.149	3.069	0.065	0.983	6.72	2.65	-162	2.84	188

In ref.4 it is shown that TW are stable at threshold in a vicinity of the CT point. Above threshold they become unstable at the Rayleigh-number

$$R_u = R_c^h + (R_c^s - R_c^h)/(1-2f_1/f_2) \tag{6}$$

and bifurcate to the stationary branch before the heteroclinic orbit is reached. The frequency at that R_u is

$$\omega_u^2 = \omega_c^2 - \delta\cdot(R_c^s - R_c^h)\cdot(f_2 - f_1)/(f_2 - 2f_1) \tag{7}$$

ω_u is always different from zero near the CT-point and at the heteroclinic orbit one would expect $\omega = 0$. To have some impression on the Ψ-dependence of ω_u and R_u one can use the linear dependence $R_c^s - R_c^h \approx \sigma\cdot(\Psi_{CT} - \Psi)$ near the CT point, where σ is given by $\sigma = 6.75\cdot 10^4$ for L = 0.04 and P = 0.75.

ACKNOWLEDGEMENTS

It is a pleasure to thank L. Kramer for helpful discussions.

REFERENCES

[1] H. R. Brand, P. C. Hohenberg and V. Steinberg, Phys.Rev.A, 30 (1984) 2548.
G. Dangelmayr and E. Knobloch, Proc.R.Soc.London, Ser.A, 322 (1987) 243.

[2] S. J. Linz and M. Lücke, Phys.Rev.A, 35 (1987) 3997.
E. Knobloch and D.R. Moore, Phys.Rev.A, 37 (1988) 860.

[3] M.C. Cross and K. Kim, Phys.Rev.A, 37 (1988) 3909.

[4] W. Schöpf and W. Zimmermann, Europhys.Lett. (in press).

[5] W. Zimmermann, D. Armbruster, L. Kramer and W. Kuang, Europhys.Lett. 6 (1988) 505.

TRAVELLING WAVES IN AXISYMMETRIC CONVECTION

Laurette S. Tuckerman and Dwight Barkley

Department of Physics and Center for Nonlinear Dynamics
University of Texas
Austin, Texas 78712

The interaction of time-dependence and spatial patterns can lead to a rich variety of interesting phenomena. In the particular case of Rayleigh-Benard convection in a cylindrical container, axisymmetric travelling waves have been predicted from theories of wavelength selection (Pomeau and Manneville, 1981), but have eluded experimental observation. Here we describe observations of such travelling waves in numerical simulations of the full Boussinesq equations, using a time-dependent code which has additional capabilities for linear stability analysis and steady-state continuation (Tuckerman 1989).

We have studied the evolution of the axisymmetric pattern, for Prandtl number $Pr = 10$ and an aspect ratio (radius/height) Γ of 5. As the reduced Rayleigh number $\varepsilon \equiv (Ra - Ra_c)/Ra_c$ is increased, the four non-central rolls of the initial five-roll state expand at the expense of the central roll (Fig. 1a,b). When the sidewalls are perfectly conducting, transition occurs abruptly to a large-amplitude *traveling wave* state (Fig. 2). The central roll grows smaller and is annihilated, while new rolls are continually created at the sidewall. Heuristically, the system alternates between four and five rolls in an irreconcilable wavelength selection conflict.The schematic phase portraits in Fig.1 illustrate the sequence of bifurcations leading to an oscillatory state as ε is increased. In phase portrait (c), the conductive state is stable. In (d), a supercritical bifurcation at $\varepsilon = 0$ from the conductive state has given rise to a pair of symmetrically related five-roll states. In (e) a second pair of (unstable) states has arisen via another supercritical bifurcation at $\varepsilon = 0.05$. The stable and unstable states approach one another and disappear via the saddle node bifurcation at $\varepsilon_* = 1.38$ in (f). This results in a heteroclinic orbit which becomes the limit cycle in (g).

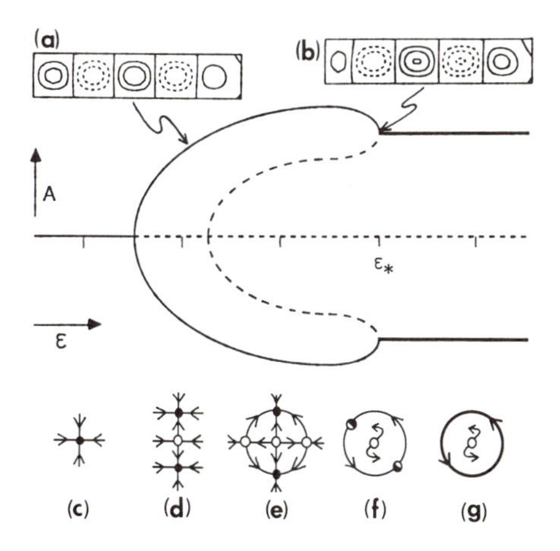

Fig. 1. Schematic bifurcation diagram for the case of conducting sidewalls. ε is the reduced Rayleigh number, and A is a coordinate which distinguishes between different states. (a) and (b) are numerically calculated stream-function contours of representative five-roll steady states at two values of ε. (c)–(g) are phase portraits at the five values of ε denoted by tick marks. Stable (unstable) states are denoted by solid (dashed) lines in the bifurcation diagram, and by solid (hollow) circles in the phase portraits. The traveling wave state (see Fig. 2) is denoted by bold lines.

New Trends in Nonlinear Dynamics and Pattern-Forming Phenomena
Edited by P. Coullet and P. Huerre
Plenum Press, New York, 1990

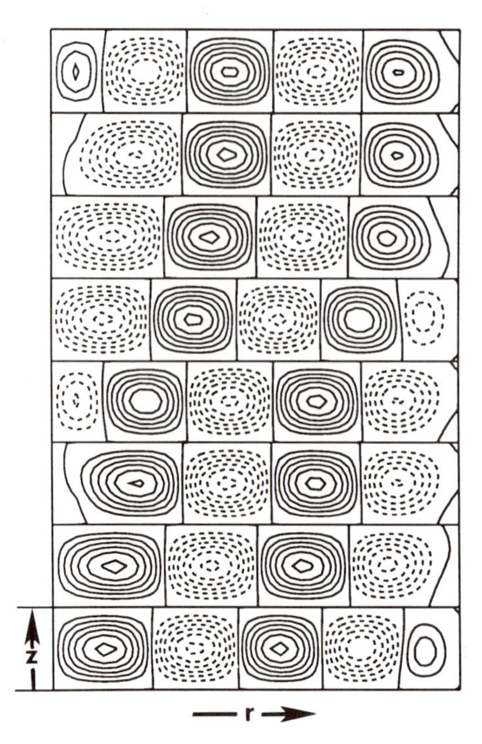

Fig. 2. Instantaneous streamfunction contours (at eight unequally spaced times) in the r–z plane of the traveling wave state at $\varepsilon = 1.39$. Solid (dashed) contours denote clockwise (counter-clockwise) flow.

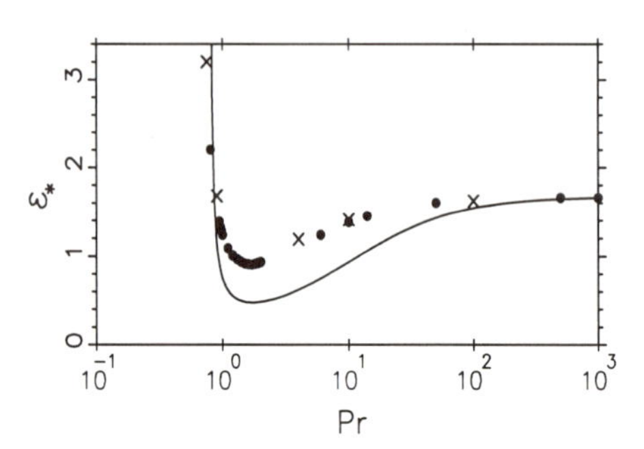

Fig. 3. The value ε_* of the saddle-node bifurcation terminating the five-roll branch plotted as a function of Pr. The circles and squares refer to conducting and insulating boundaries, respectively. The solid curve is the solution to the equation $\lambda(\varepsilon_*', Pr)$ = constant, given by the wavelength selection criterion of Manneville and Piquemal (1983), and the constant is chosen to match the curve to our data at Pr = 1000.

We have been able to interpret $\varepsilon_*(Pr)$, the Prandtl number dependence of the transition threshold using a proposed wavelength selection mechanism for axisymmetric convection (Pomeau and Manneville, 1981; Cross, 1983). These authors argue that, for axisymmetric convection to be steady, the large-scale radial flow normally generated by roll curvature must vanish and that this occurs for a unique wavelength (roll size) $\lambda(\varepsilon, Pr)$. The agreement shown in Figure 3 between $\varepsilon_*(Pr)$ and an analytic expression derived from Manneville and Piquemal's (1983) formula for $\lambda(\varepsilon, Pr)$ suggests that the transition occurs when a critical λ is reached.

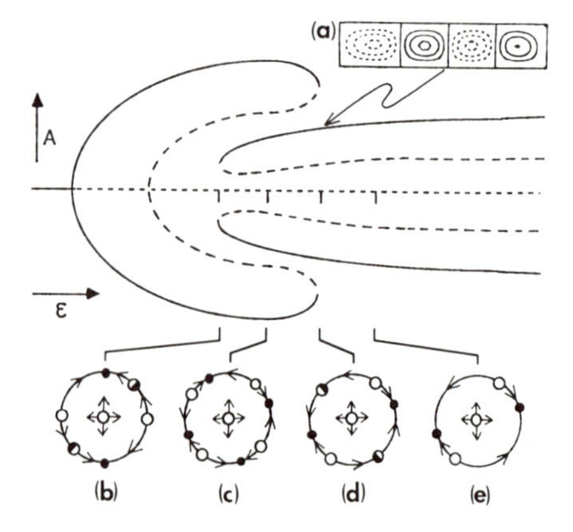

Fig. 4. Schematic bifurcation diagram for the case of insulating sidewalls. Conventions are the same as in Fig. 1. The diagram differs from Fig. 1 by the absence of traveling waves and by the presence of additional steady four-roll states (a). Phase portrait (b) shows the emergence of two pairs of four-roll states via saddle-node bifurcation, leading to the bistable situation with eight convective steady states shown in (c). (d) depicts the saddle-node bifurcation causing the disappearance of the five-roll states, leaving only the four-roll states (e).

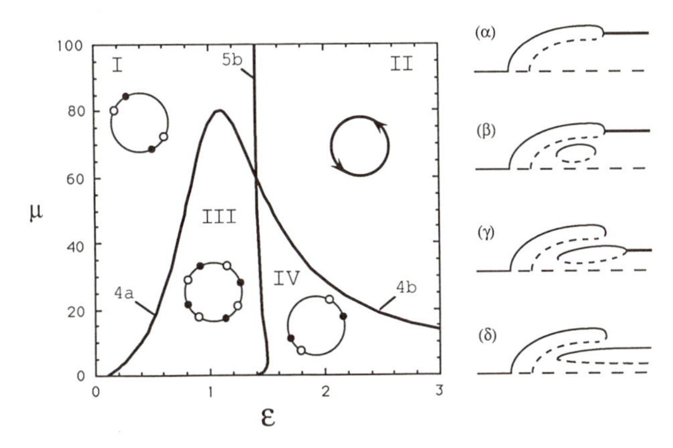

Fig. 5. Phase diagram showing the behavior of the system as a function of the two parameters μ and ε. The curves 4a and 4b mark the numerically computed saddle-node bifurcations at the low-ε and high-ε ends of the four-roll branch; curve 5b marks the saddle-node bifurcation terminating the five-roll branch. These curves delimit four regions (I – IV) with different dynamics. Each region contains a schematic phase portrait similar to those of Figs. 1 and 4. In region I, only five-roll states exist. Region II is the traveling wave regime. In region III, four- and five- roll states coexist, with separate basins of attraction. Region IV contains only four-roll states. $\alpha - \delta$ are bifurcation diagrams for four different values of μ. The conventions for the diagrams are those of Figs. 1 and 4, but only the upper half of each is shown.

We turn now to the case of insulating sidewalls (Fig. 4). The central roll again disappears via a saddle-node bifurcation at $\varepsilon_* = 1.42$, this time marking the transition to a *steady four-roll* state. The four-roll state, which intercepts the traveling wave solution, is on a disconnected branch, or *isola*, whose range of existence in ε decreases as μ, the ratio of sidewall to fluid conductivity, is increased (Fig. 5). For $\Gamma = 5$, $Pr = 10$, *traveling waves should be experimentally observable only if* $\mu \geq 60$. Numerically calculated phase portraits confirm these scenarios.

A more detailed exposition is given in Tuckerman and Barkley (1988) and Barkley and Tuckerman (1989). We hope that our quantitative analysis will stimulate experimental investigations.

Acknowledgments

This research was supported by the U.S. Office of Naval Research Nonlinear Dynamics Program, and by the NSF Mathematical Sciences Division through a Postdoctoral Research Fellowship. Computing resources for this work were provided by the University of Texas System Center for High Performance Computing. The manuscript was prepared at the Institute for Theoretical Physics, supported by NSF Grant No. PHY82-17853, supplemented by funds from the National Aeronautics and Space Administration.

References

Barkley, D. and Tuckerman, L.S., 1989, *Physica D*, in press.
Cross, M.C., 1983, *Phys. Rev. A*, 27:490.
Manneville, P. and Piquemal, J.M., 1983, *Phys. Rev. A*, 28: 1774.
Pomeau, Y. and Manneville, P., 1981, *J. Phys. (Paris)*, 42: 1067.
Tuckerman, L.S., 1989, in "Proceedings of the 11th Int'l. Conf. on Num. Meth. in Fluid Dynamics", D.L. Dwoyer, M.Y. Hussaini, and R.G. Voigt, ed., Springer-Verlag, Berlin.
Tuckerman, L.S. and Barkley, D., 1988, *Phys. Rev. Lett.*, 61: 408.

ABOUT CONFINED STATES,

NOISE-SUSTAINED STRUCTURES AND SLUGS

Helmut R. Brand

FB 7, Physik
Universität Essen
D 43 Essen 1
West Germany

INTRODUCTION

In the field of pattern formation in nonequilibrium systems it has turned out over the last couple of years to be a fruitful concept to study the onset of the first instability in systems such as binary fluid convection (Ahlers et al., 1987, Brand et al., 1984, Steinberg et al., 1987, 1989) or the electrohydrodynamic instability (EHD) in nematic liquid crystals (Joets and Ribotta, 1988). Before the emphasis in the physics community had been on the onset of thermal convection in simple fluids and on the Taylor vortex flow which arises as a first instability when the fluid in the gap between two concentric cylinders is subjected to a torque by rotating the inner cylinder. While one has learned a great deal from these studies, the onset of the first instability which is most easily accessible to theoretical treatment has always been stationary in nature. Phenomena such as oscillatory instabilities (Brand and Steinberg, 1983, Knobloch and Proctor, 1981), traveling waves, modulational instabilities etc. occurred in those systems only as higher instabilities which are harder to handle theoretically. By paying the prize of having to deal with a slightly more complex system, one has succeeded experimentally in bringing down to onset many exciting phenomena such as traveling waves in binary fluid convection (Walden et al., 1985), a co-dimension two point in binary fluid convection in a porous medium (Rehberg and Ahlers, 1985), Benjamin-Feir type behavior (Steinberg et al., 1987) etc.

In addition we have learned to view the traveling waves in a finite container as being a bridge to the understanding of open flow systems such as plane Poiseuille flow, the Goertler instability, channel flow etc. as they are of utmost importance for engineering applications including the design of aircraft wings and the body of ships (Huerre and Monkewitz, 1985). In this note we focus on the analysis of two particularly exciting phenomena as they have been seen experimentally over the last two years: the formation of confined states in binary fluid mixtures, both in rectangular (Heinrichs et al., 1987, Moses et al., 1987) and in annulus-shaped containers (Kolodner et al., 1988) and the observation of localized states near the onset of electroconvection in nematic liquid crystals (Joets and Ribotta, 1988). In all these cases parts of the experimental cells were free of thermal or electro-convection and the parts filled with a convective pattern consisted in the long time limit of traveling waves, where we will focus in the following on the patterns, which comprise propagating waves traveling in one direction only.

New Trends in Nonlinear Dynamics and Pattern-Forming Phenomena
Edited by P. Coullet and P. Huerre
Plenum Press, New York, 1990

REVIEW OF THE EXPERIMENTAL SITUATION

Concerning the localized states in the EHD all experimental observations have been reported by Joets and Ribotta. They can be summarized as follows. For voltages which are slightly lower than those for the onset of Williams domains (electro-convective rolls) filling the whole large aspect ratio cell (the aspect ratio is of the order 1000), Joets and Ribotta observe states which are localized in both directions of the plane of the sample and which contain rolls traveling in one direction. These patches have an envelope whose amplitude grows continuously from zero, i.e. a continuous transition prevails. There is no electro-convection between the patches meaning that the nematic liquid crystal is in a quiescent state. Reflections are reported to be irrelevant for the formation of the patches due to the very large aspect ratio. The same conclusion is applicable to the boundary conditions at the vertical boundaries of the sample. We note here already that traveling waves are not expected from theoretical considerations in this regime of parameter space.

For the onset of thermal convection in binary fluid mixtures a number of different localized states has been observed in different geometries.

In rectangular cells various groups have reported 'football' or 'confined' states (Heinrichs et al., 1987, Moses et al., 1987). In those, one part of the cell is free of convection within experimental resolution whereas other parts, typically at one end of the cell, are filled with a football-shaped pattern of waves traveling to the near end of the cell. The rolls are bent and there seems to be no pattern very close to the vertical boundary at the near end of the cell. The nature of the bifurcations reported ranges from forward over weakly to strongly inverted. The frequency of the traveling waves is found to be different from the onset frequency expected from a linearized theory. These states are occurring over a range of Rayleigh numbers close to onset and the boundary of the football closest to the far end of the cell stays fixed as the Rayleigh number is increased whereas the boundary closer to the near end tends to move towards the vertical boundary.

To eliminate possible effects due to reflections from the end walls of the container similar experiments were carried out for an annulus shaped cell (Kolodner et al., 1988). It was found for the separation ratio (dimensionless crosscoupling between temperature and concentration variations) studied in this experiment that the bifurcation to the patch in the annulus was strongly inverted. This confined state which was stable for long times also consisted of waves traveling in one direction whereas the patch as a whole stayed at a fixed position and the left and the right boundaries of this localized state did not move. It was also reported that the frequency of the traveling wave in the patch is vastly different from the onset frequency of the linearized theory. As the football states in rectangular containers, the confined states in the annulus also arise for a band of Rayleigh numbers close to convective onset with the length of the confined state not being constant from one experiment to the next.

Thirdly there was also in binary fluid mixtures, namely in normal fluid $^3He/^4He$ mixtures, the observation (Sullivan and Ahlers, 1988) of a phenomenon which might be related to the confined states in the room temperature mixtures. Sullivan and Ahlers reported the occurrence of irregular spikes in the Nusselt number very close to convective onset with a frequency which is several orders of magnitude different from that expected for the onset frequency in a linearized theory. The influence of reflections on these observations is unclear due to the lack of the possibility of visualization of the pattern in low temperature mixtures.

OVERVIEW OF THEORETICAL CONCEPTS

Naturally the question arises to what extent the experimental observations sketched in the last section can be interpreted with the currently available theoretical concepts, which we briefly summarize in this part of the note.

To describe an oscillatory instability to traveling or standing waves two prototype equations emerge very frequently from an analysis using envelope equations (Brand et al. I,II, 1986, Coullet et al., 1985, Newell, 1974, Newell and Whitehead, 1969, Nozaki and Bekki, 1983)

a) The Ginzburg-Landau equation with complex coefficients (CGL)

$$A_t - vA_x = aA + bA_{xx} - c \mid A \mid^2 A \tag{1}$$

where we have written down for simplicity the 1 D special case (compare Brand et al., I,II, 1986 and Coullet et al., 1987 for the 2 D case).
b) The Kuramoto-Sivashinsky (KS) equation which can be derived either as an amplitude or as a phase equation and which is also known to be the phase equation associated with eq.(1) (Kuramoto and Tsuzuki, 1976)

$$\phi_t - v\phi_x = -\phi_{xx} - \phi_{xxxx} + {\phi_x}^2 \tag{2}$$

Both prototype equations can show weak turbulence in one and two dimensions respectively and have therefore attracted considerable attention.

To describe the localized states a number of concepts arises using the properties of these two prototype equations

a) convective versus absolute instability
b) noise-sustained structure
c) reflections
d) Benjamin-Feir-Newell instability
e) subcritical bursting and nonlinear focusing effects, and
f) slugs.

a) *Convective versus absolute instability*: This is a concept which has been successfully used in the study of open flow systems (Chomaz et al., 1988, Deissler, 1985, 1987, I, 1989, Huerre and Monkewitz, 1985, Huerre, 1987) and which comes from plasma physics (Bers, 1975, Briggs, 1964). If a system is absolutely unstable, a perturbation grows with time when viewed at a fixed location. One perturbation in an absolutely unstable system is sufficient to produce a spatial pattern for all times. This is different in a convectively unstable system. Here the perturbation decays when viewed at a fixed location but grows when one is moving with the perturbation. As a consequence the perturbations or noise must be supplied continuously to maintain a pattern. Otherwise the spatial structure moves out of the container.

b) *Noise-sustained structure* and c) *Reflections*: There seem to be at least two ways to maintain a spatial pattern, namely by supplying noise continuously thus generating a noise-sustained structure (Deissler 1985, 1987, I) or by the growth of reflections generated by the end-walls of the container (Cross, 1986). For the case of a noise-sustained structure the pattern moves out if the noise is removed.

To make some of these remarks more explicit we consider two coupled Ginzburg-Landau equations (Deissler and Brand, 1988) with complex coefficients incorporating quintic terms to allow for a weakly subcritical bifurcation

$$\begin{aligned} A_t - vA_x &= aA + bA_{xx} - c \mid A \mid^2 A - d \mid B \mid^2 A - e \mid A \mid^4 A \\ B_t + vB_x &= aB + bB_{xx} - c \mid B \mid^2 B - d \mid A \mid^2 B - e \mid B \mid^4 B \end{aligned} \tag{3}$$

From a linear stability calculation we see that the system is convectively unstable provided that

$$| v | > 2 | b | \left(\frac{a}{b_r}\right)^{1/2} \tag{4}$$

This means the perturbation must propagate faster than it spreads to obtain a convectively unstable situation. Since a measures the distance from onset, this implies that a system showing traveling waves is always convectively unstable close to the onset of the instability. This result is reenforced by the observation that this must be generally true since sufficiently close to and above onset the velocities of both edges of a spatially localized perturbation will have the same sign as the group velocity of the perturbation (Deissler and Brand, 1988)

d) *Benjamin-Feir-Newell (BFN) instability*: Another important class of behaviour emerging from the analysis of the CGL equation is the BFN instability (Benjamin and Feir, 1967, Brand and Steinberg, 1984, Brand et al., I,II, 1986, Newell, 1974), which can be viewed as the generalization of the Eckhaus modulational instability occurring for a stationary pattern to an oscillatory instability. It is due to a competition between diffusive and nonlinear damping effects on one hand and the effects of dispersion and a nonlinear 'refractive index' on the other. This instability leads in 2 D to the occurrence of convective patches all over the cell as has also been verified experimentally (Steinberg et al., 1987).

e) *Subcritical bursting and nonlinear focusing effects*: The phenomenon of subcritical bursting (Brand et al., I,II, 1986) can occur when one is close to but below the onset of an instability in a subcritical situation. Focusing effects due to dispersion and nonlinear 'refraction' can then trigger the instability and drive the system above threshold and to finite amplitudes. The nonlinear part of the refractive index all by itself can also lead to experimentally observable consequences such as the bending of rolls in the football states.

f) *Slugs*: For subcritical bifurcations there is the possibility to have two 'basins of attraction' which can lead to the coexistence of two different spatial patterns. E.g. as a function of space and time to the co-existence of laminar flow interrupted by regions with a regular or an irregular spatial pattern (Deissler, 1987, I).

The interplay of some these effects (a), b), d) and f)) becomes especially clear when two coupled CGL equations , as they apply at least for the initial transient stage of the onset as traveling waves, are studied. Among the phenomena arising we just mention (Deissler and Brand, 1988) i) a transition from absolutely unstable to convectively unstable for stabilizing cross-coupling, ii) the reverse transition for a destabilizing cross-coupling, iii) a partial annihilation of colliding slugs, iv) transitions from subcritical to absolutely or convectively unstable supercritical behavior for destabilizing cross-coupling.

We close this review of theoretical concepts by noting that the KS equation with damping is absolutely stable provided the damping is high enough and that it can become convectively unstable for sufficiently high group velocity (Brand and Deissler, 1989, I).

ANALYSIS OF THE EXPERIMENTAL OBSERVATIONS

For the case of the patches observed in the EHD in nematic liquid crystals the situation is rather puzzling. Assuming - as is observed - that the transition is continuous, the following conclusions emerge when the concepts sketched above are applied. The patches cannot be 'slugs' or represent a nucleation process, since those are characteristic for discontinuous phenomena or first order type transitions. The possibility of a noise-sustained structure can also be ruled out, since in this case the patches could not be localized in all directions. Since one has a very large aspect ratio, the effect of reflections can be neglected as has already been pointed out by Joets and Ribotta. The confined states can also not be a manifestation of the Benjamin-Feir-Newell instability, since then one would also expect a spatial pattern in the regions where no electroconvection is observed. Thus only two possibilities seem to remain: either there is a completely novel mechanism at work, which has yet to be discovered, or the patches observed are a consequence of inhomogeneities in the sample thickness and/or due to impurities (e.g. dust particles) in the sample which can be mobile or fixed.

Turning to binary fluid convection it seems to be completely clear that the football states reflect the fact that the system is convectively unstable (Deissler and Brand, 1988). In the wider cells two dimensional effects come in and give rise to a bending of the rolls which is reminiscent of nonlinear focusing effects in optics as already mentioned briefly in the last section. The vertical walls perpendicular to the direction of propagation tend to slow down the speed of the rolls. The difference of the observations in rectangular cells to those in the annular geometry seems not clear, especially since recent experiments by the Santa Barbara group (Niemela et al., 1988) indicate that experiments on a rectangular cell and on an annulus, but under otherwise identical conditions, lead to the formation of very similar confined states with a frequency which is in both cases about half that obtained from the linearized theory.

The question of the relative magnitude of the influence of reflections (Cross, 1986) versus that of the influence of noise (Deissler and Brand, 1988) on the various experiments in rectangular cells can be checked experimentally by varying the box size with otherwise constant parameters. For reflections the fraction of the container which is filled with a visible convective pattern is constant whereas for the case of a noise-sustained structure the length over which no visible pattern occurs would be constant.

The confined states observed for the case of an annular geometry are natural candidates for a stationary slug. In this case it seems important to clarify the ψ-dependence of the phenomena observed and also to map out the range of possible lenghts of the slugs. That both fronts of the confined states are fixed is in accord with the slug picture, but the key open question is: why does the slug not move as a whole? This problem remains looking through all the numerical investigations reported so far (Deissler, 1989, Deissler and Brand, 1988, Thual and Fauve, 1988). Even taking into account the nonlinear gradient terms discussed first in Brand et al., I,II, 1986, does not seem to give a sufficiently large interval (Deissler and Brand, 1989) over which the confined states exist in the experiment. Another open question pertains to the possibility of reflection effects due to the curvature of the annulus and the finite width of the gap (it is a pleasure to thank Yves Couder for a stimulating discussion on this point).

Concerning the spikes observed in $^3He/^4He$ mixtures (Sullivan and Ahlers, 1988) it seems fair to say at this stage that they are good candidates for propagating slugs (Deissler and Brand, 1988), especially when one takes into account the fact that the average time interval between two spikes compares well with the transit time through the cell. The irregular time intervals for the occurrence of the spikes might indeed suggest that the excitations move out of the cell.

PERSPECTIVE

In conclusion it seems fair to say that some of the features observed experimentally for binary fluid convection can be classified in terms of the theoretical and numerical approaches outlined whereas some aspects remains yet to be understood theoretically or need further experimental study. Among those items the measurement of the length dependence of the football states on the box length seems to be particularly useful since it can help to sort out the relative contributions of noise and reflections from the end. It would also be very important to check whether one can get three or more footballs in a very long box. If this were the case our understanding of this phenomenon in terms of 'convective instability' would be gone. A systematic study of the length variation of the confined states in the annulus with the radius and of the spikes in ^{3}He/^{4}He mixtures with the box length would certainly help to reduce further the number of options for possible interpretations. From a theoretical point of view one of the most challenging questions is certainly how to get the slugs in the annulus to stand still.

In closing we briefly discuss some related phenomena observed in other systems. In a beautiful experiment on Argon under pressure Croquette and Williams (1988) have demonstrated convincingly the convective nature of the traveling waves for the Busse instability (Busse, 1972) in simple fluid convection, which arises as a secondary instability in low Prandtl number fluids. Probably the most impressive evidence for the importance of perturbations or noise on the pattern selected has been obtained in a systematic study of the sidebranching instability in perturbed Saffman-Taylor fingering (Rabaud et al., 1988).

So far most experiments on confined states and noise-sustained structures and slugs have been carried out in the amplitude domain. Recently Deissler and I (Brand and Deissler, 1989, I, II) have shown how these concepts can be carried over to the phase domain using a phase dynamic description. There seem to be several experiments which show the analog of slugs in the phase domain, that is a different roll or vortex diameter in some parts of the cell compared with that in the bulk part (Brand and Deissler, 1989, II). Candidates include the localized domains arising in the Taylor instability betwen co-rotating cylinders as observed recently (Baxter and Andereck, 1986) and the localized domains in large aspect ratio slot convection as reported for a long quasi one-dimensional cell by Dubois at this conference. The theoretical aspects of phase slugs will be elucidated in detail in a longer paper in the near future (Brand and Deissler, 1989, II).

Acknowledgements: It is a pleasure to thank Guenter Ahlers, Yves Couder, Bob Deissler, Monique Dubois and Alan Newell for stimulating discussions.
Support of this work by the Deutsche Forschungsgemeinschaft is gratefully acknowledged.

References

Ahlers, G., Cannell, D.S., and Heinrichs, R., 1987, Nucl.Phys. B2, 77
Baxter, G.W. and Andereck, C.D., 1986, Phys.Rev.Lett.57, 3046
Benjamin, T.B. and Feir, F.T., 1967, J.Fluid Mech.27, 417
Bers, A., 1975, in 'Physique des Plasmas', C. de Witt and J. Peyraud, Eds., Gordon and Breach, N.Y.
Brand, H. and Steinberg, V., 1983, Phys.Lett.A93, 333
Brand, H.R. and Steinberg, V., 1984, Phys.Rev.A29, 2303
Brand, H.R., Hohenberg, P.C., and Steinberg, V., 1984, Phys.Rev.A30, 2584 and references cited therein
Brand, H.R., Lomdahl, P.S., and Newell, A.C., 1986, I, Phys. Lett.A118, 67
Brand, H.R., Lomdahl, P.S., and Newell, A.C., 1986, II, Physica D23, 345
Brand, H.R. and Deissler, R.J., 1989, I, Phys.Rev.A39, xxx
Brand , H.R. and Deissler, R.J., 1989, II, to be published
Briggs, R.J., 1964, Electron-Stream Interaction with Plasmas, MIT Press, Cambridge, MA
Busse, F.H., 1972, J.Fluid Mech.52, 97
Chomaz, J.M., Huerre, P., and Redekopp, L.G., 1988, Phys.Rev. Lett.60, 25
Coullet, P. and Spiegel, E.A., 1983, SIAM J.Appl.Math.43, 776
Coullet, P., Fauve, S., and Tirapegui, E., 1985, J.Phys.Lett. (Paris) 46, 787
Coullet, P., Elphick, C., Gil, L., and Lega, J., 1987, Phys.Rev. Lett.59, 884
Croquette, V. and Williams, H., 1988, submitted for publication
Cross, M.C., 1986, Phys.Rev.Lett.57, 2935
Deissler, R.J., 1985, J.Stat.Phys.40, 371

Deissler, R.J., 1987, I, Phys.Lett.A120, 334
Deissler, R.J., 1987, II, Physica D25, 233
Deissler, R.J. and Brand, H.R., 1988, Phys.Lett.A130, 293
Deissler, R.J., 1989, J.Stat.Phys.54, Nos.5,6
Deissler, R.J. and Brand, H.R., 1989, to be published
Dubois, M., 1988, talk presented at this conference
Heinrichs, R., Ahlers G., and Cannell, D.S., 1987, Phys.Rev.A35, 2761
Huerre, P. and Monkewitz, P.A., 1985, J.Fluid Mech.159, 151
Huerre, P., 1987, p.141, in 'Instabilities and Nonequilibrium Structures', E. Tirapegui and D. Villaroel, Eds., Reidel, Dordrecht
Joets, A. and Ribotta, R., 1988, Phys.Rev.Lett.60, 2164
Knobloch, E. and Proctor, M.R.E., 1981, J.Fluid Mech.108, 291
Kolodner, P., Bensimon , D., and Surko, C.M., 1988, Phys.Rev.Lett.60,1723
Kuramoto, Y. and Tsuzuki, T., 1976, Prog.Theo.Phys.54, 687
Moses, E., Fineberg, J., and Steinberg, V., 1987, Phys.Rev.A35, 2757
Newell, A.C., 1974, Lectures Appl.Math.15, 157
Newell, A.C. and Whitehead, J.A., 1969, J.Fluid Mech.38, 209
Nozaki, K. and Bekki, N., 1983, Phys.Rev.Lett.51, 2171
Niemela, J., Ahlers, G., and Cannell, D.S., 1988, Bull.Am.Phys.Soc.33, 2261 and to be published
Rabaud,M., Couder, Y., and Gerard, N., 1988, Phys.Rev.A37, 935
Rehberg, I. and Ahlers, G., 1985, Phys.Rev.Lett.55, 500
Steinberg, V., Moses, E., and Fineberg, J., 1987, Nucl.Phys.B2, 109
Steinberg, V., Fineberg, J., Moses E., and Rehberg, I., 1989, Physica D, in print
Sullivan, T. and Ahlers, G., 1988, Phys.Rev.A38, 3143
Thual, O. and Fauve, S., 1988, J.Phys.(Paris) 49, 1829
Walden, R.W., Kolodner, P., Passner A., and Surko, C.M., 1985, Phys.Rev.Lett.55, 496

LOCALIZED STRUCTURES GENERATED BY SUBCRITICAL INSTABILITIES : COUNTERPROGATING WAVES

O. Thual* and S. Fauve**

* CERFACS, 42 av. Coriolis, 31057 Toulouse, France

** ENS de Lyon, 46 allée d'Italie, 69364 Lyon, France

Abstract : In the vicinity of an inverted Hopf bifurcation, we describe stable pulse-like solutions for the envelope of isolated or interacting counterpropagating waves. These localized structures correspond to droplets in first order phase transition. We show that their stabilization is a non variationnal effect, associated with the generation of constant phase gradients for the waves.

1. Amplitude equations model

Localized structures in convection experiments in binary fluid motions are widely reported in this workshop [1-3]. These structures are observed in the vicinity of a subcritical Hopf bifurcation, and are sucessfully described in the frame of the amplitude equations of the instability [4-7].

In this spirit, we consider the one dimensional complex amplitudes $W_R(x,t)$ and $W_L(x,t)$ of the left and right propagating waves governed by the equations :

$$\partial_t W_R = \mu W_R + \alpha \partial_{xx} W_R + \beta |W_R|^2 W_R + \gamma |W_R|^4 W_R + \delta |W_L|^2 W_R - c\partial_x W_R \tag{1}$$

$$\partial_t W_L = \mu W_L + \alpha \partial_{xx} W_L + \beta |W_L|^2 W_L + \gamma |W_L|^4 W_L + \delta |W_R|^2 W_L + c\partial_x W_L \tag{2}$$

where μ is the distance from criticality, $\alpha, \beta, \gamma, \delta$ are complex coefficients and c is the group velocity of the waves. We assume that the bifurcation is subcritical ($\beta_r > 0$) which explains the introduction of a quintic term to saturate ($\gamma_r < 0$) the instability. There is an interval $[\mu_c, 0]$ of the control parameter for which both the homogeneous null and bifurcated states are stable. For simplicity we restrict ourself to the case $\alpha = 1$, and consider periodic boundary conditions on the interval $[0, L]$ for the numerical simulations.

2. Isolated right moving pulse

When there is no interaction between the left and right propagative waves, say a region where $W_L = 0$, the complex amplitude W_R is thus governed by Equation (1) without the two last terms (in a frame moving at the group velocity c). We have observed numerically [7] structurally stable pulse-like solutions $W_R = R_0(x) exp\ i[\theta_0(x) + \Omega t]$ on an interval range of the control parameter μ (Figure 1).

To explain the stability of this localized structure, we consider it as a droplet of the homogeneous bifurcated state, inside the basic null state, as for a first order phase transition (e.g. the nucleation of a liquid in a supersaturated vapor). We first observe that the absolute value of $\partial_x \theta_0$ is approximatively constant, even inside the droplet. The modulus of $W_R = R(x,t) \exp i\phi(x,t)$ obeys to the equation :

$$\partial_t R = \mu_{eff} R + \beta_r R^3 + \gamma_r R^5 + \partial_{xx} R \tag{3}$$

where the effective control parameter $\mu_{eff} = \mu - (\partial_x \phi)^2$ is approximatively constant. For constant μ_{eff} Equation (3) is variationnal and has a Lyapounov functionnal $\mathcal{L}$, a "free energy", minimized by the stationnary modulus profiles :

$$\mathcal{L}\{R\} = \int_0^L \left[\frac{1}{2}(\partial_x R)^2 - V(R)\right] dx \tag{4}$$

New Trends in Nonlinear Dynamics and Pattern-Forming Phenomena
Edited by P. Coullet and P. Huerre
Plenum Press, New York, 1990

$$\text{with} \quad V(R) = \frac{1}{2}\mu_{eff} R^2 + \frac{1}{4}\beta_r R^4 + \frac{1}{6}\gamma_r R^6 \tag{5}$$

There exists only one value $\mu_p = 3\beta_r^2/16\gamma_r$ of the control parameter μ_{eff} for which the basic null state and the bifurcated state has the same local energy $-V$, and thus may coexist spatially. Around this value, one state is metastable and is slowly replaced by the stable state. One the interval of the parameter μ where we have observed stable pulse-like solutions, the computed value of μ_{eff} was closed to the equilibrium value μ_p.

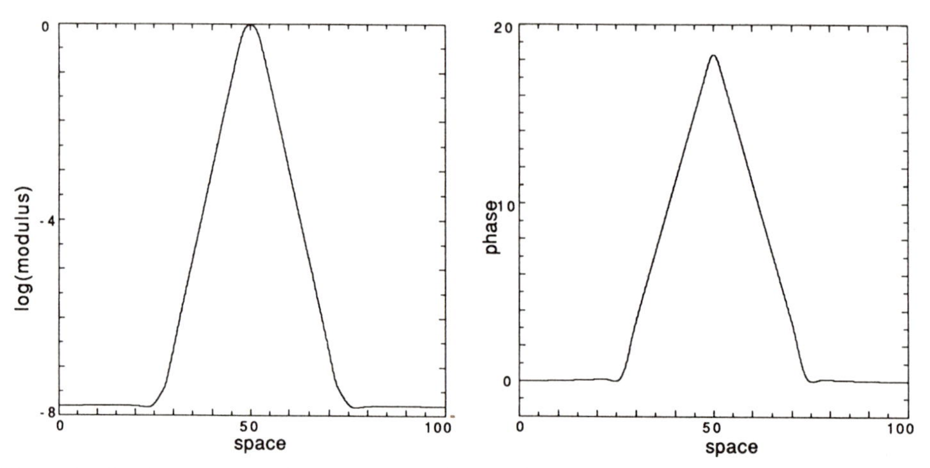

Figure 1. Stable pulse-like solution $W_R = R_0(x) exp\ i[\theta_0(x) + \Omega t]$ when $W_L = 0$, in a right group velocity frame. Parameters are $\mu = -.1$, $\alpha = 1$, $\beta = 3+i$, $\gamma = -2.75+i$, $L = 100$. We have measured $\Omega = 1.30$. **A)** Logarithm of the modulus profile $R_0(x)$, **B)** Phase profile $\theta_0(x)$.

We conclude that phase gradients are responsible for the stabilization of a droplet of bifurcated state inside the basic state.

3. Collision between right and left moving pulses

We now examine the interaction of two counterpropagating pulse-like solutions on a particular numerical experiment (parameters are those of Figure 1 plus $c = 1$ and $\delta = 1 + i$). A right pulse centered at $x_L = L/4$ and a left one at $x_R = 3L/4$ are generated by appropriate initial conditions (with a $sech[10x/L]$ shape). As long as the distance between their centers is large, there is no interaction, and each pulse moves with its own group velocity c and $-c$ (Figure 2.A).

But when the two pulses meet, they no longer propagate and a new localized structure is observed : a droplet of standing waves $|W_R| = |W_L|$ inside the basic null state. This droplet expands at a small speed compared to the group velocity c (Figure 2.B).

The phase gradients for the localized structure are nearly constant (Figure 3). We can thus consider that the dynamics of the moduli R and S of the complex amplitudes W_R and W_L is variationnal. The local energy of a homogeneous state is now $-V(R) - V(S) + 1/2\delta_r R^2 S^2$. The standing waves state have the same energy when μ_{eff} is equal to the new value $\mu_p = 3(\beta_r + \delta_r)^2/16\gamma_r$. This is indeed the case as computed from the numerical simulation.

As for the isolated pulse, we conclude that phase gradients stabilizes the droplet of standing waves.

4. Conclusion

We have shown that localized structures were characterized by constant phase gradients for the complex amplitudes W_R and W_L of the counterpropagating waves. The moduli are thus governed by a variationnal dynamics for which the localized structures are droplets of the homogeneous bifurcated state inside the null state. But contrary to ordinary first order phase transition, these droplets are structurally stable, thanks to the phase gradients which adjust themselves to maintain the equilibrium between the two phases.

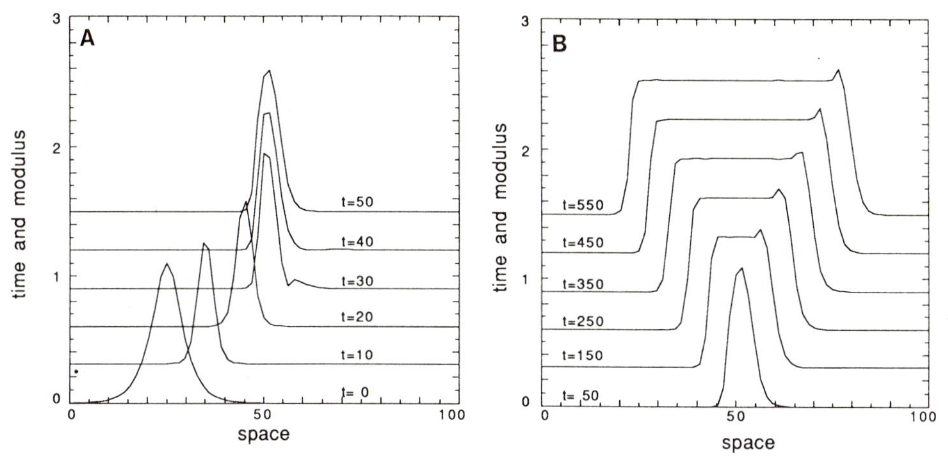

Figure 2. Time evolution $R(x,t)$ of the modulus of the right wave amplitude W_R. The evolution of the modulus of W_L is symetric around the center of the box. The two pulses start at $x_R = L/4$ and $x_L = 3L/4$, meet at $t = c/(x_L - x_R) \sim 30$, stop moving and form a localized structure slowly expanding at a speed of order 0.1 c. **A)** $R(x,t)$, modulus of W_R, **B)** $\phi(x,t)$, phase of W_R.

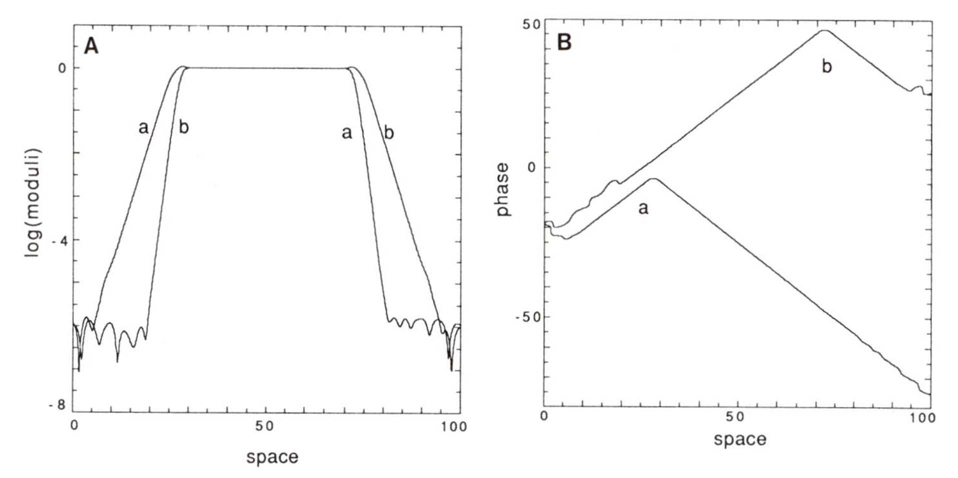

Figure 3. Localized structure (W_R, W_L) at $t = 450$. **A)** Logarithms of the modulus profile of W_R (b) and W_L (a) , **B)** phase profile of W_R (b) and W_L (a).

References

[1] E. Moses, J. Fineberg and V. Steinberg, Phys. Rev. **A 35**, 2757 - 2760 (1987).
[2] R. Heinrichs, G. Ahlers and D. S. Cannel, Phys. Rev. **A 35**, 2761 - 2764 (1987).
[3] P. Kolodner, D. Bensimon and C. M. Surko, Phys. Rev. Lett. **60**, 1723(1988).
[4] M. C. Cross, Phys. Rev. Letters **57**, 2935-2938 (1986).
[5] Y. Pomeau, Physica **23D**, 3-11 (1986).
[6] R. J. Deissler, H. R. Brand, Phys.Lett. A **130**, 293-298 (1988).
[7] O. Thual, S. Fauve, Journal de Physique, to appear (1988).

FROM THE CHAOS TO QUASIREGULAR PATTERNS

A.A. Chernikov, R.Z. Sagdeev, G.M. Zaslavsky

Space Research Institute
Profsoyuznaya 84/32, Moscow 117810, USSR

ABSTRACT

Phase portraits of dynamical systems produce certain kinds of patterns in their phase space. With help of web mapping it is possible to cover the plane with tiling of arbitrary quasicrystal symmetry. The connection between web–mapping and stationary Beltrami flows is established. New type of flows with quasicrystal symmetry is introduced. These flows have chaotic streamlines which display in real space different paterns with q–fold symmetry. Stochastic web is the region of space which separates the meshes of the pattern and inside of which Lagrangian turbulence of admixed particles is realized.

During last few years there has been progress in understanding of some features of transition from laminar flows to turbulent ones. The routes to turbulence in some cases, like thermal convection, pass through different flow patterns, their changing, deformations, etc[1]. Hydrodynamic patterns may possess some symmetry as solid state does. This permits us to approach by some new way the hydrodynamical patterns and their properties.

First, introduce the notion of the phase space pattern. Let $\tilde{R}$ be a coordinate vector of the point in the phase space of a hamiltonian dynamical system. Its equation of motion is

$$\tilde{R}_{t+T} = \hat{T}\tilde{R}_t$$

where $\hat{T}$ is the time shift operator. Then there exist invariant curves of regular motion and invariant manifolds of chaotic motion which tile the phase space and produce a phase portrait of the system.

It is usually said that the tiling produces some pattern in the phase space.

As an example let us consider a kicked oscillator[2]

New Trends in Nonlinear Dynamics and Pattern-Forming Phenomena
Edited by P. Coullet and P. Huerre
Plenum Press, New York, 1990

$$\ddot{x} + \alpha_q^2 x = - \alpha_q K \sin x \sum_{n=-\infty}^{\infty} \delta(t-n) \tag{1}$$

where α_q is the oscillator frequency, K is the parameter and the period of kicks is equal to one. The problem (1) is equivalent to the problem of particle motion in a constant magnetic field (along the z axis) and normally propagating (along the x axis) wave packet. The equation (1) can be reduced to the so—called web—mapping.

$$\hat{M}_q: \begin{cases} \bar{u} = (u + K \sin v) \cos\alpha_q + v\sin\alpha_q \\ \bar{v} = -(u + K \sin v) \sin\alpha_q + v\cos\alpha_q \end{cases} \tag{2}$$

where $u = x/\alpha_q$, $v = -x$ and $(\bar{u}, \bar{v})$ differs from (u, v) by the time shift T=1. If $\alpha_q = 2\pi/q$ and q is integer, then we have resonance condition between oscillator frequency and kicks frequency.

For small K << 1 mapping $\hat{M}_q$ produces in phase space (u, v) an invariant manifold of infinite size — the stochastic web[2,3]. The web realizes some tiling in phase space and the tiling is of q—fold symmetry pattern. For q=4 (Fig.1) and q=3,6 (Fig.2) it is periodic crystal—like symmetry. For $q \neq 1,2,3,4,5,6$ it is quasicrystal—like symmetry (Fig.3 for q=5 and Fig.4 for q=7). All patterns shown in at Figs.1—4 are produced by the only particle orbit. Its stochastic motion covers some area of finite measure inside of the thin stochastic web. The thickness of the web is exponentially small in parameter K. It is the reason to consider $\hat{M}_q$ as generator of tiling with q—symmetry.

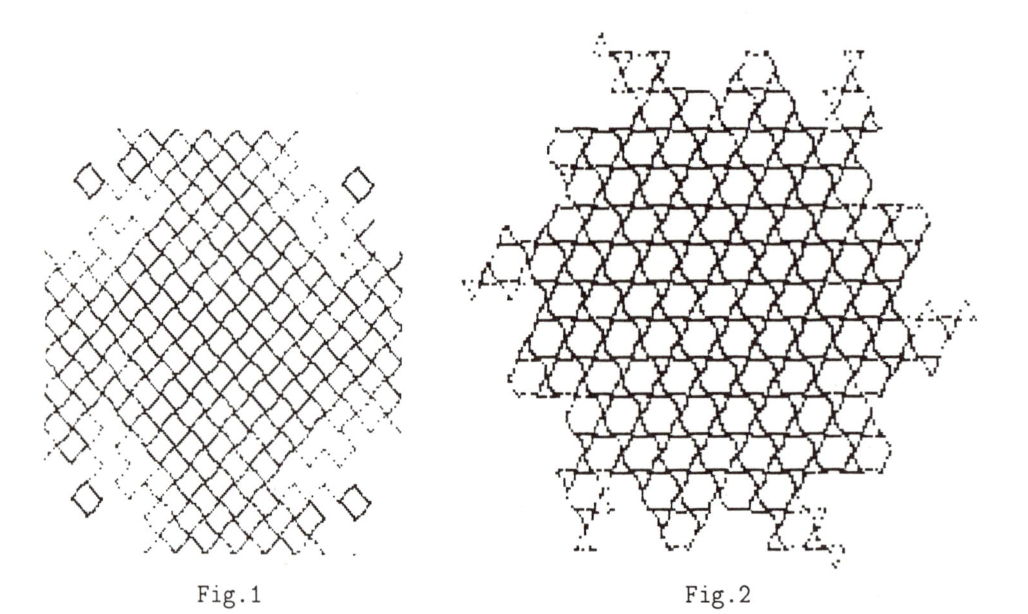

Fig.1 Fig.2

There is some another way to describe the skeleton of the stochastic web. After some transformation the dynamical system (1) or (2) may be reduced exactly to the system with hamiltonian[4]

$$H = H_q(u,v) + V(u,v,t) \tag{3}$$

where

$$H_q(u,v) = \sum_{j=1}^{q} \cos(\underset{\sim}{R}\, \underset{\sim}{e}_j), \tag{4}$$

$\underset{\sim}{R}=(u,v)$, $\underset{\sim}{e}_j=(\cos 2\pi j/q, \sin 2\pi j/q)$ are the unit vectors which form a regular star), and $V(u,v,t)$ is nonstationary perturbation.

The unperturbed dynamical system (4) is integrable. Its phase portrait depends on the value of energy $H_q=E$. For different slicings, E, we have different tilings of the phase plane (u,v) by invariant curves. But symmetry of tiling doesn't depend on a particular slice. There exists some specific energy level E_c such that the pattern on the slice $E \in (E_c-\Delta E, E_c+\Delta E)$ with small thickness $2\Delta E$ coincides with the stochastic web pattern (see Fig.5 for q=5 and Fig.6 for q=7). Thus, the hamiltonian H_q has the phase portrait with q–fold symmetry pattern in the phase space. The same function (4) can generate patterns in hydrodynamics[5].

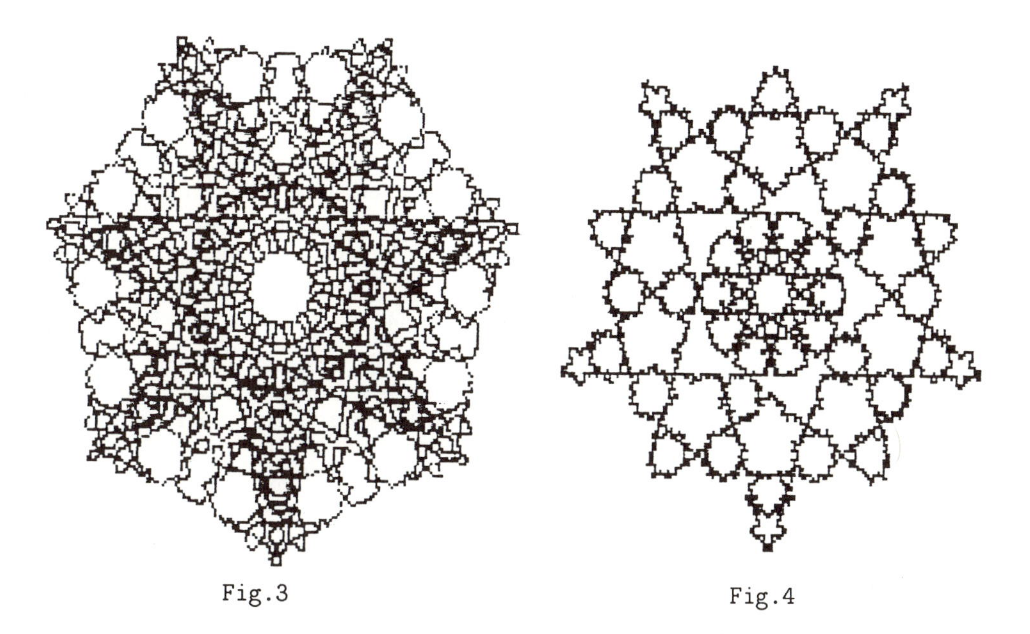

Fig.3 Fig.4

Consider 2D inviscid flow

$$\frac{\partial}{\partial t}\Delta\psi + \frac{\partial(\psi,\Delta\psi)}{\partial(x,y)} = 0 \tag{5}$$

where Δ is the 2D Laplace operator and ψ is the stream function of the flow. It can be shown that the function

$$\psi = \psi_q = H_q \tag{6}$$

becomes the stationary solution of (5) after the transformation ($u \to x$, $v \to y$). A stronger result can be obtained in 3D. Let us introduce a Q–flow:

$$\begin{aligned} V_x &= -\frac{\partial\psi}{\partial y} + \epsilon \sin z \\ V_y &= \frac{\partial\psi}{\partial x} - \epsilon \cos z \\ V_z &= \psi \end{aligned} \tag{7}$$

where ϵ is the parameter and ψ is the solution of the 2D equation

$$\Delta\psi + \psi = 0 \tag{8}$$

Then for such a flow the Beltrami property

$$[\underset{\sim}{v}, \operatorname{rot}\underset{\sim}{v}] = 0$$

exactly holds and div $\underset{\sim}{v}=0$. Therefore, Eq.(7) is the stationary solution of 3D Euler equation. Consider the particular case (6). Then the q–fold symmetry of the flow appears.

If q =4, then $\psi_4 = 2(\cos x + \cos y)$ and Eq.(7) gives the well known ABC–flow[6]. For q=3 or 6 we obtain the stream function of hexagonal pattern: $\psi_3 = \cos x + \cos(x/2 + 3^{1/2}y/2) + \cos(x/2 - 3^{1/2}y/2)$. For another q pattern of the flow has quasicrystal symmetry.

It has been shown[6] in that the ABC–flow possesses a chaos of streamlines, which are governed by equations

$$\frac{dx}{V_x} = \frac{dy}{V_y} = \frac{dz}{V_z} \tag{9}$$

The analysis[7] shows that the region covered by chaotic streamlines has certain special form close to the cubic one. Q–flows are characterized by a common property of stream lines[5]. There exists stochastic web in the real space (x,y,z), which is filled by stramlines chaotically distributed in space. The web forms the spacial pattern with ϵ for $\epsilon \ll 1$. Many properties of the hydrodynamical patterns can be obtained[8] from the properties of the dynamical system (4). Fig.5 shows the pattern produced by a single streamline for q=3. A similar picture for q=5 shown in Fig.6.

Our results indicate a universal property of the preturbulent state characterized by a 3D quasiperiodical symmetry. Streamlines of the corresponding pattern are considered as the certain dynamical system (9) for which one of the variables (x,y,z) plays the role of time. This system has the invariant manifold – stochastic web which is a carrier of hydrodynamical pattern. Admixed particles undergo unlimited transport when they are embedded within the channels of the web. This phenomenon is known as the Lagrangian turbulence[7], and such kind of turbulence is typical of Q–flows.

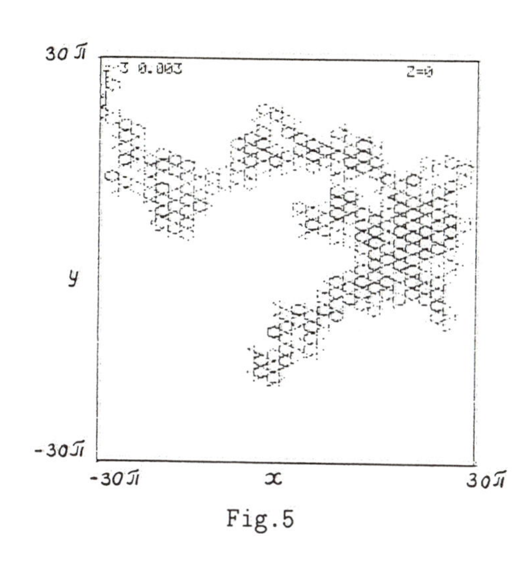

Fig.5

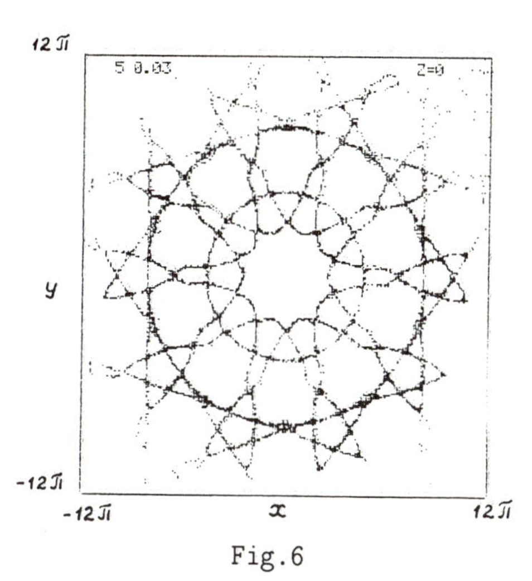

Fig.6

References

1. "Cellular Structures in Instabilities", T.E.Wesfreid and S.Zaleski, eds., Springer—Verlag, Berlin (1984).
2. G.M.Zaslavsky, M.Yu.Zakharov, R.Z.Sagdeev, D.A.Usikov and A.A.Chernikov, Stochastic web and diffusion of particles in a magnetic field, Sov.Phys.JETP, 64:294 (1987).
3. A.A.Chernikov, R.Z.Sagdeev, D.A.Usikov, M.Yu.Zakharov and G.M.Zaslavsky, Minimal chaos and stochastic webs, Nature, 326:559 (1987).
4. G.M.Zaslavsky, N.Yu.Zakharov, R.Z.Sagdeev, D.A.Usikov and A.A.Chernikov, Generation of ordered structures with a symmetry axis from a Hamiltonian dynamics, JETP Lett., 44:451 (1987).
5. G.M.Zaslavsky, R.Z.Sagdeev and A.A.Chernikov, Chaos of streamlines in stationary flows, Zh.Eksp.Teor.Fiz., 94:102 (1988).
6. V.I.Arhol'd, Sur la topologie des ecoulements stationnaires des fluides parfaits, Compt. Rendus, 261:17 (1965).
7. T.Dombre, U.Frisch, J.M.Green, M.Henon, A.Mehr and A.M.Soward, Chaotic streamlines in the ABC flows, J.Fluid Mech., 167:353 (1986).
8. A.A.Chernikov, R.Z.Sagdeev, D.A.Usikov and G.M.Zaslavsky, The Hamiltonian method for quasicrystal symmetry, Phys.Lett., 125A:101 (1987).

INTERFACIAL INSTABILITIES AND WAVES

IN THE PRESENCE OF A TRANSVERSE ELECTRIC CURRENT

René J. Moreau[α],
Sylvain L. Pigny[α],
and Sherwin A. Maslowe[β]

[α] Laboratoire MADYLAM,
ENSHMG BP 95, 38402, Saint Martin D'Hères Cedex, France
[β] McGill University, Montréal, Canada, H3A 2K6

ABSTRACT: In metals processing there are a number of situations where an electric current passes through a fluid interface (slag or electrolyte above and molten metal below, for instance). An experiment using mercury and salt water has shown that instabilities and waves may develop on such interfaces. Neutral curves deduced from linear analysis , which may have one, two, or three minima, show that different modes may become unstable. A first attempt to study the non-linear behaviour of these instabilities, based on multiple scales technique, is also presented.

1) INTRODUCTION

In metals production, or refining, there are a number of situations characterized by the passage of an electric current through an interface. This current may be alternative or direct current. It usually comes from a solid anode, which may be either rather localized or have large dimensions, and it passes first through some poor electrical conductor, like an electric arc, or a slag, or an electrolyte. Then it crosses the interface between that fluid and the molten metal (steel, or aluminium, for instance) before being collected by a cathode. A variety of examples, such as electroslag refining or welding, arc welding, arc furnaces, aluminium reduction cells, were briefly sketched and commented during the presentation of this paper. The reader who is not familiar with metals processing might find some background material in Szekely [1] or Moreau [2]. The interface is usually almost flat and horizontal, the slag being above the metal, and of course the arc still above when it is present. The relative difference of densities may vary from 0.1 to 1, and gravity has some stabilizing influence, as has surface tension. The electric current density $\mathbf{J}$ usually has values of a few $A.cm^{-2}$ in electrolytes, and its value in the metal follows from conservation of the total electric current. Some magnetic field is always present and contains two parts: the near field (induced by the internal current distribution), which varies according to $curl\mathbf{B} = \mu\mathbf{J}$, and the far field, (induced by the external conductors), whose curl is zero. The two (or three) fluids are therefore submitted to Lorentz forces $\mathbf{J} \times \mathbf{B}$, whose curl is not necessary zero. Then the two fluids are stirred and their motion obeys the Navier-Stokes equations. Because the magnetic Reynolds number is much smaller than unity in metals at the scale of these devices, the reaction of motion on the distribution of the electromagnetic field is negligible, and these data may be deduced from Maxwell's equations and Ohm's law exactly as if fluids were at rest. Furthermore in many circumstances it may be assumed that, at least along the vertical direction, a hydrostatic approximation is valid, the motion being essentially horizontal and parallel to the interface. We prefer to forget the details of a given apparatus in order to define a fundamental stability problem; the basic state of equilibrium whose stability properties should be analysed with many potential applications is the following.

New Trends in Nonlinear Dynamics and Pattern-Forming Phenomena
Edited by P. Coullet and P. Huerre
Plenum Press, New York, 1990

The undisturbed interface is plane, horizontal and parallel to the infinite surfaces of the solid electrodes. The vertical component of the electric current density is uniform along the anode surface. However, some horizontal component may nevertheless exist in the fluids, when the cathode collectors are located on the side of the device. In the next section a simple experiment is described briefly, demonstrating the existence of instabilities in a very simple situation. Section 3 presents a particular version of the typical linear stability analysis (other version have been presented earlier [3], [4]); and section 4 is a first attempt to develop a convenient theoretical tool to study the non-linear behaviour of the instability.

2) THE EXPERIMENT

A small cell (1m long, 2.5 cm wide) contains two superposed liquids: mercury below, salt water above. The parallel vertical walls are made of plexiglas in order to allow observation of the interface The anode, made of carbone, is held from above and may be plunged more or less in the electrolyte layer whose thickness is therefore an adjustable parameter. In that particular situation, with direct current densities low enough to avoid disturbing chemical reactions, gravity is always predominant. With an alternative current power supply current, densities of the order of 1 $A.cm^{-2}$ may be obtained without any instabilities before the appearance of chemical reactions. However, if an extra horizontal current is added in the mercury layer to that coming from the interface by means of electrodes located at both ends of the cell, then hydrodynamic instabilities may be observed as shown on fig.1. Clearly, there is no flow at equilibrium and therefore the instability comes only from the disturbance of the Lorentz force which results itself from some deformation of the interface geometry. This instability apppears to be, generally, a travelling wave. This is the case for the conditions of fig.1. However with a particular arrangement of the ends of the apparatus (no free surface at the end of the anodes) the instability may be forced to be a standing wave. This clearly suggests a dispersion relation with two roots of opposite signs, which is quite typical for interface problems.

3) LINEAR STABILITY ANALYSIS

3-1) Undisturbated state

In the domain $-L < x < L, -H < z < H$ of the (x,z) plane, the electric potential is modelled by the following expressions, which satisfy the Laplace equation and the conditions of continuity of Φ and $\frac{\partial \Phi}{\partial z}$ at the interface (subscripts 1 or 2 respectively refer to cryolite $0 < z < H$ or to aluminium $-H < z < 0$):

$$\begin{cases} \Phi_1 = \frac{J}{\sigma_1}\left(z + \frac{a}{H}\frac{\sigma_1}{\sigma_2}\left(x^2 - z^2\right)\right), \\ \Phi_2 = \frac{J}{\sigma_2}\left(z + \frac{a}{H}\left(x^2 - z^2\right)\right). \end{cases} \tag{1}$$

The factor a is a non-dimensional parameter allowing variation in the shape of the electric current lines. Depending on its value the horizontal component of the electric current may be positive as when the cathodic bars collect partially near the border of the cell, or negative, as when the current lines are distorted by a peripheral ledge of frozen cryolite.

The magnetic field $\mathbf{B}$ is a sum of two contributions. One, the field induced by the external conductors, is a pure gradient. As recognized by Sneyd [3] it is important to take into account its variations; it is therefore modelled by $\mathbf{B} = (-\gamma y, B_0 - \gamma x, 0)$. The other, the field of the internal current, is easily derived from Eqs.(1) through Ampère's law $curl\,\mathbf{B} = \mu\,\mathbf{J}$. Finally, we model the magnetic field as:

$$\begin{cases} B_1 = \left(-\gamma y, B_0 - \gamma x - \mu J x, 0\right), \\ B_2 = \left(-\gamma y, B_0 - \gamma x - \mu J x\left(1 - \frac{2az}{H}\right), 0\right). \end{cases} \tag{2}$$

The undisturbed forces $\mathbf{J} \times \mathbf{B}$ are rotationnal and drive some motion in the (x,y) plane, as well as some deformation of the interface. This deformation, which could have been calculated as in Moreau and Evans [5], is assumed to be small enough to be neglected.

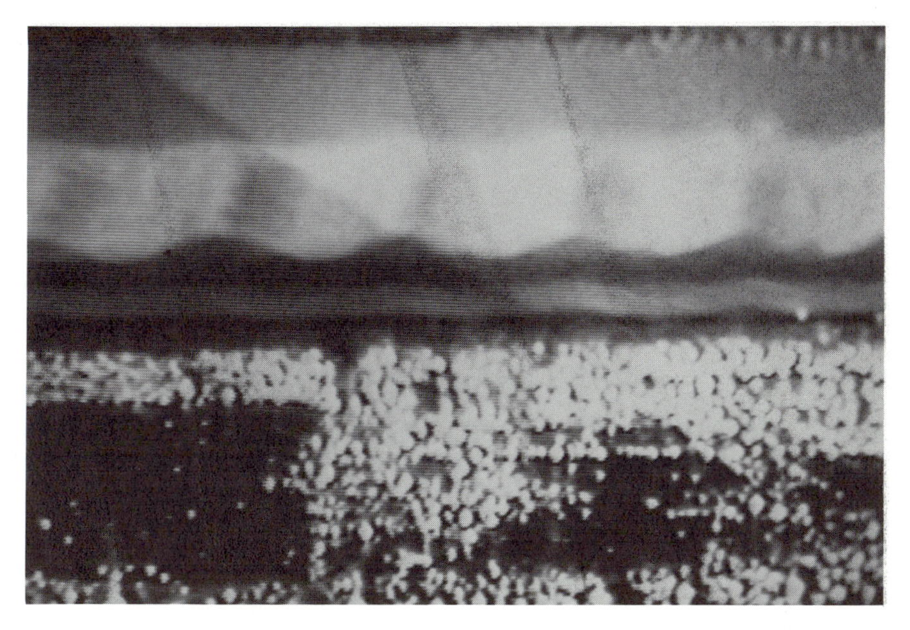

Fig. 1a. Instabilities at the interface between mercury and salt water.

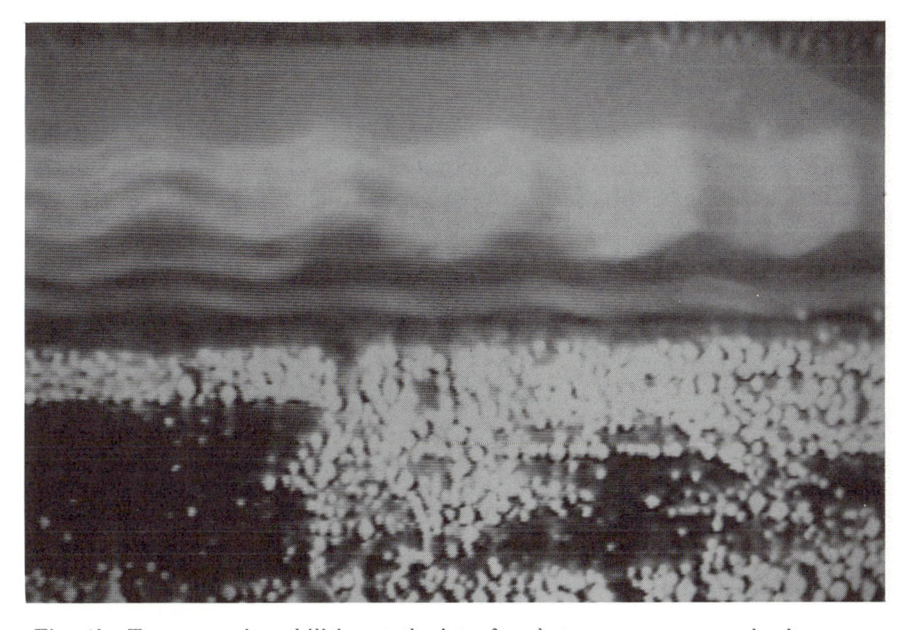

Fig. 1b. Transverse instabilities at the interface between mercury and salt water.

However, the stirring motion, as well as the preexisting turbulence, is represented via the friction coefficient κ. In the framework of this two-dimensional analysis, the main horizontal motion of the two liquids cannot be represented. However, in order to allow some interaction between Kelvin-Helmholtz instability and electromagnetically driven instabilities, uniform velocities U_1 and U_2, both in the x-direction, are assumed to be present. They are considered as independent parameters, whereas in actual cells the two horizontal velocity fields are controlled by the $\mathbf{J} \times \mathbf{B}$ forces (see [5]).

3-2) Disturbances

Because the relevant magnetic Reynolds number is much smaller than one, the disturbances of the electromagnetic quantities are independent of the velocity field. They depend only on the geometry of the interface and they instantaneously adjust themselves to any change in that geometry. Introducing a disturbance of the interface of the form:

$$\eta = A \exp\left[i\,(kx - \omega t)\right], \tag{3}$$

expressions are easily obtained for perturbations of the electric potential satisfying Laplace's equation $\Delta\phi = 0$ and the following boundary conditions:

$$\phi_1(H) = \phi_2(-H) = 0,$$

$$\Phi_1(\eta) + \phi_1(0) = \Phi_2(\eta) + \phi_2(0), \tag{4}$$

$$\sigma_1\left(\frac{\partial \Phi_1}{\partial n}(\eta) + \frac{\partial \phi_1}{\partial z}(0)\right) = \sigma_2\left(\frac{\partial \Phi_2}{\partial n}(\eta) + \frac{\partial \phi_2}{\partial z}(0)\right),$$

n being the normal to the interface.

It is straighforward to deduce expressions for the disturbances of the current density $\mathbf{j}$, the magnetic field $\mathbf{b}$ and for the electromagnetic forces $\mathbf{f} = \mathbf{j} \times \mathbf{B} + \mathbf{J} \times \mathbf{b}$.

It is remarkable that these forces are irrotational, exept for the terms involving the gradient γ (see Eq.(2)) of the far magnetic field. This suggests expressing them, as well as the velocity field, as ($i =$ ',2):

$$\begin{cases} \mathbf{f}_i = -\nabla F_i + \mathbf{f}_{ir}, \\ \mathbf{u}_i = \nabla V_i + \mathbf{u}_{ir}. \end{cases} \tag{5}$$

Then, because of continuity

$$\Delta V_i = 0, \qquad \nabla.\mathbf{u}_{ir}, \tag{6}$$

and the motion equation may be written

$$\nabla p_i^* = -\rho_i\,(r_i + \kappa)\,\mathbf{u}_{ir} + \mathbf{f}_{ir}, \tag{7}$$

where $p_i^* = p_i + F_i + \rho_i\,(r_i + \kappa)\,V_i$ and $r_i = (ik.U_i - \omega)$. The classical boundary conditions on the electrodes and on the interface are easily expressed in terms of p_i, and it is straightforward to get the distribution of p_i by solving the equation:

$$\Delta p_i^* = \nabla.\mathbf{f}_{ir}. \tag{8}$$

3-3) Characteristic equation and neutral curve

Substituting expressions for $p_1(0)$ and $p_2(0)$, obtained from Eq.(8), into the condition of continuity of the pression at the interface, where Γ stands for surface tension, we obtain:

$$p_1(0) - p_2(0) = -\eta[g\,(\rho_2 - \rho_1) + \Gamma k^2], \tag{9}$$

which gives the characteristic equation. The latter may be written as follows when $\sigma_1 << \sigma_2$ (in actual cells $\frac{\sigma_1}{\sigma_2} \approx 10^{-4}$):

$$\rho_1 r_1(r_1+\kappa)+\rho_2 r_2(r_2+\kappa)+g^* kt(\rho_2-\rho_1)+\Gamma k^3 t+J\frac{\gamma}{2}(1+M+iN)=0, \quad (10)$$

with

$$\alpha = kH,$$

$$t = \tanh kH,$$

$$M = 2a\alpha t\left(t^2 - t^{-2} - \alpha^{-2}\right),$$

$$N = 2a\frac{x}{H}\alpha t\left(2t - 2t^{-1} + \alpha^{-1}\right).$$

The first four terms in Eq.(10) represent the usual characteristic equation, exept that g^* is now the effective gravity taking into account the irrotational part of the electromagnetic forces, i.e:

$$g^* = g - 2a\frac{x}{H}.\frac{J\gamma x}{\rho_2 - \rho_1}. \quad (11)$$

This effective gravity differs from g only when a is non-zero in the aluminium. This means that the horizontal component of the current density might produce Rayleigh-Taylor instabilities when $a\gamma > 0$, or enhance the stabilizing gravity effect when $a\gamma < 0$.The last term of Eq.(10) represents the influence of the rotational part of the electromagnetic forces. The reader should notice that it has an imaginary part. This means that it is analogous to the velocity difference $U_1 - U_2$, and that this effect is destabilizing. Consequently, when $U_1 = U_2 = 0$, friction appears necessary to counterbalance the imaginary part of this term. Clearly, even without any shear of the interface some instability may develop, driven only by electromagnetic mechanisms.The marginal condition such that ω be real follows from (10). With our simplying assumptions it may be written:

$$F^2 = \frac{t}{\alpha} + T\alpha t + \frac{P\left(1 + M - QN^2\right)}{\alpha^2}, \quad (12)$$

with the following definition of the non-dimensional parameters

$$\begin{cases} F^2 = \frac{\rho_1\rho_2}{\rho_2^2-\rho_1^2}\frac{(U_1-U_2)^2}{g^* H}, & P = \frac{J\gamma H}{2g^*(\rho_2-\rho_1)}, \\ T = \frac{\Gamma}{g^* H(\rho_2-\rho_1)}, & Q = \frac{J\gamma}{2k^2(\rho_2+\rho_1)}. \end{cases} \quad (13)$$

For the sake of brevity, we limit the analysis to conditions such as $a << 1$ and $M << 1$ and $a\frac{x}{H} = 0(1)$, which seem to correspond fairly well to actual cells, where $\frac{x}{H}$ may be greater than 10. Then Eq. (12) becomes

$$F^2 = \frac{t}{\alpha} + T\alpha t + P\left(\frac{1}{\alpha^2} - Q^* G^2(\alpha)\right), \quad (14)$$

with $Q^* = 4Q.\frac{a^2x^2}{H^2}$, and $G(\alpha) = t\left(2t - \frac{2}{t} + \frac{1}{\alpha}\right)$. Figure 2 shows the variation with α of G^2 and $\frac{1}{\alpha^2}$. One may notice that these two quantities become equal when α is large enough (say $\alpha > 3$), but that $\frac{1}{\alpha^2}$ is of course predominant when $\alpha << 1$. The difference $\frac{1}{\alpha^2} - Q^*G^2(\alpha)$ may be negative only if $Q^* > 1$. If $1 < Q^* < 9.47$, it si negative for α sufficiently large and it has a minimum for $\alpha \approx 2.2$ to 2.5. If $Q^* > 9.47$, another band of wave numbers makes the difference negative around $\alpha \approx 0.5$. Of course, since $G(\alpha) = 0$ in the vicinity of $\alpha = 1$, wave numbers of order of one can never become unstable. Therefore, two different kinds of disturbance may be excited by electromagnetic mechanisms; those with

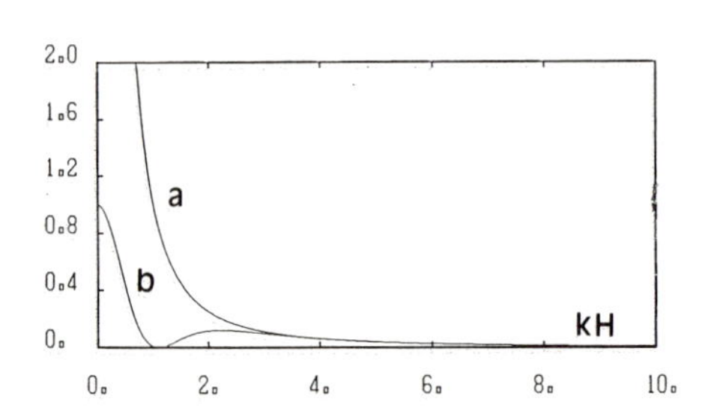

Fig. 2. Variations of $\frac{1}{\alpha^2}$ (a) and G^2 (b) as functions of $\alpha = kH$.

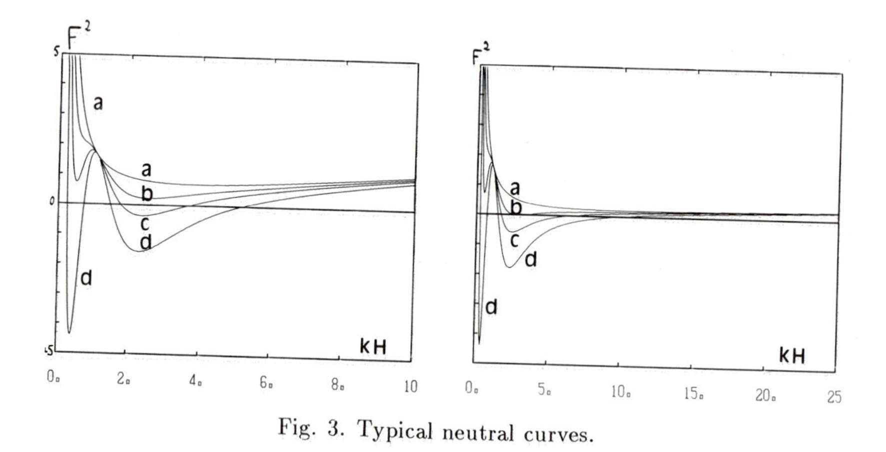

Fig. 3. Typical neutral curves.

$\alpha = 2.2$ to 2.5 (corresponding to wave-length of order of 50 cm in cells), and those with $\alpha \approx 0.5$ (corresponding to wave-length of order of 2.5 m in cells). But the second instability requires a current approximately ten times greater than the first. Figure 3 shows typical neutral curves. Of course, the ordinary minimum of the neutral curve, depending only on gravity and surface tension may still exist if it concerns a band of wave numbers significantly larger than those where electromagnetic effects are dominant (right side of figure 3); then the neutral curve may have three distinct minima. On the contrary the usual minimum may disappear when it concerns the same wave-numbers as the electromagnetic effects (left side of figure 3).

4) AMPLITUDE EVOLUTION EQUATION

In this section, we outline the derivation of an amplitude equation describing the weakly non-linear evolution of a wave packet. To indicate the nature of the expected result, an amplitude equation by Nayfeh and Saric [6] for the classical Kelvin-Helmholtz flow in the absence of surface tension is given below. Provided that we are near the cut-off number (corresponding to the onset of instability), the disturbance evolves accordingly to:

$$\frac{\partial A}{\partial \tau} - \frac{1}{2} i \omega'' \frac{\partial^2 A}{\partial \xi^2} = \mu A^2 A^*, \tag{15}$$

where τ and ξ are slow scales that will be defined later. For a conservative system, $\omega''(k)$ is real, μ is imaginary, and Eq.(15), often called the cubic Schrödinger equation, can be solved exactly by the inverse scattering method. A particularly interesting solution involves envelope solitons, but this is not the only possibility. Moreover other amplitude equations are obtained when perturbing away from the stability boundary and in this case invicid flows (including the Kelvin-Helmholtz) may lead to amplitude equations that are second- order in time. The generalisation of greatest interest in the present study concerns the treatment of instability arising from Lorentz forces. It develop that a linear term must be added to Eq.(15) and that the constants ω'' and μ will both be complex in general. To obtain an appropriate balance between these terms, as indicated earlier, the Lorentz forces are scaled such that they are of order of ϵ^2. The development will be carried out by employing a version of the multiple-scales method which requires only one slow scale in each of the variables x and t. We begin by expanding the dependent variables in series of powers of ϵ and, as usual in such studies, a Taylor series expansion about $z = 0$ is also employed. The interface displacement, velocity potential and pressure are expanded as follows:

$$\begin{cases} \eta &= \eta_0 + \epsilon\eta_1 + \epsilon^2\eta_2 + \cdots \\ \Phi &= \Phi_0 + \epsilon\Phi_1 + \epsilon^2\Phi_2 + \cdots \\ p &= p_0 + \epsilon p_1 + \epsilon^2 p_2 + \cdots \end{cases}$$

In the $O(\epsilon^0)$ problem, the classical characteristic equation is recovered. However, instead of taking the amplitude constant, as in the usual linear theory for monochromatic waves we write, for example,

$$\eta_0 = A(X,T)\, e^{i\theta} + A^*(X,T)\, e^{-i\theta},$$

where $\theta = kx - \omega t$ and the slow scales are defined by $X = \epsilon x$ and $T = \epsilon t$. The general procedure for deriving the equation satisfied by $A(X,T)$ in hydrodynamic stability problems is described in Benney and Maslowe [7], where a number of illustrative examples are given. If the group velocity ω' is finite, as this is usually the case, one obtains

$$A_T + \omega' A_X = \epsilon\left(\beta A_{XX} + \delta A + \mu A^2 A^*\right), \tag{16}$$

where it can be verified a posteriori that $\beta = \frac{i\omega''}{2}$. Equation (16) is the appropriate form in our study, which is interesting, because in the absence of Lorentz forces it was shown in [5] that the amplitude equation at marginal stability is second order in time. A discussion of how the form of the amplitude equation is dictated by the linear dispersion relation can be found in sections 7-3-2 and 7-4-1 of Swinney and Gollub [8].

Proceeding now to the $O(\epsilon)$ problem, the variables can be separated by noting that $A_T = -\omega A_X$ to this order and the time-dependent variables are all taken to be proportional to A_X. Imposing the boundary conditions at interface determines the value of ω'. At $O(\epsilon^2)$, the value of the coefficient δ in Eq.(17) is determined. This time, to separate variables and solve the resulting non-homogeneous system, the $O(\epsilon)$ terms must be included in the amplitude equation. The form of a typical term is illustrated by the way of example by writing the pressure as:

$$p_2 = A_{XX} P_1(z) e^{i\theta} + A P_2(X,z) e^{i\theta} + A^*_{XX} P_1(z) e^{-i\theta} + A^* P_2(X,z) e^{-i\theta}.$$

Because of the way we have scaled the Lorentz forces, the Landau constant μ and the dispersion coefficient β are not changed and have the the same value as in capillary-gravity waves. The non- zero value of δ is a new result due entirely to the Lorentz forces:

$$\epsilon^2 \frac{\delta}{\omega} = -\frac{J\gamma H}{g(\rho_2 - \rho_1)} a \frac{x}{H} \left(\frac{1}{sc} + \frac{1}{\alpha} \right)$$
$$+ i \left(2a \frac{x}{H} \frac{J(B_0 - \mu J x)}{g(\rho_2 - \rho_1)} - \frac{a}{\alpha} \frac{J\gamma H}{g(\rho_2 - \rho_1)} \left(\frac{1}{\alpha^2} + \frac{s^2 + c^2}{s^2 c^2} 2a \frac{x}{H} kx \right) \right). \tag{17}$$

As a final observation, we note that the linearized version of Eq.(16) can be solved exactly by employing a Fourier transform in X if δ is assumed to be constant. It is natural however to rewrite this equation by employing the coordinates $\tau = \epsilon T$ and $\xi = x - \omega' T$, corresponding to a frame of reference moving at the group velocity. This leads to the equation

$$\frac{\partial A}{\partial \tau} = \beta \frac{\partial^2 A}{\partial \xi^2} + \delta A, \tag{18}$$

whose solution is given by:

$$A(\xi, \tau) = \frac{e^{\delta\tau}}{\sqrt{4\pi\beta\tau}} \int_{-\infty}^{+\infty} A(w, 0)\, e^{-[\frac{(\xi - w)^2}{4\beta\tau}]} dw.$$

A number of special cases, e.g. self-similar solutions, can be obtained easily from the above general solution by an appropriate choice of initial condition. Rather than write this down, we discuss these properties in common. For example, there is a shortening of wave length with distance and a phase that varies slowly in time and space. The dominant effect, of course, is the exponential amplification of A with time which will eventually be modified by nonlinearity.

References

[1] SZEKELY J., "On Heat and Fluid Flow Phenomena in Electric Melting and Smelting Operations", in "Metallurgical Appl. of M.H.D", eds Moffatt H.K. and Proctor M.R.E, (1982), The Metals Soc. of London, pp.260-271.

[2] MOREAU R., "Applications Métallurgiques de la Magnétohydrodynamique", in "Theoretical and Applied Mechanics", Proc. of the 15th. IUTAM Congress, Toronto, 17-23 August 1980, eds: Rimrott F.P.J. and Tabarrok B., North-Holland Pub. (1980), pp.107-118.

[3] SNEYD A.D., "Stability of Fluid Layers Carrying a Normal Electric Current", J.Fluid Mech. (1985), vol 156, pp. 223-236.

[4] MOREAU R. and ZIEGLER D., "Stability of Aluminum Cells. A New Approach",Light Metals (1986), pp. 359-364.

[5] MOREAU R. and EVANS J.W., "An Analysis of the Hydrodynamics of Aluminum Reduction Cells",J.Electrochem.Soc, Electrochemical Science and Technology (1984), vol. 131, n^0 10, pp. 2251-2259.

[6] NAYFEH A.H. and SARIC W.S., "Non-Linear Waves in a Kelvin-Helmholtz Flow", J.Fluid Mech.(1972), vol 55, pp. 311-328.

[7] BENNEY D.J. and MASLOWE S.A, "The Evolution in Space and Time of Non- Linear Waves in Parallel Shear Flows", Studies in Appl. Math. (1975), vol.54, pp. 181-205.

[8] MASLOWE S.A.,"Shear Flow Instability and Transition", in "Hydrodynamic Instability and the Transition to Turbulence", eds. Swinney H.L. and Gollub J.P. (1982), Springer, pp. 181-228.

MODULATED TAYLOR VORTEX FLOW

Manfred Lücke, Dieter Roth, and Hendrik Kuhlmann

Institut für Theoretische Physik
Universität des Saarlandes
D-6600 Saarbrücken, FRG

We have studied the rotating Couette system, i.e., the flow between two concentric cylinders of radii R_1 and R_2, with the outer cylinder at rest. The rotation rate of the inner cylinder is modulated periodically $\Omega(t)=\Omega(1+\Delta\cos\omega t)$. We use a finite difference numerical simulation and a four-mode Galerkin model to investigate both the threshold shift of the onset of Taylor vortex flow (TVF) and the fully nonlinear behaviour of TVF (Kuhlmann et al., 1989). Our calculations were done for a radius ratio $\eta=R_1/R_2=0.65$ but we found that the results presented here are practically independent of η.

We considered only low frequency modulation. Then the response of the TVF to the modulation is like a standing wave; only the amplitude oscillates. In Fig. 1 we show the mean squared axial velocity $\overline{w^2(t)}$ at a position where w has a maximum in the absence of modulation. Modulation enhances the order parameter $\overline{w^2(t)}$ and the onset of TVF is shifted to smaller mean driving by a small amount. These effects are considerable only if $\Delta=O(1)$. We found the destabilisation of the basic circular Couette flow in good agreement with the theoretical predictions by Hall (1975) and Riley and Laurence (1976).

In Fig. 2 we compare for $\omega=4$ the time behaviour of our numerically calculated TVF with experimental results obtained by Ahlers (1988). Both agree perfectly except for the slightly premature growth of TVF in experiment (a). This might be caused by the large influence of imperfections when the mean driving is small.

The variety of dynamical behaviour under modulation with different amplitudes Δ and mean driving Ω most significantly appears for low frequencies. If the driving $\Omega(t)$ is small there is a range of exponential growth and decay (Fig. 3a and left half of Fig. 3c) where the flow is governed by linear terms in the equations. For strong supercritical rotation rates the flow follows the driving adiabatically (Fig. 3b and right half of Fig. 3c). Between both regions some oscillations can appear (Fig. 3c).

New Trends in Nonlinear Dynamics and Pattern-Forming Phenomena
Edited by P. Coullet and P. Huerre
Plenum Press, New York, 1990

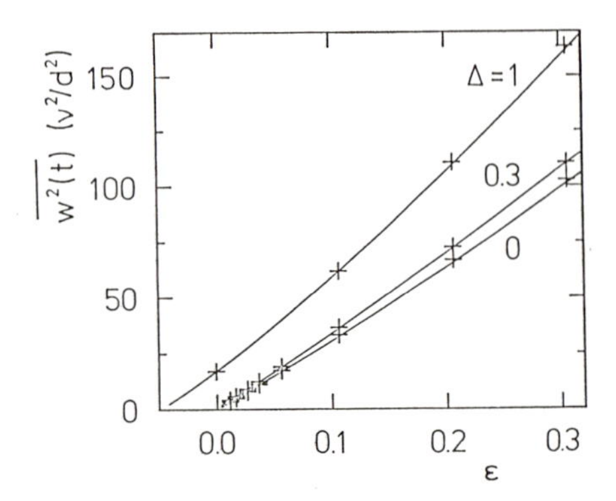

Fig. 1. Mean squared TVF amplitude $\overline{w^2(t)}$ versus $\epsilon=\Omega/\Omega_c(\Delta=0)-1$ for $\omega=\pi/2$ (numerical results: plusses, Galerkin model: lines). $\Omega_c(\Delta=0)$ is the critical rotation rate for onset of TVF in the absence of modulation.

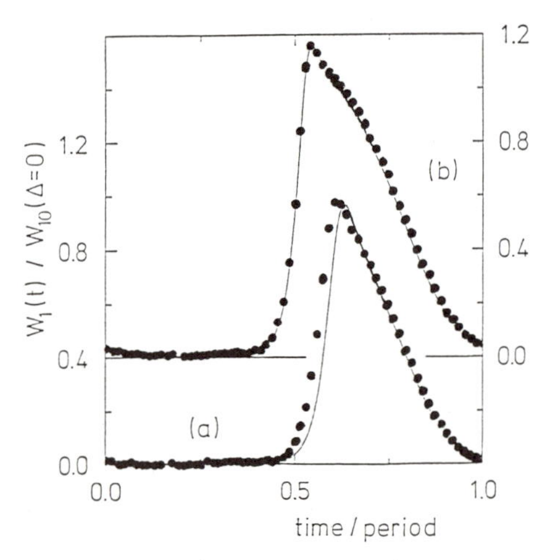

Fig. 2. Time behaviour of the reduced axial Fourier component $W_1(t)$ in experiment (Ahlers (1988); $\eta=0.75$, dots) and numerical simulation (lines) for $\omega=4$, $\epsilon=0.1$ (a), $\epsilon=0.2$ (b) and $(1+\epsilon)\Delta=0.7$. W_{10} is defined by $W_1(\Delta=0)=\sqrt{\epsilon}\cdot W_{10}(\Delta=0)+O(\epsilon)$. The time origin is given by the minimum of the inner cylinder's rotation rate.

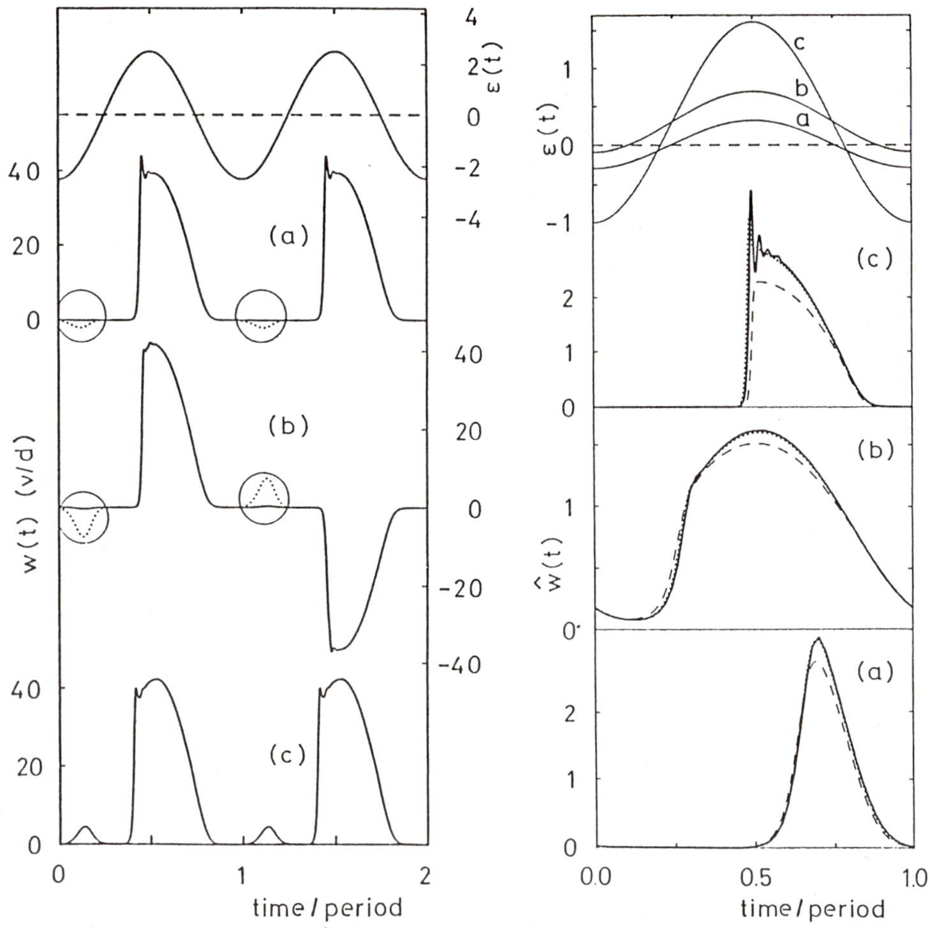

Fig. 3. Axial velocity w(t), normalized by its (Δ=0)-values as function of time for $\omega=\pi/2$, η=0.65 and (a) ϵ=0.016, Δ=0.3, (b) ϵ=0.306, Δ=0.3 (c) ϵ=0.306, Δ=1. Solid line: Galerkin model, dotted line: numerical simulation, dashed line: amplitude equation (Hall, 1975). The upper part shows the respective control parameters $\epsilon(t)=\Omega(t)/\Omega_c(\Delta=0)-1$.

Fig. 4. Axial velocity w(t) obtained by the numerical simulation for large modulation amplitude Δ=2.5 (a), 2.675 (b) and 2.8 (c) with ϵ=0, $\omega=\pi$ over two periods. The driving $\epsilon(t)$ for Δ=2.5 is shown in the top. Curves inside the circles are magnified by a factor of 20.

For modulation amplitudes so large that $\Omega(t)$ becomes also supercritical in the opposite rotation direction we found subharmonic response for a small Δ-band located between two different ranges of harmonic response (Fig. 4).

ACKNOWLEDGEMENT

This work was supported by Stiftung Volkswagenwerk.

REFERENCES

Ahlers, G., private communication (1988).
Hall, P., J. Fluid Mech. 67:29 (1975).
Kuhlmann, H., D. Roth, M. Lücke, Phys. Rev. A 39 (1989).
Riley, P. J., and Laurence, R. L., J. Fluid Mech. 75:625 (1976).

RECENT RESULTS ON THE NON-LINEAR

DYNAMICS OF CURVED PREMIXED FLAMES

Guy Joulin

U.A. 193 CNRS, ENSMA
Rue Guillaume VII
86034 Poitiers, France

INTRODUCTION

Any theoretical study of problems of flame propagation in turbulent flows encounters difficulties of various origins: instabilities, non-linearity, non-locality, evolving curved geometries, wide spectra... This extended summary reports on recent analytical models which tackle them.

FREE FLAMES, POLE DECOMPOSITIONS

The simplest model to describe the displacement $\phi(x,t)$ of a spontaneously wrinkled flame is the M.S. equation (Sivashinsky, 1977; Michelson and Sivashinsky, 1977) viz:

$$\phi_t + \phi_x{}^2/2 = \mu\phi_{xx} + I(\phi, x) \qquad (1)$$

Its LHS measures the changes in absolute normal flame velocity. $I(.,x)$, defined as the multiplication by $|k|$ in Fourier space conjugate to x, represents the changes of flow velocity due to wrinkling and causes the flame instability (Landau, 1944). The stabilizing term $\mu\phi_{xx}$ ($\mu > 0$), is due to the dependence on curvature of the local burning rate (Markstein, 1964; Clavin and Joulin, 1983). The dispersion relation $\omega = |k|(1 - \mu|k|)$ corresponding to the linearized form of (1) and to $\phi \sim \exp(ikx + \omega t)$, shows the two antagonist effects. Thual et al (1985) noticed that (1) belongs to a class for which exact solutions are available (Lee and Chen, 1982). Specifically (1) admits real $2\pi/K$-periodic solutions corresponding to:

$$\phi_x = -\mu K \sum_\alpha \cot\{K(x - x_\alpha)/2\} \qquad (2)$$

provided the N pairs of complex conjugate poles of ϕ_x satisfy:

$$\dot{x}_\alpha = \mu K \sum_{\beta\neq\alpha} \cot\{K(x_\beta - x_\alpha)/2\} - i\,\mathrm{sign}(\mathrm{Im}(x_\alpha)) \qquad (3)$$

Similar pole-decompositions of ϕ_x, with the cot's replaced by $1/x$ functions, hold in the non-periodic cases. As shown by Thual et al (1985) the poles tend to align parallely to the imaginary axis. Once this is achieved ϕ reads $-Ut + F(x)$; it represents a steadily propagating, wrinkled flame. Upon averaging (1) over a period one can then show (Joulin, 1987) that the instability-induced increase in flame speed $U = \langle F_x^2/2\rangle$ reads $2NK\mu(1 - NK\mu)$,

New Trends in Nonlinear Dynamics and Pattern-Forming Phenomena
Edited by P. Coullet and P. Huerre
Plenum Press, New York, 1990

where N must satisfy $N \leq N^* = \mathrm{Int}\{0.5(1 + 1/\mu K)\}$ (Thual et al, 1985). $N = N^*$ and $\mu K = 1/2n$ yield $U = 1/2$ in agreement with Denet's results (1988); $N = N^*$ and $\mu K = 1/(2n + 1)$ give $U < 1/2$.

FORCED DYNAMICS

To mimic a weak incoming turbulence, a forcing term $u(x,t)$ may be added to the RHS of (1). In general (1) must then be studied numerically, as Denet (1988) did recently. The following, shot-noise-like $u(x,t)$ may be an exception (Joulin, 1988a):

$$u(x,t) = \sum_m \delta(t - t_m)(\psi_m(x) - \langle\psi_m\rangle) \tag{4}$$

$$\psi_m(x) = -4\mu \,\mathrm{Re}\{\mathrm{Log}\sin(K(x - a_m - ib_m))\} \tag{5}$$

Since the action of $u(x,t)$ is then to implant pairs of poles $(a_m \pm ib_m)$ of ϕ_x at given times (t_m) without changing the dynamics (3) for $t \neq t_m$, (3) $\div$ (5) in principle enable one to study the response of (1) to a random additive noise while still accounting for the Landau instability. When the pole density $\rho(x,y,t)$ in the complex $(x + iy)$ -plane gets large enough, one may think of combining (3) - (5) in the continuous equation:

$$\rho_t + (\rho U)_x + (\rho V)_y = \Omega(x,y,t) \tag{6}$$

where $U + iV$ is the continuous analog to the RHS of (3) and Ω is the local rate of pole sprinkling. From (6), one recovers the limiting form of $F(x)$ obtained by Thual et al (1985) for $\mu K \to 0$. It is the subject of current works, along with Monte-Carlo simulations based upon (3) - (5).

NEARLY PARABOLIC FLAMES

Writing $\phi = -Ut + F(x) + \Phi(x,t)$ in (1) and approximating the convex tips of F as $Sx^2/2$ lead to a model equation to study the non-linear stability of nearly-parabolic flames (Joulin, 1988b):

$$\Phi_t + Sx\Phi_x + \Phi_x{}^2/2 = \mu\,\Phi_{xx} + I(\Phi,x) \tag{7}$$

Its linearized form has solutions in the form $\Gamma(k,t)\exp(iq(t)x)$, with $q = |k|\exp(-St)$ and $d\mathrm{Log}\,\Gamma/dt = q(1 - \mu q)$; they display the disturbance stretch by the pseudo-convective term $Sx\Phi_x$ and a most dangerous initial wavenumber: $k = 1/\mu$. Following Zel'dovich et al (1980) or Pelcé and Clavin (1987) one may define a critical initial amplitude Γ_*: for $\Gamma(1/\mu,0) \geq \Gamma_*$, $\Gamma(1/\mu,\infty) = O(1)$ obtains (hence a noticeable instability). With S given by the shape of $F(x)$ found in the limit $\mu K \to 0$, one gets $\Gamma_* = O(\exp(-\pi/2\mu K))$, in full accordance with Denet's numerical results (1988). In numerical studies of (1) Γ_* may be exceeded by the truncation errors, thereby causing spurious patterns; this is certainly what happened in a work of Sivashinsky and Michelson (1983), who argued that chaos spontaneously appears if $\mu K \ll 1$. Equation (7) also admits pole-decomposable solutions (Joulin, 1988b); in the non-periodic case the structure of Φ_x is the same as in (2), with the cotangents replaced by $1/x$ functions. A new term Sx_α then appears in the RHS of (3): it accounts for the geometry-induced causes of stretch, shift and damping of the pulse-like solutions to (7). The barycenter $\bar{x}$ of the x_α's generically satisfies $\bar{x} \sim e^{St}$, suggesting that moderate, peaked disturbances of $\phi(x,t)$ cannot stay at tips, in agreement with Lee and Chen's conclusions (1982). Periodic solutions are found upon use of new variables ($z = xe^{-St}$, $\psi = \Phi(xe^{-St},t)$) which convert (7) into:

$$\psi_t + e^{-2St}\,\psi_z{}^2/2 = \mu\, e^{-2St}\,\psi_{zz} + e^{-St}\,I(\psi,z) \tag{8}$$

ψ_z also has a pole-decomposition similar to (3); its poles z_α satisfy:

$$\dot{z}_\alpha = \mu K e^{-2St} \sum_{\beta \neq \alpha} \cot\{K(z_\beta - z_\alpha)/2\} - i e^{-St} \operatorname{sign}(\operatorname{Im}(z_\alpha)) \quad (9)$$

Non-linear analogs of the results of Zel'dovich et al (1980) and of Γ_* can then be obtained. We currently study how to adapt (4) (5) to analyze the influence of a shot-noise on nearly-parabolic flames.

EXPANDING FLAMES

The analysis leading to (1) can be extended to expanding flames (Joulin, 1988c). The model equation then looks like (8), now in terms of an angle variable z, provided exp(St) is replaced by t (this accounts for the nearly-linear flame perimeter growth). ψ_z is again pole-decomposable, the analog of (9) being readily written. First decaying, the initial disturbances of expanding flames in quiet flows finally form isolated ridges at fixed angular locations. We currently study whether generalizations of (4) (5) enable one to also understand the influence of turbulence upon expanding flames: adding poles to a front of ever-increasing perimeter is likely to account for the permanent cell-splitting seen in experiments (Palm-Leis and Strehlow, 1969).

REFERENCES

Clavin, P. and Joulin, G., 1983, Premixed flames in large-scale, high-intensity turbulent flows, J. Phys. Lettres, 1: L1.

Denet, B., 1988, Simulation numérique d'instabilités de fronts de flammes, Thèse de l'Université de Provence.

Joulin, G., 1987, On the hydrodynamic stability of flat burner flames, Comb. Sci. Tech., 53: 315.

Joulin, G., 1988a, On a model for the response of unstable premixed flames to turbulence, Comb. Sci. Tech., in press.

Joulin, G., 1988b, On the hydrodynamic stability of curved premixed flames, J. Phys., submitted.

Joulin, G., 1988c, On the non-linear theory of hydrodynamic instabilities of expanding flames, in preparation.

Landau, L. D., 1944, On the theory of slow combustion, Acta Phys. Chim., 19: 77.

Lee, Y. C. and Chen, H. H., 1982, Non-linear dynamical models of plasma turbulence, Phys. Scripta, T2: 41.

Markstein, G. H., 1964, "Non-steady flame propagation", Pergamon, N.Y.

Michelson, D. M. and Sivashinsky, G. I., 1977, Nonlinear analysis of hydrodynamic instabilities in laminar flames, Acta Astro., 4: 1207.

Michelson, D. M. and Sivashinsky, G. I., 1983, Thermal-expansion-induced cellular flames, Comb. Flame, 48: 211.

Palm-Leis, A. and Strehlow, R. A., 1969, On the propagation of turbulent flames, Comb. Flame, 13: 111.

Pelcé, P. and Clavin, P., 1987, Curved front stability, Europhys. Lett., 3: 907.

Sivashinsky, G. I., 1977, Nonlinear analysis of hydrodynamic instabilities in laminar flames, Acta Astro., 4: 1177.

Thual, O., Frisch, U., and Henon, M., 1985, Application of pole decomposition to an equation governing the dynamics of wrinkled flame fronts, J. Phys., 46: 1485.

Zel'dovich, Y. B., Istratov, A. G., Kidin, N., and Librovich, V. B., 1980, Flame propagation in tubes: hydrodynamics and stability, Comb. Sci. Tech., 24: 1.

PATTERNS AND DEFECTS IN LIQUID CRYSTALS

E. Bodenschatz, M. Kaiser, L. Kramer,
W. Pesch, A. Weber and W. Zimmermann

Physikalisches Institut
Universität Bayreuth
8580 Bayreuth, Fed. Rep. of Germany

Abstract

The nonlinear behavior of electrohydrodynamic convection in planarly aligned nematic liquid crystals is reviewed. Based on two-dimensional, universal envelope descriptions we present new results on undulated roll structures in the vicinity of a normal-to-oblique roll transition, wavelength selection by the Eckhaus processes, dynamics and ordering of defects in travelling roll patterns (also applicable to homogeneous Hopf bifurcations), and mean-flow induced turbulence.

1. Introduction

Whereas the investigation of two-dimensionally extended pattern-forming systems that are intrinsically isotropic in the plane has been well-developed during the last 20 years much less work was done for anisotropic systems. Quite recently, however, the case of axial anisotropy has attracted increasing attention. The reason for this is mainly that planarly aligned nematic liquid crystal layers which are driven electrically (or sometimes thermally) into convection constitute such a beautiful example. The preferred axis is here given by the surface anchoring of the director along the x-direction (the layer will always be chosen in the x-y-plane).

The linear stability analysis of electrohydrodynamic convection (EHC) in nematics within the conventional description is by now fairly complete [1-6]. The conventional description entails the balance of torques (two components) and balance of momentum (three components) together with incompressibility, Poisson's equation for the electric potential and charge conservation. The electric properties of the material are described by linear and anisotropic polarisabilities (dielectric and sometimes also flexoelectric [3-6]) and conductivities.

The linear modes are characterized by their wavevector $\vec{q} = (q,p)$ which for given external conditions fixes the time-exponent $\sigma = \sigma' + i\sigma''$. For the usual case of an applied sinusoidal electric field the useful external control parameters are the voltage amplitude V, frequency ω, and possibly an additional magnetic field. The threshold condition $\sigma' = 0$ leads for each frequency ω to two-dimensional neutral surfaces $V = V_0(\vec{q};\omega)$ whose minima with respect to $\vec{q}$ determine the voltage threshold $V_c(\omega)$ and critical wavevector $\vec{q}_c$.

New Trends in Nonlinear Dynamics and Pattern-Forming Phenomena
Edited by P. Coullet and P. Huerre
Plenum Press, New York, 1990

In the case of a Hopf bifurcation one would also have a nonzero critical frequency σ_c''. Within the conventional theory no Hopf bifurcation was found up to now, at least for materials with negative dielectric anisotropy (which is the relevant case). This is in contrast to some experiments on thin layers where Hopf bifurcations are observed in some frequency range, unfortunately not in a systematically reproducible fashion [7,8].

Until quite recently the wavevector of the convection rolls was always assumed to be parallel to the undistorted director at threshold ($p_c = 0$, "normal rolls"). Ribotta and Joets were the first to show experimentally that under some conditions oblique rolls ($p_c \neq 0$) occur below a certain frequency ω_z [9]. This is by now understood fairly well [1-6]. Due to the degeneracy with respect to the sign of p one usually observes a "zig-zag" structure.

The experiments as well as the existing weakly nonlinear calculations [2,10] show that the bifurcation at threshold is supercritical. Then important aspects of the behavior slightly above threshold can be described in terms of universal envelope (or amplitude) equations. This paper will be devoted to several applications of these Ginzburg-Landau type equations for a two-dimensional geometry. Actually the very simplest type of Ginzburg-Landau equation

$$\partial_T A = \left[\partial_X^2 + \partial_Y^2 + 1 - |A|^2 \right] A \tag{1.1}$$

with complex envelope function A(X,Y,T) can be used for these anisotropic systems for the case of stationary patterns away from the normal-oblique transition [11,12,2]. The slow variables are

$$X = \epsilon^{1/2}\, x/\xi_1 \quad , \quad Y = \epsilon^{1/2}\, y/\xi_2 \quad , \quad T = \epsilon\, t/T_0 \tag{1.2}$$

with $\epsilon = (V^2 - V_c^2)/V_c^2$ for EHC, coherence lengths ξ_1, ξ_2 and relaxation time T_o. In the normal-roll case the physical quantities u_i are of the form

$$u_i(\vec{r},t) = \epsilon^{1/2}\, [U_i\, A(X,Y,T)\, e^{iq_c x} + cc]\, \tilde{u}_i(z,t) + O(\epsilon) \tag{1.3}$$

with $|U_i|=1$. After a rotation of the coordinate system equation (1.1) can also be used in the oblique-roll regime for slow modulations around one of the oblique directions. If both directions contribute (zig and zag) one has to use two coupled amplitude equations [5]. At the normal-oblique transition point ξ_2 goes to zero, which characterizes a so-called "Lifshitz-point". In its vicinity a different scaling for y has to be used. Measuring ϵ (> 0) from the normal-roll threshold we write

$$Y = \epsilon^{1/4}\, y/\xi_3 \quad , \quad \omega - \omega_z = \epsilon^{1/2}\, W\, \omega_0 \tag{1.4}$$

where W is a dimensionless and blown up version of the additional control parameter (= frequency ω for EHC) that brings the system across the Lifshitz point ($\omega = \omega_z$). Otherwise (1.2) and (1.3) remain unchanged and the envelope equation becomes

$$\partial_T A = \left[\partial_X^2 - iZ\partial_X\partial_Y^2 + W\partial_Y^2 - \partial_Y^4 + 1 - |A|^2 \right] A. \tag{1.5}$$

with $|Z| \leq 2$. For Z = 2, W = 0 the famous Newell-Whitehead-Segel equation for isotropic systems is recovered from (1.5) [13]. In previous work the case Z > 0 was assumed and the parametrization $S = 2W/Z$, $b = (2/Z)^2 - 1$ was chosen [11,12,2]. For $\epsilon < 0$ one must replace in the square bracket of eq.(1.5) the +1 by -1.

In Fig.1 the quantities $\omega_z\tau_o$, Z and $\omega_o\tau_o$ ($\tau_o = \epsilon_o\epsilon_\perp/\sigma_\perp$, charge-relaxation time, $\sigma_\perp$= conductivity perpendicular to director) are given as a function of $\sqrt{\sigma_\perp}$ d (d = layer thickness) for the standard nematic material MBBA (solid curves, see [6]) and a modified material (broken curves, MBBA I of ref. 2 with $\sigma_\parallel/\sigma_\perp=2$ and $\epsilon_a=-0.2$) that has a Lifshitz point even in the limit $\sqrt{\sigma_\perp}$ d $\to \infty$ where the flexoeffect drops out.

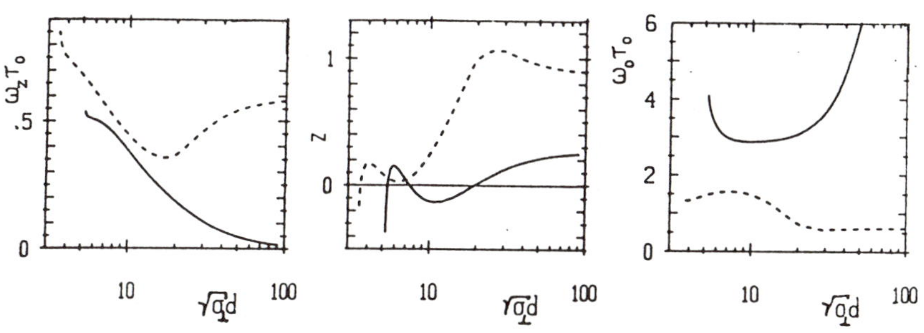

Fig.1. Reduced normal-to-oblique transition frequency $\omega_z\tau_0$ (Lifshitz point) and parameters $\omega_o\tau_o$ and Z for eq.(1.5) are shown as a function of $\sqrt{\sigma_\perp}$ d in units of $\sqrt{10^{-8}\Omega^{-1}m^{-1}}\mu m$ for EHC in MBBA like materials (see text). For small values of $\sqrt{\sigma_\perp}$ d the curves terminate when the dielectric threshold becomes lowest at ω_z.

In Sec. 2 we present new results on undulated and zig-zag solutions of eq. (1.5) and study the dynamical processes by which these states are reached. We conclude that the existing experiments can presumably be explained, but more systematic experimental work is needed. In Sec. 3 we review results on eq.(1.1) and present in particular results regarding the dynamical processes by which the system recovers a stable wavenumber when it is initialized in an unstable state. In two dimensions these "Eckhaus processes" involve nucleation, motion and annihilation of dislocations.

Section 4 is devoted to an investigation of the extension of eq.(1.1) for a Hopf bifurcation where the coefficients become complex and one may write:

$$\partial_T A = \left[(1 + ib_1)\,\partial_X^2 + (1 + ib_2)\,\partial_Y^2 + 1 - (1 + ic)\,|A|^2\right]A. \qquad (1.6)$$

For simplicity we take $b_1 = b_2 = b$. Then eq.(1.6) also applies to a spatially homogeneous Hopf bifurcation as found in oscillatory chemical reactions. We have studied numerically the behavior of solutions with defects (spirals) in the Benjamin-Feir stable range $1 + bc > 0$, where simple periodic solutions are linearly stable [24].

At a second threshold slightly above the primary bifurcation a transition to weak turbulence with spontaneous creation and annihilation of defects is often observed in EHC [14,15]. Since this effect is observed in travelling and also in stationary patterns (in fact it was first found in the latter case [14]), Benjamin-Feir resonance is probably not responsible for it, in contrast to what is sometimes claimed [15,16]. Presumably mean-flow effects which are not included in the envelope description (1.1) - (1.6) are responsible for the transition to turbulence. A strong indication in this direction is given by the fact that with the often used but unrealistic free boundary conditions all roll solutions in EHC are unstable [10] and one has a turbulence that is very reminiscent of the experiments. These results can be obtained by coupling eq.(1.1) to the mean-flow mode and are presented in Sec.5.

2. Vicinity of the Lifshitz Point

Equation (1.5) constitutes a universal description of the vicinity of a point where a continuous transition from normal to oblique rolls takes place along the threshold curve of a stationary, supercritical bifurcation. Therefore it is of interest to find the stable solutions and their basins of attraction.

Clearly one has straight-roll solutions

$$A = F \exp\left[i(QX + PY)\right] \quad , \quad F = \left[1 - Q^2 - (ZQ + W + P^2)P^2\right]^{1/2} \tag{2.1}$$

which are stable within 2-dimensional wavevector bands [11]. For $ZQ + W > 0$ and $Q^2 < 1/3$ one has stable normal rolls ($P = 0$). For $ZQ + W < 0$ only oblique rolls are stable. For $W > 0$ the amplitude is maximal for $Q_c = P_c = 0$ ("band center") whereas for $W < 0$ (and $|Z| < 2$) the band center is at

$$Q_c = ZW/(4 - Z^2) \; ; \; P_c^2 = - 2W/(4 - Z^2). \tag{2.2}$$

This wavevector corresponds to the most stable state in the sense that the minimizing potential

$$V = \langle |\partial_X A|^2 + Z \,\mathrm{Im}\,(\partial_X A^* \partial_Y^2 A) + |\partial_Y^2 A|^2 + W|\partial_Y A|^2 - |A|^2 + \tfrac{1}{2} |A|^4 \rangle \tag{2.3}$$

is lowest at (Q_c, P_c). $\langle ... \rangle$ denotes the spatial average.

In addition eq.(1.5) has stable undulated solutions which can be classified by their undulation wavenumber L [11,12]. These solutions cannot be obtained analytically, but a simple mode approximation is possible [11]. Such undulated roll states are found experimentally [9] and there is interest in a detailed understanding. A (nonlinear) phase-diffusion description, which does, however, not catch the stability properties of the undulated solutions (they always appear to be unstable), can be deduced from eq.(1.5) for the case where the undulations are weak, so that their influence on the amplitude is small. Choosing

$$A = F \exp\left[i(QX + \phi(Y,T))\right] \tag{2.4}$$

with real F one obtains

$$\partial_T \phi = (W + ZQ)\, \partial_Y^2 \phi - 4/Z^2 \partial_Y \left[\partial_Y^3 \phi - 2(\partial_Y \phi)^3\right]. \tag{2.5}$$

For the stationary solutions of (2.5) one has

$$\partial_Y^2 f = - \partial_f V \quad , \quad V = -\tfrac{1}{2} f^4 - (Z^2/8)(W + ZQ) f^3 - Cf \tag{2.6}$$

with $f = \partial_Y \phi$ and C an arbitrary constant. Equation (2.6) is analogous to the one-dimensional motion of a particle with unit mass in a potential $V(f)$. For $W + ZQ < 0$ and $|C| < 2/3\, [16/3\, Z^2\, (W + ZQ)]^{1/2}$ the potential has two maxima and a minimum corresponding to straight-roll solutions with $P = f$ (= constant). The solutions oscillating periodically in the minimum correspond to the undulated states.

For $C = 0$ one has undulated **normal** rolls. From (2.6) they are given by

$$\phi(Y) = \sqrt{m} \int^{aY} dt\, \mathrm{sn}(t|m) \quad , \quad a^2 = \frac{|W + ZQ|}{1 + m} \tag{2.7}$$

where sn is a Jacobian elliptic function with parameter m [17]. The period of the undulation is $4K(m)/a = 2\pi/L$, where K is the complete elliptic integral of the first kind [17]. The amplitude of the undulation ϕ is $\ln((1-m)^{1/2}/(1-\sqrt{m}))$ and the amplitude of the slope f (= maximum tilt of the rolls) is $a\sqrt{m}$.

In order to determine the most probable modulation wavenumber L we consider the situation when the system is initialized in the (unstable) normal-roll state. Linearization of eq.(1.5) around the normal-roll solutions, shows that for W + 2Q < 0 the most critical modes are undulations with growth rate

$$\sigma = -(W + ZQ)\, L^2 - L^4 . \tag{2.8}$$

Clearly σ is maximal for $L = L_m$ with

$$L_m = [|W + ZQ|/2]^{1/2} \tag{2.9}$$

(The same result is obtained in the phase-diffusion approximation). It is very suggestive to identify the most probable L with L_m. It is further interesting to note that the oblique rolls with $P = L_m$ correspond to the most stable rolls (minimum of V) for given Q. Consequently one has $L_m = P_c$ for $Q = Q_c$. In the phase-diffusion approximation the case $L = L_m$ corresponds to $K(m) = \pi/(2(1+m))^{1/2}$, or $m = 0.472$, leading to an undulation amplitude $\phi_0 = 0.842$ and a maximum tilt $0.57\cdot|W + ZQ|^{1/2}$. Since one roll spacing corresponds to a phase change of π the "most probably undulation" amplitude is in this approximation 0.27 in units of the roll spacing. Full numerical solutions of eq.(1.5) show that this value, which strictly holds for $L_m \ll 1$, increases up to about 0.5 in the range of stable undulations. The published experimental results on undulated rolls appear to be consistent with this result [9].

The stability of undulated rolls with respect to nonperiodic fluctuations was investigated numerically using the amplitude equation (1.5). In Fig.2 the range of stable existence of undulations with $L \gtrsim L_m$ (other stable regions with smaller L exist, see below and ref. [11,12]) is shown in the L-Q-plane for various values of W and Z = 0.9. The value of Z corresponds to modified nematic material MBBA for not too thin and clean specimens (see Fig.1 and ref. [2]). The long-dashed curve represents $L_m(Q)$. For sufficiently negative W one has substantial stable regions with $L_m(Q)$ passing through them. For increasing W the stable regions shrink and move to $L > L_m$ (for $W \gtrsim -0.12$ there is no overlap). For W = 0 only a very small island remains which disappears for positive values of W. The stable undulation wavenumbers L are between about 0.2 and 0.33 for Z = 0.9 and $W \gtrsim -0.5$. We point out that for larger values of

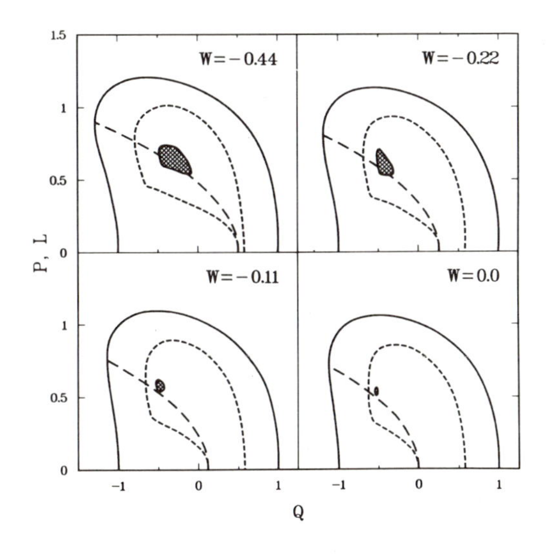

Fig.2.

Range of stable existence of undulations (shaded) together with fastest-growing undulations from normal rolls (long-dashed) is shown in the L-Q plane for Z = 0.9 and various values of W. Also shown is the range of existence (solid) and stability (short-dashed) for oblique rolls in the P-Q plane.

$|Z|$ the stable islands persist to larger values of W. Also included in Fig.2 is the range of existence (solid) and stability (short-dashed) for straight rolls in the P-Q-plane.

In our theory the "soft" undulations with $L \gtrsim L_m$ are never stable near the point where they bifurcate from normal rolls ($W + ZQ = 0$). This seems to be a fairly general feature since it is also borne out by the phase-diffusion approximation (2.5) which, in the vicinity of the bifurcation and in the absence of mean-flow effects, should be valid more generally. Thus, according to our theory there cannot be a **continuous** transition from normal to undulated rolls, although such a transition would not be forbidden by symmetry arguments. The scenario would presumably be changed only when undulations bifurcate without oblique rolls competing. Possibly finite ϵ-corrections like mean-flow effects lead to such changes.

Nevertheless our theory does permit discontinuous transitions from normal to undulated rolls by a sufficiently rapid decrease of W and possibly q (for $Z > 0$). The experimental method used to obtain undulated rolls [9] seems to be consistent with this result: At frequencies slightly above ω_z (normal regime) the voltage V was increased in small steps. Above a second threshold the undulations occurred. Taking the most rapidly growing mode (q_f, p_f) out of the unconvective state as a criterion, an increase of V shifts q_f to higher values and brings about a transition from $p_f = 0$ to $p_f \neq 0$ [2]. Thus an appropriate jump of V into the regime with $p_f \neq 0$ would seem to initialize the system in a normal-roll state with q smaller than the optimal one. Maybe undulated rolls can also be obtained for frequencies slightly below ω_z by initializing the system in an appropriate normal-roll wavenumber with the use of structured electrodes as used in different context [18]. A nonlinear theoretical treatment that includes finite ϵ-corrections explicitly seems unavoidable at this point.

We now consider undulated normal-roll states with L well below L_m. In the phase-diffusion approximation (2.7) for decreasing L the solutions go smoothly over into zig-zag-type solutions with the straight parts corresponding to oblique rolls with $P = L_m(Q)$ ($m \to 1$ for $L \to 0$). The full eq. (1.5) exhibits similar behavior, except that for intermediate L one has additional wiggles [12]. Stable islands appear to exist for $L \sim L_m/n$, $n = 1,2,...$. As pointed out before [2] one also expects the existence of non-periodic wiggled solutions. For large n the stable islands presumably merge so that zig-zag states with arbitrary wavelength exist. However, zig-zags with $P \neq L_m$, whose existence was postulated before for eq.(1.5) [12], appear to be numerically unstable.

Since the undulated solutions with decreasing L (fixed Q) have decreasing potential (2.2) (the oblique rolls have the lowest potential), it will usually be easy to have the system evolve in that direction, i.e. towards zig-zag and hard to keep the system in a soft undulation with $L \approx L_m$. This is presumably one reason why soft undulations are rarely observed. Changing some external parameter but staying within the oblique-roll regime will usually drive the system towards zig-zag.

3. Wavelength Selection Far From the Lifshitz Point

The one-dimensional version of eq.(1.1) is applicable to quasi-one-dimensional systems with or without intrinsic anisotropy. There exist periodic solutions $A = \sqrt{1-Q^2}\exp(iQX)$ which are linearly stable in the range $Q^2 < 1/3$ bounded by the Eckhaus instability. One also has unstable quasiperiodic solutions which bifurcate from the periodic ones in the unstable range and which can be expressed in terms of elliptic functions [19,20]. They play the role of saddle-points (or separatrices) separating the regions of attraction of the stable solutions. In the general context of wavelength selection one is interested in the wavenumber Q which is found after long times starting from random initial conditions. It is clear that for very small and random initial fluctuations a wavenumber Q near zero will result. The fact that Q is not necessarily exactly zero may have interesting consequences for an analogous problem in cosmology [21]. The nonlinear evolution of the system initialized in a modulationally unstable periodic state ($Q^2 > 1/3$) was investigated experimentally [18] and theoretically [22].

Equation (1.1) has in two dimensions periodic solutions

$$A = \sqrt{1-Q^2-P^2}\ \exp\left[i(QX + PY)\right] \tag{3.1}$$

which are linearly stable for $Q^2 + P^2 < 1/3$ (generalized Eckhaus instability). In this stable region evolution towards the bandcenter $Q = P = 0$ can occur by motion and annihilation of dislocations if a mechanism for their creation by fluctuations or perturbations (e.g. boundaries) is present [23]. The nucleation, and especially the motion of dislocations has been studied in detail [23,5]. Ideal boundaries that are perpendicular to the direction of normal rolls do not appear to provide spontaneous nucleation of dislocations in the stable region (they only reduce the barrier), and they accelerate the nucleation process in the unstable range [20].

We have studied the nonlinear evolution of the system when it is initialized in a modulationally unstable periodic state ($1 > Q^2 + P^2 > 1/3$) by simulating eq.(1.1) numerically in two dimensions. Without loss of generality one can choose $P = 0$ because all other cases can be obtained by a rotation of coordinates. In Fig.3 a time series for a rectangular region with periodic boundary conditions is given. Equations (1.2), (1.3) were used to obtain a rapidly varying quantity u, whose amplitude is represented on a grey scale. Parameters roughly appropriate for a liquid crystal were taken ($q_c = 0.42$ defines the length scale; $\xi_1 = 0.27$, $\xi_2 = 0.14$, $\epsilon = 0.05$). In slow units the periodicity interval is $L_Y = 128$ and $L_X = 256$. The initial state corresponds to $Q_i = 0.7$. Small random noise was added to the solution. After some time a modulation with wavenumber $K \approx 0.6$ in X-direction develops (upper graph), which corresponds approximately to the most rapidly growing unstable mode K_m (see. e.g. [22]). The modulation breaks up into dislocations which move and annihilate pairwise, and finally a state within the stable band remains (the dislocations in the lowest graph eventually disappear). The final wavenumber corresponds approximately to $Q_f = Q_i - K_m$. We have not yet been able to determine systematic deviations from this formula, as found in the one-dimensional case, especially for Q_i near to $1/\sqrt{3}$ [22]. Work in that direction is in progress.

Fig.3.

Wavelength selection process when the system is started in an Eckhaus-unstable wavenumber state. For details see text.

When one implements boundary conditions with $A = 0$ at the "upper" and "lower" boundaries of the cell (i.e. perpendicular to the rolls) instead of periodicity conditions, the initial modulation and the nucleation of dislocations occur predominantly near these boundaries, as expected from previous work [20]. Otherwise there are no important changes.

If one would initiate the system in a fully unstable periodic state with $Q^2 + P^2 > 1$ (outside the neutral curve) then the pattern would first vanish and a new pattern at (or near) band center would evolve subsequently. This is different from the above case where the initial pattern is modulationally unstable. The distinction seems to have been observed by Kai (priv. comm.) in EHC by suddenly applying a magnetic field in the y - direction, which tends to rotate the rolls: For strong fields first the old pattern decayed and then the new one evolved, whereas for weaker fields the transition occured via defects.

4. Defects in the Complex Ginzburg-Landau Equation

The Ginzburg Landau Equation (1.6) with complex coefficients has a rich variety of interesting solutions. For $1 + bc > 0$ ("Newell criterion" [24]; for simplicity we consider $b_1 = b_2 = b$) there exists a band of linearly stable roll solutions which in most cases correspond to travelling rolls in the physical system. For $1 + bc \to 0$ the stable band shrinks to zero and for $1 + bc < 0$ there do not appear to exist simple stable solutions. The determination of the stability limits of the band for $1 + bc > 0$ is in general a somewhat cumbersome task [25] which can be solved quite elegantly by using the Hurwitz criterion [26]. For $1 + bc < 0$ starting with random initial conditions the system appears to settle down in a turbulent state which has nonzero average defect density and is essentially independent of the initial conditions [27,28] (a defect is a zero of the complex field A). The density does not appear to tend to zero when $1 + bc \to 0$ [27,28].

We have investigated the parameter region $1 > 1 + bc > 0$ (the numerical studies were restricted to cases with $b = -c$; there is a symmetry with respect to simultaneous changes of b and c). The spatially constant solution $A = \exp(-icT)$ (band center) is stable in the full range. There exist isolated defect solutions of the form [29,30]

$$A(r,\phi,T) = D(r) \exp i(\phi + Qr - \Omega T) \tag{4.1}$$

where (r,ϕ) denote polar coordinates in the X-Y-plane, and $\Omega = c + (b-c) Q^2$. The (single-valued) complex function $D(r)$ tends to $(1-Q^2)^{1/2}$ for $r \to \infty$. Clearly (4.1) represents a rigidly rotating, approximate Archimedean spiral. The wavenumber Q of the spiral is apparently fixed for given b and c ($Q \to 0$ for $b,c \to 0$), and this has been investigated by Hagan analytically and numerically [29]. Defining $z = (1 + ib)/(1 + ic)$ one can show that spirals with fixed values of arg(z), which is essentially equivalent to fixed $(b - c)/(1 + bc)$, can be transformed into each other by

$$(b_1, c_1, Q_1) \to (b_2, c_2, Q_2 = Q_1 [(1 - Q_1^2)|z_2/z_1| + Q_1^2]^{-1/2}). \tag{4.2}$$

Clearly isolated defect solutions cannot exist stably in the full Benjamin-Feir stable range $1 + bc > 0$ since the state with nonzero wavenumber Q, which is built up asymptotically, loses stability before the band-center solution ($Q = 0$) does (for $b = -c$ this stability limit is about $b \sim 0.55$, or $1 + bc \sim 0.7$).

In the following we report on observations seen in numerical experiments. We used a pseudospectral code [31] which simulated eq.(1.6) on a square cell of sidewidth L with periodic boundary conditions. Usually initial conditions with two well-separated zeros of A (defects of opposite polarity) at $(0, l_1/2)$ and $(0, -l_1/2)$ were chosen ($X = Y = 0$ at the cell center). The spiral pair developes initially without translation of the centers. Eventually the whole cell is filled with the outgoing waves. The waves collide at the boundaries of the cell and also at the symmetry line Y = 0 between the spirals. Here well-defined line defects (sinks) are formed and the amplitude exhibits crests with $|A|$ rising from about $\sqrt{1-Q^2}$ to a value near 1 (see Fig.4 , there is a different arrangement shown, see below).

For well-separated spirals at $b = -c = 0.5$ this situation persists for very long times (the simulations extended up to about $T \sim 10^4$). If the inital distance l_i is decreased below some value $l_2 \approx 30$ the spirals eventually attract each other pairwise and tend to a stable separation l_s ($l_s \approx 18$). This state is also approached from below when l_i is chosen smaller than l_s but larger than some lower critical distance l_1 ($l_1 \approx 11$). For $l_i < l_1$ the spirals approach each other and collapse. Parallel to the stabilization of the distance an overall drift of the whole pattern with constant velocity along the X-direction is initiated. These results do not depend much on the cell size as long as L is sufficiently large.

Long-time simulations and runs with the addition of random noise at intermediate times showed that the above drifting pattern is not really stable. Eventually one of the spirals stays behind and settles down at a fixed position (X_1, Y_1). The other spiral performs a fairly complicated motion and settles down at (X_1 - L/2, Y_1 - L/2). In this way the simplest "antiferromagnetic" face-centered square lattice with $l = L/\sqrt{2}$ is obtained. Curiously, the spiral that stayed behind grows and in the final state the waves emitted by this spiral propagate all the way to the core of the other spiral. The final state appeared stable against random noise.

Additional runs were made by initializing the system directly in the above mentioned antiferromagnetic arrangement. In a large range of distances this arrangement proofed to be stable (including relaxation after applying random noise). However, in that case the spirals with opposite sign remained equivalent and the outgoing waves collide on lines that form a square lattice with the lattice constant L/2. In Fig.4 the amplitude $|A|$ and contours of equal phase are shown. Below some distance l_3, however, and above some minimal distance l_m the spirals pair up at a distance $l_s < L/\sqrt{2}$, and the pattern drifts again. Below l_m the spirals collapse and the homogeneous state results. In Fig. 5 we have plotted numerical results for l_m as a function of $b = -c$. Presumably l_m characterizes the maximal defect density $1/l_m^2$.

(a)

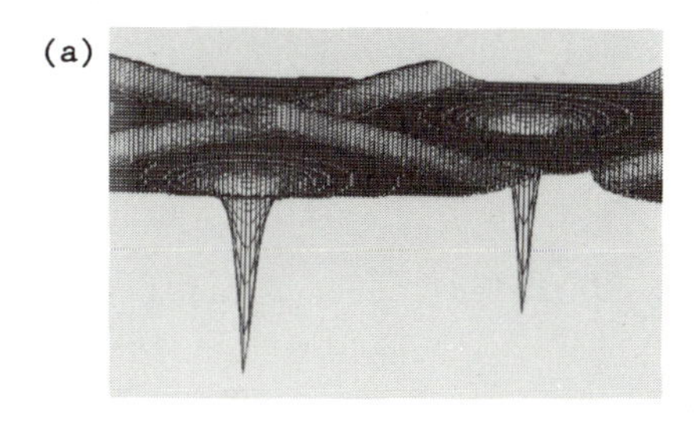

Fig.4. Amplitude seen from above (a) and phase plots (b) for a unit cell of a spiral-antispiral lattice. The steps in the amplitude plot are artifacts of the plotting routine.

Apparently l_m diverges for decreasing b. It is at this point not clear whether this divergence is at $b = 0$ or at a nonzero value of b. This can only be decided analytically. At $b \approx 0.55$ we find a break in the curve and a somewhat unsystematic behavior for $b > 0.55$. Although there is presumably some connection with the approach of the stability limit discussed above, the solutions remain well-ordered.

We conclude that at least in some range of b and c between zero and the Benjamin-Feir limit one has a continuum of stable states with a defect density between zero and an upper limit that increases as the Benjamin-Feir limit is approached. However, we point out that we have not performed tests for stability with respect to fluctuations that have not the periodicity of the lattice. Presumably, as the Benjamin-Feir limit is approached a nonzero lower boundary of the defect density is also established (in our case presumably for $b = -c > 0.55$). At not too low densities defects tend to order and, if the arrangement is not symmetrical with respect to spirals and antispirals, to drift. Both these phenomena are observed in the EHC instability of liquid crystals when chevron patterns form in the dielectric regime [32]. Although there may be an analogy it is quite clear that eqs.(1.6) do not describe that system realistically.

5. Mean-Flow Effects

The weakly nonlinear analysis of stationary roll patterns not too near to the normal-oblique transition has up to now been based on the simple amplitude equation (1.1). The stability boundaries derived from this approach are of the Eckhaus type. For isotropic fluids it is well known, that perturbations involving large scale flow can qualitatively change the result of such an analysis. One gets new instabilities of the skewed-varicose type [33]. They appear naturally within the stability analysis of normal rolls. Their description in terms of (generalized) amplitude equations, which is necessary for the analysis of more complicated patterns observed in experiments, has only been achieved approximately. It has been done convincingly for the often used somewhat artificial free boundary conditions [34]. It is believed, that the scenario found at threshold carries over to rigid boundaries slightly above threshold. Free boundary conditions simplify the analysis substantially and so it is natural that the inclusion of large scale flow (or mean-flow) effects in EHC should start with this simplification. Even then the analysis is much more involved than in isotropic fluids (e.g. Rayleigh-Bénard convection [35]). As pointed out in the introduction mean-flow effects are also expected to be responsible for the weak turbulence observed in experiments.

We have first gone through the linear and weakly nonlinear analysis as before, but with stress free and torque free boundary conditions allowing for a stabilizing magnetic field in the x-direction. A nonzero magnetic field is necessary for a consistent roll solution in the oblique roll-range. As in the

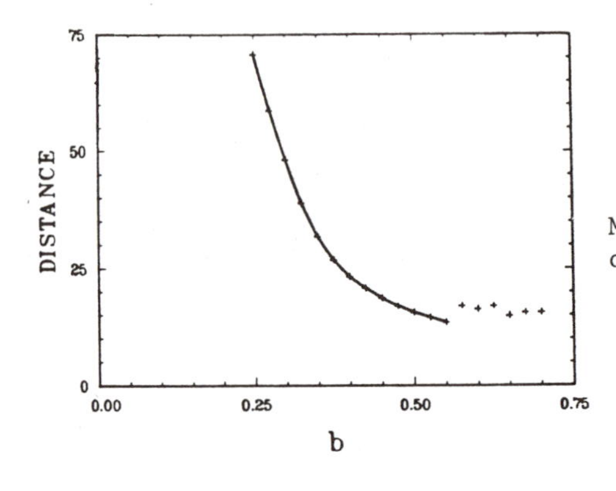

Fig.5.

Minimal distance l_m as a function of $b = -c$.

isotropic case there also exists in the linear analysis a slowly decaying shear mode with wavenumber α and growth rate $\sigma = O(\alpha^2)$. In the presence of a magnetic field the existence of this mode with nonvanishing vertical vorticity is somewhat surprising. It can be traced back to the existence of a solution of the full nonlinear equations, corresponding to an undamped rotation of the whole system with the director staying aligned. In a second step the stability of the nonlinear roll solution has to be examined. Above threshold the vertical vorticity mode can be changed into a growing one. As a result of this analysis we find that normal rolls are always unstable at threshold against perturbations with a wavevector perpendicular to the roll axis. That means physically, that a bending of the roll induces a long range, slowly varying flow, which couples back and enforces the initial distortion of the roll.

We will not present the results of the stability analysis here because they are also contained in the generalized amplitude equations given in what follows. We have adapted the concept introduced in the work of Bernoff [36] in the context of Rayleigh-Benard convection. The solution vector $\vec{u}$ of the full nonlinear equations, which contains the electric potentials, the velocity potentials and the angles of the director is expanded with respect to the linear modes with small growth rates. One uses the ansatz:

$$\vec{u} = \int dq\, A_q(t)\, \vec{V}_0(q) + \int d\alpha\, B_\alpha(t)\, \vec{V}_{MF}(\alpha) \tag{5.1}$$

Here $\vec{V}_0$ denotes the modes nucleating the pattern at threshold with a wavenumber q of the order of q_c. $\vec{V}_{MF}$ contains the vertical vorticity mode with a small wavenumber α. Inserting this into the full nonlinear equations one obtains at leading order in ϵ for the expansion coefficients A_q and B_α :

$$a_1 \frac{d}{dt} A_q(t) = a_2 A_q + \int dq \int dq'\, a_3 A_{q'} A_{q''} A_{q-q'-q''} + \int d\alpha\, a_4 B_\alpha A_{q-\alpha} + \ldots$$

$$b_1 \frac{d}{dt} B_\alpha(t) = b_2 B_\alpha + \int dq'\, b_3 A_{q'} A_{\alpha-q'} + \ldots \tag{5.2}$$

The factors a_i and b_i depend on the wave numbers (besides the material parameters) and have been calculated.

Focussing on long wavelength modulations, the system for the Fourier coefficients (5.2) can be transformed into partial differential equations. We introduce the slow variables as in eq.(1.2) and take out a factor $\epsilon^{1/2}$ from the amplitudes A and B. We give only the one dimensional form, because we are mainly interested in modulations along the y-direction:

$$\partial_T A(Y,T) = \left(1 + \partial_Y^2\right) A - A|A|^2 + i s_1 A \partial_Y B$$

$$\partial_Y^2\, \partial_T B = q_1\, \partial_Y^4\, B + q_2\, \partial_Y\, (A^*\, i\partial_Y^2 A - c{\cdot}c.) + \text{higher order terms} \tag{5.3}$$

A typical set of constants in the normal roll range of MBBA-type material is $\xi_2 = 0.14$ d, $s_1 = 0.7$, $q_1 = 210$, $q_2 = -110$ and $T_0 = 0.2\ d^2[\mathrm{msec}/\mu m^2]$. Using eq. (5.4) it is easy to investigate the stability of the normal roll at band center i.e. of the solution $A = 1$ and $B = 0$. Linearising around this solution one finds for the growth rate λ of perturbations with wavenumber l the following expression:

$$\lambda = -\tfrac{1}{2}(l^2 + q_1 l^2) \pm \sqrt{\tfrac{1}{4}(l^2 + q_1 l^2)^2 - 2 s_1 q_2 l^2 - q_1 l^4} \tag{5.4}$$

One sees directly that the solution is unstable for small l if $s_1 q_2 < 0$, which turns out to be the case. This is unlike Rayleigh-Benard convection, where the skewed-varicose mode is tilted with respect to the roll axis. It can be seen that an adiabatic approximation ($\partial_T B = 0$) is questionable because the resulting growth rate remains finite at $l=0$.

In order to see where the system settles down starting from a (unstable) normal roll we have solved the eqs. (5.3) numerically on a periodicity intervall $L_Y \gg \xi_2$ with slightly disturbed normal rolls as initial conditions. One finds irregular excitation and depression of the weight of the Fourier modes. In Fig.6 we have plotted as a representative example the modulus of the Fourier coefficients $A_n(T)$ of the amplitude $A(Y,T) = \Sigma A_n \exp(inqY)$ ($q = 2\pi/L_Y = 0.0065$; $-512 \le n < 512$) for consecutive time intervals $20 \cdot \Delta T$ and time step $\Delta T=0.01$ as used in the time integration. One observes a concentration of the modes at large wavelenghts (small n). As a function of time the position of the spikes changes permanently. In real space one has periodic bending and reconnection of the rolls, triggered by the linear instability. During this process an extended defect, a one dimensional phase slip is generated . In Fig.7 we show a typical picture of the local wavenumber as function of Y for an arbitrarily chosen time. The sharp maxima signalize phase slips. If one also allows for variations in the x-direction one expects that these defects remain localized and appear as dislocations in the roll pattern. We have solved the two dimensional extension of eqs.(5.3). In Fig.8 a representative snapshot of the resulting pattern is shown. As time goes on we find spontaneous generation and annihilation of point defects. Note that the defects have a distortion field, which is extended in the x-direction typical for the experiments in EHC. It is interesting to note that structures reminiscent of those presented here, have been found in modelling instabilities of free shear layers [37].

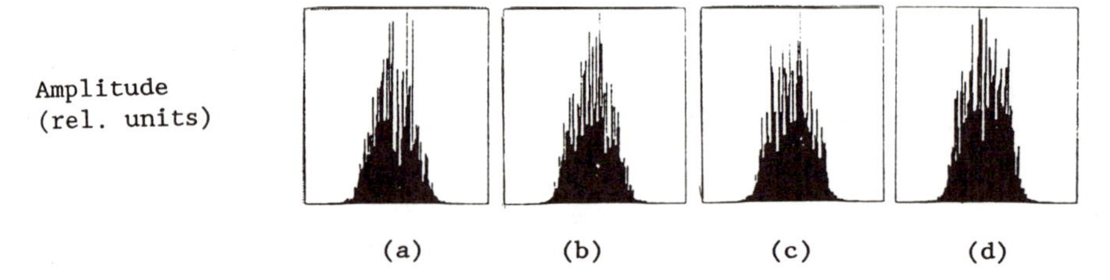

Fig.6. The Fourier spectrum of the amplitude A(Y,T) for consecutive times (see text)

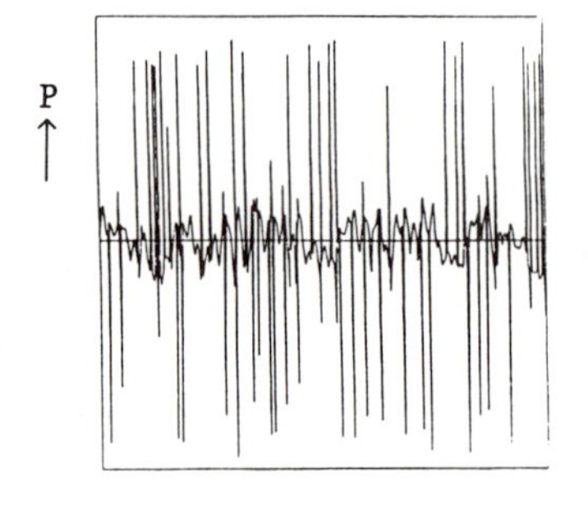

Fig.7.

The local wavenumber P, i.e. derivative of the phase of A(Y,T) corresponding to Fig. 5(a).

Fig.8.

Two dimensional characteristic scenario of roll patterns with defects triggered by mean flow effects.

Acknowledgement

We wish to thank A.C. Newell and W. Thom for useful discussions and contributions. One of the authors (E.B.) would like to thank P. Coullet for his kind hospitality in Nice and in particular L. Gil and J. Lega who introduced him to their pseudo-spectral-code. This work was supported by the Deutsche Forschungsgemeinschaft (SFB 213,Bayreuth), HLRZ,KFA Juelich, MATFO-Programm of the BMFT and by Fonds der chemischen Industrie.

This contribution covers material presented in lectures given by E. Bodenschatz, L. Kramer and W. Pesch.

References

1. W. Zimmermann and L. Kramer, Phys. Rev. Lett. **55**, 402 (1985).
2. E. Bodenschatz, W. Zimmermann and L. Kramer, J. Phys. France **49**, 1875 (1988).
3. N. V. Madhusudana, V. A. Raghunathan and K. R. Sumathy, Pramana-J. Phys. **28**, L 311 (1987); V. A. Raghunathan and N. V. Madhusudana, Pramana-J. Phys. **31**, L 163 (1988); W. Thom, W. Zimmermann and L. Kramer, Liquid Crystals (in press).
4. N. V. Madhusudana and V. A. Raghunathan, Mol. Cryst. Liq. Cryst. Lett. **5**, 201 (1988).
5. L. Kramer, E. Bodenschatz, W. Pesch, W. Thom and W. Zimmermann, submitted to Liquid Crystals.
6. W. Zimmermann and W. Thom, to be published.
7. S. Kai and K. Hirakawa, Suppl. Progr. Theor. Phys. **64**, 212 (1978); Mem. Fac. Engin. Kyushu Univ. **36**, 269 (1977).
8. A. Joets and R. Ribotta, Phys. Rev. Lett. **60**, 2164 (1988); I. Rehberg, S. Rasenat and V. Steinberg, preprint 1988.
9. A. Joets and R. Ribotta, in "Cellular Structures in Instabilities", J. E. Wesfreid and S. Zaleski, Eds. (Springer, Berlin) 1984, p. 294; R. Ribotta, A. Joets and Lin Lei, Phys. Rev. Lett. **56**, 1595 (1986); A. Joets and R. Ribotta, J. Phys. France **47**, 595 (1986).

10. M. Kaiser, Diploma thesis, Bayreuth, 1988; M. Kaiser, W. Pesch and E. Bodenschatz, to be published.
11. W. Pesch and L. Kramer, Z. Phys. **B63**, 121 (1986).
12. L. Kramer, E. Bodenschatz, W. Pesch and W. Zimmermann, in "The Physics of Structure Formation", W. Güttinger and G. Danglmayr, eds. (Springer-Verlag, Heidelberg, 1987) p. 136.
13. A. C. Newell and J. A. Whitehead, J. Fluid Mech. **38**, 279 (1969); L. A. Segel, J. Fluid Mech. **38**, 203 (1969).
14. S. Kai in "Noise in Nonlinear Dynamical Systems", P.V. E. Mc Clintock and F. Moss, eds. (Cambridge University Press, 1987); H. Yamazaki, S. Kai and K. Hirakawa, J. Phys. Soc. Japan **56**, 1 (1987).
15. V. Steinberg, J. Feinberg, E. Moses and I. Rehberg, preprint 1988.
16. P. Coullet, L. Gil, and J. Lega, preprint 1988 and this conference.
17. See e.g. M. Abromowitz and I. Stegun, "Handbook of Mathematical Functions" (Dover, New York, 1965) Chapts. 16 and 17.
18. M. Lowe and J. Gollub, Phys. Rev. Lett. **55**, 2575 (1985).
19. L. Kramer and W. Zimmermann, Physica **D16**, 221 (1985).
20. E. Bodenschatz and L. Kramer, Physica **D27**, 249 (1987).
21 W.H. Zurek, Nature **317**, 505 (1985).
22. L. Kramer, H. Schober and W. Zimmermann, Physica **D31**, 212 (1988).
23. E. Bodenschatz, W. Pesch and L. Kramer, Physica **D32**, 135 (1988).
24. A.C. Newell, Lect. appl. Math. **15**, 157 (1974).
25. J.T. Stewart and R.C. DiPrima, Proc. R. Soc. London **A362**, 27 (1978).
26. W. Zimmermann, Doctoral thesis, Bayreuth, 1987.
27. P. Coullet and J. Lega, Europhys. Lett. **7**, 511 (1988).
28. L. Gil, J. Lega and J.L. Meunier, preprint 1988.
29 P.S. Hagan, SIAM J. Appl. Math, **42** 762 (1982).
30. Y. Kuramoto "Chemical Oscillations, Waves and Turburlence" (Springer, Berlin, 1984) P 106.
31. L. Gil and J. Lega, private communication.
32. S. Kai (private communication) and M. de la Torre, I. Rehberg (private communication).
33. See e.g. F.H. Busse ,Rep.Prog.Phys.,**41**,1931,(1978).
34. E.Siggia and A.Zippelius Phys.Fluids **26**,2905,(1983).
35. F.H. Busse and E.W. Bolton J.Fluid Mech. **146**,115,(1984).
36. A.J. Bernoff ,Univ.of Arizona ,Tuscon (Preprint 1986).
37. P.Coullet,P.Huerre and R.Yang (This conference).

DEFECTS AND TRANSITION TO DISORDER

IN SPACE–TIME PATTERNS OF NON-LINEAR WAVES

A. Joets and R. Ribotta

Laboratoire de Physique des Solides, Bât.510
Université de Paris–Sud
91405 Orsay Cedex, France

INTRODUCTION

The traveling–wave convection found in a liquid crystal gives an example of nonlinear waves. The basic state is a uniform progressive wave which is a perfectly ordered structure in space–time. We present an experimental study of some elementary mechanisms that trigger the nucleation of defects in waves and we show that these defects mediate a transition to a chaotic state. It is found that a homogeneous progressive wave can become unstable against local perturbations of the phase which generally have a shock structure. The shocks give rise to topological singularities that are defects of the space–time ordering. It is shown that the structure and the role of these defects in the evolution to a space-time disordering are reminiscent of that of the dislocations and the grain boundaries in stationary convective structures. Some numerical simulations of defects nucleation by use of a Landau-Ginzburg equation, are also presented.

TRAVELING WAVE CONVECTION IN A NEMATIC LIQUID CRYSTAL

A nematic liquid crystal is an anisotropic fluid which can develop convective instabilities when driven by an AC electric field (as a reference see de Gennes' book, 1974). A layer of nematic becomes unstable to convection above some well defined voltage threshold V_{th} which is frequency dependent and diverges at a cutoff f_c. The range of frequency limited by f_c has been named the "conduction regime". The anisotropy raises the degeneracy of the direction of the wavevector, contrary to the case of isotropic convection. Therefore, the ordered homogeneous pattern consists of equally spaced rolls which are aligned along the perpendicular (hereafter **y**) direction to the anisotropy axis (**x**). The layer thickness d is an important parameter and it is found that for values of d typical of usual samples $5\mu m < d < 100\mu m$ the periodic molecular alignement associated to the convective pattern is never stationary but propagates along **x** with a uniform velocity u (Joets and Ribotta, 1988, Proceedings of the Liquid Crystal Conference, Freiburg). It is found that u decreases as the thickness d increases and increases significantly when the frequency is either close to DC or approaches the cut–off of the conduction regime. A typical value for $d \simeq 50\mu m$ is $u \simeq 4.10^{-2}\mu m/s$ in the middle frequency range of the conduction regime, i.e. typically from 20 Hz to 100 Hz, the cut–off being at $f \simeq 120$ Hz. Close to the cut–off and close to DC, a typical value for the velocity is $u \simeq 10\mu m/s$. For smaller d ($d \simeq 5\mu m$) u increases by an order of magnitude all over the conduction regime. At high frequencies the pattern

New Trends in Nonlinear Dynamics and Pattern-Forming Phenomena
Edited by P. Coullet and P. Huerre
Plenum Press, New York, 1990

is not usually homogeneous in space : the envelope is strongly modulated in space and the convective rolls form isolated domains separated by a steady (non–convective) state. In some cases the domains can be quite elongated in the **x** direction with a rather small extension L_y along the roll axis ($L_y \simeq 1$ to 2 d). Thus, one has an example of quasi 1-D traveling-wave pattern. Another way to obtain a 1-D pattern is to restrict the size of the electrode along **y** while keeping it very large along **x**. We shall describe hereafter the defects which are spontaneously created in quasi 1-D patterns propagating at relatively high velocity (Joets and Ribotta preprint 1988), i.e. either for quite small or large values of d (the thickness does not seem to be an important parameter in the nucleation mechanism of defects). Our main result is that it is the localized modulations of the phase that lead to the formation of these defects which in turn can mediate a transition to a chaotic state, as in the quasi–stationary states (Yang et al. 1986, 1988; Ribotta , 1988).

ORDERED STRUCTURES OF TRAVELING WAVE CONVECTION

Propagating patterns are an example of non-linear waves and the theoretical study of some of their defects is quite recent (Coullet *et al.*, 1988).

A pattern can move in either direction ±**x**, and the space–time evolution of a local variable, say a, that defines the convective flow represents a progressive wave with ± **u** velocity. It is a space-time ordered structure of amplitude

$$a_{R,L} = A_{R,L}(x,t)\, e^{i(\omega t \pm kx)},$$

where the subscripts stand for a wave traveling either to the right (R, –kx) or to the left (L, kx).

For a pure (non–perturbed) progressive wave $A(x,t) = A_0$ and the phase is $\varphi = (\omega t \pm kx)$ with a wavector $\mathbf{k} = \frac{\partial \varphi}{\partial x}$ and a frequency $\omega = \frac{\partial \varphi}{\partial t}$. The velocity (here the phase velocity) is $u = \omega/k$. Therefore, the ordering involves both the space through k and the times through ω, and any perturbation either in space or in time are reflected by phase changes. Typical experimental values are $\lambda \simeq$ 60–70 μm (i.e. $k \simeq 10^3 cm^{-1}$) and $u \simeq 10$ μm/sec. (i.e. $\omega \simeq 1$ rad/sec.or $T \simeq 6.3$ sec.).

It is experimentally, quite convenient to adopt the space-time representation for the propagative rolls. In a nematic liquid crystal which is strongly birefringent the periodic modulation in space of the molecular alignment correlated to the convective flow induces an optical pattern of parallel focal lines for light transmitted accross the layer, on a plane parallel to the layer. The position of the focal lines is related to that of the upwards motion (up and down).

An optical intensity profile recorded along **x** at some position y_0 consists then of a series of peaks separated by the wavelength λ, on a flat baseline. The successive profiles recorded at equal time intervals (here $\delta t \simeq 0.04$ sec.) for the same $y = y_0$ are plotted equally spaced one above each other along the t axis. One obtains a space-time diagram on which the peaks are aligned on oblique lignes, the slope of which is $dt/dx = u^{-1}$. For a uniform motion (progressive wave), the space–time diagram is a perfectly ordered 2–D structure of oblique lines (Fig.1). Such space–time diagrams can be recorded over times up to some hours and are then particularly convenient to characterize states in which the phase is localized in space and slowly varying in time. The accuracy of the measurements is limited by the spatial resolution of the individual peaks and by the spatial homogeneity of the structure (absence of moving defects). In our experiments we have been able to measure velocities u as low as $1.5\ 10^{-5}\lambda$/sec.($\simeq$ 10 Å/sec.) in uniform motion, for times up to 10^4 sec (Joets and Ribotta, 1988).

In such a diagram any pure traveling wave forms an ordered spatio-temporal structure characterized by its phase $\varphi = (\omega t \pm kx)$ and a direction of invariance $x = ut$. There is a symmetry between the left and the right traveling rolls ($k \Leftrightarrow -k$, or $u \Leftrightarrow -u$), which in the $\{x,t\}$ diagram corresponds to a symmetrical tilt angle α of the peak lines over the t axis (α : arctan.u). This symmetry is formally equivalent to the $\theta \leftrightarrow -\theta$ symmetry of the Oblique Rolls structure (Joets and Ribotta, 1986) with respect to the **y** axis and we shall hereafter compare the two types of structures.

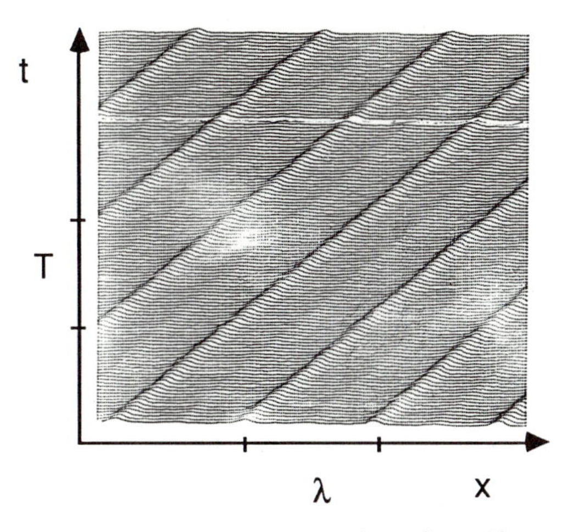

Fig. 1. Space-time representation of a uniform wave propagating to the right.

LOCALIZED MODULATIONS OF THE PHASE AND NUCLEATION OF DEFECTS

An initially uniform progressive wave i.e. a fully ordered wave with a unique wavevector k may become unstable against long wavelength disturbances as it is the case for modulational Benjamin-Feir instabilities of water waves (Benjamin and Feir, 1967). Experimentally, the conditions for such perturbations are not yet well understood although they are found to be effective relatively close to the threshold for the basic wave. We usually find that if $\epsilon = (V^2 - V_{th}^2)/V_{th}^2$, where V_{th} is the threshold value for the basic state, is higher than some 2 - 3.10^{-2}, then the basic wave may become unstable. Localized modulations of the phase $\varphi = (\omega t \pm kx)$ develop, apparently at random in space and in time. They are evidenced by a sudden change Δu in the phase velocity u over some wavelengths λ and might have a large relative amplitude $\Delta u/u \simeq 0.2$. Typically, two types of phase perturbations are found: a) the velocity amplitude jumps suddenly at some point and its profile along **k** is kink-like (Fig.2a) ; b) the profile of the velocity jump is symmetrical and has a lump shape (Fig.2b). The phase variation $\Delta\varphi = \varphi - (\omega t \pm kx)$ has a kink–shape in the first case and a lump–shape in the latter one (Fig. 2). One can also deduce the variation of the local "wavevector" as $\delta k = \partial(\Delta\varphi)/\partial x$. It has a lump-shape in the first case while it is antisymmetrical in the latter one.

The Nucleation of Space–Time Dislocations

In the first case (a) the wavetrain is separated into two halves with a different velocity and at the front (the interface) a shock occurs as in usual waves. The effect of the shock is either to change the magnitude of the local wavevector k_l so as to drive it beyond its stability limits. In a compressive shock k_l increases and when it hits the upper limit of the stability domain k_b, right below the upper branch of the Eckhaus-Benjamin-Feir (E.B.F.) instability domain, restabilization is obtained by expelling one spatial period. In an expansive shock a new period is created when the local wavevector k_l hits the lower limit k_a of the stability domain (above the lower branch of EBF unstable domain). Thus, the local convective period if pulled into the E.B.F. instability domain will suddenly, either disappear if $k_l \geq k_b$ or give rise to a new period if $k_l \leq k_a$, within a time of the order the period of the wave $T = 2\pi/\omega$. The creation or annihilation of a wave-period is equivalent to a dislocation creation or annihilation respectively, and indeed, it is how it appears in the {x,t} diagram. If the two wavetrains keep traveling with a different velocity for a time long compared to T, the creation or annihilation of dislocations occurs periodically in time (Fig. 3). However, such a case seems not to be a typical one even if it may sometimes be observed.

Sources and Sinks, Space–Time Grain–Boundaries

The case (b) is a more typical one, since it corresponds to a localized symmetric variation of the velocity (or of the total phase difference $\Delta\varphi$), most likely to occur inside a initially homogeneous wave which keeps unaltered at large distances. Then, two shocks, a compression and a dilation ones, simultaneously break the wave and trigger both the creation and annihilation of one period. This corresponds to the appearance of two opposite singularities: a sink S_i and a source S_0. In the space between the two defects, the local wave velocity is strongly decreased. A source is a topological singularity where two fronts of counterpropagating waves a_R and a_L meet in such a way that a_L is on the right side and a_R on the left side (Fig. 4). The relative position of the two half waves a_L and a_R is reversed for a sink. It is found that the two singularities are stable in time, i.e. they remain for a time usually much larger than the natural period T. They are in fact, topological defects that limit a counterpropagating wavetrain (say left-going a_L) inside the initial one (right-going a_R). In the {x,t} diagram this is represented as a domain embedded in the initial one and separated by two grain–boundaries as in the Oblique Rolls structure (Fig. 4, 5).

A source, or a sink, is composed of a core which extends over about one period λ and of an intermediate region where the phase undergoes a damped oscillation over some periods until the wave is uniform. The core can be simply understood as the point where two half waves a_L and a_R with a kink-like profile connect (a_L and a_R mutually exclude). It represents the area over which a topological constraint forces the two waves a_L and a_R to be simultaneously present with equal amplitudes. Thus, it represents a local standing wave. In fact, it is restricted to the point where the amplitude oscillates around zero and this fact can be confirmed by considering either the structure of the flow or the local variable (the velocity component u_z for instance). The intermediate zone is a damped modulation of the basic progressive wave (say a_L) wave under the decreasing coupling with the its opposite (a_R) since only the wave a_L is a stable state in this region.

The representation of the sink and source in {x,t} is quite similar to the grain boundaries found separating two domains of stationary Oblique Rolls symmetrically tilted with an angle $\pm\theta$ over the direction **y** (Fig. 4b). There, the core is found to represent locally the bimodal state (of lower symmetry) while the intermediate zone is the skew-varicose structure (Joets and Ribotta, 1984). It can also be shown that there too, the bimodal is a superposition of two solutions $+\theta$ and $-\theta$ with equal amplitude while the skew-varicose structure is a superposition of the same basic states but with different amplitudes. As the time goes the wave a_L "domain" may develop or disappear. This raises the problem of the motion of the sink and source. Preliminary studies indicate that a topological defect is unstable in space and it can move in either direction by steps of half a period. This motion is ascribed to long wavelength modulations of at least, one of the waves velocities u_L or u_R, which change locally the amplitude of that wave. Whenever this modulation is continuous in time, it makes the defect move with an average velocity that can be higher than the velocity of anyone of the waves a_L or a_R. In that case, the moving defect is similar to the front of a shock wave (Joets and Ribotta, preprint 1988).

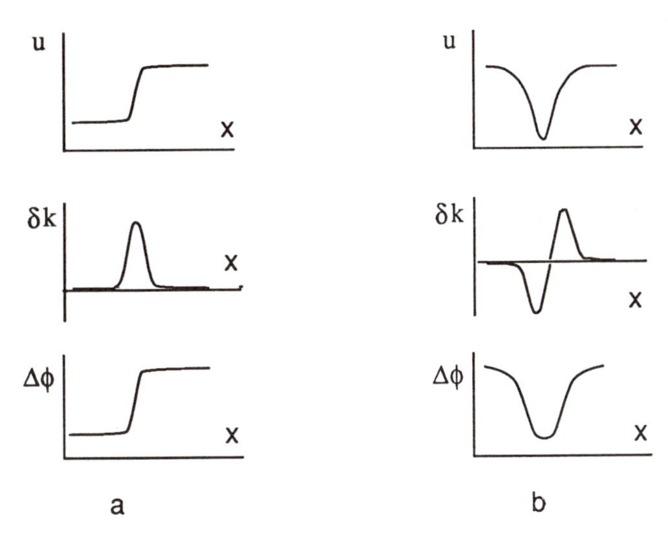

Fig. 2. Typical localized modulations of the phase. The velocities are u, the local wavevector variation is δk and the phase difference is $\Delta\varphi$.

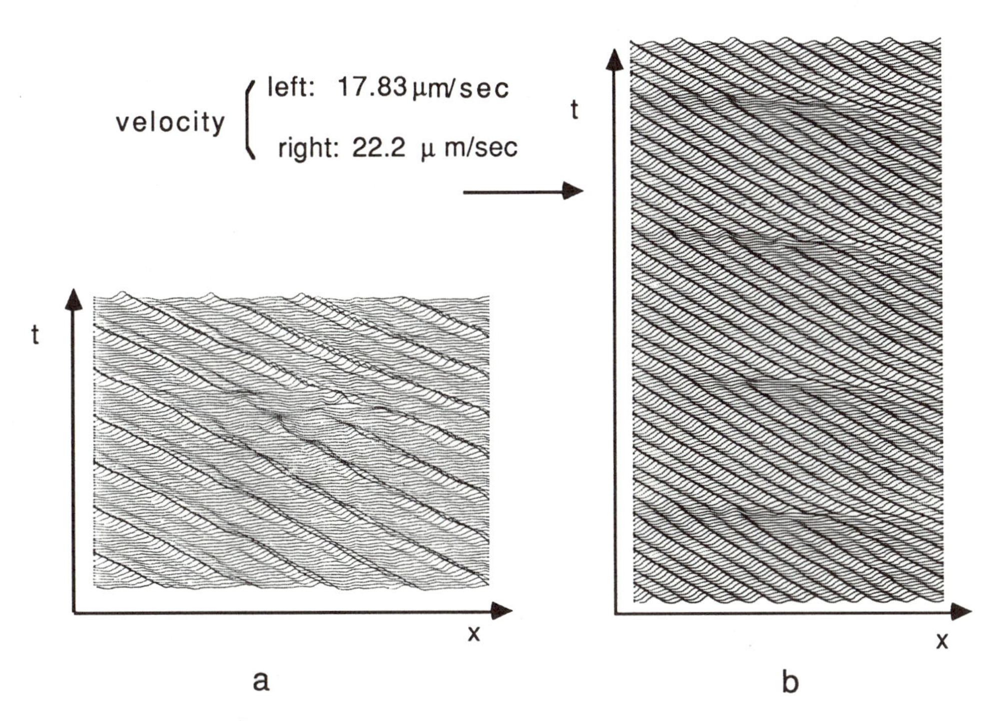

Fig. 3. a) Dislocation in a progressive wave.
b) A periodic array of dislocations results from a continuous shock between two waves with different velocities.

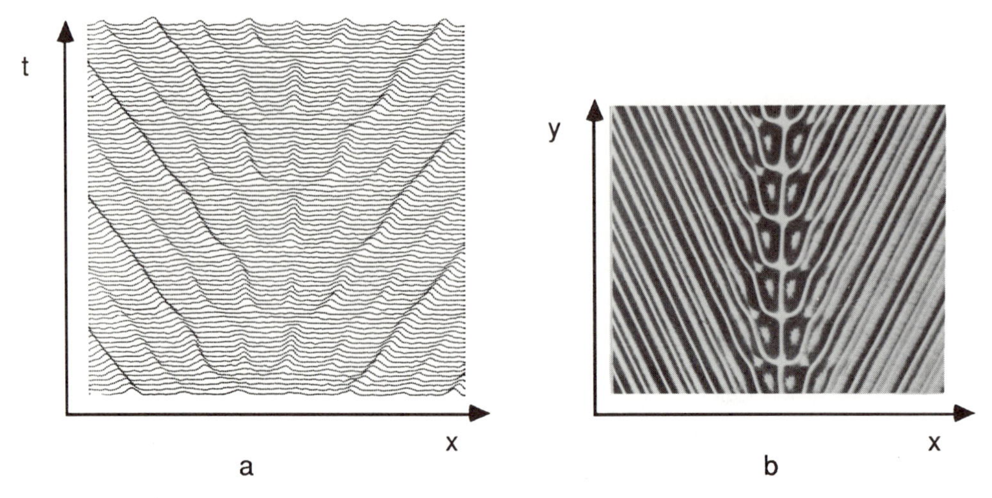

Fig. 4. a) A source separating two counterpropagating waves is represented as a grain_boundary . b) A real grain_boundary in the Oblique Rolls

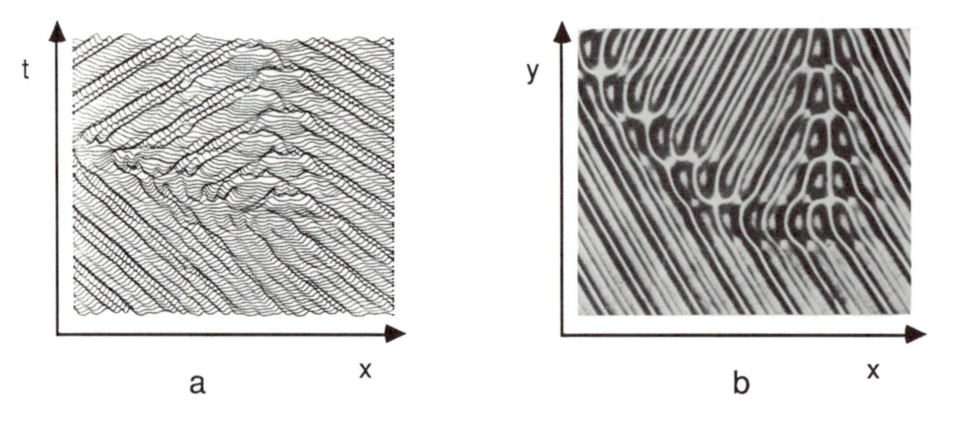

Fig. 5. a) The source and sink enclose a counterpropagating wave "domain".
b) Domains in the Oblique Rolls structure.

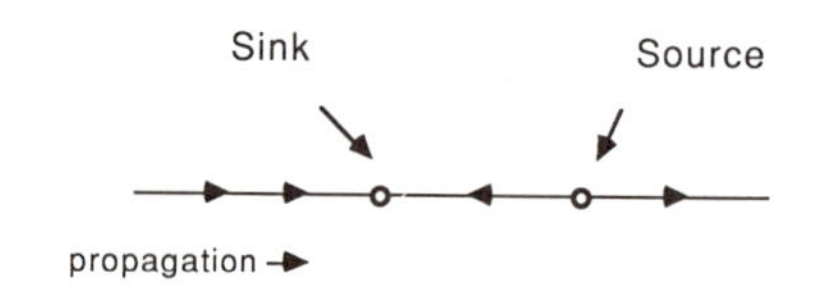

Fig. 5. c) A source and a sink are singularities created in pair in a progressive wave, following a shock.

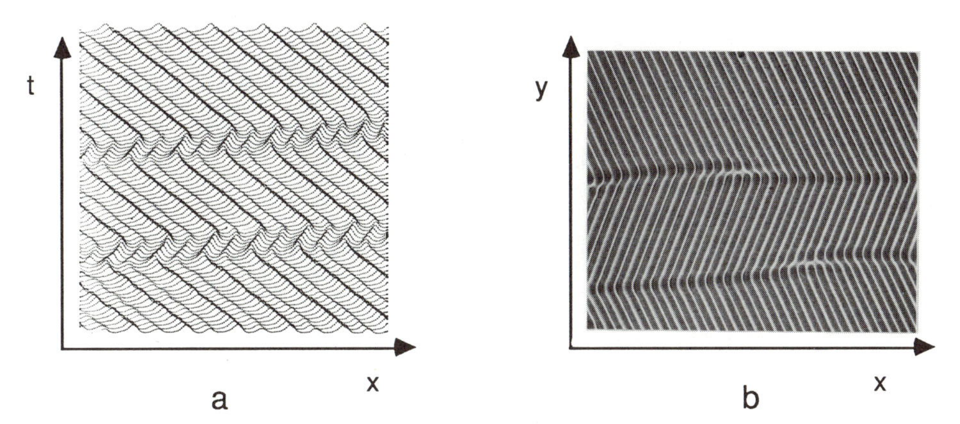

Fig. 6. a) Quasi–collective mode of velocity reversal that can be interpreted as a due to a fast moving source (see text).
b) "Twin–boundaries" in the Oblique Rolls

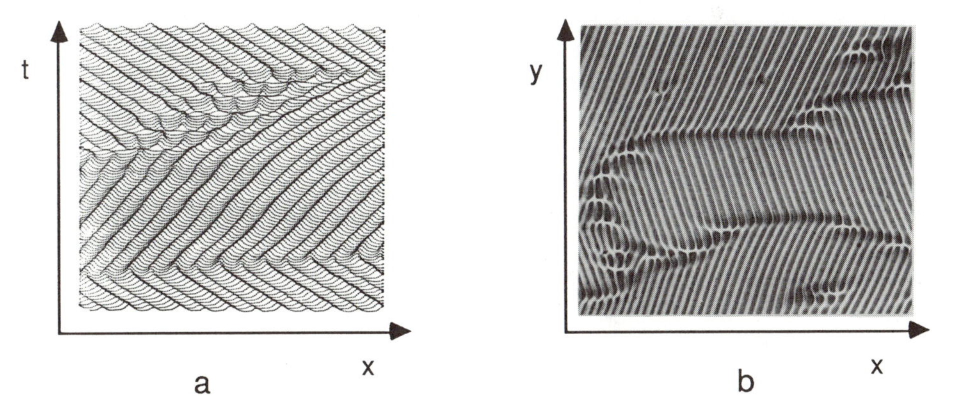

Fig. 7. a) Space–Time complexity involving the defects.
b) Defect–mediated disordering in spatial patterns.

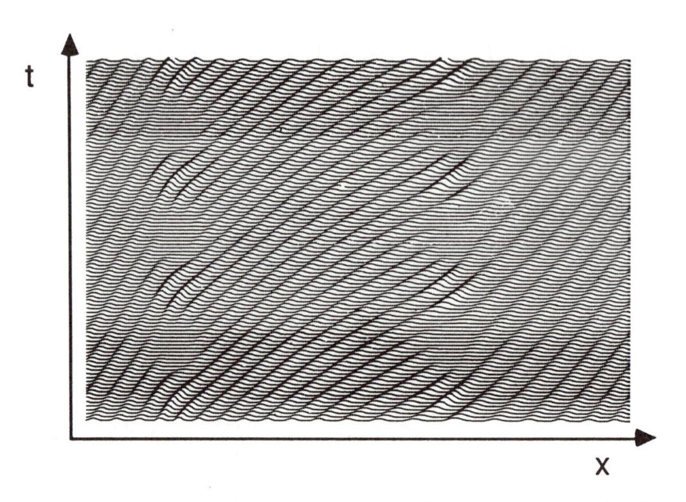

Fig. 8. Numerical simulation of a time-periodic nucleation of dislocations by the shock between two waves with different velocities (compare with Fig. 3b).

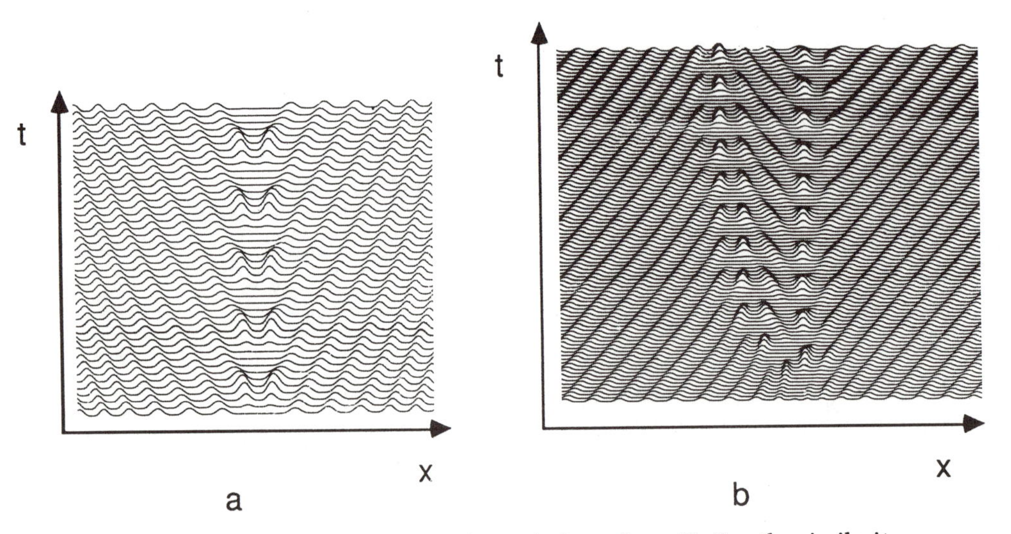

Fig. 9. a) Numerical simulation of a grain-boundary. Notice the similarity with Fig. 4
b) Numerical simulation of the nucleation of a pair of singularities (Sink and Source) that give rise to a "domain" of counter–propagating wave between the two grain-boundaries.

Large-Scale Phase Modulations and Quasi–coherent Modes

A sink or a source is a topological singularity which "separates" in space two waves a_L and a_R and it may move with a velocity w that can be higher than that of the wavetrain inside which it propagates. A limit case is when a source propagates with infinite velocity (in the x,t diagram the singular line would be an horizontal). Then at a sudden, the wavetrain reverses its propagation direction as shown in Fig. 6a. If this reversal occurs periodically in time and on a large space scale, one has a quasi–collective longitudinal mode of oscillation of the structure (bistable state). Therefore, there is an interesting apparent continuity between a pure mode of oscillation in time and the motion of a defect. The space–time representation of this case is quite analogous to that of the twin–boundaries in the Oblique Rolls structure (Fig. 6b). There exists also, under some other conditions, another longitudinal collective mode of oscillation of the structure: the periodic compression–dilation (therefore named "time–dependent Eckhaus" mode) with a wavelength that can be as low as that of the basic structure of convection. Reports on its observation will be given elsewhere (Joets and Ribotta, preprint 1988).

Space–Time Complexity

The progressive wave is the basic state which can be unstable against either localized or homogeneous perturbative modes. The complexification of the basic state (disordering both in time and space) can be thought as the result of either homogeneous modes or localized states as in the case of quasi–stationary structures. In our experiments as the control parameter is increased just above threshold, so that $\epsilon \simeq 0.05$, defects appear randomly both in time and in space and their density increases with ϵ. The grain boundaries may reverse the direction of the propagation of the basic wave over space scales that can either grow or decrease to zero whenever they mutually annihilate. Mixed states with grain boundaries, dislocations and twin–boundaries are shown in figure 7. Further increasing ϵ makes the behaviour become unpredictable and the result is a fully erratic motion with no apparent temporal correlations on small time scales. Usually the disordered state is reached for a value of ϵ around 1.

If we now consider the more general case of 2–D structures, then the progressive wave may become unstable to transverse modes that can be either propagative such as the Busse oscillatory instability (Busse, 1972) or standing waves. This latter case is most often found and as for the stationary states it seems to be the most dangerous one in the process of a rapid transition to the chaos, termed Martensitic Transformation (Ribotta, 1988; Yang and Ribotta, preprint 1988).

DISCUSSION

It is likely that the phase modulations which give rise to singularities can be accounted for by a one–dimensional envelope equation with complex coefficients (Newell, 1974). This time–dependent Landau–Ginzburg equation (TDLG) describes the amplitude evolution of instability waves (wave packets) close to the transition. It has been the subject, in the recent years, of intensive theoretical and numerical study (Newell, 1974; Kuramoto, 1978; Moon et al., 1983; Nozaki and Bekki, 1983, 1984; Keefe, 1986, Coullet et al., 1988). In most cases the results would show the presence of chaotic behavior. Indeed, Bretherton and Spiegel (1983) have shown that localized states of phase instability fluctuating in time are also solutions when the initial data is a noise. Also Nozaki and Bekki (1983) who studied the natural evolution in the absence of initial noise have proposed that soliton–like solutions as well as shocks may be obtained from a Landau–Ginzburg equation. However it is only recently that the nonlinear phase dynamics of a propagating pattern were studied by Kuramoto (1984) who showed that the main nonlinearity of the phase equation comes from local velocity changes. Coullet et al (1988) have studied the sinks and sources as topological defects of waves and have numerically shown the essential feature of these defects using a set of two coupled TDLG equations. The existence of shocks inside nonuniform wavetrains has been demonstrated and also numerically evidenced (Bernoff, 1988).

However, up to now the important question of the effects of strongly localized states resulting from long wavelength disturbances has not been raised yet in the case of periodic propagative structures in hydrodynamics. We believe that the development of topological singularities, which in our case, are triggered after a local phase modulation may be interpreted as a nucleation mechanism following an Eckhaus–Benjamin–Feir instability. Up to now a complete equation for the envelope, which could account for heterogeneous time-dependent states has not yet been derived in the case of liquid crystals from the basic microscopic system of equations. Nevertheless, we have found that it is possible to reproduce numerically most of the spatio–temporal effects hereabove described by use of a set of coupled envelope equations (TDLG) for the two counterpropagative waves. Some of the results obtained with adequate initial conditions for the phase and the amplitude are shown on figures 8 and 9 (Ribotta and Joets, preprint 1988).

CONCLUSION

The traveling–wave state of convection inside a layer of an anisotropic fluid offers quite rich examples of localized states and of singularities in 1–D non–linear waves. Even right above the threshold for the basic state, localized modulations of the envelope occur on scales comparable to the spatial period of the wave. In this case, the localized phase perturbations give rise to shocks. Because of the finite bandwidth of the unstable state, topological singularities are created. These singularities correspond to defects in the space–time ordering. A fully disordred state may be rapidly reached through the random nucleation of the defects and their incoherent motion inside the homogeneous structure. We also have shown the striking similarity of the transition to a fully disordered state both in time–dependent and in quasi–stationary structures, mediated by the defects that are nucleated from localized modulations of the phase. Up to now numerical simulations show that these elementary defects can simply be described by the use of a complex Landau–Ginzburg equation.

Acknowledgements

This work was supported by the Direction des Recherches et Etudes Techniques (DRET). We thank P. Coullet, L. Gil and J. Lega (C.G.L.)for making available to us a numerical code and C.G.L. and A.C. Newell for useful discussions.

References

Benjamin, T. B. and Feir, J. E., 1967, The disintegration of wave trains on deep water, J. Fluid Mech., 27:417.

Bernoff, A., 1988, Slowly varying fully nonlinear wavetrains in the Ginzburg–Landau equation, Physica D, 30:363.

Brand, H.R., Lomdahl, P. and Newell, A., 1986, Evolution of the order parameter in situations with broken rotational symmetry, Phys. Let.A 118:67.

Bretherton, C. S. and Spiegel, E. A., 1983, Intermittency through modulational instability, Phys. Lett., 96A:152.

Busse, F. H., 1972, The oscillatory instability of convection rolls in a low Prandtl number, J. Fluid Mech., 52:97.

Coullet, P., Elphick, C., Gil, L. and Lega, J., 1987, Topological defects of wave patterns, Phys. Rev. Lett., 59:884.

Coullet, P. and Lega, J., 1988, Defect–mediated turbulence in wave patterns, Europhys. Lett.,7:511.

Eckhaus, W., 1963, "Studies in nonlinear stability theory", Springer, Berlin.
de Gennes, P. G., 1974, "The Physics of Liquid Crystals", Clarendon, Oxford.
Joets, A. and Ribotta, R., 1986, Hydrodynamic transitions to chaos in the convection of an anisotropic fluid, J. Phys. (Paris), 47:595.
Joets, A. and Ribotta, R., 1988, Propagative structures and localization in the convection of a liquid crystal, in: "Propagation in Systems far from Equilibrium", J.E. Wesfreid, H.R. Brand, P. Manneville, G. Albinet, and N. Boccara, eds.,Springer, Berlin.
Joets, A. and Ribotta, R., 1988, Localized time–dependent state in the convection of a nematic liquid crystal, Phys. Rev. Lett., 60:2164.
Joets, A. and Ribotta, R., 1988, Propagative patterns in the convection of a nematic liquid crystal, in: "Proceedings of the 12th international liquid crystals conference, Aug. 1988, Freiburg), Liquid Crystals, London.
Joets, A. and Ribotta, R., 1988, Structure of defects in nonlinear waves, preprint
Joets, A. and Ribotta, R., 1988, Nucleation of defects in traveling–wave convection, preprint
Kawasaki, K.and Ohta, T., 1982, Kink dynamics in one dimensional nonlinear systems, Physica, 116A:573.
Keefe, L.R., 1986, Dynamics of perturbed wavetrain solutions to the Ginzburg–Landau equation, Phys. Fluids, 29:3135.
Kuramoto, Y., 1978, Diffusion induced chaos in reaction systems, Prog. Theor. Phys. Suppl., 64:346.
Kuramoto, Y., 1984, Phase dynamics of weakly unstable periodic structures, Prog. Theor. Phys. 71:1182.
Moon, H. T., Huerre, P., Redekopp, L. G., 1983, Transitions to chaos in the Ginzburg–Landau equation, Physica, 7D:135.
Newell, A.C., 1974, Envelope equations, Lect. Appl. Math., 15:157.
Nozaki, K. and Bekki, N., 1983, Pattern selection and spatiotemporal transition to chaos in the Ginzburg–Landau equation, Phys. Rev. Lett., 51:2171.
Nozaki, K. and Bekki, N., 1984, Exact solutions of the generalized Ginzburg–Landau equation, J. Phys. Soc. Jap.,53:1581.
Ribotta, R. and Joets, A., 1984, Defects and interactions with the structures in EHD convection in nematic liquid crystals, in "Cellular Structures in Instabilities", J. E. Wesfreid and S. Zaleski, eds, Springer, Berlin.
Ribotta, R., Joets, A., and Lin Lei, Oblique roll instability in an electroconvective anisotropic fluid, 1986, Phys. Rev. Lett., 56:1595.
Ribotta, R., 1988, Solitons, defects and chaos in dissipative systems, in "Non–linear phenomena in materials science", L. Kubin, G. Martin, eds., Trans Tech Publications, Switzerland.
Stuart, J.T. and DiPrima, R.C., 1978, The Eckhaus and Benjamin–Feir resonance mechanisms, Proc. R. Soc. Lond. A, 362:27.
Whitham, G.B., 1974, "Linear and Nonlinear Waves", John Wiley & Sons, New York.
Yang, X. D., Joets, A. and Ribotta, R., 1986, Singularities in the transition to chaos of a convective anisotropic fluid, Physica, 23D:235.
Yang, X.D., Joets, A., and Ribotta, R., 1988, Localized instabilities and nucleation of dislocations in convective rolls, in "Propagation in Systems far from Equilibrium", J.E. Wesfreid, H.R. Brand, P. Manneville, G. Albinet, and N. Boccara, eds., Springer, Berlin.
Yang, X. D., Ribotta, R., 1988, Transition to chaos mediated by defects in convective stationary structures, in preparation.

DEFECT-MEDIATED TURBULENCE IN SPATIO-TEMPORAL PATTERNS

J. Lega

Laboratoire de Physique Théorique
Université de Nice
06034 NiceCedex France

Introduction

A lot of interest has been devoted, these last few decades, to systems driven far from equilibrium by an external parameter (see for instance the book edited by Wesfreid and Zaleski, 1984). The study of such problems involving many degrees of freedom and displaying nonlinear behaviours is expected to give some clues in the understanding of complex spatio-temporal phenomena, such as hydrodynamic turbulence. Many model equations, displaying spatio-temporal complexity (Kuramoto and Tsuzuki, 1976; Pomeau and Manneville, 1979; Chaté and Manneville, 1987), have been studied. We show in this paper that, in the framework of amplitude equations (Newell and Whitehead, 1969; Segel, 1969) which describe nonlinear dynamics near a bifurcation point, a new form of turbulence may occur (Coullet et al., 1989), which is associated with topological defects. First, we consider the simplest model which leads to such behaviours, namely the amplitude equation associated with a Hopf bifurcation. Then, same considerations are applied to amplitude equations corresponding to a Hopf bifurcation where space translational invariance is also broken, and which describe the occurence of wave patterns in an anisotropic medium.

Amplitude Equation and Linear Stability Analysis

The amplitude equation which describes a Hopf bifurcation with $k_0 = 0$ in an isotropic, parity invariant medium, reads:

$$\frac{\partial A}{\partial t} = (\mu_r + i\mu_i)A + (\alpha_r + i\alpha_i)\Delta A - (\beta_r + i\beta_i)|A|^2 A,$$

where μ_r measures the distance from bifurcation threshold, μ_i corresponds to a shift in frequency, α_r, α_i, β_r and β_i are associated respectively with diffusion ($\alpha_r > 0$), dispersion, nonlinear saturation ($\beta_r > 0$), and nonlinear renormalisation of the temporal frequency. The complex order parameter A is assumed to vary slowly in time, and corresponds, near the bifurcation threshold, to the amplitude of the most unstable modes of a quantity C, such as a concentration of a given chemical species. After performing a shift on the frequency of A, together with scalings on space, time, and field A, above equation can be written in the form:

$$\frac{\partial A}{\partial t} = \mu A + (1 + i\alpha)\Delta A - (1 + i\beta)|A|^2 A. \tag{1}$$

This equation possesses a one-parameter family of homogeneous solutions, parametrized by an arbitrary phase φ: $A_0 = \sqrt{\mu}\exp(-i\beta\mu t + i\varphi)$. Linearization around one of these solutions leads to two eigenvalues:

$$\lambda_\pm = -(\mu + k^2) \pm \sqrt{\mu^2 - 2\mu\alpha\beta k^2 - \alpha^2 k^4},$$

New Trends in Nonlinear Dynamics and Pattern-Forming Phenomena
Edited by P. Coullet and P. Huerre
Plenum Press, New York, 1990

which read in the limit of small $|k| = \sqrt{k_x^2 + k_y^2}$:

$$\lambda_- = -2\mu - (1 - \alpha\beta)k^2 + \ldots \tag{2.a}$$

$$\lambda_+ = -(1 + \alpha\beta)k^2 - \frac{\alpha^2(1+\beta^2)}{2\mu}k^4 + \ldots . \tag{2.b}$$

While the amplitude perturbations of A_0 are damped under the dynamics of Eq.(1), Eq.(2.b) shows that the homogeneous solution A_0 may become unstable with respect to phase perturbations if $1 + \alpha\beta$ is negative. Such an instability (Benjamin and Feir, 1967; Newell, 1974) is known (Kuramoto and Tsuzuki, 1976; Kuramoto, 1978) to lead to complex spatio-temporal behaviours (Pomeau and Manneville, 1979). In the following, we are interested in the dynamics of Eq.(1) when a phase instability occurs.

Phase Dynamics

A first attempt in this study is to reduce the dynamics of Eq.(1) to that of the unstable mode, namely the mode associated with the phase of the order parameter A (Kuramoto and Tsuzuki, 1976). This is justified by the fact that λ_- has a negative upper bound. Writing $A = R\exp(-i\beta\mu t + i\phi)$, and expanding R as a function of spatial derivatives of ϕ, which are assumed to be small, one gets:

$$R = \sqrt{\mu} - \frac{\alpha}{2\sqrt{\mu}}\Delta\phi - \frac{1}{2\sqrt{\mu}}(\nabla\phi)^2 + \frac{\alpha^2\beta}{4\mu\sqrt{\mu}}\Delta^2\phi + \ldots \tag{3.a}$$

$$\frac{\partial\phi}{\partial t} = (1 + \alpha\beta)\Delta\phi - \frac{\alpha^2(1+\beta^2)}{2\mu}\Delta^2\phi + (\beta - \alpha)(\nabla\phi)^2 + \ldots . \tag{3.b}$$

The one dimensional analog of Eq.(3.b), known as the Kuramoto-Shivasinsky equation (Kuramoto and Tsuzuki, 1976; Shivasinsky, 1977), has been widely studied (for a review see Hyman et al., 1986), and displays smooth turbulence when $1 + \alpha\beta$ is negative. In two spatial dimensions, for small values of $|1 + \alpha\beta|$, one also observes a smooth turbulent behaviour, and phase gradients remain small enough for the derivation of Eq.(3.b) to be valid. This regime is characterized by long range phase correlations (see Fig.1), where the correlation function is defined as:

$$\mathcal{C}(r) = \frac{\mathcal{F}_1(r)}{\mathcal{F}_1(0)} \qquad \mathcal{F}_1(r) = < \sum_{x^2+y^2=r^2} \Re e[\exp i(\phi(x_0, y_0, t) - \phi(x, y, t))] > \tag{4}$$

and $< . >$ denotes time averaging.

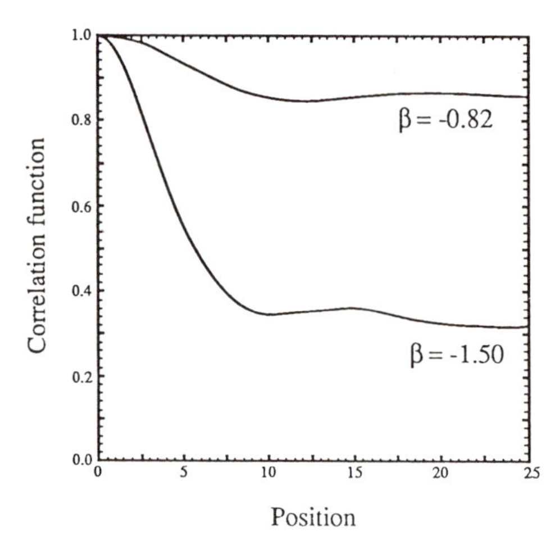

Figure 1. Numerical simulation ($\mu = 1$, $\alpha = 2$) of Eq.(3.b) showing the correlation function $\mathcal{C}$ (see Eq.(4)) as a function of position for $\beta = -0.82$ and $\beta = -1.50$.

For higher values of $|1+\alpha\beta|$, the dynamics of Eq.(3.b) leads to short range correlations (see Fig.1), but this is associated with the appearance of large phase gradients (of the order of $\sqrt{\mu}$), so that Eq.(3.b) can no longer be seen as describing the dynamics of Eq.(1). For a fixed value of the size of the box, the time required for large phase gradients to develop gets smaller when $|1+\alpha\beta|$ is increased. Moreover, for fixed values of α and β, large phase gradients appear sooner and get higher - their maximum scales linearly with the logarithm of the size of the box - when the size of the box is increased. Hence, the feeling is that whatever the negative value of $|1+\alpha\beta|$, large phase gradients always appear in infinite size boxes, and the phase approximation always breaks down.

Thus, for large values of $|1+\alpha\beta|$ or in sufficiently extended systems, Eq.(3.b) does not describe the behaviour of Eq.(1). In order to understand how the phase instability develops, we are led to study the dynamics of Eq.(1) with new variables U_R, U_ϕ so that:

$$\frac{\partial U_R}{\partial t} = \lambda_- U_R$$
$$\frac{\partial U_\phi}{\partial t} = \lambda_+ U_\phi$$

They read at second order in spatial derivatives:

$$U_R = \frac{1}{4\mu}\left[2\mu(a+\bar{a}) - i\alpha\Delta(a-\bar{a}) - 2\alpha\beta\Delta(a+\bar{a})\right] \tag{5.a}$$

$$U_\phi = \frac{1}{4\mu}\left[-2\beta\mu(a+\bar{a}) - 2i\mu(a-\bar{a}) + 2\alpha\beta^2\Delta(a+\bar{a}) + i(\alpha\beta - \frac{\alpha}{\beta})\Delta(a-\bar{a})\right], \tag{5.b}$$

where $a = A\exp(i\beta\mu t) - \sqrt{\mu}$.

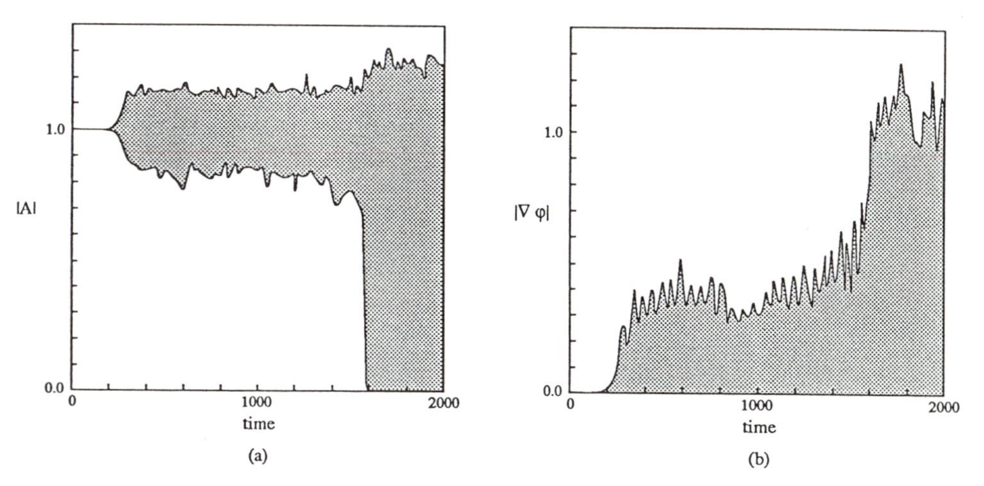

Figure 2. Numerical simulation ($\mu = 1$, $\alpha = 2$, $\beta = -0.82$) of Eq.(1) showing the spreading of (a) $|A|$ and (b) $|\nabla U_\phi|$ (see Eq.(5.b)) as a function of time. At t=1600, $|\nabla U_\phi|$ becomes of order $\sqrt{\mu}$, and topological defects appear in the system (there exist points where $|A| = 0$).

The behaviours of $|A|$ and $|\nabla\phi|$ for $\mu = 1$, $\alpha = 2$, and $\beta = -0.82$ are shown in Fig.2. One can clearly see that after some transient, gradients of U_ϕ become high enough to drive $|A|$ far from its initial value, which is $\sqrt{\mu}$, so that the phase approximation breaks down. Moreover, the existence of a strong coupling between amplitude and phase modes is confirmed by the fact that the phase approximation breakdown in Eq.(1) occurs before Eq.(3.b) develops large gradients. Hence, phase instability leads in two dimensional systems to the destabilization of slaved amplitude modes. We show in the following that this results in the spontaneous nucleation of topological defects.

Topological Defects

Topological defects are solutions of Eq.(1) which exhibit localized amplitude variations. Around the core of a defect, the phase ϕ of the order parameter $A = R\exp(i\phi)$ turns by 2π, i.e. the circulation of the phase gradient along any closed path around the core of the defect equals $\pm 2\pi$. As shown in Fig.3, a defect takes the form of a spiral wave which emanates from its core. Figure 3.a is a half tone-plot of the real part of the order parameter ($\Re e(A) \propto C - C_0$) as a function of space, and Fig.3.b shows the lines where $\Re e(A) = 0$ and $\Im m(A) = 0$ (which are close to special equipases). These two lines meet at the core of the defect, where the order parameter A vanishes. Far from the core, the amplitude of the solution is of order $\sqrt{\mu}$.

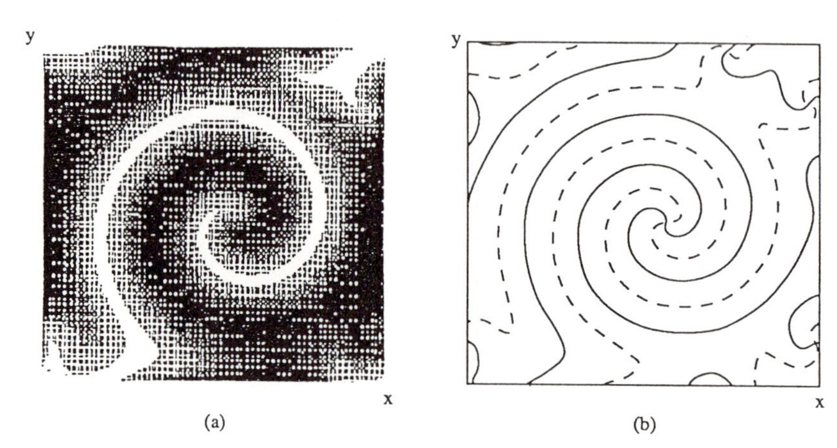

Figure 3. Numerical simulation ($\mu = 1$, $\alpha = -0.3$, $\beta = -2$) of Eq.(1) showing a spiral wave defect. (a) $\Re e(A)$ has been plotted as a function of space (white corresponds to $-\sqrt{\mu}$, and black to $\sqrt{\mu}$). (b) Lines $\Re e(A) = 0$ and $\Im m(A) = 0$ meet at the cores of defects.

Defect-mediated turbulence

We show by means of numerical simulations of Eq.(1) that a phase instability of the homogeneous solution A_0 leads to the spontaneous nucleation of topological defects. We use a spectral code with periodic boundary conditions, in a box of size 50×50 with 80 grid points in both directions. Initial conditions correspond to the slightly perturbed homogeneous solution A_0, and the external parameters are chosen so that a phase instability occurs.

We have shown above that in sufficiently extended systems, Eq.(1) develops large phase gradients, which become high enough to destabilize slaved amplitude modes. After some transient, there exist points where the amplitude of the order parameter goes to zero (see Fig.2), which are seeds of pairs of defects. Once created, spiral defects move away through the system and break the order that was induced by the temporal structure. Far from the core of each defect, the amplitude is close to the homogeneous solution, so that the complex spatio-temporal state we are describing can be seen as made of spiral wave defects evolving in a phase turbulent field.

In order to evaluate how much defects disorganize the temporal pattern, we have computed the field correlation function:

$$\mathcal{C}(r) = \frac{\mathcal{F}_2(r)}{\mathcal{F}_2(0)}, \qquad \mathcal{F}_2(r) = < \sum_{x^2+y^2=r^2} \Re e[A(x_0, y_0, t)\bar{A}(x, y, t))] > \tag{6}$$

where $< . >$ denotes time averaging. Its behaviour when there are defects in the box and when there are not is displayed in Fig.4, for fixed values of external parameters. Actually, external parameters μ, α and β, can be chosen in order that there are periods so that all the defects have annihilated, bringing the solution close to A_0. But because of phase instability, new defects appear after some lapse of time required for phase instability to stir up amplitude modes. Here, α and β have been chosen so that periods when defects exist or do not exist are long enough to compute the field correlation function. As it can be seen in Fig.4, defects are responsible for an exponential decrease of the field correlations.

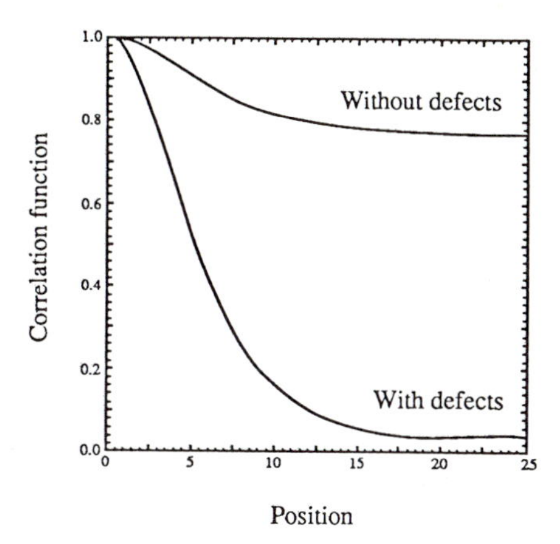

Figure 4. Numerical simulation ($\mu = 1$, $\alpha = 2$, $\beta = -0.82$) of Eq.(1) showing the correlation function $\mathcal{C}$ (see Eq.(6)) as a function of position with and without defects in the box.

Many runs have been performed in order to determine the role of the parameters α and β (Coullet et al., 1989). While the former slightly modifies the mean number of defects $< N >$, the latter rules this number. For a fixed value of the size of the box, and for a finite computation time, there exists a critical value $|\beta_c|$ beyond which no defects appear in the box. For small values of $|\beta|$ above this threshold, the system alternates between periods with defects and periods without defects, and $< N >$ is small when defects exist. For higher values of $|\beta|$, this "intermittent" feature disappears and $< N >$ gets larger when $|\beta|$ is increased. For large values of $|\beta|$, more and more amplitude modes are excited, so that "defect-mediated turbulence" merges into "amplitude turbulence" (Kuramoto and Tsuzuki, 1976).

Hence, we have seen on the simplest model equation displaying such an instability, that phase instability leads in two dimensional sufficiently extended systems to spontaneous appearance of topological defects, whose motion gives rise to complex spatio-temporal phenomena. Now, we apply same considerations to wave patterns in an anisotropic system, where our results can be qualitatively compared with experiments.

Defect-mediated turbulence in wave patterns

We assume we are dealing with an anisotropic, parity invariant medium where a Hopf bifurcation with temporal frequency ω_0 and spatial frequency $k_0 \neq 0$ occurs. The two coupled amplitude equations (Bretherton and Spiegel, 1983; Chossat and Iooss, 1985; Coullet et al. 1985; Coullet et al., 1987) describing such a bifurcation towards a normal rolls pattern read:

$$\frac{\partial A}{\partial t} = \mu A + (1 + i\alpha_x)\frac{\partial^2 A}{\partial x^2} + (1 + i\alpha_y)\frac{\partial^2 A}{\partial y^2} - c\frac{\partial A}{\partial x} - (1 + i\beta)|A|^2 A - (\gamma + i\delta)|B|^2 A \quad (7.a)$$

$$\frac{\partial B}{\partial t} = \mu B + (1 + i\alpha_x)\frac{\partial^2 B}{\partial x^2} + (1 + i\alpha_y)\frac{\partial^2 B}{\partial y^2} + c\frac{\partial B}{\partial x} - (1 + i\beta)|B|^2 B - (\gamma + i\delta)|A|^2 B. \quad (7.b)$$

where A and B are the complex amplitudes of the right and left propagating waves, μ, α_x, α_y and β have the same meaning as in Eq.(1), c is related to group velocity, γ rules the competition between travelling and standing waves and δ is associated with nonlinear renormalisation of the frequency of each wave by the counter-propagating one.

Equations (7) possess two kinds of solutions: right ($A \neq 0$ and $B = 0$) or left ($A = 0$ and $B \neq 0$) travelling waves, and standing waves where A and B are both finite. The following will focus on travelling waves, for instance a right one. The simpler one reads: $A_0 = \sqrt{\mu}\exp(-i\beta\mu t)$, $B_0 = 0$. Here again, we are interested in the dynamics of Eqs.(7) when a phase instability occurs. Initial conditions correspond to a slightly pertubed right travelling wave (A_0,B_0). The numerical simulation presented below is performed with parameters so that the amplitude of B quickly goes to zero under the dynamics of Eq.(7) and remains equal to this value during the whole run. Hence, after a change of referential, Eqs.(7) can be reduced to:

$$\frac{\partial A}{\partial t} = \mu A + (1 + i\alpha_x)\frac{\partial^2 A}{\partial x^2} + (1 + i\alpha_y)\frac{\partial^2 A}{\partial y^2} - (1 + i\beta)|A|^2 A \quad (8)$$

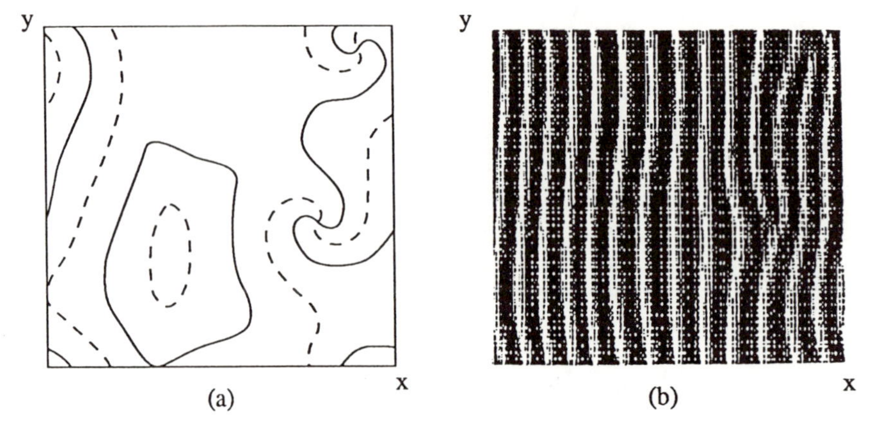

Figure 5. Numerical simulation ($\mu = 1$, $\alpha_x = 0.5$, $\alpha_y = -1$, $\beta = -1.8$, $\gamma = 1.5$, $c = 5$) of Eqs.(7) showing a pair of travelling wave dislocations. (a) Lines $\mathcal{R}e(A) = 0$ and $\mathcal{I}m(A) = 0$ take a spiral shape near the core of each defect, where they meet. (b) Half-tone plot of $C - C_0 = \Re e[A(X,Y,T)\exp(i(\omega_0 t - k_0 x)) + B(X,Y,T)\exp(i(\omega_0 t + k_0 x))]$ displaying two dislocations. Dark areas correspond to maxima of $C - C_0$.

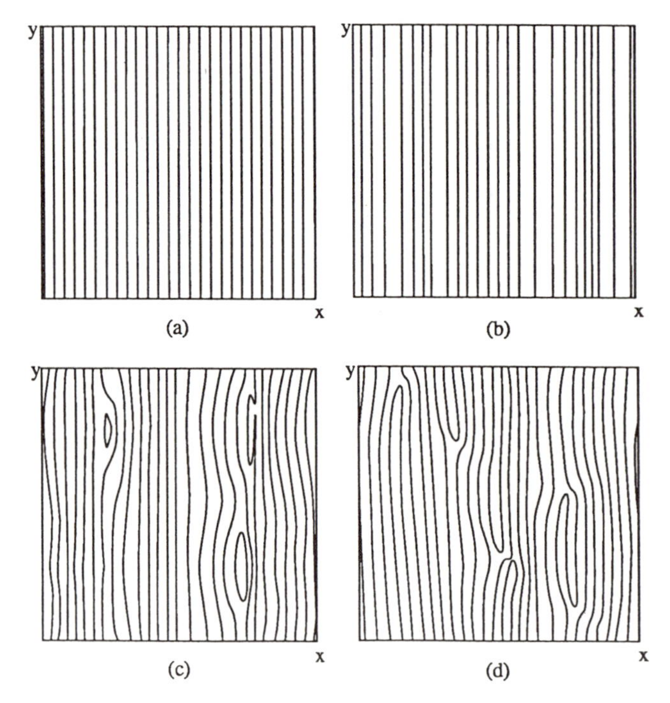

Figure 6. Numerical simulation ($\mu = 1$, $\alpha_x = 1$, $\alpha_y = -2$, $\beta = -1.8$, $\gamma = 1.5$, $c = 5$) of Eqs.(7) showing the destabilisation of a travelling wave pattern, giving rise to dislocations. Lines where $\mathcal{R}e(A \exp i(-k_0 x + \omega_0 t)) = 0$ have been plotted in the (x, y) plane, at times (a)100, (b)300, (c)400, and (d)500. The space between two lines corresponds to a travelling roll.

which is analog to Eq.(1) except that coefficients of $\frac{\partial^2 A}{\partial x^2}$ and $\frac{\partial^2 A}{\partial y^2}$ are not equal because of anisotropy. Hence, phase instability will occur if either $1+\alpha_x\beta$ or $1+\alpha_y\beta$ or both are negative. Here we have chosen $1+\alpha_x\beta < 0$ and $1+\alpha_y\beta > 0$. As Eq.(1), Eqs.(7) lead to the spontaneous nucleation of topological defects, which disorganize the system (Coullet and Lega, 1988; Lega 1989).

As shown in Fig.5, topological defects are spiral waves for the amplitude equations (7), but correspond to dislocations of $C \propto \Re e(A\exp(-ik_0x) + B\exp(ik_0x))$. The destabilization of the initial phase unstable travelling wave pattern is shown in Fig.6, where the lines $\Re e(A\exp(-ik_0x) + B\exp(ik_0x)) = 0$ have been plotted. Each line separates two travelling rolls. After some transient (Fig.6.a), large scale modulations appear (Fig.6.b) and the first pairs nucleate (Fig.6.c). "Defect-mediated turbulence" is illustrated in Fig.6.d where several dislocations are shown. Defects move through the box, some annihilate, others are created, and their existence is responsible for the disorganization of the ordered wave pattern. As described for Eq.(1), for small values of $|\beta|$, there also exist periods when all the defects have annihilated, so that the system alternates between "phase turbulence" and "defect-mediated turbulence".

Liquid crystals, where turbulent wave patterns associated with dislocations have been observed (Guazelli, and Guyon, 1981; Ribotta, 1988; Yang et al., 1988; Rehberg and Steinberg, 1988), could be relevant experimental setups to test our results. Besides, since for some range of parameters, Eqs.(5) reduce to Eq.(8), "defect-mediated turbulence" may also be observed in systems where parity invariance is broken, such as shear flows or wakes, where analogous behaviours have been reported (Browand and Troutt, 1985). Let us conclude by pointing out that more complex behaviours involving both left and right travelling waves can be observed when $1+\alpha_x\beta$ and $1+\alpha_y\beta$ are both negative.

Conclusion

We have shown in this paper that a phase instability of a temporal periodic pattern leads in sufficiently extended 2D systems to a turbulent state associated with topological defects. We have seen that phase modes destabilize slaved amplitude modes, and allow the spontaneous nucleation of defects. Once created, those defects move through the system before they annihilate, while others appear. They break the order induced by the initial periodic pattern and are responsible for an exponential decrease of the field correlations. The resulting turbulent state has been termed "defect-mediated turbulence" and we hope it has an experimental relevance since analogous behaviours have been observed, especially in systems giving rise to wave patterns. Finally, let us mention that this study may also be related to previous works (Dreyfus and Guyon, 1981; Walgraef et al., 1982) where the analogy between such phenomena in far from equilibrium systems and two-dimensional melting was emphasized.

Acknowledgements

The author is grateful to P. Coullet and L. Gil for fruitful discussions. I also thank the CCVR where the numerical simulations have been performed, the NCAR for the use of its graphic library, the DRET for financial support under contract n° 86/1511, and the EEC for financial support under contract n° 500241.

References

Benjamin T. B., and Feir J. E., 1967, J. Fluid Mech. **27**, 417.
Bretherton C. S., and Spiegel E. A., 1983, Physics Lett. **96A**, 152.
Browand F. K., and Troutt T. R., 1985, J. Fluid Mech. **158**, 489.
Chaté H., and Manneville P., 1987, Phys. Rev. Lett. **58**.
Chossat P., and Iooss G., 1985, Japan J. Appl. Maths. **2**, 37.
Coullet P., Fauve S., and Tirapegui E., 1985, J. Physique Lettres, **46**, 787.
Coullet P., Elphick C., Gil L., and Lega J., 1987, Phys. Rev. Lett. **59**, 884.
Coullet P., Gil L., and Lega J., 1989, *Defect-mediated turbulence*, Phys. Rev. Lett. **62**, 1619.
Coullet P., Gil L., and Lega J., 1988, *A form of turbulence associated with topological defects*, to appear in Physica D.
Coullet P., and Lega J., 1988, *Defect-mediated turbulence in wave patterns*, Europhysics Lett. **7**, 511.
Dreyfus J. M., and Guyon E., 1981, J. de Phys. **42**, 283.
Guazelli E., and Guyon E., 1981, C. R. Hebd. Séan. Acad. Sci. **292 II**, 141.
Hyman J. M., Nicolaenko B., and Zaleski S., 1986,Physica **23D**, 265.

Kuramoto Y., and Tsuzuki T., 1976, Prog. Theor. Phys. **55**, 356.
Kuramoto Y., 1978, Suppl. of Prog. Th. Phys. **64**, 346.
Lega J., 1989, *Defect-mediated turbulence: an example in wave patterns*, Le Journal de Physique, Colloque C3, **50**, C3-193.
Newell A. C., and Whitehead J. A., 1969, J. Fluid Mech. **38**, 279.
Newell A. C., 1974, in *Lectures in Applied Mathematics*, Vol.15, Am. Math. Society, Providence.
Pomeau Y., and Manneville P., 1979, J. Phys. Lettres **40**, 609.
Rehberg I., and Steinberg V., 1988, *Defect-mediated turbulence in travelling convection patterns of a nematic liquid crystal*, preprint submitted to Phys. Rev. Lett.
Ribotta R., 1988, in *Non linear phenomena in material science*, Eds. L. Kubin and G. Martin, Trans Tech Publications.
Segel L. A., 1969, J. Fluid Mech. **38**, 203.
Shivasinsky G.I., 1977, Acta Astronautica, **4**, 1177.
Walgraef D., Dewel G., and Borckmans P., 1982, Z. Phys. B **48**, 167.
Wesfreid J. E., and Zaleski S., 1984, *Cellular Structures in Instabilities*, Springer Verlag.
Yang X. D., Joets A., and Ribotta R., 1988, in *Propagation in Systems far from Equilibrium*, Eds. J. E. Wesfreid, H. R. Brand, P. Manneville, G. Albinet, and N. Boccara, Springer-Verlag.
Yang X. D., Joets A., and Ribotta R., 1988, *Localized instabilities and nucleation of dislocations in convective rolls*, preprint submitted to Phys. Rev. Lett.

MEAN FLOWS AND THE ONSET OF TIME-DEPENDENCE IN CONVECTION

Alain Pocheau

Laboratoire de recherche en Combustion
Université de Provence, Centre de Saint Jérome, S 252
13397 Marseille, France

We report recent progress in the understanding of the transition to time-dependence in convective patterns. These works refer to the fundamental role played by mean flows in convection and especially to their non-local character. In the following, two different situations are addressed. In the first one, the mean flow effects are forced; in the second one, they develop freely. In the latter case, we construct an analytical solution which succeeds in simultaneously handling global nonlocal modes and local unstable ones. This solution provides an understanding of the overinstability of distorted patterns in convection and of the influence of the Prandtl number on the route to turbulence.

Since the attention of physicists has been drawn on extended pattern forming systems, the Rayleigh-Bénard thermoconvective instability has been widely used as a natural ground for elaborating primary tools such as the phase or the amplitude formalisms. Recently, as an additional example of the richness of this system, the concept of mean flows first emerged in convection and has further spread among a lot of other topics, even including open flows (see this volume). Because the spatial scale of the mean flows is large compared to the roll size, convection presents itself as a system governed by a non-local interaction between two scales. In this paper, we report recent advances in the understanding of the spatio-temporal properties of such a system (Pocheau et al.,1987; Pocheau, 1988). They allow a close connection between experiments and theory to be made.

The paper is divided into five parts. In the first part, we stress the central problem raised by the low Prandtl number convection i.e. the great difference in the threshold of time dependence between straight and bent patterns. We next emphasize the nonlocal features of mean flows and we study in detail forced mean flow effects. In the fourth part, we finally derive, at the dominant orders, an exact perturbative solution of the Cross-Newell equations for stationary slightly distorted patterns in a cylindrical container. The study of the destabilization of this solution yields a wide understanding of the transition to time dependence in a cylindrical container.

I PATTERN DISTORTION AND TIME DEPENDENCE

When experimentalists avoid using artificial tricks to orientate the convective motions, the patterns which spontaneously arise in sufficiently wide containers display long range spatial disorder (i.e. patches of regular rolls linked by defects) or, at least, long range distortion (i.e. roll bending or roll compression) (fig1). A fundamental question then naturally emerges from the duality between the long range character of the distortion and the short range nature of the rolls : does the pattern dynamics relate to a local analysis or to a much more global approach ? In other words, can we understand the pattern behavior by looking at each part of the convective field separately ? If so, one has to follow the tracks already taken by Busse and Clever and thus to perform a stability analysis for the patterns which locally approximate textures i.e. for straight rolls (Busse, 1978; Busse and Clever, 1979 and 1981). Their main result may be summarized as follows : at each Prandtl number Pr, a band of stable roll wavenumbers exits for Rayleigh numbers Ra below a threshold value RaB(Pr). In the (Ra,k,Pr) space, the full domain of stability is called the Busse balloon (fig 2). Outside of this balloon, since

New Trends in Nonlinear Dynamics and Pattern-Forming Phenomena
Edited by P. Coullet and P. Huerre
Plenum Press, New York, 1990

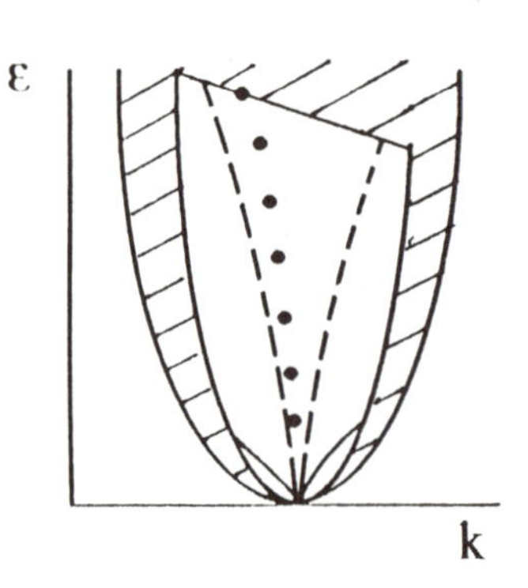

Fig. 1. Shadowgraphic picture of convection at high Prandtl number in a cylindrical container. Notice the perpendicularity of rolls at the walls and the spatial disorder.

Fig. 2. Sketch of the Busse balloon. The dashed domain indicates roll instabilities. The lines inside the stability domain refer to the solution of the Cross-Newell equation at various Prandtl numbers (see IV). a) low Prandtl number : Pr=0.7 : full line. This also corresponds to experimental measurements. b) moderate Prandtl numbers : dashed line. c) high Prandtl numbers : dotted line

straight rolls are unstable, one expects patterns to be ***locally*** unstable and thus time dependent as ***a whole*** . Our basic question then comes down to wonder in what extent such a local study of pattern stability is consistent with the pattern behaviors observed experimentally ? This question would be interesting to consider for a large range of Prandtl numbers (Pocheau, 1987). However, for the sake of conciseness, we only focus here on the low Pr regime. As reported below, the pattern dynamics then emphasizes the difference between straight and distorted patterns.

1) Straight roll patterns at Pr=0.7

A natural cell geometry for studying straight roll patterns is the rectangular one. Provided that the aspect ratio is moderate enough, visual observation at various Prandtl numbers shows that straight rolls parallel to the shortest side of the container are usually realized. Experimental studies of this situation at low (Pr=0.7) (Croquette, 1988) , moderate (Pr=5) (Kolodner et al., 1986) or high Prandtl numbers (Pocheau and Croquette, 1984, Busse and Whitehead, 1974) agree with the following behavior : As ϵ varies, straight patterns are stationary and stable until a Busse-clever instability is reached. Within the former range of Prandtl numbers, this usually occurs by increasing ϵ and then by encountering the skewed-varicose instability. When instability arises, the whole pattern is affected so that we may call such instabilities ***global*** instabilities. They result in defect nucleation but the defects are usually expelled from the pattern after having moved through the convective field. This provides a wavenumber adjustment process which enables a stationary straight pattern with a stable wavenumber to arise. Further ϵ variations yield similar evolutions provided the Busse balloon boundaries are crossed. Below the maximum Ra value of the Busse balloon, RaB(Pr), no sustained time-dependent states may thus be reached; only restabilization process involving global instabilities may occur in response to a Ra variation. Permanent dynamics can only occur above RaB(Pr), when any wavenumber is then unstable. The value of this threshold decreases with Pr. It is still high at Pr=0.7, however: $\epsilon(0.7) \approx 3$ where $\epsilon = \dfrac{Ra - Rac}{Rac}$.

For straight patterns, the Busse balloon is thus sufficient to understand the dynamical behavior: local analysis and global behavior are in agreement. This conclusion is not surprising since in a straight pattern, any part is similar to the whole. However, that property breaks down whenever non-homogeneous geometries (i.e. distortions) arise. As shown below, local analysis is then insufficient solely to understand the global behavior.

2) Distorted patterns at Pr=0.7

At any Prandtl number, experimental observations agree with the following rule : rolls tend to arise perpendicularly to the sidewalls. Instead of straight rolls, bent rolls should therefore arise in a cylindrical container. In fact, near onset, convection is usually suppressed close to the walls, because

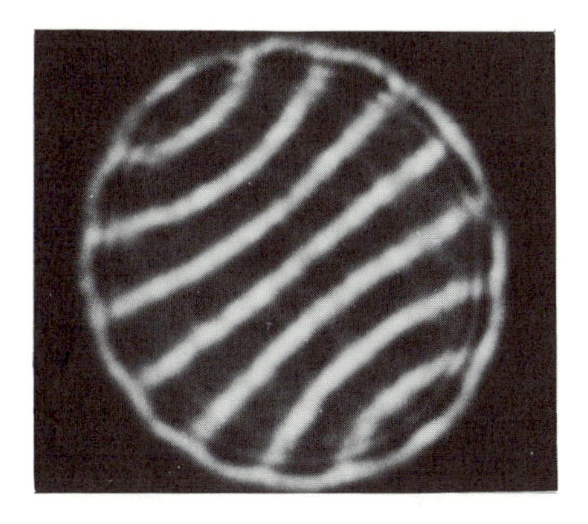
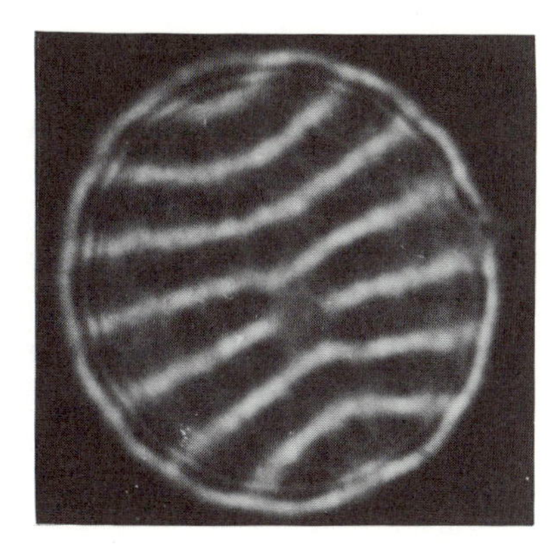

Fig. 3. Shadowgraphic picture of convection at Pr=0.7 in a cylindrical container. a) Notice the pattern bending and the slight compression at the center. b) Roll pinching at $\epsilon = 0.13$ due to compression. The pattern is then time-dependent.

of the thermal conductivity difference between the sidewalls and the fluid. This inhibits the former rule and allows straight rolls to appear. Nevertheless, as Ra increases, the subcritical layer at the walls disappears so that the expected situation is recovered for $\epsilon > 0.2$. In a cylindrical container, a whole pattern bending is then gradually but quickly forced as ϵ increases (Pocheau et al., 1985).

The pattern distortion is made of two opposite curvatures and also of a compression going from each focus to the pattern center (fig 3a). This compression is hardly seen on the pattern pictures but it is nevertheless extremely dangerous for the pattern stability. Indeed, quite close to onset ($\epsilon = 0.13$), it becomes sufficient to bring the center rolls out of the Busse balloon (fig 2). One then observes the development of an instability there : roll diameters quickly shorten more and more until a pinching occurs. Two dislocations then nucleate (fig 3b) and quickly travel all around the pattern : each defect climbs separately to a sidewall, then glides towards a focus and disappears there : the local instability drives a global evolution. At that time, a roll pair has been expelled from the pattern. Since this corresponds to a decrease of the distortion (i.e. the stress), we might expect to recover the stability. But the distortion increases again, finally leading to a sustained repetition of the previous events : distorted patterns are time-dependent for $\epsilon > 0.13$.

This statement may be rather surprising since the upper limit of the Busse balloon corresponds to $\epsilon \approx 3$: indeed, distorted patterns become time-dependent for ϵ values one order of magnitude lower than for straight rolls. This overdestabilization by distortion cannot be handled by a local analysis alone. On the other hand, we will show in the following that it may be understood by a proper nonlocal analysis of convection (Pocheau, 1988) (see part III). Surprisingly, the essential physical ingredients leading to such a huge effect will prove to be related to a very weak component of the convective flows : the mean flows. This apparent paradox comes from the fact that the mean flow effects are not only related to the amplitude of mean flows but also to their nonlocal features. For this reason, we emphasize their non-local properties in the following section.

II MEAN FLOWS : A NONLOCAL PHENOMENON

Since, at first glance at a convective picture, rolls are the most evident manifestation of convection, one might be prone to deal only with the flows which constitute the rolls i.e. with some small scale closed flows. However, such nearly periodic and single scale functions cannot be complete solutions of the Boussinesq equations, as soon as the Prandtl number is finite. Other flows must be accounted for and their features prove to be very different : these flows look like parallel flows in a Hele-Shaw cell (Siggia and Zippelius, 1981). In particular, their horizontal scale is large compared to the cell depth. In a closed cell, they must of course be closed flows. On a small scale level however, i.e. for each roll, they always look like open flows which allow the fluid to cross the rolls by means other than molecular diffusion. Near the onset of convection, their amplitude is usually quite small compared to that of the rolls. In a naive way, one could therefore be tempted to neglect them. However, they succeed in inducing the most important effects on the pattern behavior. As shown later, the origin of this surprising effect lies in the nonlocal features of mean flows i.e. in the possibility for

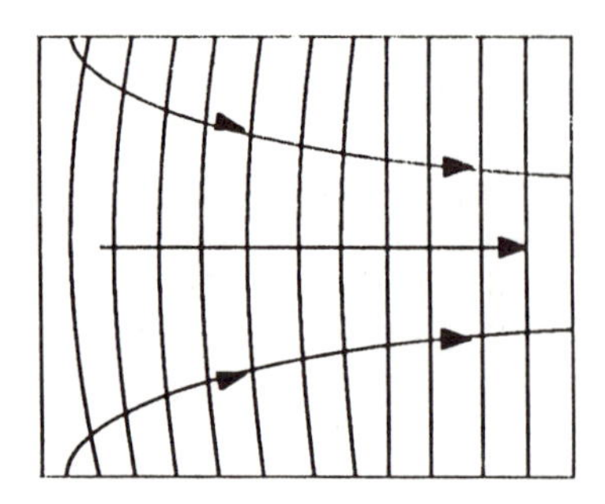

Fig. 4. A nonlocal feature of mean flows : a mean flow (sketched by the arrows) produced by roll bending is present even in a straight roll domain.

mean flows of acting over large distances. This action thoroughly counterbalances the weakness of their amplitude.

1) Mean flows generation by roll distortions

Large scale mean flows in convection arise from the advection term of the Navier-Stokes equation (Siggia and Zippelius, 1981). They are thus inversely proportional to Pr (Siggia and Zippelius, 1981). They are only produced if the pattern exhibits some distortion and their amplitude is then proportional to the distortion amplitude i.e., in a box with aspect ratio Γ, to $1/\Gamma$. The former property can be readily understood since the symmetries of a straight pattern forbid any prefered mean flow direction and thus any mean flow at all.

The mean flow generation may be understood in terms of mode interactions in the Navier-Stokes equations i.e. in terms of the excitation of a vanishing harmonic by destructive interferences among wavepackets. However, we stress that a deeper insight in the meaning of mean flows may be obtained by considering gauge invariance within the phase formalism (Pocheau, 1987). Indeed, because of the relative meaning of reference, invariance of any phase equation is required with respect to arbitrary local phase reference changes. In Rayleigh-Benard convection as well as in any other pattern forming system, this cannot be achieved unless gauge fields representing mean flow effects are taken into account in the phase formalism. This statement means that the existence of mean flows (i.e. of a large scale interaction) comes from an invariance principle, regardless of any technical property of the microscopic equations (as the Navier-Stokes for instance).

Because the spatial scales of mean flows and roll flows are quite different, the global mass conservation applies to each of them separately. For the mean flow, this condition requires the introduction of a pressure gradient which gives rise to an additional mean flow suitable for leading to a global divergence free mean flow. A further role of this pressure variable is to enable the mean flow to satisfy the boundary conditions.

In the convective medium, local roll distortions thus create local sources of mean flows by the Siggia-Zippelius mechanism. However, since the actual mean flow is divergent free, one realizes that a mean flow at a given point may well have been created by quite a far source (fig 4). That means that the mean flow field is related to the distortion field in a nonlocal way : the physics of convection is nonlocal. Following these statements, it is meaningless to compute the mean flow field by considering only a part of a pattern : the whole sources together with the full boundary conditions must be simultaneously taken into account to achieve a right determination of the mean flow field.

2) Advection forcing by mean flows

Let us superimpose a mean flow F to a nearly periodic roll flow $v \propto (Ae^{i\phi} + Ae^{-i\phi})$, where A is the roll amplitude, ϕ the phase and $\mathbf{k} = \nabla\phi = O(1)$ the wavevector. The advection term of the Navier-Stokes equation then generates a driving term which is resonnant with the roll flow spatial frequency : $iAe^{i\phi}(\mathrm{F} \cdot \nabla)\phi$. When the amplitude and the phase formalisms are further applied, this resonnant term gives rise to an advection term of the phase by the mean flow. The phase equation is thus modified according to the following transformation :

$$\phi_t \Rightarrow \phi_t + (\mathbf{F} \cdot \nabla)\phi$$

This advection term by the mean flow has been naturally introduced by a number of authors : (Siggia and Zippelius, 1981; Cross, 1983; Manneville and Piquemal, 1983; Dubois-violette et al, 1983; Cross and Newell, 1984; Coullet and Fauve, 1985; Tessauro and Cross, 1986; Pocheau et al, 1987 and lastly Brand, 1987). However, two different contexts have been considered to infer the mean flow effects. In both of them the nonlocal features of mean flow play an essential role. However, in one case, only homogeneous geometries are allowed, so that local effects produced by nonlocal mechanisms cannot be handled. On the contrary, in the second context, the attention is drawn on the occurence of non-homogeneous geometries. Additionnal mechanisms related to the local effects brought about by nonlocal phenomenons are then found. Both these kinds of mean flow effects are addressed in the following section.

III MEAN FLOW EFFECTS

1) Homogeneous geometries

In this first approach, one considers infinite straight roll patterns and one wonders whether mean flows modify their global instabilities. This may indeed occur in the following way : as instabilities involve roll distortion, some mean flows can be generated and then, either enhance or lower the instability mechanism by advecting the phase. New instabilities are hence found (skewed-varicose, oscillatory) (Cross, 1983) and previous ones are quantitatively modified (Eckauss, Zigzag) (Siggia and Zippelius, 1981; Manneville and Piquemal, 1983).

We stress that in such an analysis, the spatial distortion produced by the instability is coherent all along the pattern. For instance, the wavy undulation inherent to the skewed-varicose instability is the same in any part of the pattern and the mean flows spontaneously produced are simply a constant flow. In these global instability studies, each part of the pattern is thus similar to the whole. This means that, at a large scale level, one deals with a homogeneous distortion, a homogeneous state and a homogeneous behavior. Moreover, because of that similarity between each local part of the pattern and the pattern itself, the duality between local and global behaviors cannot be thoroughly handled. In other words, some richness of the nonlocal physics brought about by mean flows escapes this analysis.

2) Inhomogeneous geometries

In order to study the mean flow effects in a somewhat different context, capable of leading to a more developed behavior of distortions, we have chosen to separate the existence of mean flows from that of roll distortions. In other terms, we have studied a forced mean flow effect (Pocheau et al., 1987). Since the amplitude of the mean flows spontaneously generated by the Siggia-Zippelius mechanism is inversely proportional to Pr, we can inhibit the mean flow production by distortion, working with a high Prandtl number fluid (Pr=70). However, since mean flows are no longer produced internally in the convective medium, one has to generate them by an external mean : this is achieved by using a controled thermosyphon capable of producing very small parallel flows and hence of mimicking "natural" mean flows. In doing this, the mean flows become a control parameter, hence giving the opportunity of performing a quantitative analysis. For the sake of simplicity, we worked with an unidimensional roll pattern i.e. with a chain of rolls. To avoid sidewall effects, we closed the chain on itself i.e. we used an annular roll chain. The mean flows enter the convective medium at one filling hole, then split into the two opposite half circle chains and finally go out of the cell at an exhaust hole diametrically opposite to the filling hole (fig 5 a,b). The annulus therefore gives rise to two equivalent roll chains. We stress that these chains are finite so that some boundary conditions are operating on their extremities. We will see further that this is a most important feature in selecting geometrical inhomogeneous solutions of the chain behavior.

Let us now report the pattern behavior for low value of ϵ, i.e. far below the top of the Busse balloon RaB(Pr). When no mean flows cross the rolls, a uniform wavenumber is displayed all along the pattern : no distortion is present and the pattern is stationary (fig 5a). This is consistent with the usual high Prandtl number fluid behavior. However, when a mean flow is superimposed upon the roll flows, the rolls near the filling hole exhibit lower wavenumbers than the rolls near the exhaust hole : a distortion is present together with a wavenumber distribution (fig 5b). Unlike the figure 5b, the pattern is still stationary, however. When the mean flow rate is further increased, the distribution

Fig. 5. A roll chain at high Prandtl number (Pr=70) : a) No mean flow is present ; the wavenumber is homogeneous. b) A mean flow produces phase distortion : large rolls occur near the filling hole and short rolls near the exhaust hole. We even notice a roll pinching at the shortest rolls : the roll chain is dynamical.

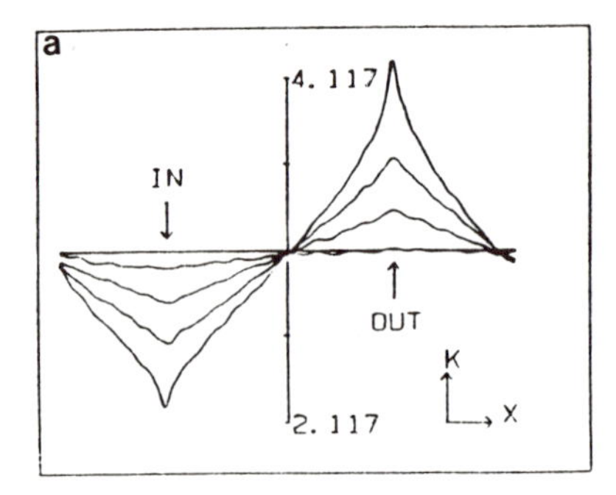

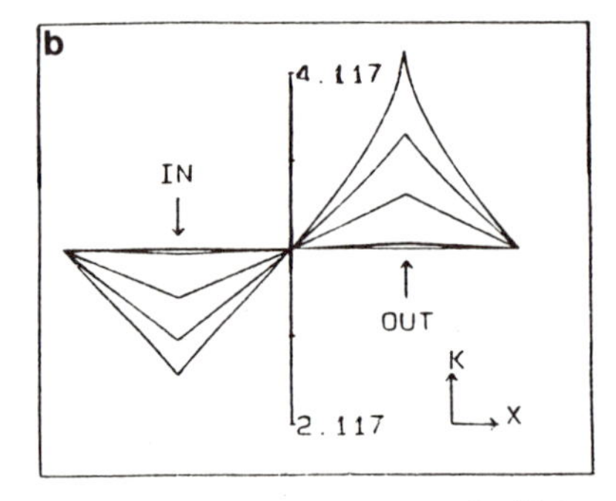

Fig. 6. Plot of the roll wavenumber along the chain for increasing flow rates : the higher the mean flow amplitude, the higher the distortion. a) Experimental measurements b) Solution of the diffusion-advection phase equation.

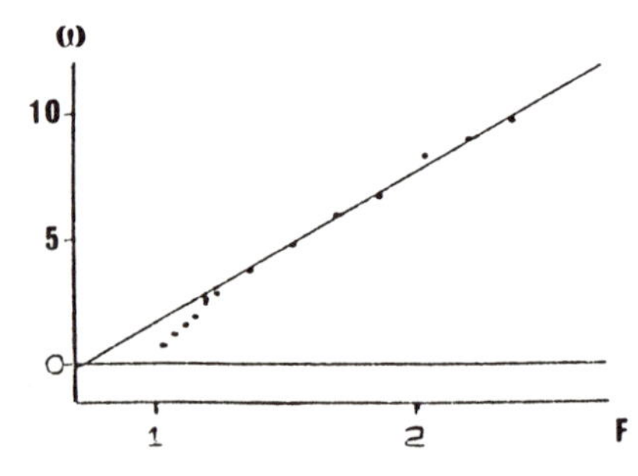

Fig. 7. Plot of the phase speed ω(unity : $2\pi 10^-4Hz$) as a function of the flow rate (unity : 1μm/s). These measurements are in agreement with the linear relation $\phi_t = \tilde{k}A = F - F_c$ (where $F_c = 0.68\mu$m/s), except near F_c where the roll chain succeeds in reaching stationary states by changing its mean wavenumber.

grows, and, at a threshold value F_c dependent on ϵ ($F_c \propto \epsilon^{1/2}$), localized instabilities occur at the rolls having a maximal or a minimal value for their wavenumber : the smaller rolls disappear and the larger ones split. At this time, the roll chain is dynamical (fig 5b).

By the sole adjunction of a mean flow onto a high Prandtl number roll chain, we have thus been able to produce, as close to the onset of convection as desired, roll distortion, localized instabilities and time dependence, i.e. all the features of the spontaneous phase turbulence inherent of low Prandtl numbers. This result is essential because it demonstrates that mean flows are the main difference between the high Prandtl number physics and the low Prandtl number one. It also draws on the ability of mean flows of producing roll distortion, which is indeed a key for understanding phase turbulence.

As the mean flow rate is a control parameter, we may get into a quantitative study of the mean flow effects, first in a stationary state and then in a dynamical one. In figure 6a we report the measurements of a stationary wavenumber distribution along the roll chain for various flow rates (the similar evolutions which arise on both the half circle chains are plotted one beside the other). We notice that the extremal wavenumbers take place at the holes and that the wavenumber distribution increases with the flow rate. To understand these measurements, we have considered, as others did previously in the homogeneous case, the usual phase equation of Pomeau-Manneville (Pomeau and manneville, 1979) completed by an advection term of the phase by the mean flow F:

$$\phi_t + (\mathbf{F}\cdot\nabla)\phi = D_{||}(\phi_x)\ \phi_{xx}$$

Indeed, a lot of solutions of this equation may be found, depending on the boundary conditions. They may be classified into three subsets :

2 a) Homogeneous advection

$\phi_t \neq 0$ and $\phi_{xx} = 0$ i.e. $\phi = (kx+\omega t)$ with k a constant and $\omega = -\mathbf{k}\cdot\mathbf{F}$. The rolls are translated at the speed F : this corresponds to a global advection by the mean flow.

2 b) Stationary distortion

$\phi_t = 0$ and $\phi_{xx} \neq 0$ so that $F\phi_x = D_{||}(\phi_x)\phi_{xx}$. We have solved analytically this equation (Pocheau, 1987), using the rational expression of $D_{||}(k)$ near onset of convection, where **k** is the phase gradient :

$$D_{||}(k) = \frac{\xi_o^2}{\tau_o}\frac{1-3Q^2}{1-Q^2} \quad \text{where } Q = (k-k_c)\epsilon^{-1/2}\xi_o \quad \text{and where } k_c \text{ is the critical wavenumber.}$$

The relation between k and x is

$$(1+2\gamma^2)\frac{\text{Log}(1+\beta Q)}{\beta} - 2\gamma^2\text{Argth}(Q) - \beta\gamma^2\text{Log}(1-Q^2) = \epsilon^{-1/2}\frac{F}{F_o}k_c x$$

$$\text{where } \beta = \frac{\epsilon^{1/2}}{\xi_o k_c} = \frac{1}{Q}\frac{k-k_c}{k_c},\ \gamma^2 = \frac{1}{1-\beta^2} \text{ and } F_o = \frac{\xi_o}{\tau_o}$$

For weak distortions, i.e. for $\beta \lll 1$, it reduces to $3Q - 2\text{Argth}(Q) = \epsilon^{-1/2}\frac{F}{F_o}k_c x$. The former solutions are displayed in the figure 6b for the flow rates corresponding to the experimental study. The agreement is remarkably good, even near the holes where the spatial variation of $D_{||}$ is great, i.e. even in the fully nonlinear domain. We stress that this success provides a clear validation of the phase diffusion-advection equation. We also emphasize the existence of a complete nonlinear analytical solution for this problem.

2 c) Distortion together with advection

Any mixed state between the two previous cases (homogeneous advection and stationary distortion) may indeed be realized. We give below two examples of such states, one displaying a global advection of a distorted phase field and the other showing a phase advection within a stationary distortion.

global advection of a distortion

$\phi(x,t) = \psi(x-At)$ where A is the advection speed.

This state may readily be understood as the result of the Gallilean invariance of the system. It may also be found from the phase equation by writting for any flow field **A** :

$$\phi_t + (\mathbf{A}\cdot\nabla)\phi = D_{||}\phi_{xx} - [(\mathbf{F}-\mathbf{A})\cdot\nabla]\phi$$

Setting each member to 0, one obtains a solution for any A $:\phi(x,t) = \psi(x - At)$ where ψ corresponds to a stationary distorted phase field under the stress of the mean flow $\mathbf{F} - \mathbf{A}$. This solution describes a global advection at speed A of a roll chain distorted by the mean flow F-A. The action of the flow F thus splits into a global advection (A) and a distortion (F-A).

phase advection within a stationary distortion

This state must correspond to the following kind of phase field $\phi : \phi = \tilde{k}(x - At) + \psi(x)$ where $\tilde{k}$ is a constant in order that the wavenumber k is stationary. Let us seek such solutions of the phase equation (Pocheau, 1987). They correspond to stationary distorted ψ-solutions of the following equation

$$(F - A)(\tilde{k} + \psi_x) + A\psi_x = D_{||}(\tilde{k} + \psi_x)\ \psi_{xx}$$

Such solutions indeed exist. In particular, let us take for $\tilde{k}$ the mean mavenumber (so that $\overline{\psi_x} = 0$) and let us perform a spatial average over the chain (denoted by a bar). We obtain $(F - A)\tilde{k} = \overline{D_{||}\ \psi_{xx}}$, so that, "in average", the phase distortion is close to that produced by the mean flow (F-A). Hereto, the action of the mean flow F splits into a phase advection at speed A and a phase distortion under (merely) the mean flow stress (F-A). However, the distortion is stationary there, so that the phase advection is decoupled from the distortion.

In the present experiment, when, by increasing the flow rate $\mathbf{F}$, the stationary distortion becomes wide enough, localized instabilities occur at the extremities of the chain, where the rolls are either too large or too short. The phase adaptation to these repetitive roll destructions or creations produce a roll translation along the cell : each roll is created at the filling hole by the splitting of the largest roll and travels to the exhaust hole where it is destroyed by roll pinching. We stress that, since the chain distortion is stationary, the wavenumber of a roll changes all along its traveling. This dynamical state thus corresponds to a phase advection within a stationary distortion i.e. to the second solution of III]2)c). In addition, we emphasize that, since the unstable wavenumbers are linked to a phase instability, their value is independent of the flow rate. At fixed ϵ, the wavenumber distribution is thus the same for any dynamical state $:\overline{D_{||}\psi_{xx}}$ is invariant among the dynamical states. We determine it by considering the marginaly stable distorted chain ($A = 0$), for which this average phase diffusion just equals the phase advection by the critical flow F_c: $\tilde{k} \cdot F_c = \overline{D_{||}\psi_{xx}}$. For the other dynamical states, the excess of F compared to F_c gives rise to the phase speed A: $\tilde{k}A = F - F_c$. This behavior is indeed recovered by the experimental measurements (Pocheau, 1987) (fig 7). The only difference arises near the threshold $F = F_c$ where the chain succeeds in reaching a stationary or a regime slower than expected by simply changing its mean wavenumber $\tilde{k}$.

Shortly after our experiment, Brand has proposed to extend our analysis (Pocheau et al., 1987) in order to describe mixed advection-distortion states (Brand, 1987). He has then completed the phase diffusion-advection equation by a driving term (i.e. a term independent of ϕ). His equation is written

$$\phi_t + F\phi_x = D_{||}\ \phi_{xx} + \alpha F$$

We stress that such an equation cannot apply here, since it does not satisfy the symmetry $\phi \Rightarrow -\phi$. Indeed, that symmetry is required because, since the phase describes a real variable v (velocity, temperature, etc ...) which is written $v = Ae^{i\phi} + Ae^{-i\phi}$, the opposite phase field $-\phi$ must be a solution of the phase equation whenever ϕ is a solution. In fact, the simple diffusion-advection equation is sufficient to grasp any of the observed behaviors. Indeed any of them correspond to either of the three families of solution III] 2) a), b) and c). The selection of the right solution is linked to the choice of the boundary conditions. For instance, in the previous dynamical state, it came from the fact that the extremal wavenumbers had a constant value.

3) A nonlocal coupling

Mean flows may be produced by distortion. However, mean flows may produce in turn distortion. This gives rise to a coupling between mean flows and distortion, which displays nonlocal features. In the roll chain study, one branch of the coupling (the mean flow generation) was inhibited by taking a high Prandtl number fluid. Nevertheless, by forcing a large mean flow amplitude, we succeeded in triggering permanent dynamics. When, on the opposite, the feedback interaction is operating, one may expect a distortion to be sustained by its own mean flows that it produces. We may even guess that a spontaneous dynamics is likely to be triggered when such self-sustained distortion becomes locally unstable. That process is indeed the fundamental mechanism for the generation of phase turbulence. In the following, we aim to demonstrate it from an analytical analysis of the system describing the nonlocal coupling.

IV ROUTE TO TIME-DEPENDENCE

The nonlocal coupling between the mean flow field and the roll distortion is described by the Cross-Newell equations (Cross and Newell, 1984):

$$\tau[\phi_t + \mathbf{k} \cdot \mathbf{F}] + \nabla(\mathbf{k}B) = 0 + o(1/\Gamma) \tag{1}$$

$$\mathbf{F} = -\gamma \mathbf{k} \nabla(\mathbf{k}A^2) + \nabla\Pi + o(1/\Gamma) \tag{2}$$

where $B(k, Ra, Pr)$ and $\tau(k, Ra, Pr)$ are suitable scalar functions and A the roll amplitude. The phase equation corresponds to the Pomeau-Manneville phase equation written in a rotationally invariant form and completed by an advection term of the phase by the flow $\mathbf{F}$. The second equation expresses in a rotationaly invariant form, the mean flow field produced by the Siggia-Zippelius mechanism. It involves a coupling coefficient γ between the amplitude of the mean flow sources and the distortion. This coefficient is mainly inversely proportional to Pr. The pressure gradient $\nabla\Pi$ enables $\mathbf{F}$ to satisfy the mass conservation and the boundary condition. This condition is similar to that of a parallel flow in a Hele-Shaw cell : $\mathbf{F} \cdot \mathbf{n} = 0$ where $\mathbf{n}$ is the wall normal. Both equations are valid at order $1/\Gamma$, where Γ is the box aspect ratio. The mean flow equation is restricted to the vicinity of the convective threshold, however.

Our goal is to understand within the phase formalism why, at Pr=0.7, the threshold of turbulence of distorted patterns is one order of magnitude lower than that of straight rolls (Pocheau, 1988). The answer is hidden inside the Cross-Newell equations but, unfortunately, these equations are presumably tremendously difficult to solve in a general way. We are thus led to select a generic situation, then to try to solve it in the framework of the Cross-Newell equations and finally to guess to extract from the solution the fundamental mechanisms of time dependence. To achieve this purpose, the chosen situation must be both simple and representative enough to be both tractable and fruitfull. A good candidate stands in the transition to time-dependence in a cylindrical container.

From the experimental observations of that transition, we notice the following features. A strong curvature is generated by the roll tendency of ending perpendiculary to the walls (fig 3a). However, the dangerous mode is compression (fig 3b). Since it has a weak amplitude compared to that of the curvature mode, one might expect that compression is a produce of curvature within the nonlocal coupling between mean flow and distortion. In order to recover this process within the Cross-Newell equations, we focus on a single generic analytical mode of curvature and we determine what mode of compression it generates and how. This leads us to face a quantitative test: is the so found compression sufficiently large to produce a loss of stability for $\epsilon = O(o.1)$ at Pr=0.7 ? Our approach indeed corresponds to the search for an exact solution describing the effects of mean flows, when these flows are spontaneously generated by distortion. It is therefore a generalization of the study of forced mean flow effects. In the course of this derivation, three classical problems of hydrodynamics will be encountered: nonlocality, closing of the scale interactions, boundary conditions. Due to the analytical simplicity of our phase structure, we will be able to solve them here.

1) Construction of an analytical solution

1 a) Analytical modelization of the phase field ϕ

Owing to the symmetries of the stationary pattern which arises in a cylindrical container (figure 3a), we look to a modelization of its phase field ϕ in the shape $\phi(x,y) = k_o\,(1+\Delta)\,y\,[1+\psi(x,y)]$ where the y axis joins the centers of curvature and where ψ is even respectively in x and y (fig 8a). Since the pattern is defectless and since its distortion displays a large scale, polynomial functions are suitable to model it. We next restrict ourselves to a single generic mode of curvature $\psi(x,y) = -ax^2/R^2 + \ldots$ (R being the cell radius) and to the most general polynomial form of compression at the fifth order in ϕ :

$$\phi(x,y) = k_o\,(1+\Delta)\,y\,[1 - ax^2/R^2 + by^2/R^2 + cy^4/R^4 + dx^2y^2/R^4] + o(\frac{1}{R})$$

This modelization leads us to work in a functional space FS=$(y, yx^2, y^3, y^5, y^3x^2)$ with five parameters p=(a,b,c,d,Δ) where the first one, a, governs curvature and where the others describe com-

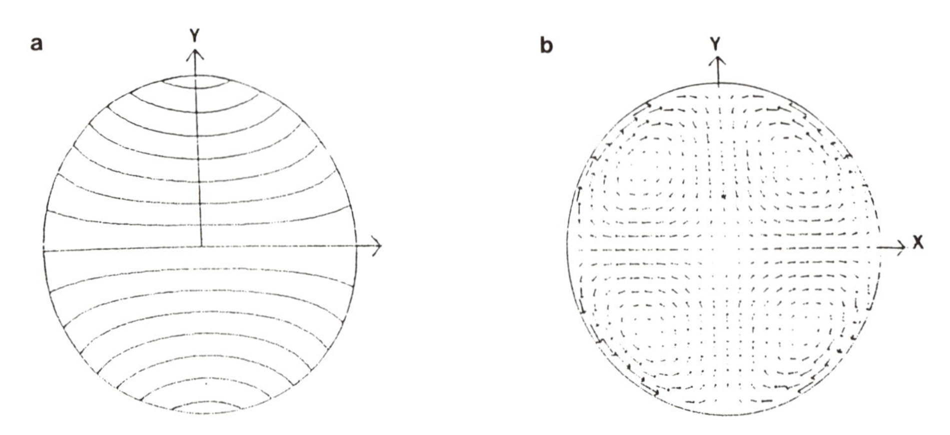

Fig. 8. An exact perturbative solution of the Cross-Newell equations. a) Phase field. Notice the compression as in figure 3a. b) Mean flow field. Notice that the four vortices produce a backflow from each foci to the pattern center.

pression. To solve the CN equations, we perform an expansion with respect to these parameters. We thus assume that the curvature is weak ($a \ll 1$). Moreover, since compression is hardly noticeable on the pattern picture compared to the curvature, we assume that the compression is second order in a. This ansatz may be justified at the course of the derivation.

1 b) Mean flow field

We wish to determine the mean flow field which is generated by the phase field ϕ and which satisfies in addition the boundary condition. We first take into account the divergence free property from the earliest stage, by introducing the stream function ξ ($\mathbf{F} = \left| \begin{array}{c} \xi_y \\ -\xi_x \end{array} \right.$). We next expand the vertical vorticity of $\mathbf{F}$ with respect to the parameter a from equation (2) and we integrate twice to obtain ξ ($\Omega_z = -\Delta\xi$). Three integration constants finally enable us to satisfy the boundary condition, hence leading to :

$$\mathbf{F} = \gamma \frac{k_o^2 A_o^2(k_o)}{3R^4} \left[a^2(1-5p) - 3d(1+p)\right] \left[(R^2 - r^2) \left| \begin{array}{c} x \\ -y \end{array} \right. + 2xy \left| \begin{array}{c} -y \\ x \end{array} \right. \right] + o(a^2/R)$$

We note that whatever ϕ is in the functional space FS, the shape of $\mathbf{F}$ is the same : it is made of four vortices. Since, as shown in the following, $[a^2(1-5p) - 3d(1+p)]$ is positive, the vortices produce a backflow from the centers of curvature to the pattern center (fig 8b). According to the study of forced mean flow effects, we expect this backflow to produce a compression. This is confirmed below by the study of the stationary states.

1 c) Stationary states

We put the determination of $\mathbf{F}$ back in the phase equation (1) and we expand it with respect to a. We emphasize that any mode belongs to FS, so that the phase dynamics is an internal problem of FS. This corresponds to a closure of the expansion and means that FS indeed contains the relevant modes of distortion. The stationarity condition gives rise to the following relation among the parameters :

$$10a^2 + 6d = \alpha[a^2(1-5p) - 3d(1+p)]$$

$$6b - 2a\Delta = -(\alpha/3)[a^2(1-5p) - 3d(1+p)]$$

$$20c = (\alpha/3)[a^2(1-5p) - 3d(1+p)]$$

where $\alpha = -\gamma \left(\frac{kA^2\tau}{dB/dk} \right)_{(k_o)}$ and $p = \left(\frac{k\, dA^2/dk}{A^2} \right)_{(k_0)}$. Close to the convective threshold, these parameters may be determined by perturbative expansion : $\alpha = 4.19\epsilon + O(\epsilon^2)$ and $p = 0.5\epsilon + O(\epsilon^2)$

An extra relation is needed to determine the compression parameters as functions of the curvature parameter a. One may get it by involving the local equilibrium near the foci. As may be seen there, the mean flows are directed along the roll axis (fig 8 a,b). Accordingly, they cannot produce phase advection so that the well known wavenumber selection by focus singularity may be invoked. This leads to $k(0,\pm R) = k_o$ and finaly to the following stationarity criterion :

$$b = -a^2[2\alpha/3]/[2 + \alpha(1+p)]$$

$$c = a^2[\alpha/5]/[2 + \alpha(1+p)]$$

$$d = a^2[1/3][\alpha(1-5p) - 10]/[2 + \alpha(1+p)]$$

$$\Delta = a^2\alpha/[2 + \alpha(1+p)]$$

As assumed previously, the compression parameters are second order in a.

2) Route to time-dependence

In the stationary states, experiments show that the distortion grows with ϵ. We may therefore wonder at what ϵ value, is the distortion so great that a local instability develops. In other words, at what ϵ does the wavenumber distribution cross the BC balloon?

Let us look for the pattern maximum wavenumber. It takes place at the pattern center (0,0) and writes $k_{max} = k_o(1 + a^2\alpha/[2+\alpha(1+p)]$. We recall that, for $\epsilon \succeq 0.1$), rolls are nearly perpendicular to the walls. That means that the parameter a has reached a large value, $a = O(2/3)$, which fits outside the domain of validity of our derivation. Nevertheless, we will assume that the extrapolation to large a gives a valuable insight into what an exact treatment should lead to. Looking to the BC balloon, we notice that for $\epsilon = O(o.1)$ it is crossed when $k > k_c + 0.5$ (fig 2). Within our determination of k_{max} and a, this arises for $1 + \frac{\alpha}{2(2+\alpha)} \approx 1 + \frac{o.5}{3}$ i.e. for $\alpha = O(1)$. Since at Pr=0.7, $\alpha = 4.19\epsilon$, this threshold of time-dependence corresponds to $\epsilon = O(0.2)$. That demonstrates that distortion decreases the threshold of time-dependence by an order of magnitude at least.

3) Mechanisms of the transition

The study of the behavior produced by a given mode of curvature in the CN equations leads us to point out the following nonlocal mechanisms on the route to time-dependence. First of all, the boundary condition is responsible for the roll curvature, by requiring the roll perpendicularity at the walls. Provided that the Prandtl number is finite, this curvature produces mean flows by the Siggia-Zippelius mechanism. The mean flow boundary condition then gives rise to a backflow, which finally enhances the compression and triggers a localized instability. In other terms, the boundary conditions appear to be at the origin of the destabilization via nonlocal phenomenons such as the backflow. We may notice that this process bears some similarity with the destabilization of homogeneous geometries. For instance, in the skewed-varicose instability, curvature produces stabilizing mean flows but, since it allows mean flows to work in compression, it finally gives rise to a net destabilizing effect. A similar feature arises here : the mean flow produced by curvature would be stabilizing in absence of the mean flow boundary condition ; however, due to it, mean flows can work in compression and overdestabilize the pattern.

How do these mechanisms work at other Prandtl numbers ? Let us first apply the same analysis to moderate Prandtl numbers: $Pr \approx 3$. The coupling coefficient γ is then smaller than at $Pr = 0.7$ so that α/ϵ is decreased. As shown in figure 2, there still exists a large wavenumber distribution, but this distribution crosses the BC balloon at a high Ra value i.e. for $Ra \approx RaB(Pr)$. The skewed-varicose instability is then encountered and gives rise to roll pinchings. At high Prandtl numbers, since no mean flows are generated ($\gamma \lll 1$), no wavenumber distribution occurs and selection mechanisms lead the pattern to cross the BC balloon at high Ra values. We thus come up with a comprehensive definition of the Prandtl number regimes :

-low Pr: there is a wavenumber distribution and a low threshold of time-dependence ($Ra \lll RaB(Pr)$)

-moderate Pr: there is a wavenumber distribution and a high threshold of time-dependence ($Ra \approx RaB(Pr)$)

-high Pr: there is no wavenumber distribution and a high threshold of time-dependence ($Ra \approx RaB(Pr)$)

V CONCLUSION

Rayleigh-Benard convection behaves as a system governed by two coupled scales which interact in a non-local way. The coupling coefficient is a function of the control parameter (Pr) and may be varied experimentally in a large range. Convection in extended geometries therefore appear as the simplest problem for studying scales interaction in hydrodynamics. We guess that understanding it might provide valuable tools for addressing the study of systems exhibiting interaction between more than two scales, in open as well as in closed flows.

In this paper, we have reported both experimental results showing the effects of forced mean flows on convective rolls and their thorough interpretation in terms of phase diffusion-advection. We have then focused on the spontaneous coupling which develops between free phase and mean flow fields at a low Prandtl number in a cylindrical container. The pattern behavior is understandable by means of mean flow effects driven by boundary conditions. It is derivable from a perturbative solution of the Cross-Newell equation capable of handling simultaneously large scale modes and local unstable modes. The stability of this solution enables us to understand the transition to time-dependence in terms of local instability modes driven by global non-local modes. The Prandtl number dependence of this mechanism is in agreement with the available experimental results on the transition to time-dependence at low, moderate and high Prandtl numbers.

We expect similar mechanisms to be responsible for the transition to time-dependence in other hydrodynamical cellular patterns. We also guess that this understanding of the onset of time-dependence may provide a good ground for addressing the dynamical regimes i.e. the transition to turbulence. As a first step, we have been able recently to identify the modes of bifurcations of the present spatio-temporal system in terms of large scale instabilities of the large scale fields and then to reconstruct its route to turbulence (Pocheau, 1988).

REFERENCES

Brand, H.R., 1987, Phase dynamics with a material derivative due to a flow field, Phys.Rev. A, 35, 4461

Brand, H.R., 1987, Phase dynamics - A review and a perspective, in "Propagation in system far from equilibrium", J.E.Wesfreid, Ed., Springer Verlag

Busse, F.H., 1978, Non-linear properties of thermal convection, Rep.Prog.Phys., 41, 1929

Busse, F.H. and Clever, R.M., 1979, Instabilities of convection rolls in a fluid of moderate Prandtl number, J.Fluid.Mech., 91, 319

Busse, F.H. and Clever, R.M., 1981, Low-Prandtl number convection in a layer heated from below, J.Fluid.Mech., 102, 61

Busse, F.H. and Whitehead, 1974, J.A., Instabilities of convection rolls in a high Prandtl number fluid, J.Fluid.Mech., 47, 305

Coullet, P. and Fauve S., 1985, Propagative phase dynamics for systems with galilean invariance, Phys.Rev.Lett., 55, 2857

Croquette, V., 1988, Convective pattern dynamics at low Prandtl number, submitted to Contemp.Phys

Cross, M.C., 1983, Phase dynamics of convective rolls, Phys.Rev. A, 27, 490

Cross, M.C. and Newell, A.C., 1984, Convection patterns in large aspect ratio systems, Physica D, 10, 299

Dubois-Violette, E., Guazzeli, E. and Prost, J., 1983, Dislocation motion in layered structure, Phil.Mag. A, 48, 727

Kolodner, P., Walden, R.W., Passner, A. and Surko, C.M., 1986, Rayleigh-Bénard convection in an intermediate-aspect-ratio rectangular container, J.Fluid.Mech., 163, 195

Manneville, P. and Piquemal, J.M., 1983, Zigzag instability and axisymmetric rolls in Rayleigh-Bénard convection : the effects of curvature, Phys.Rev. A, 28, 1774

Pocheau, A. and Croquette, V., 1984, Dislocation motion : a wavenumber selection mechanism in Rayleigh-Bénard convection, J.Phys. (Paris), 45, 35

Pocheau, A., Croquette, V. and Le Gal, P., 1985, Turbulence in a cylindrical container of argon near threshold of convection, Phys.Rev.Lett., 55, 1094

Pocheau, A., Croquette, V., Le Gal, P. and Poitou, C., 1987, Convective pattern deformations under mean flow stress, Europhys.Lett., 3, 915

Pocheau, A., 1987, Structures spatiales et turbulence de phase en convection de Rayleigh-Bénard, Thèse d'Etat

Pocheau, A., 1987, Phase turbulence and mean flow effects in Rayleigh-Bénard convection, in "Propagation in system far from equilibrium", J.E.Wesfreid, Ed., Springer Verlag

Pocheau, A., 1988, Transition à la turbulence des écoulements convectifs, C.R.Acad.Sci. Paris, 306, 331

Pocheau, A., 1988, Transition to turbulence of convective flows in a cylindrical container, J.Phys. (Paris), 49, 1127

Pocheau, A., 1988, Phase dynamics attractors in a cylindrical convective layer, submitted to J.Phys. (Paris)

Pomeau, Y. and Manneville P., 1979, Stability and fluctuations of a spatially periodic convective flow, J.Phys.Lett. (Paris), 40, 609

Siggia, E.D. and Zippelius, A., 1981, Pattern selection in Rayleigh-Bénard convection near threshold, Phys.Rev.Lett., 47, 835

Tessauro, G. and Cross, M.C., 1986, Climbing of dislocations in nonequilibrium patterns, Phys.Rev. A, 34, 1363

EXPERIMENT ON PATTERN EVOLUTION IN THE 2-D MIXING LAYER

F.K. Browand and S. Prost-Domasky

Department of Aerospace Engineering
University of Southern California
Los Angeles, CA 90089-1191

INTRODUCTORY REMARKS

The first detailed, quantitative study of the 2-D mixing layer was completed by Liepmann and Laufer in 1947. They described this technologically important flow in terms of the variation of the mean velocity and various mean fluctuation intensities. The measurements were of high quality, and have scarcely been improved upon in the intervening forty years. Yet they give very little fundamental understanding of the structure of the flow. What has changed within the past twenty years is the increased concern with *process* in turbulent flows. Today turbulent flows are perceived to contain identifiable structure. The *interaction of structure* is the turbulent *process*. Thus *process* attempts to provide a dynamical description of the flow, and is a more ambitious undertaking than a simple description of the *state* of the flow.

The identification of structure in highly turbulent flows first came about by relatively simple visualizations using dye or hydrogen bubbles. These techniques allowed the viewing of selected volumes or areas within the flow in a manner not possible with single point velocity measurements. Spatial patterns and pattern evolution are emphasized in these qualitative visualizations. We are now beginning to ask quantitative questions about pattern evolution. This has only become possible within the past ten years or so, as the result of inexpensive computers having vastly increased storage capacity.

The present experiment is an attempt to describe a high Reynolds number, highly turbulent shear flow from a geometrical point of view, and is an outgrowth of earlier work by Browand and Troutt, (1980, 1985). Here we extend these concepts to include an evolution equation model as has been shown to be effective in the study of the Rayleigh-Benard and Taylor-Couette flows. The companion analytical/numerical effort by Yang, Huerre, and Coullet is presented in the paper immediately following. The long-term goal is to provide a combined mathematical-experimental model which will quantitatively describe the evolution of the large scale structure in the highly turbulent regions of the flow. To date, we concentrate on describing the initial instability region where a small range of scales is present. The comparisons with theory are largely qualitative at this stage.

DESCRIPTION OF THE EXPERIMENTAL APPARATUS

The experiment is performed in a large wind tunnel. A side view of the test section is shown schematically in figure 1. The mixing layer is produced by the merging of two parallel streams at the termination of a splitter plate. Flow speeds are on the order of 5-20 m/sec. The inital thickness of the laminar shear layer, δ_i, is about 2.3 mm. The width of the wind tunnel -- expressed in multiples of this thickness -- is about 400. The Reynolds number, based upon

New Trends in Nonlinear Dynamics and Pattern-Forming Phenomena
Edited by P. Coullet and P. Huerre
Plenum Press, New York, 1990

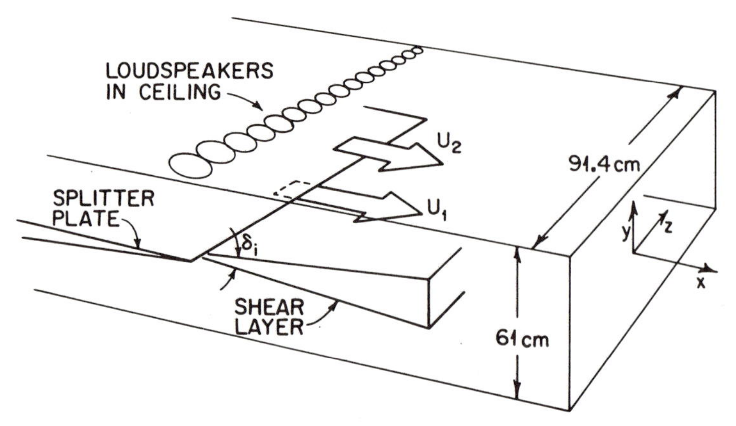

Fig. 1. Sketch of wind tunnel.

velocity difference $\Delta U = U_1 - U_2 = 18.34$ m/sec and the initial shear layer thickness is about 2600. The Reynolds number based upon a local thickness increases roughly linearly with downstream distance. The velocities of the two streams can be varied -- the relevant nondimensional parameter is the speed ratio, $R = (U_1 - U_2)/(U_1 + U_2)$. For many of these experiments, R is approximately 0.65.

The initial laminar flow is convectively unstable -- the most amplified wave is two dimensional, having a wave length of $\lambda = 1.37$ cm and a frequency just above one kilohertz. All waves near the most unstable wave travel with a wave speed very close to the mean flow speed, $\overline{U} = (U_1 + U_2)/2$, so a slightly higher frequency is equivalent to a slightly shorter wave length. Frequency estimates at many stations within the central one-third span -- and within a few wave lengths of the origin -- give an average frequency of $f_i = 1042$ Hz, with a spanwise standard deviation of $\mp$ 4 Hz. This is less than 0.5 per cent variation (just about the limit of our spectral resolution), and is an indication of the uniformity of the flow velocity and boundary layer thickness across this portion of the span. Time series of the longitudinal velocity fluctuation at multiple spanwise positions are made with hot-wires. In some cases, eleven hot-wires are simultaneously monitored; the spanwise extent of the rake is 9.3 λ_i.

The flow can either develop naturally, or can be forced acoustically by means of sixteen small loudspeakers arranged in a row across the span in the ceiling of the wind tunnel. Each speaker is independently driven -- the collection functions as a phased array, producing a small vertical velocity disturbance at the plate trailing edge. This disturbance is sufficient to affect the separation process and impose an initial disturbance field upon the developing vortical flow, Bechert (1983).

RESULTS

Natural Development of Vortex Defects (Unforced)

Figure 2 illustrates several ways the velocity fluctuation data may be displayed. The downstream position is $Rx/\lambda_i = 1.2$, approximately one wave length from the origin of the flow. The array is positioned vertically just above the plane of the plate trailing edge, at $y/\delta = 0.4$. The raw signals are shown in 2(a). In 2(b), a simple waterfall display is obtained by connecting the measured velocities with straight line segments for each time step. The horizontal origin is incremented by Δt, and the next sequence of values are plotted. Continuing the procedure results in the span (vertical axis)-time(horizontal axis) velocity display. The dark bands aligned along the span represent times of most rapid velocity increase (greatest acceleration) -- they give a good indication of the spanwise structure of the flow field. Within about a wave length of the

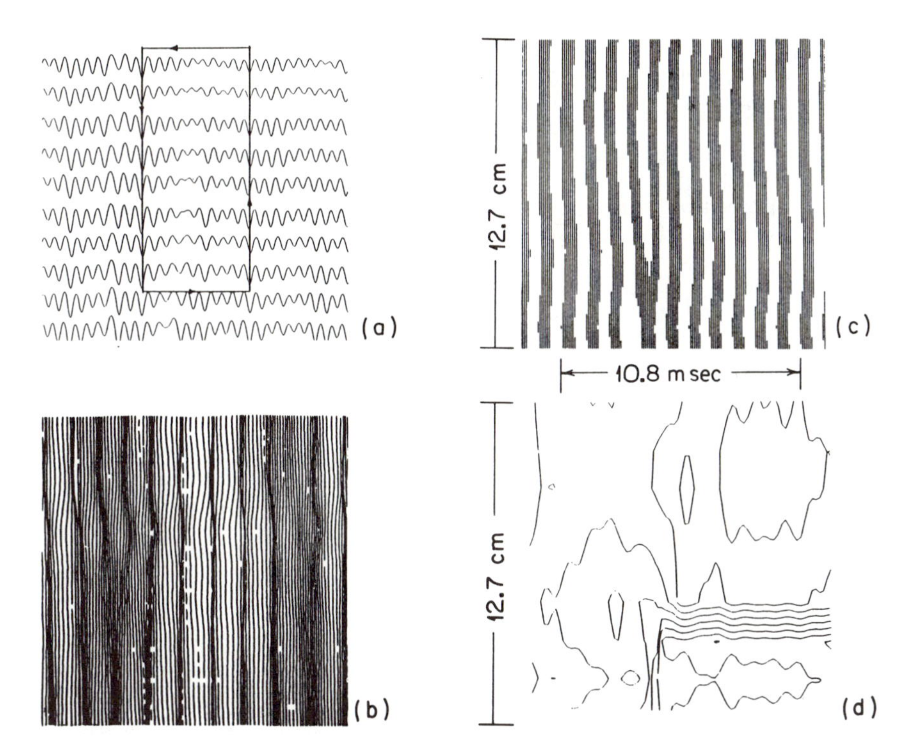

Fig. 2. Various displays of multiple hot-wire velocity signals at $Rx/\lambda_i = 1.2$. (a) raw signals, 11 positions across span; (b) same data, waterfall plot; (c) same data, two-level contour plot, positive and negative signal amplitude with respect to the time average; (d) contours of constant residual phase, where periodic portion of phase has been removed.

flow origin, the nearly periodic pattern represents the amplifying wave motion. At a downstream distance of 2-3 initial wave lengths, this growing disturbance field has concentrated the vorticity into localized vortical regions (or simply vortices). The distinction between nonlinear waves and vortices is not important here. To emphasize the nonlinear nature of the development, the terms vortex and vortex spacing will be used, but these are interchangable with wave and wave length for those preferring wave nomenclature. The most prominent feature is the interconnection of two vortex structures (waves) occurring in the left center (third and fourth dark bands). This same feature is also contained in the boxed region in 2(a). The loss of a vortex crest is evident by counting in the time direction -- first along the lower portion of the contour, and then back along the upper portion. Such defects or dislocations arise early in the development of the flow. Longer time records show the continued production of similar defects across the span of the flow. Figure 2(c) contains the same information, now expressed as a two-level contour plot. The darkened portions represent positive fluctuation, with respect to the time mean at each span position. Finally, 2(d) depicts contours of constant residual phase, θ',

$\theta'(z,t) = \theta(z,t) - \langle f\rangle t$ where

$\theta(z,t)$ = phase at any point in the field, and

$\langle f\rangle = 1/N\ \Sigma\ (f_K)$, $N = 11$

The f_k are the spectral peak frequencies at each position k, for the time interval in question, and $\langle f\rangle$ is the area average frequency. Lines of constant residual phase radiate away from the singularity. The closely spaced parallel lines represent the branch cut -- the thickness of the cut is determined by the hot-wire spacing across the span.

The total number of defects occurring in a long time record has been determined. To express this number, a physically meaningful area must be chosen. The area (span–time area) is taken to be the product $\lambda_i\lambda_i/\bar{U}$. For a time record containing several thousand defects, the number N is $1\text{-}2\times10^{-2}$. Thus in a span of ten vortex spacings (wave lengths), and during the time required for the passage of ten vortices (waves), one to two defects will appear on average. The factor of two difference is the result of choosing different spanwise locations for the hot-wire rake. Defects occur in the positive contours and negative contours (the solid or open regions in 2(c)) with equal probability. The number N was determined at two downstream stations, Rx/λ_i = 1.2 and 1.8, with no significant difference in the result. Thus the initial amplification of small disturbances in the laminar mixing layer rapidly gives rise to a topological structure containing modest numbers of defects. This condition may be compared to the results of evolution equation model calculations by, Coullet et al., (1988), who have used the descriptive term, defect-mediated turbulence. There is, of course, no firm connection between the model and the experiment, but the number N is in rough agreement.

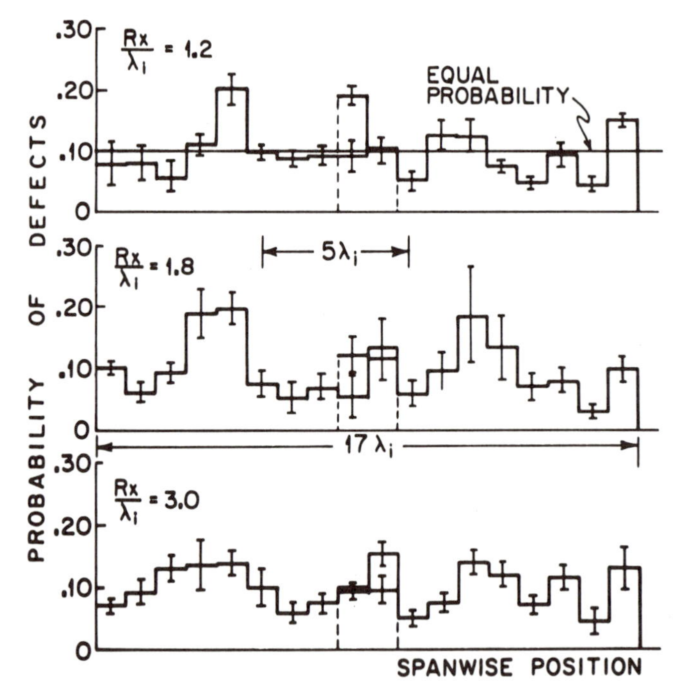

Fig. 3. Probability of defects at different spanwise positions, for three downstream locations. Error bars denote confidence in the estimates.

The probability of finding defects at different spanwise positions is shown in figure 3. The hot-wire rake defines ten bins of width 0.93 λ_i. The rake is then translated to cover a new span segment, slightly overlapping the first segment. A portion of the span of approximately 17 λ_i is covered for three downstream stations, Rx/λ_i = 1.2, 1.8, and 3.0. The plots give probabilities averaged over four separate time intervals. Error bars indicate the estimates of standard deviation. The probability of finding defects is not uniform across the span. The highest and lowest probabilities differ by about a factor of three. The positions of highest probability seem to mark the boundaries of cells having lower probability. Roughly, the cells have width of 4-5 λ_i, and cell boundaries are seen to persist with downstream distance (although the spanwise differences are somewhat diminished at three wave lengths downstream).

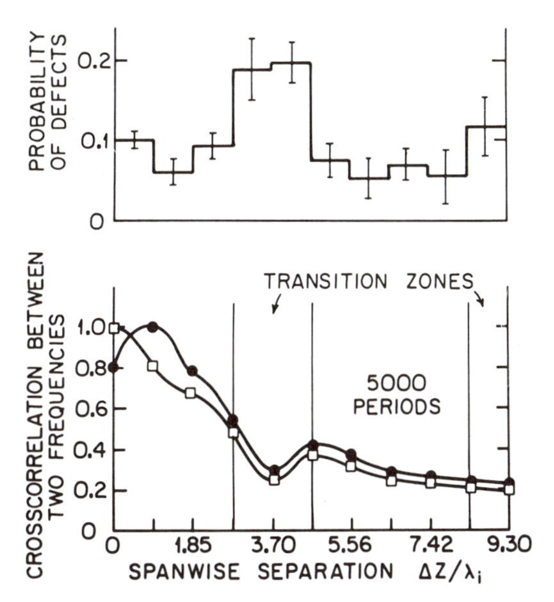

Fig. 4. Frequency correlation estimates across a portion of the span. Open symbols denote correlation of frequency at position 1 with each of the neighboring positions; filled symbols denote correlations based upon position 2.

There is additional evidence for the existence of cell-like regions. Estimates of passage frequency are made at each spanwise station using a (moving) linear fit to the instantaneous phase. The 16 point, least-squares smoothing corresponds to a time interval of approximately 1.3 periods. In figure 4, the upper plot gives the probability distribution -- the left-most rake position at $Rx/\lambda_i = 1.8$. The lower plot shows the result of correlating the frequency estimates at position one with the remaining ten span positions. (The filled circles correspond to cross correlation of frequency at position two with each of the neighboring positions.) Regions of higher-than-average defect probability are here referred to as transition zones, and separate the regions of lower-than-average probability (cells). For a time interval of 5000 periods, the trend shows decreasing correlation with increasing span separation. Correlation values in the adjoining cell are lower, but this result is to be expected since spanwise separation is greater. The special nature of the transition zone is reflected in the anomalous dip in the correlation curve. Choosing a time interval of 500 periods gives a more interesting portrait; figure 5 are the results of cross correlating positions one and seven with neighbors. Now the correlations between neighboring wires within the same cell are more uniformly high, with a rapid change in correlation across the transition zone. The most dramatic results are evident for the two separate 50 period intervals depicted in figure 6. The numbers along the horizontal axis give the average frequency at each span position for each 50 period interval. It is possible now to clearly see the reason for the diminished correlations, and the special nature of the transition zones. At certain times, as shown in the middle plot, the correlations are quite high across the entire region (with the exception of a dip within the transition zone). The average passage frequency is constant across the span. At other times (lowest plot), there is a significant jump in frequency between one cell and the adjoining cell -- in this case a difference of about 38 Hz. Almost perfect correlation exists within the cell, and a discontinuous jump occurs across the transition zone.

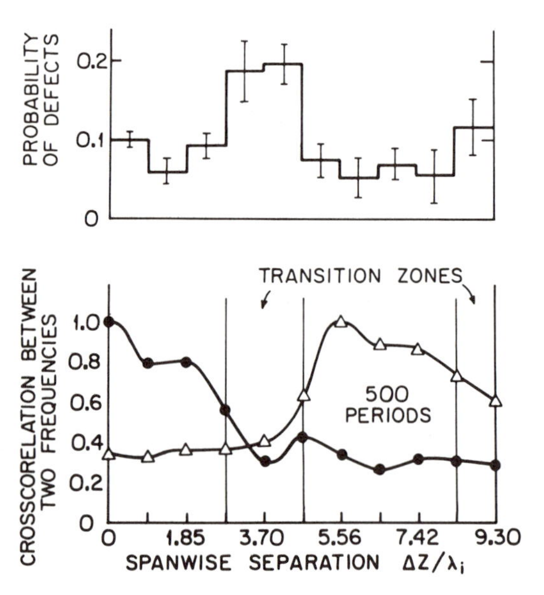

Fig. 5. Frequency correlation estimates. Filled symbols correlate position 1 with neighbors; open symbols correlate position 7 with neighbors.

The differences in frequency which may temporarily exist in different cells provide an explanation for the appearance of defects within the transition regions. Since the vortex structures travel downstream at constant speed (wave speed), the observation of frequency differences is equivalent to differences in spacing (wave length). Figure 7 is a sketch of the structure in two adjoining cells when the spacings (wave lengths) differ by about ten per cent. When the phase differences are small, an accomodating distortion of the vortex (wave front) can be made across the transition zone. But for each ten cycle period, there is an additional vortex (wave front) in the pattern on the right. This must inevitably lead to the production of a defect within the central portion of the pattern.

The reason for the persistence of a higher-than-average number of defects at certain spanwise locations is not known. Evidently, small nonuniformities exist in the oncoming flow, but we are unable to identify the cause. The results previously described are, to a certain extent,

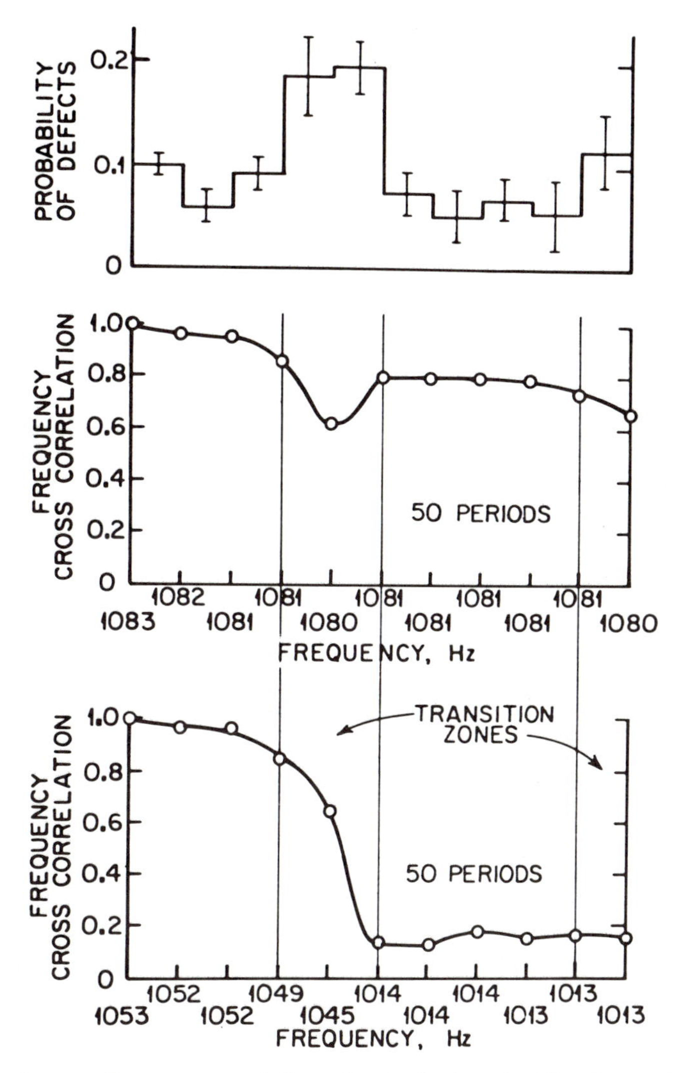

Fig. 6. Frequency correlation estimates for two short time intervals. Average frequency at each span position is shown on abscissa.

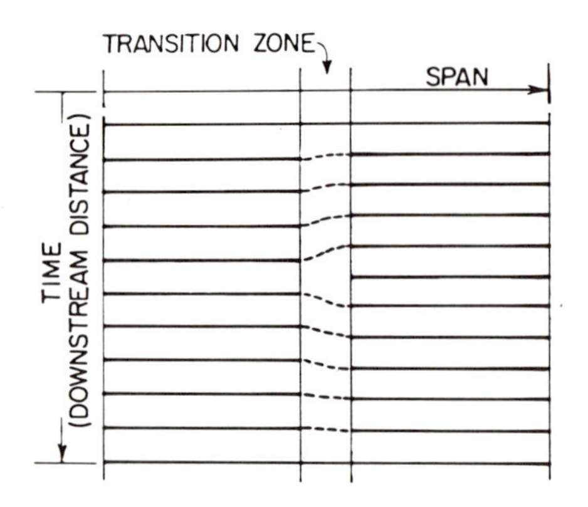

Fig. 7. Different frequencies (wave lengths) in different span cells produces defects.

dependent upon the particular wind tunnel apparatus. However, the flow quality cannot be substantially improved, and small irregularities will always exist. Also, this circumstance will be true in all other facilities. The positions marking transition zones, and perhaps the cell sizes, may differ in different flow facilities, but we are confident that a similar process of defect production would be operative. This belief is supported indirectly by the observations of similar features in the wakes of cylinders. Several of the most recent studies include the work of Gerich and Eckelmann, (1982); Mathis, et al., (1984); Williamson (1988),(1989); and Eisenlohr and Eckelmann (1989).

Acoustic Forcing Produces Artificial Vortex Defects

Naturally occurring defects can never be studied in great detail in our wind tunnel. It is simply impossible to concentrate the appropriate number of sensor probes in the required volume. Additional difficulties arise because defects appear at random positions and times, and no two natural occurrences are exactly alike. We have devised a simple way to produce artificial defects using the loudspeakers in the wind tunnel ceiling. The technique mimics the natural mechanism. Two finite-length trains of square waves are generated by the computer, and have frequencies which differ by about ten per cent. Each pulse train is fed to certain of the loudspeakers to produce spanwise cells of differing frequency (wave length). Figure 8 displays examples of the measured acoustic pressure field at the plate trailing edge as a function of time, using the two contour plotting format of figure 2(c). In 8(a) the central half of the speaker array is driven at the higher of the two frequencies, while each end (one-quarter span) is driven at the lower frequency. The output acoustic field in the central half of the span is observed to contain one additional wave crest in ten. Figure 8(b) is the inverse pattern, and 8(c) consists of a single defect produced by driving each half of the speaker array with one of the two pulse trains. Three measurement microphones (or later, hot-wires) are used to produce these plots. A pulse train is initiated at the instant t = 0, and the signals from the three measurement probes are digitized. The rake is shifted a small distance along the span, and the pulse train is initiated a second time. Figure 8 (and figure 9) are composites of twenty pulses and contain sixty measurement stations across the span. The residual phase contours in 8(d) reflect this increased spatial resolution.

Figure 9 displays the flow response to forcing at four downstream locations, $Rx/\lambda_i = 0.34$, 1.7, 2.4, and 3.4. For this case, R = .46, and λ_i for the forced disturbance is 1.35 cm. (The naturally occurring most unstable wave has λ_i = 1.8 cm.) In 9(a) at one-third wave length downstream, the defect in the velocity field is clearly present. By 1.7 wave lengths, the defect

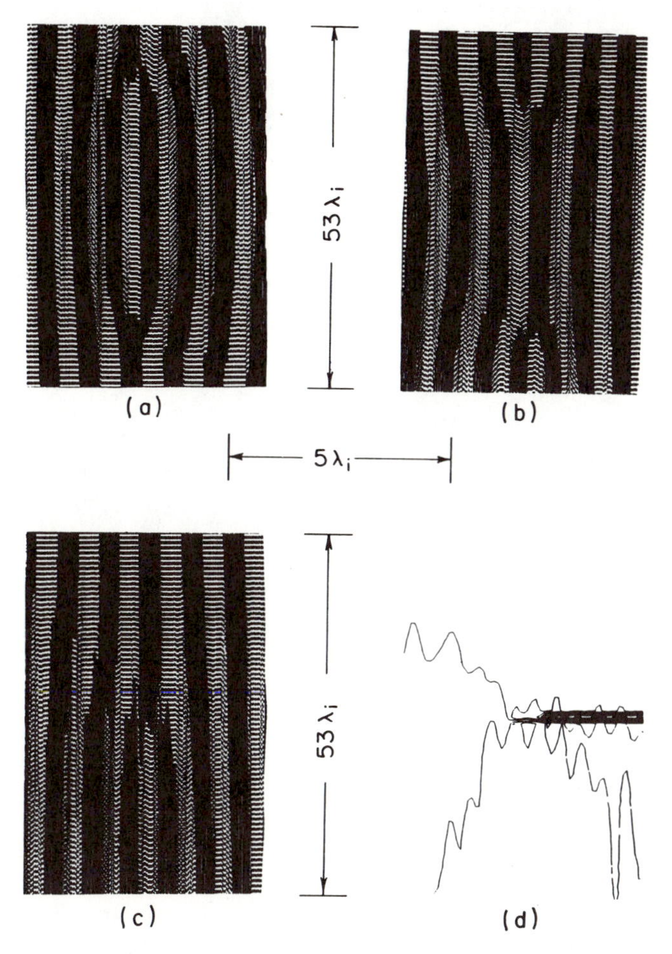

Fig. 8. Acoustic forcing signal at origin of mixing layer. (a), (b), and (c) correspond to different forcing geometries; (d) contours of residual phase for case (c).

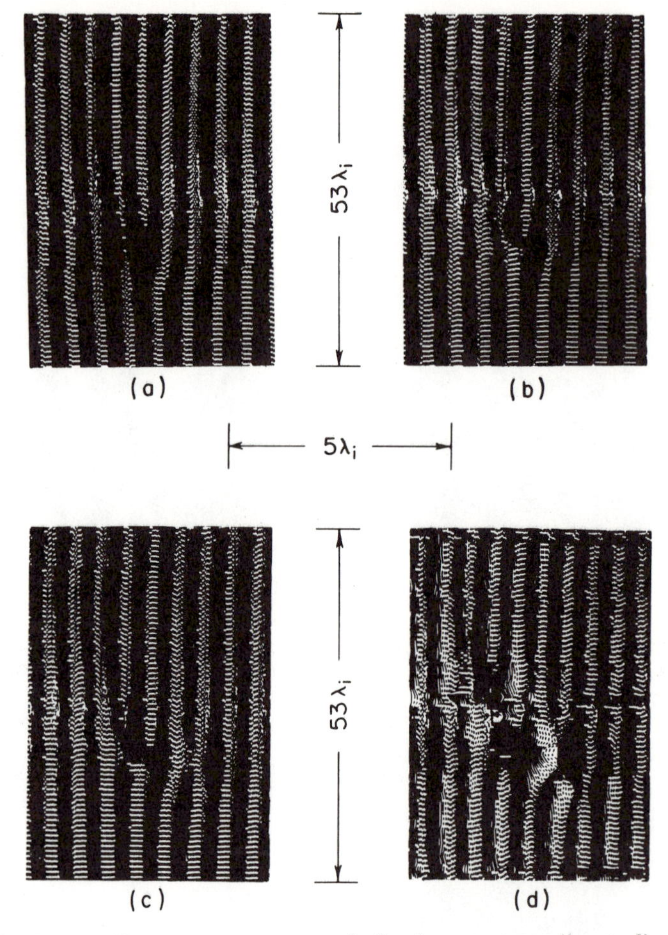

Fig. 9. Flow response to acoustic forcing corresponding to figure 8(c). (a) Rx/λ_i = .34; (b) 1.7; (c) 2.4; (d) 3.4.

region is more concentrated in span and has moved laterally (climbed). At 2.4 wave lengths, figure 9(c), two additional defects have been nucleated at a time just prior to the passage of the original defect. The horizontal time axis can be thought of as downstream distance -- at least for small displacements -- by translating at the mean speed, $\overline{U}$. Increased downstream displacements correspond to events occurring earlier in time. Finally, at 3.4 wave lengths downstream, the two previously nucleated defects have separated, and two additional defects have appeared at a later time (upstream of the initial defect). The influence of the original defect appears to spread laterally across the span as a propagating disturbance field, extending both upstream and downstream from the original defect.

CONCLUDING REMARKS

We have observed that vortex (wave front) defects, or dislocations, are the major topological feature of the spatial instability developing in the 2-D mixing layer. The number of dislocations is

$$N \simeq 1\text{-}2 \times 10^{-2} \quad \text{in} \quad \lambda_i \lambda_i / \overline{U}$$

We have also demonstrated that artificial defects can easily be created and studied. The principal conclusion to date is that single defects nucleate additional defects. Model calculations show qualitatively similar behavior, as indicated in the succeeding paper by Yang, Huerre, and Coullet (1989). In certain model parameter ranges, defects arise spontaneously and nucleate additional dislocations. Since the unstable modes are wave-like, effects propagate in both the spanwise and streamwise directions in qualitative agreement with the experiments.

ACKNOWLEDGMENTS

This research is supported by the Office of Naval Research, Fluid Mechanics Program, under the auspices of the University Research Initiative.

REFERENCES

Bechert, D.W., 1983, A Model of the Excitation of Large Scale Fluctuation in a Shear Layer, AIAA paper no. 83-0724.

Browand, F.K., and Troutt, T.R., 1980, A Note on Spanwise Structure in the Two-Dimensional Mixing Layer, J. Fluid Mech., 93, 325.

Browand, F.K., and Troutt, T.R., 1985, The Turbulent Mixing Layer: Geometry of Large Vortices, J. Fluid Mech., 158, 489.

Coullet, P., Gil, L., and Lega, J. 1988, Defect-Mediated Turbulence, preprint, University of Nice.

Eisenlohr, H., and Eckelmann, H., 1989, Vortex Splitting and Its Consequences in the Vortex Sheet Wake of Cylinders at Low Reynolds Numbers, Phys. Fluids, A, 1, 189.

Gerich, D., and Eckelmann, H., 1982, Influence of End Plates and Free Ends on the Shedding Frequency of Circular Cylinders, J. Fluid Mech., 122, 109.

Liepmann, H.W., and Laufer, J. 1947, Investigations of Free Turbulent Mixing, NACA Technical Note number 1257.

Mathis, C., Provansal, M., and Boyer, L., 1984, The Bénard-von Karman Instability: An Experimental Study Near the Threshold, J. Phys. Paris Lett., 45, 483.

Williamson, C.H.K., 1988, Defining a Universal and Continuous Strouhal-Reynolds Number Relationship for the Laminar Vortex Shedding of a Circular Cylinder, Phys. Fluids, 31, 2742.

Williamson, C.H.K., 1989, Oblique and Parallel Modes of Vortex Shedding in the Wake of a Circular Cylinder at Low Reynolds Numbers, J. Fluid Mech., in press.

Yang, R., Huerre, P., and Coullet, P., 1989, A Two-Dimensional Model of Pattern Evolution in Mixing Layers, this volume.

A TWO DIMENSIONAL MODEL OF PATTERN EVOLUTION IN MIXING LAYERS

R. Yang and P. Huerre

Department of Aerospace Engineering
University of Southern California
Los Angeles, California 90089-1191
USA

P. Coullet

Laboratoire de Physique Théorique
Parc Valrose
Université de Nice
06036 Nice
France

1 INTRODUCTION

The evolution of coherent structures in shear layers and wakes provides a particularly simple example of pattern dynamics in open spatially-developing non-equilibrium systems. As in closed flows such as Rayleigh-Bénard convection, one observes a wealth of possible flow configurations involving dislocations, periodic arrays of vortices with distinct orientation, quasi-two dimensional vortical arrangements, etc. However, in contrast with Rayleigh-Bénard convection, the presence of a basic shear lifts the orientational degeneracy: vortices tend to remain more or less perpendicular to the flow direction.

The geometry of three-dimensional patterns in mixing layers has been extensively studied from an experimental point of view by Browand and his colleagues [see for instance Browand (1986), Browand & Ho (1987) and in this volume, Browand & Prost-Domasky (1989)]. These experiments have clearly and carefully documented the generation of defects in nearly periodic arrangements of spanwise vortices. Recent experiments by Stuber & Gharib (1988) have demonstrated that complex spatio-temporal regimes result from external forcing of dislocations in wakes. Lateral boundaries can also considerably affect the direction of shedding and lead to oblique patterns as shown by Williamson (1989). The present investigation should be viewed as an attempt at a theoretical description of some of these phenomena in the specific context of Browand & Prost-Domasky's observations in mixing layers.

New Trends in Nonlinear Dynamics and Pattern-Forming Phenomena
Edited by P. Coullet and P. Huerre
Plenum Press, New York, 1990

One possible course of action would be to perform direct numerical simulations of three-dimensional mixing layers. The work of Arter & Newell (1988) provides an example of application of this approach in Rayleigh-Bénard convection. We choose instead to study pattern evolution models in two space dimensions, where the cross-stream direction is essentially disregarded. The choice of a suitable model is partially dictated by the symmetry properties of the problem and by physical intuition. This strategy, first suggested by Swift & Hohenberg (1977), has been very successful in accounting for many of the pattern transitions observed in Rayleigh-Bénard convection cells. One crucial advantage is that many of the linear stability properties of periodic arrays of vortices can be obtained explicitly from a simple analysis. Numerical experiments can then be undertaken to explore the fully nonlinear regime in various regions of parameter space [Greenside & Coughran (1986)]. Such models are also very amenable to a phase dynamics description of vortices far away from the onset of the instability. The reader is referred to Cross & Newell (1984) and Newell (1988) for detailed discussions of this approach. It is also possible to characterize, via approximate perturbation procedures, the dynamics of dislocations [Siggia & Zippelius (1981), Pomeau, Zaleski & Manneville (1983), Tesauro & Cross (1986)].

2 PHENOMENOLOGICAL MODEL

The problem of interest is to describe the evolution of Kelvin-Helmholtz vortices in a parallel mixing layer of basic velocity profile $U(y)$ such that $U(y) = -U(-y)$. A sketch of configuration space is given in Fig. 1. The streamwise, cross-stream and spanwise coordinates are denoted by x, y and z respectively, and the velocity field $\vec{v}$ has components u, v and w. This temporally evolving flow should be distinguished from the spatially-developing mixing layer studied in the preceding article of Browand and Prost-Domasky. Many qualitative aspects of the dynamics will nonetheless be reproduced in this idealized setting. Since the primary instability mechanism is inviscid and due to the inflection point of $U(y)$, the boundary layers at the walls are neglected and the horizontal walls are assumed to be stress-free. Under these conditions, the full set of governing equations and boundary conditions are invariant under the following transformations:

space translations	$x \to x + \Sigma_x, \quad z \to z + \Sigma_z$
time translations	$t \to t + \theta$
space reflections	$x \to -x, \quad u \to -u$
	$z \to -z, \quad w \to -w$
Galilean transformations	$x \to x - Ut, \quad z \to z - Wt$
	$\vec{v} \to \vec{v} - U\vec{i} - W\vec{k}.$

Furthermore, we shall postulate that the dispersion relation between wave vector $\vec{K} = \alpha\vec{i} + \beta\vec{k}$ and complex frequency ω is of the form

$$\omega = i[\mu - (1 - \alpha^2)^2 - \gamma\beta^2], \tag{1}$$

where μ is a control parameter and $\gamma > 0$ is the anisotropy parameter which ensures that two-dimensional waves are more amplified than three- dimensional waves. According to Coullet & Fauve (1985), Galilean invariance is responsible for the existence of a marginal phase mode at zero wavenumber which may be coupled with the pattern wavevector $\vec{K}$. It is assumed that this large scale phase mode admits a dispersion relation of the

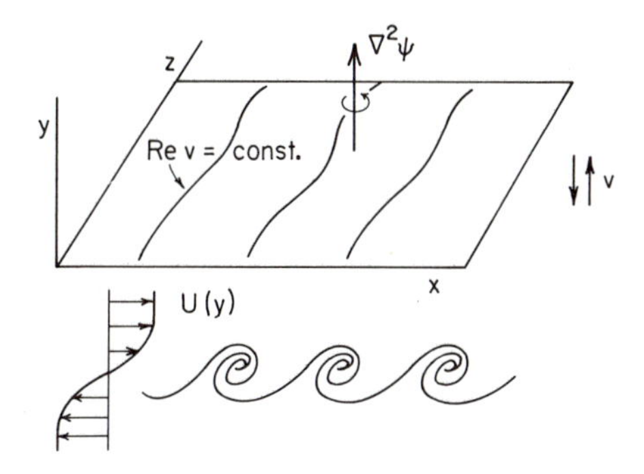

Figure 1. Sketch of mixing layer.

form

$$\omega = -i\nu(\alpha^2 + \beta^2), \tag{2}$$

with ν being a diffusion coefficient.

A phenomenological model satisfying all the above constraints is given by

$$\left[\frac{\partial}{\partial t} + \frac{\partial \psi}{\partial z}\frac{\partial}{\partial x} - \frac{\partial \psi}{\partial x}\frac{\partial}{\partial z}\right] v = \mu v - \left(\frac{\partial^2}{\partial x^2} + 1\right)^2 v + \gamma \frac{\partial^2 v}{\partial z^2} - |v|^2 v \tag{3}$$

$$\left[\frac{\partial}{\partial t} - \nu \nabla^2\right] \nabla^2 \psi = \sigma \frac{\partial^2 |v|^2}{\partial x \partial z}, \tag{4}$$

σ being a real positive coupling coefficient. The complex dependent variable $v(x, z, t)$ should be interpreted as a measure of the strength of the vortices, for instance the vertical fluctuating velocity. The linear evolution of this field is governed by (1). The real stream function $\psi(x, z, t)$ is associated with large scale motions in the $x - z$ plane, its linear properties being given by (2). Equation (4) essentially states that spatial inhomogeneities in the strength $|v|$ of the vortices induce a large scale vertical vorticity $(-\nabla^2\psi)$. In turn the field v is advected by ψ, as represented by the convective derivative in (3).

It is worth mentioning, as an additional justification for our choice of model,that the amplitude evolution equations governing the dynamics near onset at $\mu = 0$ are identical to those pertaining to stratified mixing layers near the critical Richardson number [Huerre (1987)].

Numerical solutions of the above model are obtained by making use of a pseudo-spectral method with periodic boundary conditions in both x and z directions. The pattern evolution is confined within a rectangular domain of size $L_x = 2\pi/q_x$ and $L_z = 2\pi/q_z$, for different initial conditions at $t = 0$.

This numerical study is guided by relying on analytical results which can readily be obtained from the model. The basic state $v = 0$ becomes linearly unstable within the domain $\mu > (\alpha^2 - 1)^2 - \gamma\beta^2$. The corresponding neutral stability boundary in (α, μ) space is the heavy solid line on Fig. 2. Within the unstable region, there exists finite amplitude stationary solutions of the form

$$v = Ae^{i(\alpha x + \beta z)}, \qquad \psi = \text{const.}, \tag{5}$$

with amplitude given by

$$A^2 = \mu - (\alpha^2 - 1)^2 - \gamma\beta^2. \tag{6}$$

These perfect patterns correspond to a periodic distribution of straight vortices. They are themselves subjected to two classes of secondary phase instabilities. The classical Eckhaus instability involves streamwise modulations of the periodic patterns and it prevails in the hatched region on Fig. 2. The skew-varicose instability corresponds to streamwise and spanwise deformations of perfect patterns and occurs within the dotted region on Fig. 2. This leaves a narrow band of wavenumbers within which the finite-amplitude solutions (5) are stable and observable in a domain with no wall effects.

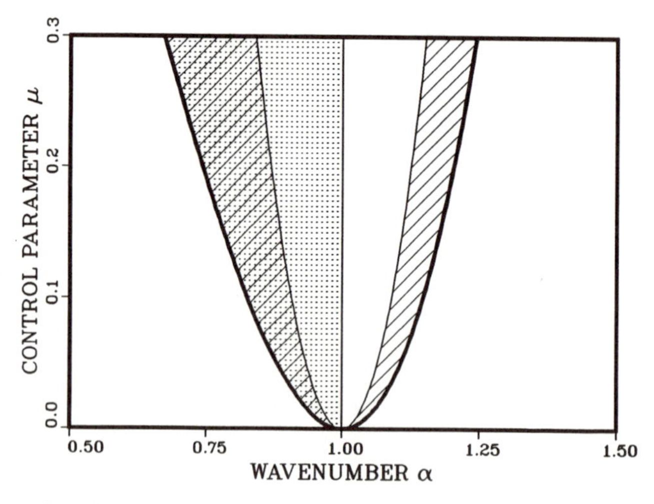

Figure 2. Neutral stability boundaries in (α, μ) space. Thick neutral curve refers to the instability onset of the state $v = \psi = 0$. Hatched region refers to the Eckhaus instability of periodic solutions at finite α and $\beta = 0$. Dotted region refers to the skew-varicose instability of periodic solutions at finite $\alpha < 0$ and $\beta = 0$. The clear region delimits the band of stable periodic solutions.

3 PATTERN EVOLUTION UNDER "NATURAL" CONDITIONS

To simulate the broadband fluctuations present in any real experiment we first explore the dynamics resulting from a random input. The initial state is taken to have a broadband spectrum in both wavevector components α and β, the number of Fourier modes in each spatial direction being $N_x = 64$ and $N_z = 64$. The box wavevector has components $q_x = 0.106$ and $q_z = 0.10$ and the parameters of the model are set at the values $\mu = 1, \gamma = 1, \nu = 0$ and $\sigma = 0.1$. Initial fluctuation levels are high, of the

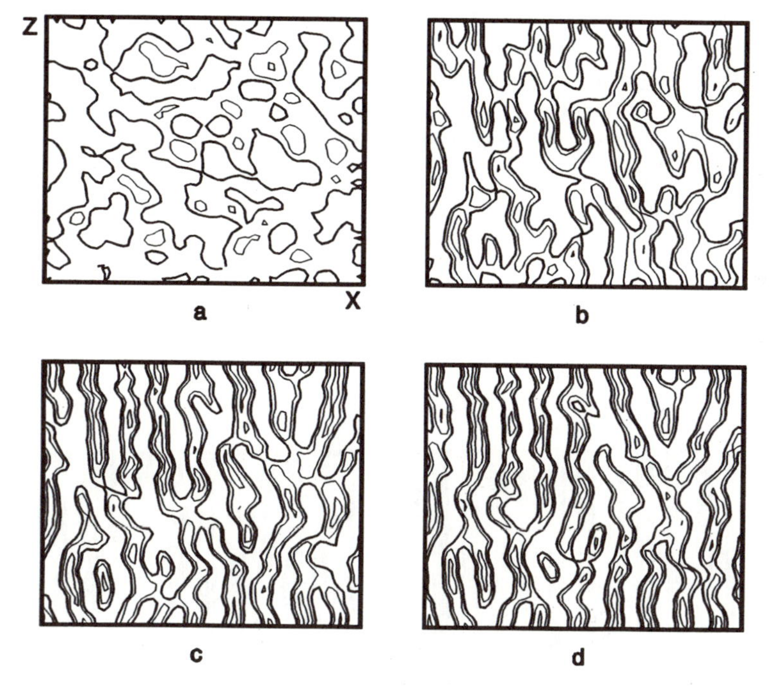

Figure 3. Pattern evolution under "natural" conditions. $\mu = 1$, $\gamma = 1$, $\nu = 0$, $\sigma = 0.1$, $q_x = 0.106$, $q_z = 0.10$. (a) $t = 0$; (b) $t = 2$; (c) $t = 4$; (d) $t = 7$.

order of 0.5. The evolution of the pattern is followed from snapshot pictures of the contours of constant *Rev* as displayed in Fig. 3. The phenomenological model is seen to act as a filter, a quasi-two-dimensional pattern gradually emerging from the random input as the simulation progresses. This behavior is not altogether unexpected since the dispersion relation (1) ensures that streamwise wavevectors are more amplified than their oblique counterparts. We note, however, that regions of slowly varying wavevector fail to completely dominate everywhere in space. Dislocations appear within the flow, in qualitative agreement with the experiments of Browand & Prost-Domasky.

4 NUCLEATION OF DISLOCATIONS BY PHASE PERTURBATIONS

In the previous numerical experiment, dislocations arose from a random initial state with no discernable periodic pattern. It is also possible to observe the spontaneous nucleation of defects when a low level of noise is added to an initially periodic array of structures. In the present case, the control parameters are chosen to be $\mu = 0.3$, $\gamma = 1$, $\nu = 0.05$, $\sigma = 5$, with box wavenumbers $q_x = q_z = 0.08$. The initial state is composed of a slightly perturbed finite-amplitude solution at $\alpha = 0.8$, $\beta = 0$, lying within the skew-varicose instability region. The initial noise amplitude is 25% of the strength of the primary array. The simulation is carried out with 64×64 Fourier components, 192 of which are unstable skew-varicose modes. The dynamics therefore involve multiple interactions between a large number of unstable disturbances superposed on the primary structures. As seen from Fig. 4, one may distinguish a regime dominated by spatio-temporal modulations of the phase of the structures which is followed, for $t > 36$, by the nucleation of dislocations within the computational domain. The first defect appears at about $t = 36$ and multiple dislocations are generated rapidly thereafter. The skew-varicose instability has given rise to a "defect-mediated" turbulent regime of the kind described in Coullet, Gil & Lega (1988). Whereas in this earlier investigation, defect nucleation ultimately resulted from the Benjamin-Feir-Eckhaus instability, here it is associated with the skew-varicose instability.

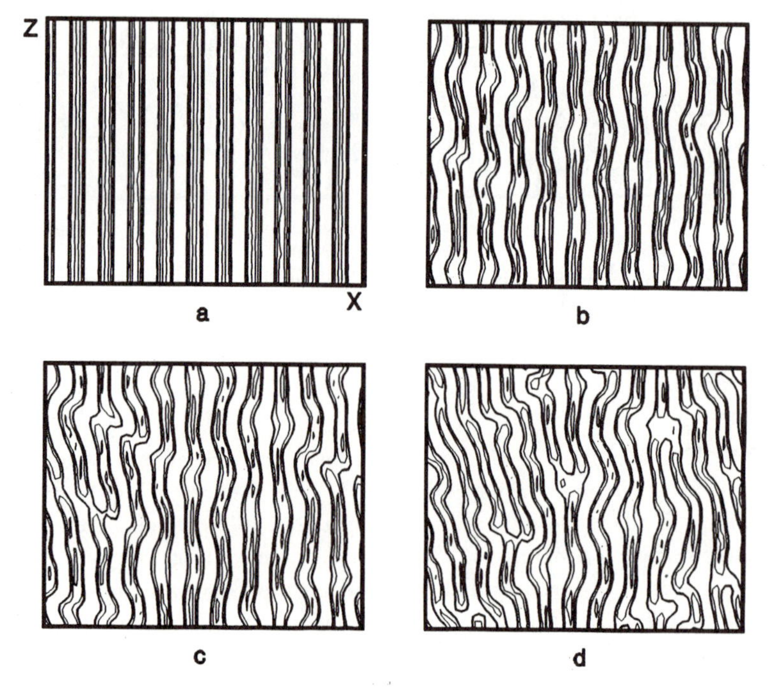

Figure 4. Nucleation of dislocations by phase perturbations. $\mu = 0.3$, $\gamma = 1$, $\nu = 0.05$, $\sigma = 5$, $q_x = q_z = 0.08$. (a) $t = 1$; (b) $t = 33$; (c) $t = 36$; (d) $t = 40$.

The spatio-temporal evolution of the vortices results from the interaction between a modulated field of the form

$$v = A(x,z,t)e^{i\theta(x,z,t)}e^{i\alpha x}$$

and of a large scale field $\psi(x,z,t)$. The spatial distributions of the modulation amplitude $A(x,z,t)$, phase $\theta(x,z,t)$ and of the large-scale horizontal stream function $\psi(x,z,t)$ are illustrated on Fig. 5 at time $t = 37$. Dislocations are seen to coincide with phase singularities of $\theta(x,z,t)$, pairs of dislocations being connected by branch cuts of θ. At these same locations, the amplitude $A(x,z,t)$, reaches zero. It is important to note that the small-scale patterns associated with $v(x,z,t)$ coexist with a horizontal large-scale flow $\psi(x,z,t)$ composed of counter-rotating recirculating regions of vertical vorticity (in the y direction).

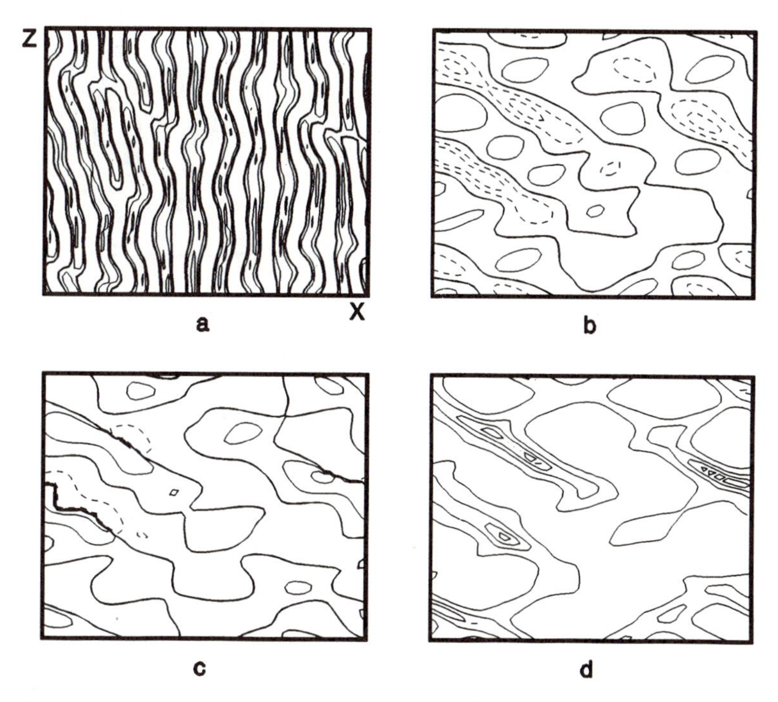

Figure 5. Same conditions as in Fig. 4, $t = 37$. (a) *Rev* contours; (b) Streamlines of large-scale field ψ; (c) lines of constant phase θ; (d) Contours of constant amplitude A.

5 NUCLEATION OF DISLOCATIONS BY DISLOCATIONS

In our last example, the parameters of the model are set at the same values as in the previous section ($\mu = 0.3, \gamma = 1, \nu = 0.05, \sigma = 5, q_x = q_z = 0.08$). As illustrated in Fig. 6, the initial state consists of a periodic array of finite-amplitude vortices with wavenumber $\alpha_1 = 0.8$ in the center of the computational box and with wave number $\alpha_2 = 0.88$ elsewhere. At time $t = 16$, two additional dislocations have been generated and this process of dislocation nucleation continues beyond the last configuration of Fig. 6d. Defect nucleation takes place more rapidly than in Fig. 4, where the first dislocation pair only appears around $t = 37$. We note that the generation of defects by defects is consistent with the experimental observations of Browand & Prost-Domasky (see Fig. 9 of the preceding paper). There are however noticeable differences: in the experiments, secondary defects are not generated symmetrically with respect to the primary one. This might be due to the fact that only one dislocation was introduced in the flow whereas a dislocation pair was initially produced in the numerical simulations.

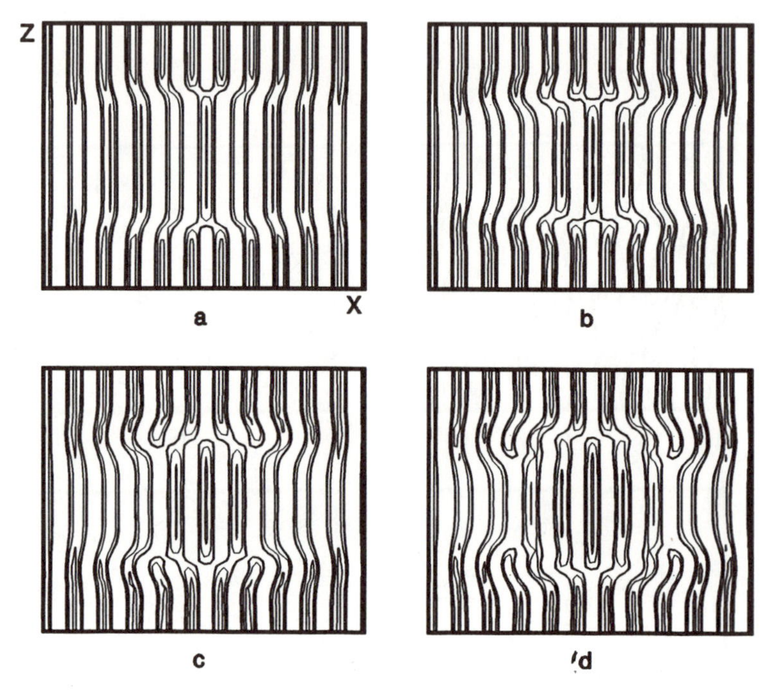

Figure 6. Nucleation of dislocations by dislocations. $\mu = 0.3$, $\gamma = 1$, $\nu = 0.05$, $\sigma = 5$, $q_x = q_z = 0.08$. (a) $t = 4$; (b) $t = 14$; (c) $t = 16$; (d) $t = 20$.

6 CONCLUDING REMARKS

The non-potential pattern evolution model specified by equations (3) and (4) has been shown to reproduce certain of the features observed in mixing layers. If the initial state is chosen to be turbulent, one observes experimentally a downstream evolution towards an ordered quasi-two-dimensional pattern. Imperfections in the pattern lead to the formation of dislocations. The model results reveal a similar behavior: random input gives rise to organized spanwise vortices in the long-time evolution, with the presence of dislocations.

When a mixing layer is acoustically perturbed at slightly different frequencies along the span, controlled dislocations can be produced experimentally, which generate additional defects further downstream. Dislocation pairs are also produced by suitably choosing the initial state in the model. The results of numerical simulations indicate that new defects are then nucleated.

The last type of input, which has so far been analysed on the model, involves noisy random infinitesimal perturbations superimposed on a perfect two-dimensional array of vortices. Remarkably, dislocations are found to be spontaneously generated by the flow as a result of intrinsic phase instabilities. This result may provide an important clue to the mechanism of dislocation nucleation in shear flows.

ACKNOWLEDGMENTS

We have benefitted from many stimulating discussions with F. K. Browand, L. G. Redekopp, and D. Walgraef. This research is supported by a University Research Initiative of the Office of Naval Research under contract # N0001486-K-0679 and by ATT under contract # ATT-Redekopp.

References

Arter, W.D., & Newell, A.C., 1988, Numerical simulation of Rayleigh-Bénard convection in shallow tanks, Phys. Fluids, 31:2474.

Browand, F.K., 1986, The structure of the turbulent mixing layer, Physica, 18D:135.

Browand, F.K., & Ho, C.M., 1987, Forced, unbounded shear flows, Nuclear Physics B (Proc. Suppl.), 2:139.

Browand, F.K., & Prost-Domasky, S., 1989, Experiments on pattern evolution in the 2-D mixing layer, this volume.

Coullet, P. & Fauve, S., 1985, Propagative phase dynamics for systems with Galilean invariance, Phys. Rev. Lett., 55:2857.

Coullet, P. Gil, L., & Lega, J., 1988, Defect mediated turbulence, preprint, University of Nice.

Cross, M.C., & Newell, A.C., 1984, Convection patterns in large aspect ratio systems, Physica, 10D:299.

Greenside, H.S., & Coughran, W.M., 1984, Nonlinear pattern formation near the onset of Rayleigh-Bénard convection, Phys. Rev., A30:398.

Huerre, P. 1987, Evolution of coherent structures in shear flows: a phase dynamics approach., Nuclear Physics, B (Proc. Suppl.), 2:159.

Newell, A.C., 1988, The dynamics of patterns: a survey, in: "Propagation in systems far from equilibrium", J.E. Wesfreid, H.R. Brand, P. Manneville, G. Albinet, & N. Boccara, eds., Springer, Berlin.

Pomeau, Y., Zaleski, S., & Manneville, P., 1983, Dislocation motion in cellular structures, Phys. Rev., A27:2710.

Siggia, E.D., & Zippelius, A., 1981, Dynamics of defects in Rayleigh-Bénard convection, Phys. Rev., A24:1036.

Stuber, K., & Gharib, M., 1988, Transition from order to chaos in the wake of an airfoil, preprint, University of California at San Diego.

Swift, J., & Hohenberg, P.C., 1977, Hydrodynamic fluctuations at the convective instability, Phys. Rev., A15:319.

Tesauro, G., & Cross, M.C., 1986, Climbing of dislocations in non-equilibrium patterns, Phys. Rev., A34:1363.

Williamson, C.H.K., 1989, Oblique and parallel modes of vortex shedding in the wake of a circular cylinder at low Reynolds numbers, submitted to J. Fluid Mech.

SOME STATISTICAL PROPERTIES OF DEFECTS IN COMPLEX GINZBURG-LANDAU EQUATIONS

J.-L. Meunier

Laboratoire de Physique théorique
Parc Valrose 06034
Nice Cedex, France

The following is devoted to a part of the systematic study of defects properties, in particular in the Complex Ginzburg-Landau Equation, which have been developped in Nice [1]. The numerical work has been performed on the Cray II of the Ecole Polytechnique. For details, see J.Lega contribution to these procedings.

Let us first recall the C.G.L. equation :

$$\partial_t A(x,t) = A(x,t) + (1+i\alpha)\Delta A - (1+i\beta)A|A|^2 \tag{1}$$

Where A is a complex field in two dimensions, which is supposed to describe some physical quantity (chemical concentration for exemple).

When $1+\alpha\beta$ is positive, we are in the stable phase regime, i.e. the equation admits a stable homogenious solution :

$$A_h(x,t) = \exp(-i\beta t) \tag{2}$$

else, this solution is unstable, and the linear analysis provides us the number of unstable modes :

$$N_0 = -\frac{2\pi(1+\alpha\beta)}{\alpha^2(1+\beta^2)} \tag{3}$$

which gives the caracteristic length in the sample :

$$\lambda c = \alpha\sqrt{\frac{-(1+\beta^2)}{2(\alpha\beta+1)}} \tag{4}$$

In all the following, for a sake of brievety, we shall describe an intermediate zone of parameters defined by $\beta > 1$, $\alpha = -2$ and focus ourself on a very specific trend of the field, namely the defects statistical properties .

We first verifiy that, in the unstable phase regime of the equation, the field correlation :

$$C(r) = \mathrm{Re}(\langle A(r)A^*(0)\rangle / \langle |A(0)|^2\rangle) \tag{5}$$

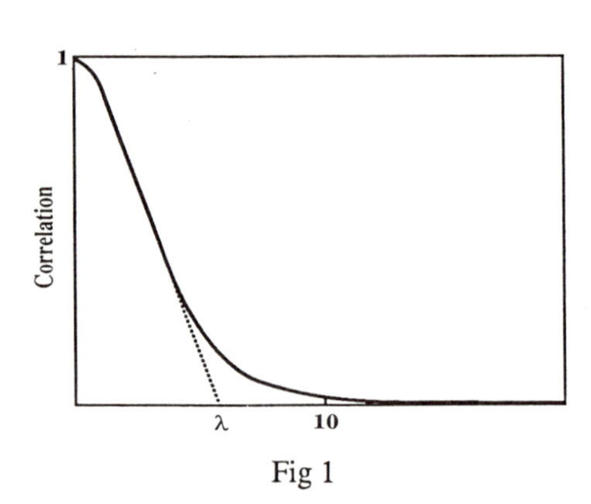

Fig 1

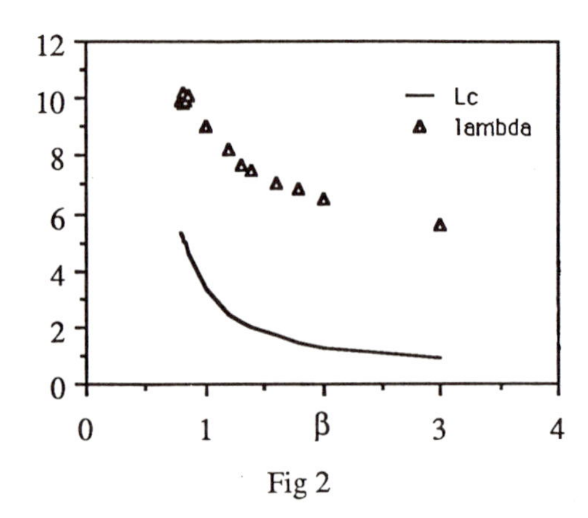

Fig 2

decrease exponentialy Fig (1). This means that there is a well defined correlation length, λ, in the sample which roughly behaves like the caracteristic length (dimentionnality) while somewhat greater Fig (2).

This correlation length is in general much less than the size of the sample (50 units of length). This first observation indicates that, in the unstable phase regime, the C.G.L. equation develops a spatial disorder which can be temptatively studied in the numerical sample we have used.

Let us now concentrate on our subject, namely the defects statistical properties of the C.G.L equation.

The G.L. equation (α=β=0) can develop, as stable solutions in two dimensions, some topological defects (vortices) caracterised by :

(6)
$$A(r=r_{def}) = 0.$$
$$\Delta\phi = \pm\, 2\pi$$

where $\Delta\phi$ is the phase circulation of the field around the defect position. In this case, the size of the defect is of the order of 1 units of length, which is the length over which the G.L. amplitude relaxes to the homogeneous solution.

In the unstable region of the C.G.L. equation, we observed the presence in the sample of several moving defects, appearing and desappearing by pairs, and living isolated one from the others, during a non negligeable amount of time. Figures (3) and (4) show the spatial structure of the field and the time dependance of the number of defects in the box.

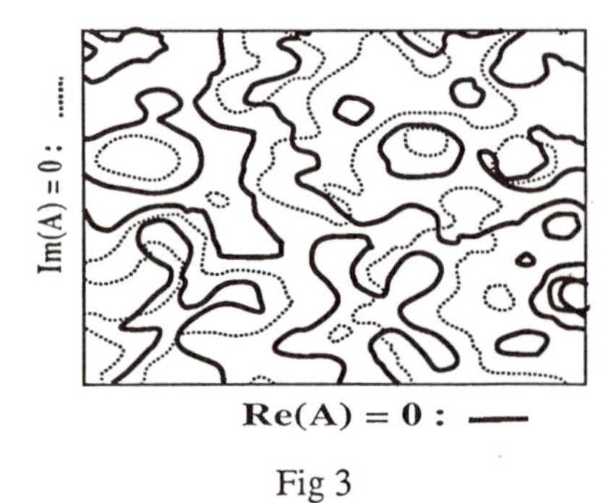

Fig 3

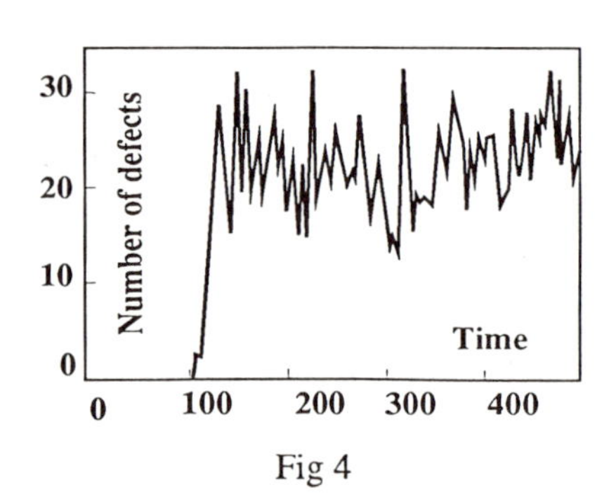

Fig 4

The defects are created by the occurence, at a given place in the sample, of a strong phase gradient which sudently relaxes by creating a pair of defects. As there is a relatively low number of defects in the box (25 in average in a surface of 2500 units square) the creation rate, a_+ must be rather independant of the number of defects in the box :

(7) $$a_+ = c_+$$

Now, the discociated defects (antidefects) travel independently in the box (periodic boundary conditions) and annihilate when they meet another antidefect (defect). Thus the anihilation rate, a_- must be essentialy proportional to the number n of pairs in the sample :

(8) $$a_- = c_- n^2$$

This can be temptatively compared to an electron anti-electron gaz in a very hot plasma.

Using equations (7) and (8) one can show that the probability of having n pairs of defects in the box is :

(9) $$P_n = k \frac{\gamma^n}{(n!)^2}$$

where $k = 1/I_0(2\sqrt{\gamma})$ is the normalisation constant (I_0 is the usual modified Bessel function), and $\gamma = c_+/c_-$ is a parameter which can be fited on the mean value of the number of pairs of defects in the box :

(10) $$<n> = \sqrt{\gamma}$$

Now, the unambigous prediction of the model is that $\sigma^2 = <n^2>-<n>^2 = <n>/2$ for the pairs of defects, that is

(11) $\Sigma^2 = <N>$ for the defects.

These predictions are compared with numerical results in figures (5) and (6) and are in very good agreement with the Rehberg and Steinberg's experimental results [2] (see De la Torre Juarez's contribution to these proceedings)

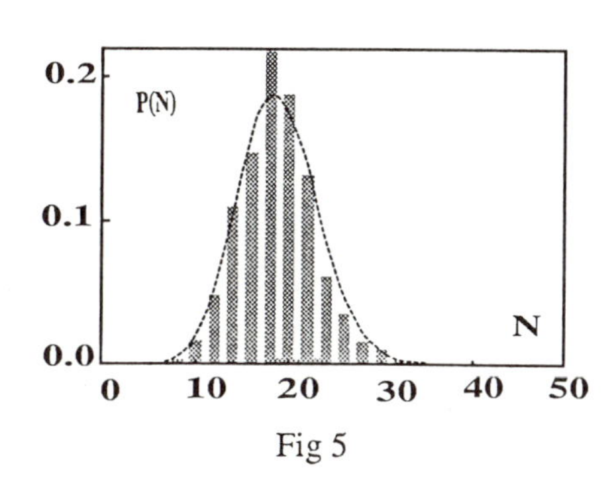

Fig 5

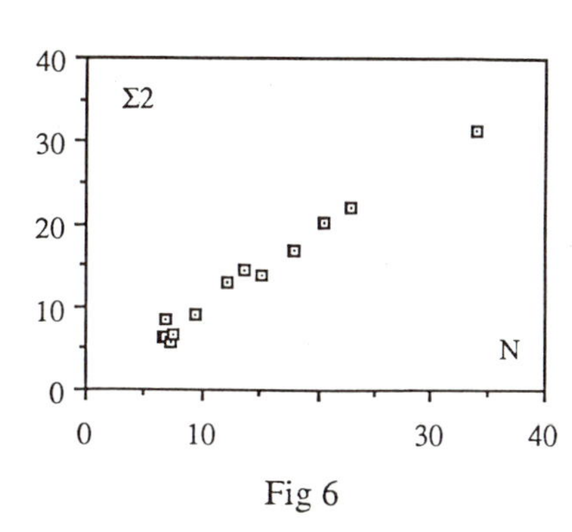

Fig 6

References

1 P. Coullet, L. Gil, J. Lega," defect-mediated turbulence" submited to Phys-Rev Lett. & L. Gil, J. Lega and J.-L. Meunier," Statistical .." submited to Phys-Rev. Lett.

2 Rehberg and Steinberg, "Defect-mediated turbulence in traveling convection paterns.." submited to Phys. Rev. Lett.

VORTEX DYNAMICS IN A COUPLED MAP LATTICE

Tomas Bohr and Anders W. Pedersen

The Niels Bohr Institute
Blegdamsvej 17
DK-2100 Copenhagen

Mogens H. Jensen

NORDITA
Blegdamsvej 17
DK-2100 Copenhagen

David A. Rand

Mathematics Institute
University of Warwick
Coventry CV4 7AL

Abstract

We present a new method for investigating the behaviour of partial differential equations, specifically the complex Ginzburg-Landau equation, by approximating them as systems of coupled map lattices. The method is very efficient and well suited to an investigation of possible universal results about the phase diagram and the transition to turbulence. Preliminary results on vortex structure, dynamics and occurrence are given and we note the existence of turbulent states well below the linear stability threshold from which we argue that one must investigate the dynamics in the regime of very low vortex density in order to gain useful insight into the onset of turbulence.

Introduction

In many interesting nonequilibrium systems it is possible to have a transition into a temporally periodic, spatially uniform state. In the time dependent state there will usually be strong fluctuations tending to diminish the spatio-temporal coherence and possibly even drive the system into turbulent motion. Often the situation close to the transition is well-discribed by an *order parameter,* which, in the spirit of Landau theory, can vary both spatially and temporally and which represents the strength of the critical mode. The order parameter satisfies an *amplitude equation* [1,2] whose form is insensitive to the precise dynamics of the problem, but determined only by the dimensionality and symmetries of the order parameter and of the surrounding medium.

In this work we shall study the case of a two component order parameter in a two-dimensional isotropic medium. Here one gets very interesting behaviour caused by *topological defects* in the form of *vortices* like the ones that drive the so-called Kosterlitz-Thouless phase transition in the planar XY-model and in superfluid He-films [3].

The prototypical amplitude equation for this situation is the complex Ginzburg-Landau equation [1,2]

New Trends in Nonlinear Dynamics and Pattern-Forming Phenomena
Edited by P. Coullet and P. Huerre
Plenum Press, New York, 1990

$$\dot{A} = \mu A - (1+i\alpha)|A|^2 A + (1+i\beta)\nabla^2 A \quad (1)$$

where the order parameter, A, is a complex field and μ, α and β are real numbers. The parameter μ is the usual Landau coefficient: Negative μ implies a quiescent state (A=0) whereas positive μ gives nonzero values to the order parameter. In fact there is a homogeneous solution

$$A = \sqrt{\mu} e^{-i\alpha\mu t} \quad (2)$$

for positive μ and it is seen that the frequency of this periodic state is $\omega = \alpha\mu$. The parameter β is related to the presence of anisotropy of the diffusivity in the complex A-plane (the local configuration space). It is important to note that both α and β introduce a preferred sense of rotation in the complex A-plane so we anticipate that the relation between their signs will play a significant role. On the other hand, complex conjugation of (1) leads to an equation for A^* of the same form, but with opposite signs of α and β, and therefore the simultaneous change of sign of these two parameters doesn't affect the dynamics. In the following we can thus, without loss of generality, restrict our attention to the case $\beta \leq 0$.

A simple physical realization of the above system is provided by a class of chemical reactions between two species in a shallow dish [2]. Here the parameter α determines (with μ) the period of the state after the Hopf bifurcation and β is proportional to the difference between the diffusivities of the two species. Another interesting example is provided by Rayleigh-Benard convection in binary mixtures, although the appearence of running modes complicates the situation [4].

The linear stability of the homogeneous state (2) can be investigated by standard techniques [1]. If we look at weak perturbations of (2) in the form

$$A = (\sqrt{\mu} + \rho(x,y,t)) e^{i(-\alpha\mu t + \phi(x,y,t))} \quad (3)$$

and treat (1) to linear order in ρ and ϕ we find that ρ and ϕ will decay exponentially as long as $1 + \alpha\beta > 0$, whereas long wavelenth modes ($|\vec{k}| < k_c \propto |1+\alpha\beta|$) will be exponentially enhanced if $1 + \alpha\beta < 0$. In particular, if α and β have the same sign, (2) is always linearly stable. For later convenience we shall denote the lowest unstable value of α for a given β as α_0 (thus $\alpha_0(\beta) = -1/\beta$ for (1)).

It has been noted in several recent studies that the loss of coherence and the transition to turbulence in this system is closely linked to the appearence of *vortices* [2,5,6]. Vortices are defined in this model, like in the planar XY-model, as singularities of the angle field. The total variation of the angle over a closed loop, i.e. $\Delta\phi = \oint d\phi$, doesn't necessarily vanish if the loop encloses vortex centers. Instead $\Delta\phi = 2\pi n$, where the integer n is the total vorticity of the region enclosed. In the vortex center the angle is not defined so for the order parameter itself ($A = re^{i\phi}$) to remain well-defined requires that it (i.e. r) vanishes in the center.

If the system is followed above the linear instability threshold, an initially almost uniform state will eventually - through the creation of vortex-antivortex pairs - evolve into a seemingly turbulent state [2,5,6]. In ref. [6], where equation (1) was solved numerically, it was argued that the turbulent state is characterized by a certain *density* of vortices. The vortex density was determined for different parameters and it was found to exhibit a jump as the linear instability point α_0 was crossed. This jump, however, does not necessarily imply any fundamentally singular behaviour of the system at α_0. It is due to the special (smooth) initial conditions chosen.

In fact, what we find in the present work is that the *vortex state* - which can be reached by applying *random* initial conditions instead of smooth deformations of the uniform state - is turbulent already below the linear stability threshold. Our simulations show turbulent states for α-values well below α_0 and even for β=0, where $\alpha_0 = \infty$, and thus we do not believe that the primary onset of turbulence (for random initial conditions) has anything to do with linear instability of the uniform state or any other simple state. To understand the onset of turbulence one must follow the vortex state all the way down into the regime where the vortices form a dilute gas.

The Coupled Map Lattice

In order to simplify our understanding of the system and to save computer time we have replaced the PDE (1) by a coupled map lattice. It can be viewed as a rough approximation to the complex Ginzburg-Landau equation (which can be made exact by taking certain limits of the parameters) or as an interesting dynamical system in its own right, which we hope to be in the same universality class.

We split the coupled map lattice into two parts: a *local* map $A' = F(A)$ representing the two first terms of (1) and a *nonlocal* part representing the complex heat equation which results from omitting the local terms. The latter part has the solution

$$A(t+\tau_0) = e^{\tau_0(1+i\beta)\nabla^2} A(t)$$

On the lattice we approximate the Laplacian by an average ΔA over neighbors. Thus on a two-dimensional hexagonal lattice $(x,y)=(i+j/2, \sqrt{3}/2 j)$ with $i,j = 1,\dots,N$ we take

$$\Delta A(i,j) = \frac{2}{3} \sum_{i',j'} A(i',j') - A(i,j)$$

where the sum is over the six nearest neighbors (i',j'). The nonlocal map is then given by $\tilde{A} = (1 + \frac{\tau_0}{M}(1+i\beta)\Delta)^M A$. Here M is an integer that determines the range of the effective interaction. The limit $M \rightarrow \infty$ reproduces the exponential above (except, of course, that Δ and ∇^2 are not the same). We take M somewhere between 1 and 5, large enough to ensure that short wavelength instabilities do not occur.

The properties of the local map F are very simple. In contrast to most of the literature on coupled map systems they are completely non-chaotic. If we look at (1) without the last term it can be written as

$$\dot{r} = \mu r - r^3 \tag{4a}$$

$$\dot{\phi} = -\alpha r^2 \tag{4b}$$

The general structure of this can be easily reproduced by maps

$$r_{n+1} = f(r_n) \tag{5a}$$

and

$$\phi_{n+1} = \phi_n - \tau\alpha r_n^2 \tag{5b}$$

where the map f has an unstable fixed point in 0 and a stable one in $r=\sqrt{\mu}$. Specifically, one can integrate (4a) as

$$r(t+\tau) = \frac{\sqrt{\mu}\, r(t)}{\sqrt{\lambda\mu + (1-\lambda) r(t)^2}} \tag{6}$$

where $\lambda = e^{-2\mu\tau}$, which fixes the map f. The full map lattice can now be written

$$A_{n+1}(\vec{r}) = F(\tilde{A}_n(\vec{r})) \tag{7}$$

and its properties closely resemble equation (1). We have mostly worked with periodic boundary conditions, but in special cases like a single vortex (see below), we have chosen "free boundary" conditions meaning that terms in $\Delta A\,(i,j)$ extending outside the boundary are omitted. Most of our simulations have been done with lattice size N=50 with occasional excursions up to N=100.

One can again ask for linear stability of the homogeneously rotating state $A_n = \sqrt{\mu} e^{-i\alpha\mu\tau n}$ and the resulting criterion (replacing $1+\alpha\beta>0$) turns out to be

$$1+2\frac{\mu}{1-s}\tau\alpha\beta > 0 \tag{8}$$

where s is the stability parameter for the stable fixed point of f, i.e. $s=f'(\sqrt{\mu})$. For the map (6) $s=e^{-2\mu\tau}$, and it is seen that the new stability criterion approaches the old one as $\tau\to 0$. Again, instability occurs at α_0, where now $\alpha_0=\frac{1-s}{2\mu\tau|\beta|}$.

Results of simulations

As in the XY-model, vortices come in two different species: vortex and antivortex, whose structure are shown in fig.1. For these configurations $\Delta\phi=\pm 2\pi$. The evolution of a single vortex is shown on fig.2 (note that free boundary conditions are chosen in this case). When α and β have oppopsite signs it develops into a spiral, which is rotating more or less uniformly. At larger values of α one sees strong depressions in the modulus $|A|$, in fact the vortex begins to look like a *target pattern* periodically emitting spherical waves [2]. It should be noted that, contrary to other discriptions of target patterns, we do not put in any inhomogeneity at the center. The pattern arises simply from the vortex initial condition and thus could also arise spontaneously.

Pairs of opposite vortices attract. At low values of α they approach each other and annihilate (if they are not too far apart). The annihilation process has (at least for positive α) an interesting aftermath in the from of a "splash" accompanied by a phase-shift: The collision of the two vortices creates a little, growing island in which (roughly) all vectors have been rotated some angle with respect to the bulk. The island keeps growing until it covers the whole lattice.

On increasing α an interesting phenomenon occurs, which we call *entanglement.* Vortex pairs far enough from each other will approach each other for some time, after which they get stuck. As shown in fig.3 they seem to get entangled in each others' spiral arms. If we start them off further apart they also end up further apart, so there is no definite final distance characterizing the situation. One might suspect that this behaviour is due to pinning from the lattice, but we have convinced ourselves that this not the case by varying the parameters μ, which changes the size of the vortex core.

The phenomenon of entanglement shows clearly the difference between the equilibrium XY-model and a non-equilibrium theory like (1). In the XY-model one can integrate out all small fluctuations in the field variable, which leaves a gas of vortices interacting by logarithmic pair-potentials (Coulomb gas [7]). Now, if entanglement occurs, a similar procedure will certainly be very hard to carry out for the nonequilibrium system: Vortex motion isn't simply determined by the coordinates of the vortices, one needs to know the configuration of the spiral arms as well, so in a sense vortex dynamics becomes "history dependent".

When α is increased, the entangled state can break loose and a seemingly turbulent state occurs. This happens below the linear instability threshold. As an example, we have performed simulations with $\tau_0=0.2$, $\beta=-1$, $\tau=1$, $\mu=0.2$. From (8) we find $\alpha_0=0.82$ and we see turbulence from $\alpha\approx 0.70$.

To study the "vortex state" we have made a series of runs with random initial conditions. Here turbulence is seen at much lower values of α - with the parameters above from at least around $\alpha=0.3$. The density of vortices fluctuates a lot both in time and with respect to different initial conditions and it is therefore hard to determine exactly where it vanishes. From our preliminary results we believe that the vortex density (defined as an ensemble average) is finite for *any positive* α (for $\beta<0$). This is in accordance with extrapolations of our simulations of vortex-antivortex dynamics: For any, positive, α, vortex pairs far enough from each other seem *not* to annihilate, but get entangled. Further work is clearly needed to clarify the situation and a quantitative investigation of the turbulent states is currently being undertaken.

Discussion

We have presented a new method and a preliminary study of a two-dimensional system of coupled simple maps, which allows us to simulate votex motion in a system with a Hopf bifurcation into a coherent periodic state. We have stressed the importance in trying to understand the dynamics of the "vortex state" as opposed to focussing on the linear instability point.

In studies of this kind one has to keep in mind that the discretization leading to the coupled map lattice can create several problems. Firstly, complete *isotropy* is lost on the lattice. We have tried to

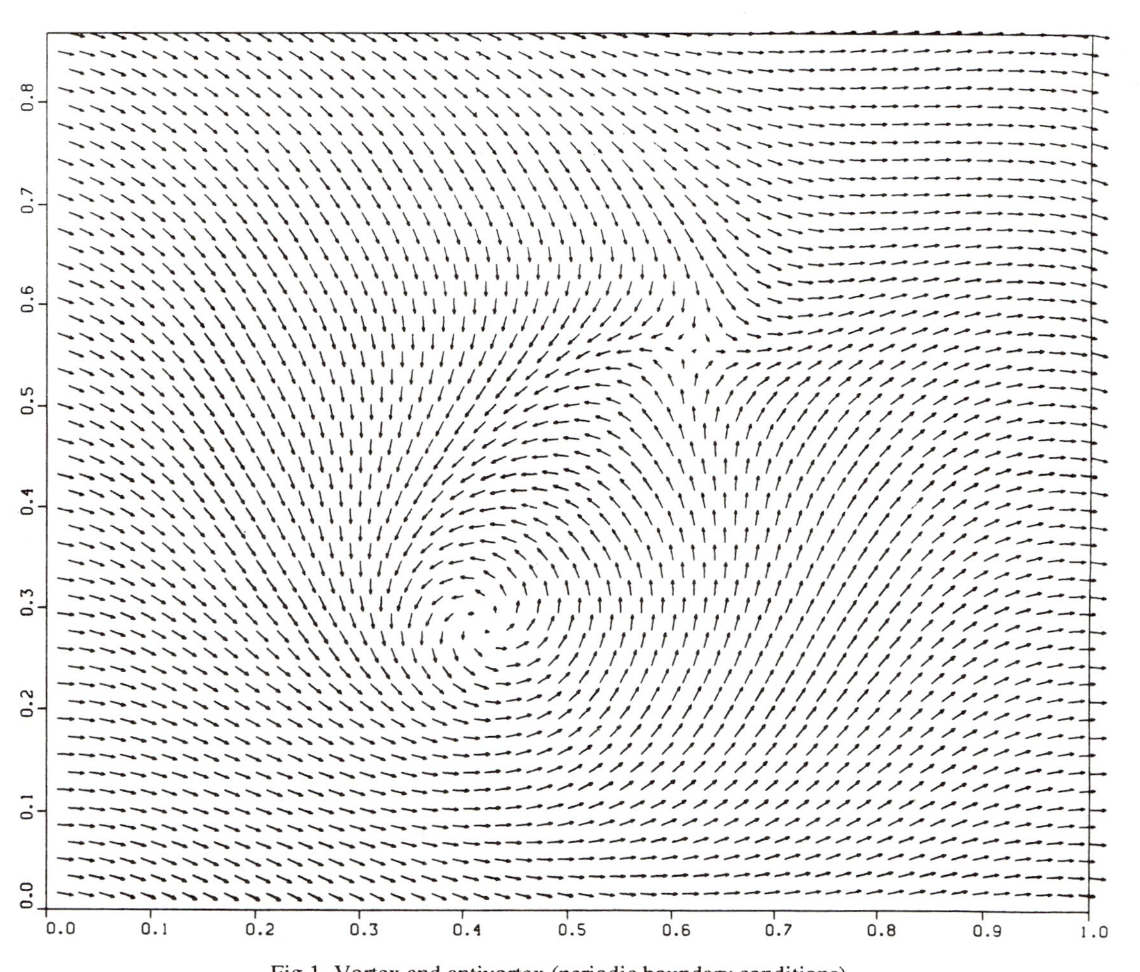

Fig.1. Vortex and antivortex (periodic boundary conditions).

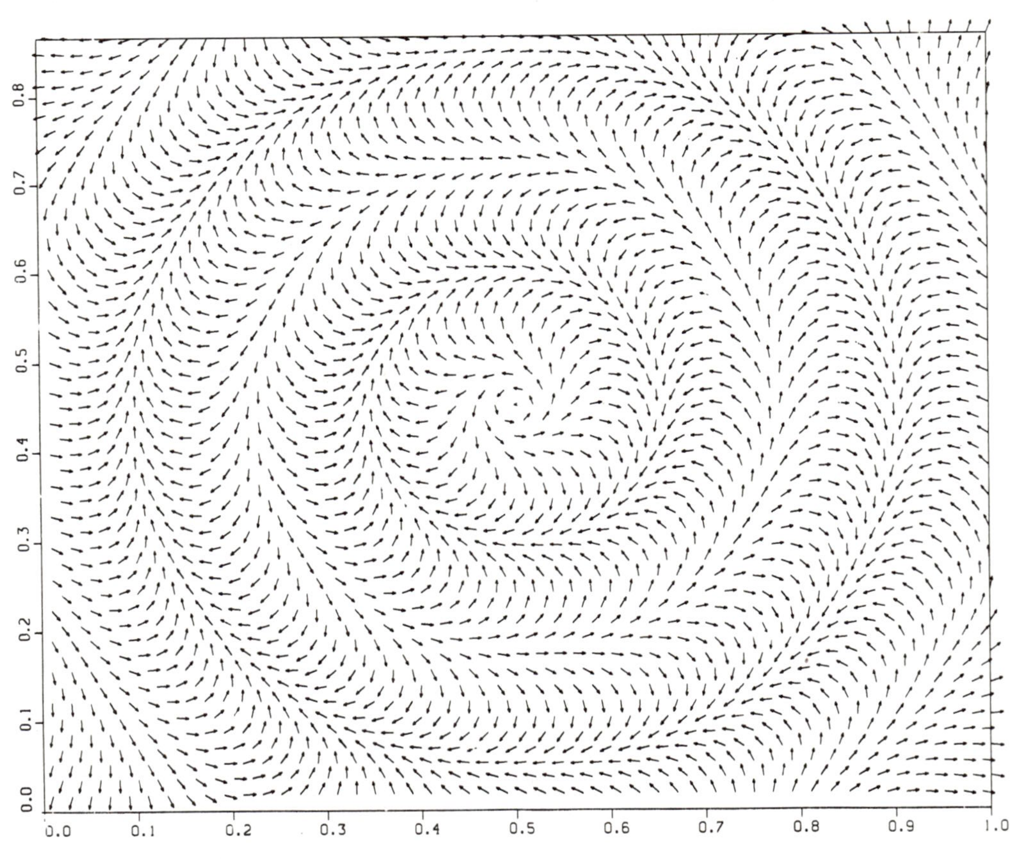

Fig.2. Single vortex (free boundary conditions). The initial state was $A=(z-z_0)/|z-z_0|$ and the figure shows the 300'th iterate. At later times the vortex basically performs a rigid rotation. Parameters are: α=0.5, β=−0.25, τ_0=0.2, μ=0.2, τ=1.

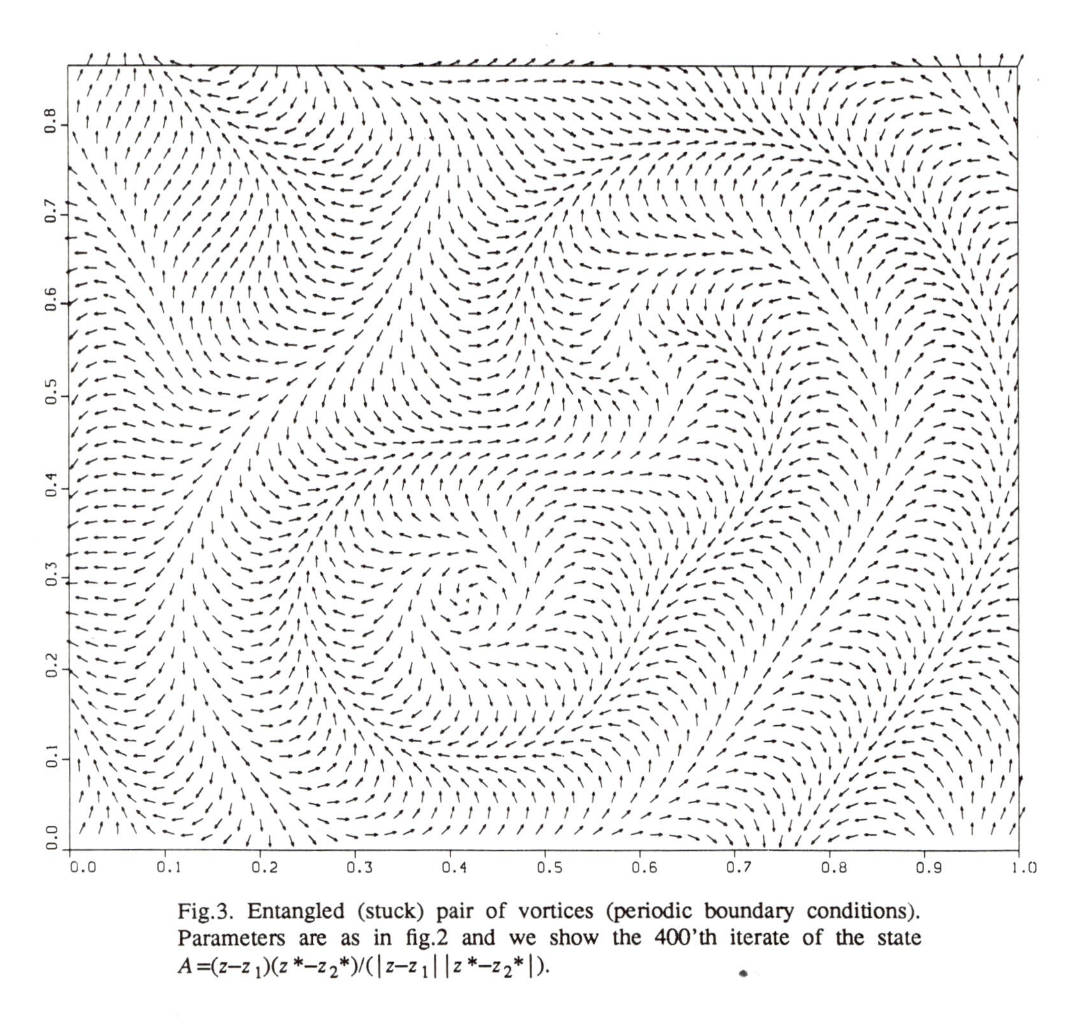

Fig.3. Entangled (stuck) pair of vortices (periodic boundary conditions). Parameters are as in fig.2 and we show the 400'th iterate of the state $A=(z-z_1)(z^*-z_2^*)/(|z-z_1||z^*-z_2^*|)$.

reduce this problem by using a hexagonal lattice. Secondly, the lattice introduces *pinning effects* since vortices maybe prefer some positions in the unit cell over others. We have tried to asses these problems by varying our parameters (especially μ). Finally, our aim is to say something about the *infinite* system, since that is the limit in which one might expect to see universal features. Thus it is very important to combine simulations at different lattice sizes and hope that some form of "finite size scaling" will work.

Acknowledgements

We are very grateful to P. Coullet, G. Grinstein, J. Lega, H. Rugh, L. Kramer and E. Bodenschatz for many useful discussions. DAR thanks the Mathematical Sciences Research Centre at the University of Arizona for its hospitality when part of this work which was supported by ONR grant N0001485K042 and AFOSR grant FQ8671-8601551 was carried out. We aknowledge support from the Carlsberg Foundation through a scholarship for AWP.

References

1 A.C.Newell and J.A.Whitehead J.Fluid Mechanics 38 , 279 (1969); A.C.Newell in *Lectures in Applied Mathematics,* vol. 15, Am. Math. Society, Providence (1974).

2 Y.Kuramoto, *Chemical Oscillations, Waves and Turbulence* Springer, Berlin (1980)

3 See e.g. J.M.Kosterlitz in "Nonlinear Phenomena at Phase Transitions and Instabilities" ed. T.Riste (Plenum 1982) p.397, or D.R.Nelson in "Phase Transitions and Critical Phenomena" vol. 7 ed. C.Domb and J.L.Lebowitz (Academic Press 1983) p.1.

4 H.R.Brand, P.S.Lomdahl and A.C.Newell, Physica 23D , 345 (1986).

5 A.V.Gaponov-Grekhov and M.I.Rabinovich, Sov.Phys.Usp. 30 , 433 (1987). A.V.Gaponov-Grekhov, A.S.Lomov, G.V.Osipov and M.I.Rabinovich, "Pattern formation and dynamics of two-dimensional structures in nonequilibrium dissipative media". Gorky preprint (1988).

6 P.Coullet, L.Gil and J.Lega, "Defect mediated turbulence". Nice preprint (1988).

7 For a recent review see P.Minnhagen, Rev.Mod.Phys. 59 , 1010

SPATIO TEMPORAL INTERMITTENCY IN RAYLEIGH-BENARD CONVECTION IN AN ANNULUS

S. Ciliberto

Istituto Nazionale Ottica
Largo E.Fermi 6-50125 Firenze-Italy

1) Introduction

Spatio temporal intermittency is a problem of great current interest that has been theoretically studied in system of coupled maps [Kaneko; Chate'-Manneville], partial differential equations[Chate'-Manneville;Nicolaenko] and in some cellular automata [Chate'-Manneville; Bagnoli et al.; Livi]. It consists of a fluctuating mixture of laminar and turbulent domains with well defined boundaries . Such a behaviour appears also in Rayleigh Benard convection[Ciliberto-Bigazzi; Berge'] and in boundary layer flows[Van Dyke; Tritton]. We report here a statistical analysis of the onset of spatiotemporal intermittency done in an experiment of Rayleigh-Benard convection. Our results display features typical of phase transitions similar to those obtained by Chate'-Manneville and Kaneko. In what is following we describe very briefly the experimental apparatus. In section 3) we will show the spatial patterns that preceed the spatiotemporal intermittency. In section 4) we descibe the reduction of the space time evolution to a symbolic dynamic and we report only the main results of the statistical analysis whose details have been the object of a previous paper[Ciliberto-Bigazzi].

2) Experimental apparatus

The system of interest is an annular fluid layer confined between two horizontal plates and heated from below.When the temperature difference ΔT exceeds the threshold value ΔTc a steady convective flow consisting of radial rolls arises.With the annular geometry the spatial pattern has periodic boundary conditions. The inner and outer diameters of our annulus are 6 cm and 8 cm respectively and the depth of the layer is 1 cm. The other details of the experimental apparatus have been described elsewhere[Ciliberto-Bigazzi;Ciliberto et al.;Ciliberto-Rubio]. The fluid under study is silicon oil with a Prandtl number of about 30. The critical value of ΔT at the onset of convection is $\Delta Tc = 0.06°C$. The spatial patterns produced by the convective motion have been studied qualitatively by shawdowgraph. For a more quantitative analysis we measure,on the circle of radius r_o =3.5 cm the component of the temperature gradient perpendicular to the roll axis. This component, averaged along the vertical , will be called u(x,t),with $x = \theta/2\pi$ that is the position on the circle of radius r_o. The function u(x,t) is sampled at 128 points in space and in time dependent regimes for at least 5000 times at intervals of 1 sec,that is about 1/10 of the main oscillation period of our system.

New Trends in Nonlinear Dynamics and Pattern-Forming Phenomena
Edited by P. Coullet and P. Huerre
Plenum Press, New York, 1990

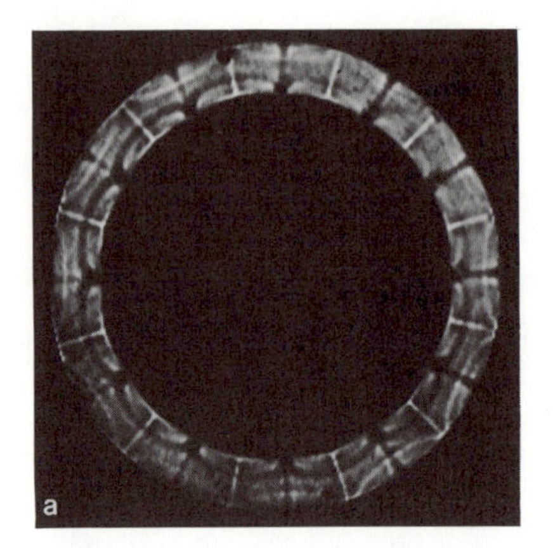

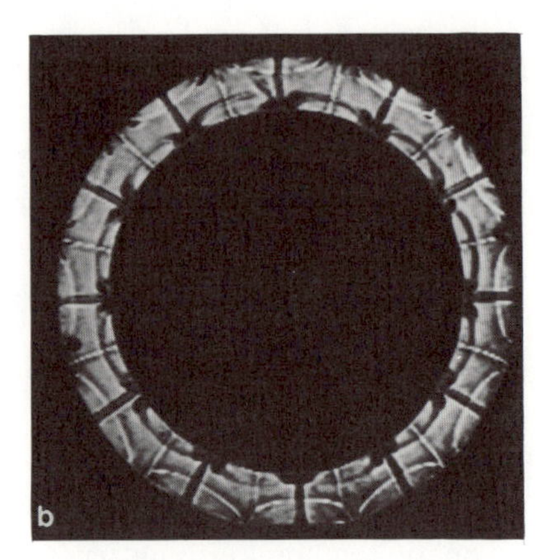

Figure 1. a)Shadowgraph of a stationary spatial patterns at $\eta = 100$. b) Snapshot of the spatial pattern at $\eta = 190$ in a time dependent regime.

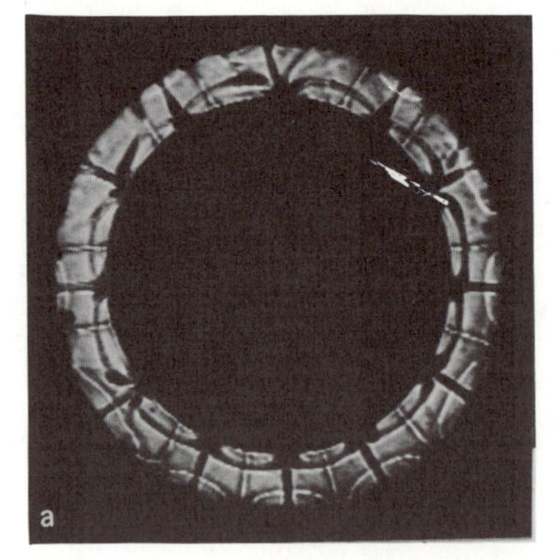

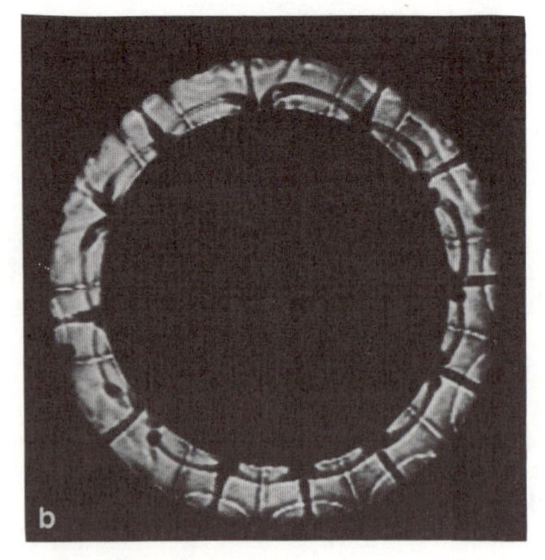

Figure 2. Spatial patterns at $\eta = 250$ and t=0,30 Sec.

3) Spatial patterns

In Fig.1a we show the shadowgraph of the spatial pattern at $\eta = 100$, where $\eta = \Delta T/\Delta Tc$. Dark and white regions correspond to hot and cold currents respectively. We observe that our geometry constrains the spatial structure to an almost one dimensional chain of rolls.

The spatial structure remains stationary for $\eta < 183$ where a subcritical bifurcation to the time dependent regime takes place.For $\eta > 183$ the time evolution is chaotic but, reducing η, the system presents either periodic or quasiperiodic oscillations, and at $\eta = 152$ it is again stationary.In the range $152 < \eta < 200$ the time dependence consists of rather localized fluctuations that slightly modulate the convective structure, which mantains its periodicity. This is clearly seen in Figs.1b) where a snapshot of the spatial structures at $\eta = 190$ is reported. The presence of hot and cold currents transverse to the main set of rolls merit a special comment. Such a two dimensional effect certainly influences the dynamics. However considering that the ratio between the length and the width of the annulus is roughly 22 we realise that the system can be considered almost one dimensional for what concerns the propagation time of thermal fluctuations along the circle. Furthermore the transition to spatiotemporal intermittency in two dimensional systems of coupled maps presents features similar to those in one dimension [Chate'-Manneville].

The spatiotemporal intermittency appears at $\eta = 200$. Typical spatial patterns at $\eta = 230$ are shown in Fig.2 for two different times. They present, at the same time several domains where the spatial periodicity is completely lost (we will refer to them as turbulent) and other regions (that we call laminar) where the spatial coherence is still mantained. On a very long time scale (order of minutes) the disordered and ordered regions exchange their position on the annulus.

4) Reduction to a symbolic dynamic and statistical analysis

The space time evolution of u(x,t) shows that in the turbulent domains the time evolution is characterised by the appearence of large oscillatory bursts.Instead in laminar regions the oscillations remain very weak. Thus the two regions can be identified by measuring the local peak to peak amplitude,for a time interval comparable with the mean period of the oscillation . Choosing a cutoff α, and making black all the points where the oscillation amplitude is above α, we can easily represent the dynamics of turbulent and laminar regions. As an example of

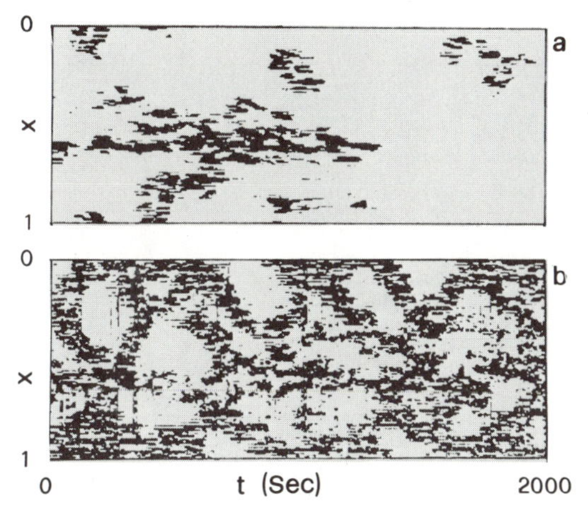

Figure 3.Binary representation,at $\alpha = 1.5°C/cm$, of the spacetime evolution of u(x,t) at $\eta = 216$ a) and $\eta = 248$ b). The dark and white area correspond to turbulent and laminar domains respectively.

such a code we show the spacetime evolution of u(x,t) at $\eta = 216$, in Fig. 3a, and $\eta = 248$ in Fig.3b. We remark that the qualitative features of these pictures are rather independent of the precise value of the cutoff. At $\eta = 216$, Fig.3a), a wide laminar region surrounds completely the turbulent patches that remain localized in space, after their appearence. Furthermore, the nucleation of a turbulent domain has no relationship with the relaxation of another one. On the contrary, at $\eta = 248$ Fig.3b),the turbulent regions migrate and slowly invade the laminar ones. This regim that sets in for $\eta > 245$ is very similar to those obtained in theoretical models. The change from the regime of Fig.3a) to that of Fig.3b) is reminescent of a percolation [Pomeau], that, indeed, has been proposed as one of the possible mechanisms for the transition to spatiotemporal intermittency. Following the method used by Kaneko and Chate'-Manneville, we quantitatively characterize such a behaviour by computing, over a time interval of $10^4 sec$, the distibution P(x) of the the laminar domains of length x. For $\eta < 248$ P(x) decays with a power law.The exponent does not depend within our accuracy, either on α or on η . Its average value is $\rho = 1.9 \pm 0.1$. On the other hand, for $\eta > 248$,the decay of P(x) for $x > 0.1$ is exponential with a characteristic length $1/m$.The existence of two different regimes is clearly seen in Figs.4a),4b) which display $P(x)$ versus x at $\eta = 241$ and $\eta = 310$.

We find that the dependence of m on η and α is the following:

$$m(\alpha, \eta) = m_o(\eta) exp(-\alpha/\alpha_o) \tag{1}$$

with $\alpha_o = (0.87 \pm 0.06)°C/cm$ independent of η.

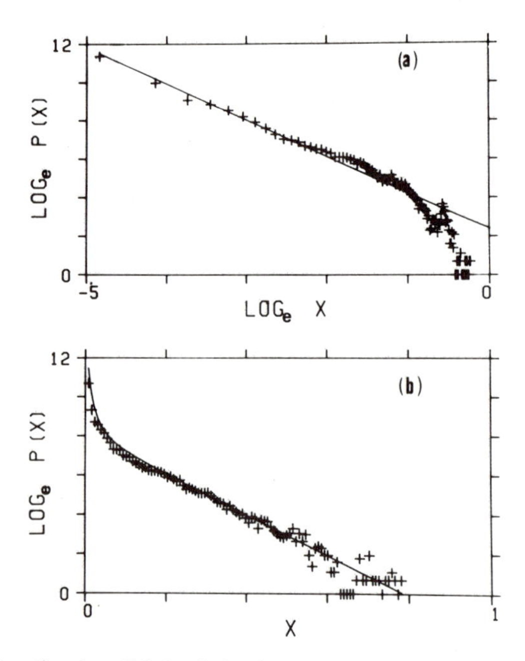

Figure 4. Distribution P(x) of the laminar regions of length x. (a) $\eta = 241$, the algebraic decay with exponent 1.9; (b) η=310, and $\alpha = 1.6°C/cm$, exponential decay with a characteristic length $1/m = 0.10$. The solid lines are obtained from Eq.3).

The dependence of m_o versus η is reported in Fig.5) . The linear best fit for $\eta > 246$ of the points of Fig.5) gives the following result:

$$m_o(\eta) = m_1(\eta/\eta_s - 1)^{\frac{1}{2}} \tag{2}$$

with $\eta_s = 247 \pm 1$ and $m_1 = 117 \pm 2$. This equation shows existence of a well defined threshold η_s for the appearence of an exponential decay in P(x).Besides we see that the characteristic length $1/m_o$ diverges at $\eta = \eta_s$. In the range $200 < \eta < 400$, P(x) is very well approximated by the following equation:

$$P(x) = (Ax^{-\rho} + B)exp[-m(\alpha, \eta)x] \tag{3}$$

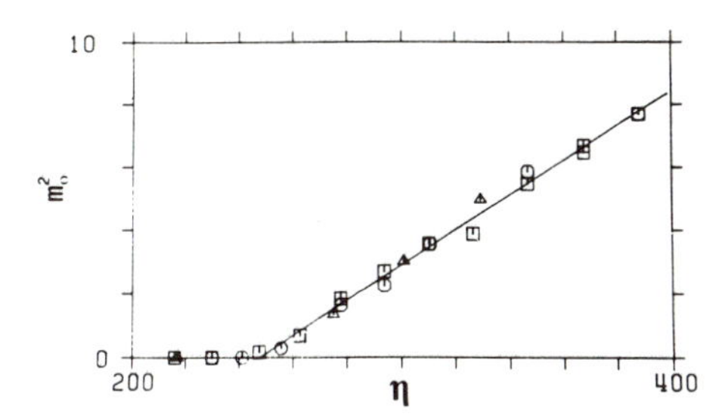

Figure 5. Dependence of ${m_o}^2$ on η ,the different symbols pertain to different sets of measurements done either increasing or decreasing η.The solid line is obtained from Eq.2.

where $m(\alpha, \eta)$ is given by 1) and ρ has the previous determined value.A,B are instead free parameters that can be very easily determined. It is possible to fit our experimental P(x),in the range $0.4°C/cm < \alpha < 3°C/cm$, with $A = 10$, $B = 4 \cdot 10^3$ for $\eta > eta_s$ and B=0 for $\eta < \eta_s$. The features of P(x) displayed by equations 2),3) are typical of phase transitions . Therefore, being the transition point η_s very close to the point where the behaviour like that of fig.3b) sets in, we conclude that the transition to this behaviour may be a phase transition [Mueller-Krumbhaar]. The main features of P(x) for $\eta > \eta_s$ qualitatively agree with those obtained in coupled maps [Kaneko; Chate'-Manneville] and P.D.E. [Chate'-Manneville; Nicolaenko] in spatiotemporal intermittent regimes. Of course these models do not reproduce the values of the non-universal exponents in Eqs.2),3) [Kaneko].

The transition may also be characterised by measuring p_0 that is the probability of finding a laminar point [Bagnoli]. If we suppose that a laminar site is generated at a certain time with space-time independent probability p_0, the probability of finding a laminar region of length x is given by $P(x) = exp[x \log(p_0)/l_o]$, where l_0 is a suitable characteristic length. Therefore we may conclude that $m = \log(p_0)/l_o$. We can verify this hiphotesis by computing directly p_0 on the experimental data. By following the same procedure used to rescale m as a function of α , we find that $\log p_0$ extrapolated at $\alpha = 0$ has the following dependence on η :

$$|\log p_0| = const.(\eta/\eta_c - 1)^{\frac{1}{2}} \tag{4}$$

with $\eta_c = 216$. So we conclude that it has the same exponent of m but a different critical threshold. This means that the appearence of a laminar site is an independent process for $\eta >> \eta_s$, and that a certain correlation exists between laminar and turbulent sites near the critical value of η. The presence of a power law decay of P(x) for $\eta_c < \eta < \eta_s$ may be due either to finite size effects [Muller-Krumbhaar] or to defects [Nelson]. This aspect of the problem is not yet very well understood and further investigation is in progress to clarify this point.

5) Conclusion

Summarizing, the onset of spatiotemporal intermittency, in Rayleigh-Benard convection of a medium Prandtl number fluid in a annular cell, displays features of a phase transition that is reminescent of a percolation. Although many aspects of this phenomenon are still to be investigated,the analogy of the behaviour of our system with that observed in coupled maps,P.D.E. and some cellular automata, suggests that these models may be very useful to understand the general features of spatiotemporal intermittency. This work has been partially supported by G.N.S.M.

References

Bagnoli F., Ciliberto S., Francescato A., Livi R., Ruffo S., in "Chaos and complexity", M.Buiatti, S.Ciliberto, R.Livi, S.Ruffo eds., (World Scientific Singapore 1988).

Berge' P., in " The Physics of Chaos and System Far From Equilibrium", M.Duong-van and B.Nicolaenko, eds. (Nuclear Physics B, proceedings supplement 1988).

Chate' H., Manneville P., Phys.Rev.Lett. 54, 112 (1987); Europhysics Letters 6,591 (1988) ; Physica D in press; see also these proceedings.

Ciliberto S.,Bigazzi P.,Phys.Rev.Lett.60,286(1988).

Ciliberto S.,Francini F., Simonelli F., Optics Commun. 54, 381 (1985).

Ciliberto S., Rubio M.A., Phys.Rev.Lett. 58, 25 (1987).

Crutchfield J., Kaneko K. in "Direction in Chaos", B.L.Hao (World Scientific Singapore 1987).

Dubois M.,in these prooceedings.

Kaneko K.,Prog.Theor.Phys.74,1033(1985).

Livi R., in these prooceedings.

Muller-Krumbhaar H. in 'Monte Carlo Method in Statistical Physics", edited by K.Binder (springer Verlag,New York 1979).

Nicolaenko B., in " The Physics of Chaos and System Far From Equilibrium", M.Duong-van and B.Nicolaenko, eds. (Nuclear Physics B, proceeding supplement 1988). See also these proceedings.

Pomeau Y., Physica 23D, 3 (1986).

Tritton D.J., Physical Fluid Dynamics (Van Nostrand Reinold, New York, 1979), Chaps.19-22.

Van Dyke V.,An Album of Fluid Motion (Parabolic Press,Stanford,1982).

Nelson D.R. , 'Phase transitions and critical phenomena' edited by C.Domb and J.L.Lebowitz (Academic,London 1983)

Transition to Turbulence *via* Spatiotemporal Intermittency: Modeling and Critical Properties

Hugues Chaté and Paul Manneville

Institut de Recherche Fondamentale
DPh-G/PSRM, CEN-Saclay
91191 Gif-sur-Yvette Cedex, France

INTRODUCTION

When studying the transition to turbulence in closed flow systems, it has become traditional to make a distinction between them according to confinement effects. These effects can be measured by *aspect-ratios*, i.e. ratios of the lateral dimensions of the physical system to some internal length linked to the basic instability mechanism. Small aspect-ratio systems are strongly confined so that the spatial structure of the modes involved can be considered as frozen. As such, they experience a transition to turbulence according to scenarios understood in the framework of low dimensional dynamical systems theory. The general procedure in this field involves the reduction to center manifold dynamics by elimination of stable modes, the Poincaré surface of section technique and the iteration of maps. Disorder is then interpreted rather as *temporal chaos* arising from the *sensitivity of trajectories to initial conditions and small perturbations*.

When the aspect-ratio increases, confinement effects become weaker and the spatial structure can no longer be considered as frozen. Therefore, disorder is no longer strictly temporal but regains spatial features which have to be understood along different lines. Reference is then made to an *ideal pattern* and to its modulations described by a *complex envelope*, say $A = |A|\exp(i\phi)$. Sufficiently far from the instability threshold, a further reduction derives from the fact that the modulus $|A|$ is slaved to the phase ϕ, which allows the elimination of the former. A *phase equation* is obtained, which in the simplest case is simply a diffusion equation (Pomeau and Manneville, 1979): $\partial_t\phi = D\partial_{xx}\phi$. A phase instability takes place when the diffusion coefficient D, which

New Trends in Nonlinear Dynamics and Pattern-Forming Phenomena
Edited by P. Coullet and P. Huerre
Plenum Press, New York, 1990

is a function of control parameters, becomes negative. Further stages of the transition to turbulence can then be understood in terms of *phase modes* whose evolution is governed by the diffusion equation adequately completed to include higher order terms (a much studied specific model is given by the Kuramoto-Sivashinsky equation, for a review and references see: Manneville, 1988).

In fact, the reduction process sketched above for large aspect-ratio systems implies a somewhat progressive growth of space-time disorder, even though the precise nature of the asymptotic regimes (the attractors) may depend in a complicated and discontinuous way on the control parameters. Accordingly, the associated scenarios can be called "supercritical" in an enlarged sense. It turns out that the transition to turbulence in weakly confined systems can be much "wilder", specifically when the increase of the number of degrees of freedom cannot be controlled easily. We can call this situation "subcritical", again in an enlarged sense (see Pomeau, 1986). For usual subcritical instabilities, the stable branch of bifurcated states remains at "finite distance" of the basic state, this distance being usually measured by a scalar quantity. Here, we examine a case where this is no longer possible and where even slowly modulated envelopes remain useless. In the first part of this review, we present an illustration of the transition to turbulence *via spatiotemporal intermittency* (STI) from simulations of a partial differential equation (for accounts of other concrete examples from convection experiments, see: Ciliberto, these proceedings, or Bergé, 1987). We then give some hints on how the transition can be interpreted in terms of discrete-time discrete-space systems, coupled map lattices and cellular automata. Such systems are best approached within the framework of conventional statistical physics, so that we shall be led directly the second part devoted to the discussion of their critical properties and the relation between STI and directed percolation conjectured by Pomeau (1986).

MODELING: FROM CONTINUOUS TO DISCRETE SYSTEMS

Transition to Turbulence *via* Spatiotemporal Intermittency in a Partial Differential Equation

We have studied the following one-dimensional variant (Pomeau and Manneville, 1980):

$$\frac{\partial w}{\partial t} = \varepsilon\, w + \left(\frac{\partial^2}{\partial x^2} + 1\right)^2 w - w\,\frac{\partial w}{\partial x} \tag{1}$$

of a convection model due to Swift and Hohenberg (1977). For ε positive and sufficiently small, this partial differential equation possesses steady (i.e. time independent) cellular (i.e. periodic in space with period of the order of $\lambda_c = 2\pi$) solutions as in convection but, at variance with the original model whose nonlinear term (w^3) derives from a potential, the present one can display a transition to chaos as ε is increased. Indeed, equation (1) can be rewritten in the form:

$$\frac{\partial u}{\partial t} + \frac{\partial^2 u}{\partial x^2} + \frac{\partial^4 u}{\partial x^4} + u\frac{\partial u}{\partial x} + \eta\, u = 0 \tag{2}$$

which appears as the Kuramoto-Sivashinsky model with an additional damping term

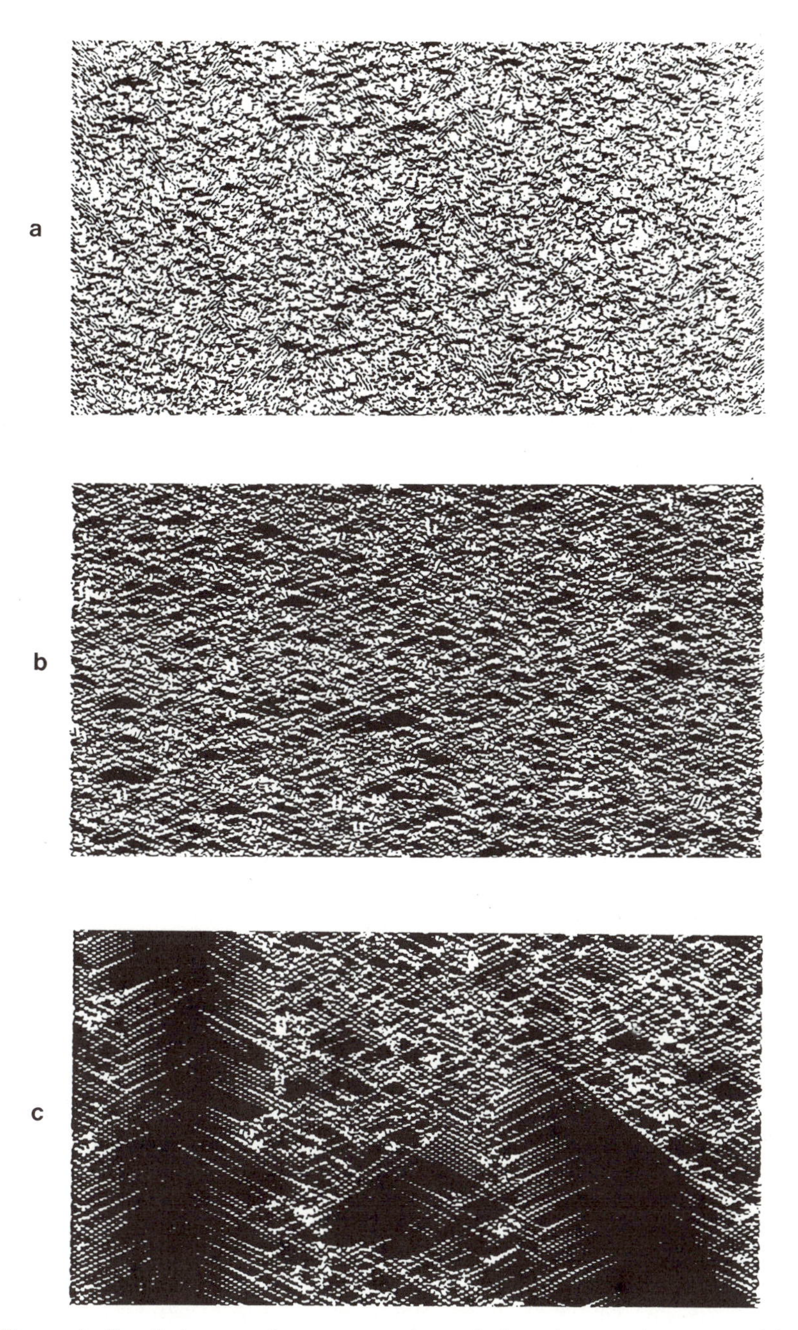

Figure 1. *Spatiotemporal representations of the solution of equation (1) in a box of length $L = 1600$ with rigid boundary conditions. Laminar cells are in black, turbulent ones in white. For a description of the reduction process, see Chaté and Nicolaenko, these proceedings. Time is running upwards for $t \in [0, 1500]$ after long transient periods. a: pure Kuramoto-Sivashinsky model ($\varepsilon = 1$). b: $\varepsilon = 0.84$. c: at the STI threshold, $\varepsilon = \varepsilon_c \simeq 0.68$.*

of coefficient $\eta = (1 - \varepsilon)/4$ (see also Chaté and Nicolaenko in these proceedings). For $\varepsilon = 1$, this term vanishes, so that the "pure" Kuramoto-Sivashinsky model is recovered, which is well known for exhibiting turbulent regimes. In this limit, a single control parameter remains: the length L of the interval at which boundary conditions are imposed to w. Here we have two parameters at our disposal to "unfold" the dynamics: ε and L. Usually L is kept constant and ε is increased. For L small, vis. $\sim 2\lambda_c$, confinement effects are strong and the transition to chaos follows a route typical of low dimensional systems. The intermediate region, vis. $L \sim 20\lambda_c$, already exhibits essentially spatiotemporal dynamics, but the situation is extremely complicated due to still sizable end effects. On the other hand, at very large L's, say $L > 200\lambda_c$, the transition to turbulence occurs systematically *via spatiotemporal intermittency* (see e.g. Kaneko, 1985).

As illustrated in fig. 1, the spatiotemporally intermittent regime is a fluctuating mixture of regular patches here called *laminar* and chaotic or *turbulent* domains. This binary reduction is based on the observation that the local structure may be very close to the perfect cellular solution (for details see Chaté and Nicolaenko, these proceedings). The proportion of laminar and turbulent regions is a function of the control parameter and the size of laminar patches increases as ε is decreased (fig. 1b). At a well defined (in the large L limit) value $\varepsilon_{\mathrm{STI}}$, some laminar patches reach a "macroscopic" size, i.e. of the order of the width of the system, see fig.1c. For $\varepsilon < \varepsilon_{\mathrm{STI}}$, these patches slowly invade the system at the expense of the domain left spatiotemporally intermittent (for an illustration, see Chaté and Manneville, 1987). Upon increasing ε from zero beyond $\varepsilon_{\mathrm{STI}}$, a direct transition to turbulence from a regular pattern to spatiotemporal intermittency is observed and, as far as can be inferred from the numerical experiments reported, the process keeps a continuous character from a statistical viewpoint since global average quantities evolve gently around the threshold.

The observed transition to turbulence thus displays features of a "contamination" of the laminar state by turbulence and it has been suggested by Pomeau (1986) that stochastic processes of the class of directed percolation (for a review see Kinzel, 1983) could offer a particularly relevant interpretation framework. In spite of some support to this viewpoint coming from the evolution of statistical properties strongly reminiscent of the critical behavior observed at a second order phase transition, steps are obviously lacking in the reduction from the original deterministic fully continuous system to a probabilistic fully discrete system. The first one, from continuous to discrete space-time is examined in the next section.

From Partial Differential Equations to Coupled Map Lattices and Cellular Automata

In spite of the complexity of the spatiotemporal disorder shown by the continuous model above, it can be seen as the interplay of a few elementary local dynamical events, mainly space-time dislocations (cell creation and cell annihilation) and deformation waves (for details see Chaté *et al.*, 1988a). In particular, the basic cellular structure and its (local) temporal oscillations provide natural space and time units for a discretization scheme. This reduction to local processes rests on the remark that the transition is not supercritical in a strict sense so that no divergence of a coherence length or relaxation time is expected. Here both quantities have to be understood at a "microscopic" level in much the same way as the lattice spacing, the interaction range or any other atomic characteristics in a macroscopic system, e.g. a crystal. The continuous-space continuous-time system can then be reduced, at least at a formal level, to a discrete-space discrete-time system, i.e. a coupled map lattice (CML). The next crucial step is then the choice of the local map(s) and the kind of coupling(s) between maps.

The minimal requirement to be fulfilled by the map is a splitting of the local phase space into two main regions, one corresponding to the direct basin of attraction of a regular asymptotic state to be assimilated with the laminar regime, the other with complex dynamics, either a true chaotic attractor with sizable basin of attraction, or more probably a chaotic transient, to be understood as the turbulent regime. Usually a single local one-dimensional map f is considered:

$$X^{n+1} = f(X^n)$$

where the superscripts denote time and X is a single real variable. Figure 2a shows an example chosen to fulfill this requirement in a simple manner. This "minimal" map is piecewise linear with a chaotic domain modeled by a tent map:

$$\begin{aligned} f(X) &= rX && \text{for } X \in [0, 1/2] \\ f(X) &= r(1-X) && \text{for } X \in [1/2, 1] \end{aligned} \tag{3a}$$

with $r > 2$, and a laminar domain where the dynamics is a mere relaxation toward a fixed point, e.g.:

$$f(X) = k(X - X^*) + X^* \quad \text{for } X > 1 \tag{3b}$$

with $X^* = (r+2)/4$ and $|k| \le 1$ (when $k = 1$ all points $X > 1$ are fixed points of the local map so that there is no local time-scale).

Using the vocabulary of directed percolation, the laminar state is said to be *absorbing*, the turbulent one *active*. The principal constraints put on the choice of the

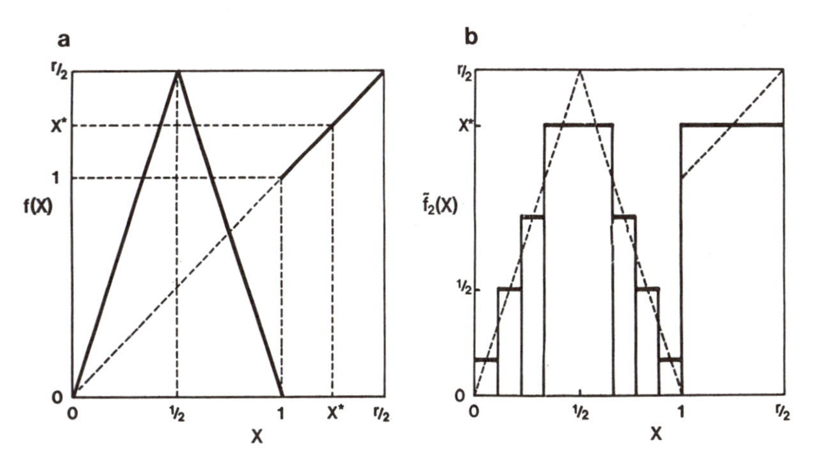

Figure 2. Local maps. a: minimal map f for the minimal CML. b: step-function approximation $\tilde{f}_2$ of f at order $p = 2$ from which the 4-state DCA rules approximating the minimal CML are derived.

coupling are that it must be compatible with the absorbing character of the laminar state, that is to say it must preserve its stability with respect to infinitesimal delocalized fluctuations but allow the growth of finite localized perturbations. The simplest realization obviously corresponds to the discrete approximation of a diffusive coupling:

$$D\,\nabla^2 Y \longrightarrow g\,(Y_{i+1} - 2Y_i + Y_{i-1}) \tag{4}$$

where the subscripts denote the spatial position, here on a one-dimensional lattice but this extends to any space dimension d, Y stands for the quantity which is to be smoothed, usually simply $f(X)$, and g a coupling constant. The coupling step can be generalized into:

$$X_i^n = \sum_{j\in\mathcal{V}_i} W_{ij} Y_j^n$$

where $\mathcal{V}_j$ is some neighborhood of site i and W_{ij} a weight factor depending of the relative position of sites i and j.

For the final expression of the CML, several equivalent formulations are possible. The global evolution can be decomposed into two steps, the local evolution and the coupling:

$$\cdots \overset{coupling}{\longrightarrow} X^n \overset{f}{\longrightarrow} Y^n \overset{coupling}{\longrightarrow} X^{n+1} \overset{f}{\longrightarrow} Y^{n+1} \overset{coupling}{\longrightarrow} \cdots$$

Assuming that the system is observed after the coupling step and before the next iteration step yields:

$$X_i^{n+1} = \sum_{j\in\mathcal{V}_i} W_{ij}\, Y_j^{n+1} \quad \text{with} \quad Y_i^{n+1} = f(X_i^n)$$

that is to say:

$$X_i^{n+1} = \sum_{j\in\mathcal{V}_i} W_{ij}\, f(X_j^n) \tag{5a}$$

If, on the contrary, we assume that the system is observed after the iteration step and before the coupling step, we obtain:

$$Y_i^{n+1} = f(X_i^n) \quad \text{with} \quad X_i^n = \sum_{j\in\mathcal{V}_i} W_{ij}\, Y_j^n$$

that also reads:

$$Y_i^{n+1} = f\Big(\sum_{j\in\mathcal{V}_i} W_{ij}\, Y_j^n\Big) \tag{5b}$$

In any case, underlying this modeling is the assumption that the transition process derives from the conversion of a local *transient* chaos into global *sustained* turbulence as a result of the spatial coupling only.

Statistical properties of the class of CMLs displaying a transition to turbulence *via* STI will be reviewed in the next section. For the moment, let us describe two further reduction steps aiming at a thorough theoretical account of the process. First of all, the definition of CMLs especially in the form of equation (5*b*) is strongly reminiscent of that of *deterministic cellular automata* (DCA) which are discrete-space discrete-time systems with a finite number of states, i.e. a discrete phase space evolving according to rules defined on some local neighborhood. DCAs are the simplest systems capable of displaying complex spatiotemporal behavior. They have been the subject of intensive research recently (see Wolfram, 1986) so that this further reduction step is particularly appealing. In fact, the effective restriction to a finite number of states can easily be obtained even in a system with continuous state variables if the local evolution is governed by step functions.

The problem of giving a dynamical meaning to the reduction of the continuous dynamics described by model (3) in terms of step functions is sketchily solved in fig. 2b: backwards iterates of the laminar domain and complementary intervals define the steps of the function, while the interaction rule remains given by the diffusive coupling. Depending on the depth of the backward iteration, increasingly refined DCA approximations to the CML can be constructed (for a detailed account of this approximation, see Chaté and Manneville, 1988a).

In striking similarity with their continuous counterpart, DCAs constructed in that way display a transition to STI which is now interpreted as a transition in the space of rules. This prepares us to the last step which is a reduction to *probabilistic cellular automata* (PCA) of which *directed percolation* is the best known example. One way to understand these systems is to consider them as the result of the stochastic interpolation between two or more *deterministic* cellular automata rules. Phase transitions between macroscopically different regimes are then observed akin to the laminar-intermittent transition, the control parameter(s) being in this case the interpolation coefficient(s) between the rules. This connects directly to the whole field of phase transitions and critical phenomena, or, at least, to the statistical mechanics type of approach of macroscopic physics, which opens the way for a mean-field type theory of STI.

In fact, most steps described above remain at a formal level and difficulties can emerge at every stage in practice. For example, in the reduction from the space-time continuous model to the discrete model, it was tacitly assumed that decomposition into localized low dimensional dynamical systems was legitimate. However, it can be learnt from the study of medium aspect-ratio systems that spatiotemporal disorder can appear by crises involving subsystems of various sizes so that the large aspect-ratio system should be viewed more as a complicated, perhaps hierarchical, arrangement of such nested subsystems than as a simple array of identical elements. In addition, the linearly stable character of the laminar basic state was judged important to justify the existence of the absorbing state. This assumption, while valid for the model used here, may well have to be critically examined if the same reduction scheme is to be applied to other cases (in particular, in situations where nonlocal hydrodynamic interactions are expected to play a role) However, a weakly phase-unstable or drift-sensitive system could appear stable enough to allow the definition of the required absorbing state. In any case, spatiotemporal intermittency presents itself as an original scenario specific to very large aspect-ratio systems and the principal merit of the approach described above is to build a new bridge with statistical mechanics, thus justifying the recourse to concepts familiar in the theory of phase transitions as will be shown below.

SPATIOTEMPORAL INTERMITTENCY AND CRITICAL PHENOMENA

In this section, the critical properties of the minimal CML defined above by the local map (3) and the coupling (4) in one and two space dimensions are first reviewed, keeping the conjectured equivalence with directed percolation as a general guideline. Then, coming back to the approximation of CMLs by cellular automata, a possible explanation of the non-universality of the transition to turbulence *via* spatiotemporal intermittency is given, together with a discussion of a mean-field approach of CMLs. This last point is strengthened by the report of new results concerning the transition in three and four space dimensions and an experimental mean-field treatment of the minimal CML.

The Minimal CML: Precise Setting of the Problem

The minimal CML based on the local map f defined by (3) and a nearest-neighbors diffusive coupling given by (4) is expressed in the usual form of equation ($5a$) by:

$$X_i^{n+1} = (1-\varepsilon)\, f(X_i^n) + \frac{\varepsilon}{\mathcal{N}_d} \sum_{j \in \mathcal{V}_i} f(X_j^n) \tag{6}$$

where $\mathcal{V}_i$ is now the set of the $\mathcal{N}_d$ nearest neighbors of site i in d space dimensions (in all the following we used hypercubic lattices so that $\mathcal{N}_d = 2d$) and ε is the coupling strength.

The existence of a threshold for STI is easily understood: for small values of the coupling, sites behave as if uncoupled so that the whole lattice quickly evolves to an homogeneously laminar state for any initial condition. On the contrary, when starting from random initial conditions, the lattice may remain disordered if ε is large enough. As a matter of fact, thanks to the local finite amplitude perturbations introduced by the coupling, a site may escape the laminar state and be "re-injected" in the chaotic part of the local map f, a mechanism at the origin of the sustained regimes of STI. At this point, and although other choices are possible, a natural order parameter is the fraction f_t of sites in the turbulent state, a quantity that goes to zero when decreasing ε below the transition threshold.

This transition when ε is varied may have different characteristics for different values of the two other parameters left, r and k, which define precisely the local map f (fig. 2a). Basically (see below for a confirmation), two cases have to be considered for each parameter. Slope r governs the turbulent state and the escape process from it: for $r \simeq 2$, local chaotic transients may be very long while they are much shorter for larger r's. Coefficient k rules the attracting character of the laminar state: for $k \neq 1$ there is a single stable fixed point X^* while for $k = 1$ all $X > 1$ are fixed points of f so that the laminar state is in a way only "marginally" attracting. Although many cases have been studied, in all the following numerical results we restrict ourselves to typical parameter values chosen along these guidelines.

Transitions in One Space Dimension

Figure 3 shows spatiotemporal representations of the minimal CML near threshold for four typical cases. Note that the patterns have very different shapes depending on the values of r: the $r = 3$ case is characterized by propagating structures delimiting triangular embeddings while in the $r = 2.1$ case no particular local structure is evidenced (as is the case for directed percolation). We will see in the following that this is not a mere aesthetic remark.

For $k = 1$ (fig. 3a and 3b), the transition is *continuous*: the order parameter f_t goes continuously to zero at a well defined threshold value ε_c of the coupling, no

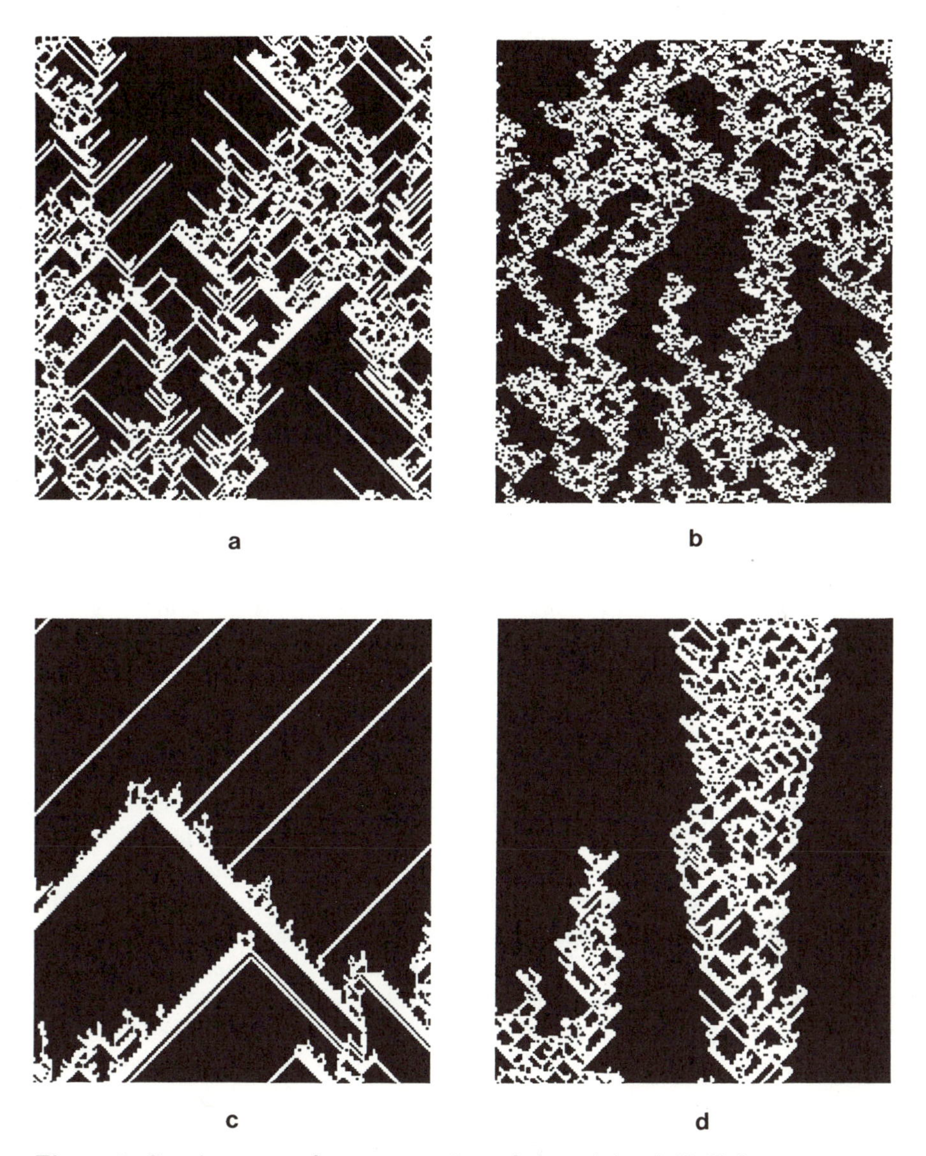

Figure 3. Spatiotemporal representation of the minimal CML in one space dimension near the STI threshold for a lattice of size $N = 200$ with periodic boundary conditions. Time is running upwards during 250 iterations after long transients. Laminar sites ($X > 1$) are in black, turbulent ones in white. (a): $k = 1$, $r = 3$, $\varepsilon_c = 0.360$ *(continuous transition).* (b): *same as* (a) *but for* $r = 2.1$ ($\varepsilon_c = 0.0047$). (c): $k = 0.5$, $r = 3$, $\varepsilon_c = 0.402$ *(threshold masked by defects).* (d): $k = 0.8$, $r = 3$, $\varepsilon_c = 0.397$.

hysteresis is observed, and various critical exponents can be determined. Among them, the exponent $\varsigma_\perp$ governing the distribution of sizes of the laminar clusters at threshold is easy to measure. This distribution, while exponential away from threshold, indeed shows a power-law behavior near ε_c. The corresponding critical exponent can be related, in the field of directed percolation, to other ones through: $\varsigma_\perp = 1 + d_f = 2 - \beta/\nu_\perp$. Exponent β can also be measured, although at a greater numerical cost. The table summarizes the measured exponents and compare them to the case of directed percolation. Two important remarks immediately follow: the measured values for the CML all show significant discrepancies with directed percolation, so that the transition to turbulence *via* STI is not in the universality class of this model and moreover they differ for different values of r, so that the existence of such a class is forbidden, at least in the usual acceptance of the term in statistical mechanics.

Table 1. Summary of the measured critical exponents for the minimal CML in the case of a continuous transition when $k = 1$ for space dimension $d \leq 2$. The corresponding values for directed percolation are given between parentheses for comparison. For details on the numerical procedure, see Chaté and Manneville (1988b,c).

d	r	β	ς
1	2.1		1.78 (1.75)
1	3	0.25 (0.28)	1.99 (1.75)
2	3	0.50 (0.60)	

For $k \neq 1$ and $r = 3$ (fig. 3c and 3d), this continuous phase transition picture is broken. For $k < 0.70$, the transition region is masked by the linearly stable propagating structures appearing in fig. 3c. The order parameter f_t is no longer well defined, since it then depends on the number of such *defects* which itself has no statistical mean. For $0.70 < k < 1$, the transition region is not in the domain of stability of these defects (Chaté and Manneville, 1988d), and f_t keeps its meaning as an intensive quantity. Preliminary results in this case show that f_t does *not* go smoothly to zero at threshold, but with a discontinuity that gradually decreases when k increases and vanishes for $k = 1$ (recovering thus the continuous transition described above). This is characteristic of a *discontinuous* phase transition, although this term has to be taken with great care in one space dimension where one lacks the usual confirmation by surface effects.

Transitions in Two Space Dimensions

Results in two space dimensions are very similar to those in one space dimension except that discontinuous transitions are now clearly evidenced. Due to the large numerical cost, only $r = 3$ cases have been studied in detail (Chaté and Manneville 1988c).

For $k = 1$, the transition is still continuous. We measured the exponent β and again found it different from the corresponding values for directed percolation (see Table).

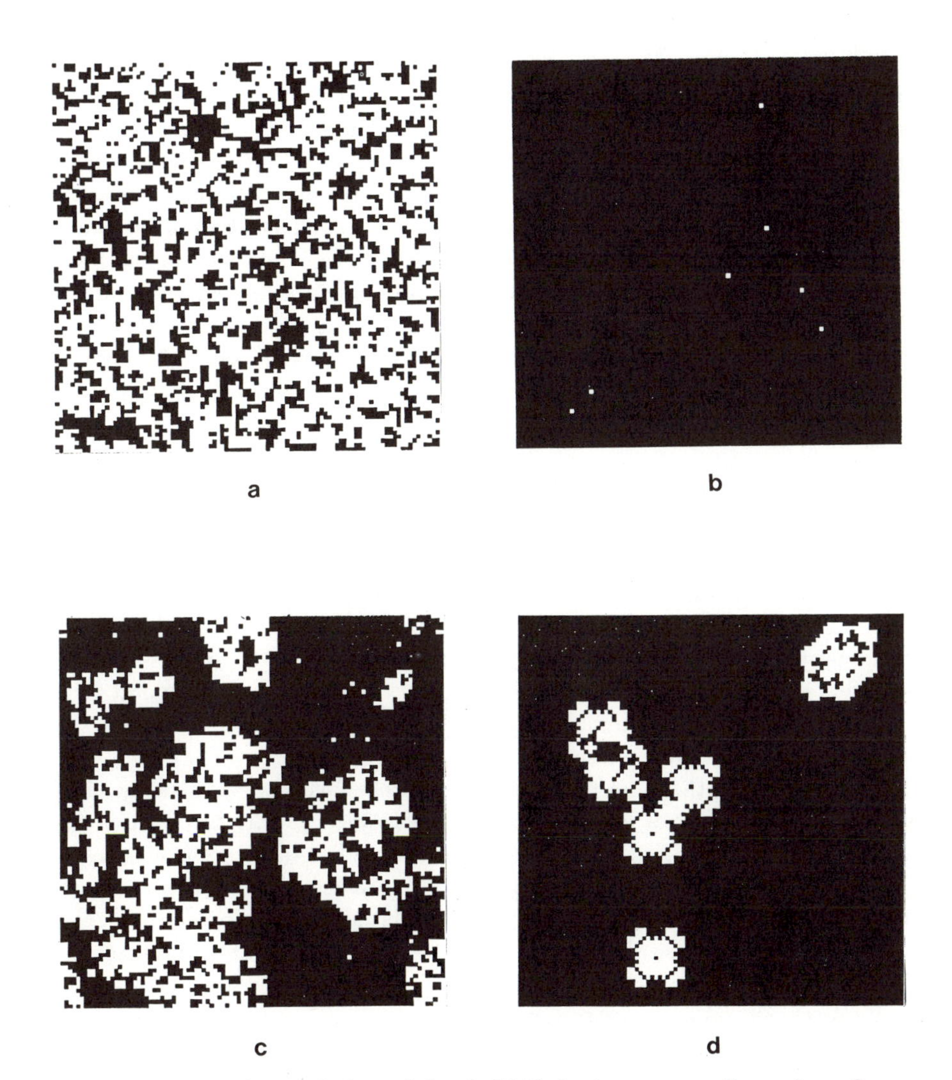

Figure 4. Snapshots of the minimal CML in two space dimensions for a lattice of linear size $N = 100$ with periodic boundary conditions for $r = 3$ and $k = 0.5$. Laminar sites are in black, turbulent ones in white. (a): intermittent phase at $\varepsilon = 0.33$. (b): quiescent state with defects at $\varepsilon = 0.315$. (c): during the melting transient between the intermittent phase and a quiescent state. (d) during the growing transient when $\varepsilon > \varepsilon_s$, the upper limit of stability of the defects.

For $k = 0.5$, the transition is *discontinuous* with practically all the usual characteristics of first order phase transition: decreasing the coupling, there is a clear jump of the order parameter at threshold when the system falls from the intermittent phase (fig. 4a) to the laminar phase (fig. 4b) during a "melting transient" (fig. 4c). No finite-size effects have been observed, in contrary to the case of continuous transitions. Surface effects were shown by studying planar fronts between the intermittent phase and the laminar phase whereas, in the continuous case, such interfaces break under the evolution of the CML. The front velocity goes to zero at threshold, a definition which coincides with the divergence of the critical size for a bubble of one phase to invade the other phase (note that there is no critical size effect at all in the continuous case). Finally, point-wise defects which consists in linearly stable pinched structures appear in the transition region and their stability domain in ε delimits the hysteresis loop (Chaté and Manneville, 1988d). Indeed, increasing ε passed the upper limit ε_s of the stability of the defects, the system leaves the laminar phase to "jump back" to the intermittent phase in a very regular growth transient (fig. 4d).

The Minimal CML as a Cellular Automaton: Origin of the Non-Universality

Results obtained in one and two space dimensions show that the transition to turbulence *via* STI is not in the universality class of directed percolation. In one dimension, the critical exponents in the case of a continuous phase transition vary with the parameters of the local map f. The cases $r = 3$ and $r = 2.1$ which already appear very different visually (fig. 3), are also very different from the point of view of the approximation of the minimal CML by deterministic cellular automata sketched in the last part of the previous section.

Let us summarize the main properties of this scheme described at length in Chaté and Manneville (1988a): at every order p of the approximation, the CML is approximated by 2^p-state, three-site neighborhood, legal rule DCAs with one absorbing state and the transition when ε is varied is translated into a discrete sequence of rules describing a path in the set of possible automata. This sequence progressively recovers a continuous character when the approximation is better and better refined (large p's). In this limit, the minimal CML itself, formally, is a DCA with a huge number of states when implemented on a digital computer.

For $r = 3$, the sequence of rules reproduces the STI threshold phenomenon: for small values of ε, the equivalent rules are trivial (class 1 or 2) while they are complex (class 3 or 4) for large values of the coupling. This explains the propagative structures and the triangular embeddings exhibited by the CML above threshold (fig. 3a). For $r = 2.1$, the situation is very different: all the rules of the sequence are trivial; there are no complex rules, hence no local structures (fig. 3b).

The physical processes involved in the $r = 3$ and $r = 2.1$ cases are of very different nature. For $r = 3$, the local turbulent transients of the local map f are very short and the sustained spatiotemporal disorder does not derive from a local stochastic process but from a quasi-deterministic one of the type exhibited by class 3 and class 4 DCAs. On the contrary, the local turbulent transients for f may be very long for $r = 2.1$. Thus, although the Lyapunov exponent associated with the chaotic part of the phase space of f is smaller than for $r = 3$, the local mixing (typical of a Bernouilli shift) is the dominant process at the origin of disorder while it was negligible for $r = 3$.

STI derives then from the spatial translation of this local stochastic process and the underlying local deterministic dynamics has practically no influence on the resulting spatiotemporal disorder. It is remarkable that this translates itself, under the approximation of the CML, by trivial DCA rules all along the sequence corresponding to the transition, fostering the irrelevance of any deterministic local process and stressing the increasing similarity of the minimal CML with directed percolation (a "fully probabilistic" automaton) when $r \to 2$.

This important difference not only applies to the visual aspect of the problem, but, as the above discussion of the physical processes indicates, is at the origin of the non-universality of the transition to STI in the CML from both viewpoints of the order of the transition (continuous/second order vs discontinuous/first order) and the quantitative discrepancies observed in the critical exponents.

Moreover, coming back to the problem of the universality of the phase transitions of PCAs, there are no *a priori* reason for a cellular automaton with more than two states to be in the universality class of directed percolation, the number of possible states per site being a relevant parameter to distinguish classes. Also, the usual status of the universality class of directed percolation has also to be revised itself: it is widely believed that this class is formed of all PCAs with two states one of which is absorbing (Kinzel, 1985). Now, a recent work (Bidaux *et al.*, 1988) has shown that this is not true for a particular PCA which can be seen as a "probabilized" class 3 DCA, by opposition to directed percolation which is a probabilized class 1 DCA. The phase transition of this automaton is discontinuous for space dimension $d \geq 2$ and is continuous with critical exponents significantly different from those of directed percolation for $d = 1$ (note incidentally that the transfer matrix scaling analysis of Kinzel (1985) fails near class 3 DCAs). The observed non-universality for the minimal CML probably relies on the same grounds as the discrepancy between these cellular automata, especially in the light of the approximation discussed above which presents CML as CAs.

Results in Higher Space Dimensions and Mean-Field Approach of CMLs

Preliminary results (a detailed account will be reported in Chaté *et al.* (1988b)) in three and four space dimensions for $r = 3$ indicate that the transition to STI of the minimal CML is then very different from the previous cases. The intermittent phase is characterized by a *chaotic collective behavior* of the lattice evidenced by the time series of the instantaneous density of turbulent sites (fig. 5). We observe irregular long period oscillations of large amplitude during which the population of turbulent sites may vary from a few percent to over fifty percent. The threshold is reached when the instantaneous density incidentally goes below a certain limit concentration. Similar behavior is also observed when performing an "experimental mean-field" treatment of the minimal CML by e.g. coupling each site to a large number of random "neighbors".

This comes in contradiction with a recent result obtained by Bohr *et al.* (1987) which claims that no chaotic collective behavior should be observed in extended dynamical systems. This is probably due to the very different nature of spatiotemporal disorder in our case: local chaos is only transient and the spatial coupling is essential to the occurrence of sustained spatiotemporal intermittency while in their case the coupling is not a condition to sustained spatiotemporal disorder since the local map is already chaotic. In other terms, the existence of an absorbing state (the laminar

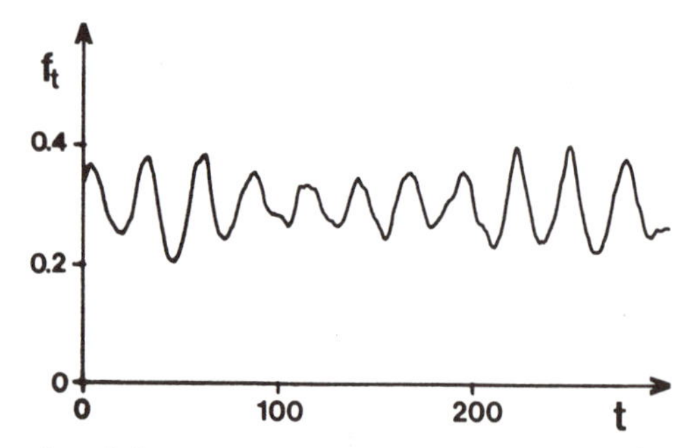

Figure 5. Time series of the concentration of turbulent sites ($X > 1$) for the minimal CML with $r = 3$, $k = 1$ and $\varepsilon = 0.183$ on a four dimensional cubic lattice of linear dimension $N = 10$ with periodic boundary conditions.

state) is at the origin of both spatiotemporal intermittency and the chaotic collective behavior.

All this suggests that the global behavior of the CML may be described, in large space dimensions, by an effective iterative map for the concentration of turbulent sites. This is exactly what one expects from a simple (one-body) mean-field analysis of two-state cellular automata which replaces the interaction with actual neighbors by an effective field (Bidaux *et al.*, 1988). Now, such an analysis is difficult to apply directly to CMLs, because of the continuous nature of the local phase space. But the approximation described above provides a natural framework in which to solve this problem, at least at a formal level, by providing an automatic procedure for a coarse-graining of the local phase space. The minimal CML approximated by a 2^p-state automaton at order p, can be reduced to a two-state PCA by distinguishing the absorbing state from the others, so that the mean-field analysis is straightforward at this stage.

CONCLUSION

The detailed study of the minimal CML has shown that the seminal conjecture of Pomeau (1986) does not hold *stricto sensu*. Indeed, the transition to turbulence *via* spatiotemporal intermittency is not in the universality class of directed percolation nor does it form a class by itself. But the essential point remains: this type of transition, which, depending on the model, is more or less directly related to hydrodynamics, is well described by the critical phenomena concepts familiar to statistical mechanics. In this respect, it seems a very general scenario typical of large aspect-ratio systems which present the "subcritical" characteristics discussed in the first section (i.e. the possibility of local abrupt transition to disorder).

But the question of the non-universality of the transition has risen several important questions in the field of cellular automata itself. The approximation scheme described above for the minimal CML has given strong indications that the status of the universality class of cellular automata with two states one of which is absorbing should be reconsidered. Moreover, results in high space dimensions d clearly exhibited the existence of chaotic collective behavior and suggested that a mean-field analysis reducing the system to an effective iterative map for a global concentration may work and be qualitatively correct as soon as $d \geq 3$.

Acknowledgements

We want to thank R. Bidaux and Y. Pomeau for their constant interest in all the parts of this work, and for the fruitful discussions we had with them.

REFERENCES

Bergé, P., 1987, From temporal chaos towards spatial effects, Nucl.Phys.B (Proc. suppl.) 2:247.

Bidaux, R., Boccara, N., and Chaté, H., 1988, Order of transition vs space dimension in a family of cellular automata, submitted to Phys. Rev. A.

Bohr, T., Grinstein, G., Yu He, and Jayaprakash, C., 1987, Coherence chaos and broken symmetry in classical many-body dynamical systems, Phys. Rev. Lett. 58:2155.

Chaté, H. and Manneville, P., 1987, Transition to turbulence via spatiotemporal intermittency, Phys. Rev. Lett. 58:112.

Chaté, H. and Manneville, P., 1988a, Coupled map lattices as cellular automata, submitted to J. Stat. Phys..

Chaté, H. and Manneville, P., 1988b, Spatiotemporal intermittency in coupled map lattices, to appear in Physica D.

Chaté, H. and Manneville, P., 1988c, Continuous and discontinuous transition to spatiotemporal intermittency in two-dimensional coupled map lattices, Europhys. Lett. 6:591.

Chaté, H. and Manneville, P., 1988d, Role of defects in the transition to turbulence via spatiotemporal intermittency, to appear in Physica D.

Chaté, H., Manneville, P., Nicolaenko, B., and She, Z.S., 1988a, in preparation.

Chaté, H., Manneville, P., and Rochwerger, D., 1988b, Mean-field approach of coupled map lattices, in preparation.

Kaneko, K., 1985, Spatiotemporal intermittency in coupled map lattices, Prog. Theor. Phys. 74:1033.

Kinzel, W., 1983, Directed percolation, in *Percolation Structures and Processes*, G. Deutscher *et al.* ed., Annals of the Israel Physical Society 5:425.

Kinzel, W., 1985, Phase transitions of cellular automata, Z. Phys. B 58:229.

Manneville, P., 1988, The Kuramoto-Sivashinsky equation: a progress report, in *Propagation in Systems Far from Equilibrium*, J.E. Wesfreid *et al.* ed., Springer.

Pomeau, Y., 1986, Front motion, metastability and subcritical bifurcations in hydrodynamics, Physica D 23:3.

Pomeau, Y., and Manneville, P., 1979, Stability and fluctuations of a spatially periodic convective flow, J. Phys. (Paris) Lett. 40:609.

Pomeau, Y., and Manneville, P., 1980, Wavelength Selection in Cellular Flows, Phys. Lett. A 75:296.

Swift J., Hohenberg P.C., 1977, Hydrodynamic fluctuations at the convective instability, Phys. Rev. A 15:319.

Wolfram, S., 1986, *Theory and Applications of Cellular Automata*, World Scientific, Singapore.

Phase Turbulence, Spatiotemporal Intermittency and Coherent Structures

Hugues Chaté [†‡] and Basil Nicolaenko [†]

[†]Center for Nonlinear Studies and Theoretical Division
Los Alamos National Laboratory
Los Alamos, NM 87545, USA

[‡]Institut de Recherche Fondamentale
DPh-G/PSRM, CEN-Saclay
91191 Gif-sur-YvetteCedex, France

GENERAL SETTING

The essentially spatiotemporal dynamics of extended physical systems is still poorly understood. Indeed, in spite of the success of dynamical systems theory to explain the *deterministic chaos* occurring in confined situations, this approach is unable to handle the complex behaviors appearing when the spatial structure is not frozen. From this point of view, *turbulence* may be seen as the ultimate complex system, and it seems interesting to study models of extended systems somewhat "simpler" than the Navier-Stokes equations but nevertheless retaining some of the essential physics of the problem.

In this work, we study the following one dimensional partial differential equation (PDE), named the Kolmogorov-Spiegel-Sivashinsky (KSS) model:

$$\frac{\partial \psi}{\partial t} + \frac{\partial^2 \psi}{\partial x^2} + \frac{\partial^4 \psi}{\partial x^4} + \left(\frac{\partial \psi}{\partial x}\right)^2 + \eta\psi - \delta\frac{\partial}{\partial x}\left(\frac{\partial \psi}{\partial x}\right)^3 = 0 \tag{1}$$

where $x \in [0, L]$. The KSS model can be seen as a more general version of the modified Swift-Hohenberg model of convection studied by Chaté and Manneville, known as "model b" (Pomeau and Manneville, 1980), which was shown to exhibit, in the large

New Trends in Nonlinear Dynamics and Pattern-Forming Phenomena
Edited by P. Coullet and P. Huerre
Plenum Press, New York, 1990

size limit, a transition to turbulence *via spatiotemporal intermittency* (for a review see the work of Chaté and Manneville in this volume). We report here a similar investigation of the spatiotemporal dynamics exhibited at large sizes by the KSS model.

We first briefly review the physical background of the model and its mathematical properties (for a more detailed account see Nicolaenko, 1987). We then describe qualitatively the spatiotemporal dynamics exhibited by careful numerical simulations in the whole region of parameter space of physical interest. The last part is devoted to more quantitative approaches like the possibility of a description in terms of cellular-automata and the statistical characterization of the transitions exhibited by the model.

PHYSICAL BACKGROUND OF THE KSS MODEL

Spiegel *et al.* derived a slightly more general version of equation (1) in the context of solar convection (Depassier and Spiegel, 1981; Poyé, 1983). They performed a systematic multiple-scale analysis of a compressible solar convective layer zone. Variable ψ of equation (1) then stands for the mean horizontal temperature field. Note that the quadratic nonlinear term in (1) is the usual Burgers nonlinearity if we rewrite the equation for $u = \partial\psi/\partial x$. It is the main destabilizing term in the model, transporting energy from large to small scales.

Equation (1) was also derived by Sivashinsky in an asymptotic analysis of the large scale dynamics of the two-dimensional Kolmogorov flow (Sivashinsky, 1985). This flow, reproduced in the laboratory by Bondarenko *et al.* (1979), is a viscous flow induced by a unidirectional external force field periodic in the transverse coordinate. It was proposed by Kolmogorov to test the formation of large scale structures after the basic laminar shear flow breaks down (Kolmogorov, 1960). Sivashinsky used the extreme anisotropy introduced by the forcing to perform an asymptotic expansion of the large scale stream function (having removed the mean flow). It was shown that a *negative viscosity* instability is generated at large scales which is believed to be at the origin of the *coherent structures* (large eddies) observed in the turbulent flow (She, 1987a; Yakhot and Sivashinsky, 1987).

This negative local viscosity is easily visualized by rewriting the second and the last term of (1) under the form of the divergence of a current J_ψ with an effective local viscosity ν_{eff}:

$$\nabla J_\psi = \frac{\partial}{\partial x}\left(-\nu_{\text{eff}}\frac{\partial\psi}{\partial x}\right) \quad \text{with} \quad \nu_{\text{eff}} = -1 + \delta\left(\frac{\partial\psi}{\partial x}\right)^2 \tag{2}$$

Feeding energy from small to large scales, it can be thought to be at the origin of the inverse cascade phenomenon observed by She (1987a). However, the dominant nonlinear interaction for unstable large scales may generate a locally *positive* effective viscosity as seen in (2). Indeed, for δ big enough and for rapid (local) variations of ψ, the effective (local) coefficient of the second order term can become negative (yielding a positive viscosity), so that the cubic term governed by parameter δ is in fact a stabilizing factor. This process can be understood as the saturating effect of the $(\partial\psi/\partial x)^2$ factor appearing both in the cubic term and in the quadratic nonlinearity which have opposite influence on the stability of large scales.

Stabilizing also is the damping term of coefficient η, which can be seen, in the original problem of the Kolmogorov flow, as the linear friction at the boundaries of the experimental apparatus.

Finally the role of the fourth order term $\partial^4\psi/\partial x^4$ is well known since the numerous studies of the Kuramoto-Sivashinsky model (for a recent review, see Manneville, 1988). In this case, which corresponds here to $\eta = \delta = 0$, it is the only stabilizing term, dissipating energy at small scales.

QUALITATIVE DYNAMICS AT LARGE L

We now report a careful numerical exploration of the KSS model. All simulations have been performed with periodic boundary conditions, using pseudo-spectral methods with a number of modes roughly equal to ten times the number of unstable modes for $\eta = 0$ at given L (typically 512 modes for 50 unstable modes). For most simulations, we used an adaptive time step, adaptive order (at least fourth order) code developed by J.M. Hyman in Los Alamos to insure the precision of the results (Hyman *et al.*, 1986). Long transient periods were allowed after each change of parameter values. The mean drift ($k = 0$ mode) was usually subtracted to keep ψ between constant values.

Detailed Setting of the Problem

We are interested here in the spatiotemporal dynamics of the KSS model in the large size limit. The dynamics of the KSS model in small boxes $L < 50$, best described in the framework of low dimensional dynamical systems theory, was studied in detail by Nicolaenko (1987). While the situation becomes very intricate for intermediate L, one hopefully recovers a certain simplicity for very large boxes ($L > 500$) where a statistical description is better suited.

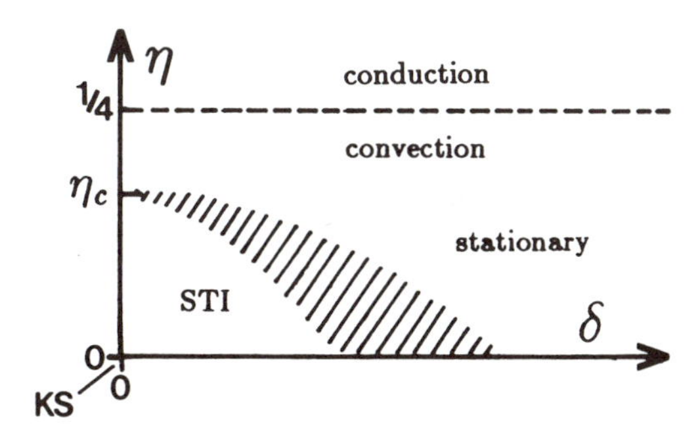

Figure 1. Overview of the different regimes of the KSS model in the (η, δ) parameter plane at large L.

In the following, we thus kept the length L of the box fixed (and large) while varying the two control parameters η and δ. Given the above discussion of the relative physical importance of the different terms in (1), the $\eta = \delta = 0$ case appears as the "most turbulent" state in the (η, δ) parameter plane. The KSS model then reduces to the well known "pure" Kuramoto-Sivashinsky (KS) equation for which a similar study is impossible since it has no other control parameter than the length of the box itself. For η and/or δ large, the KSS model exhibits stationary solutions, so that decreasing η and/or δ, the system undergoes transitions to a spatiotemporal disorder characteristic of the pure KS equation at large L. This crude description is summarized in fig. 1.

Summary of the $\delta = 0$ Case

When $\delta = 0$, the KSS model reduces to the modified Swift-Hohenberg model studied by Chaté and Manneville in the same spirit. The linear stability analysis (valid of course for the full KSS equation) helps to understand the observed transitions. The dispersion relation reads:

$$s(k) = k^2 - k^4 - \eta \tag{3}$$

For $\eta < \eta_0 = 1/4$ the quiescent state $\psi(x) = 0$ (conductive regime) is unstable with respect to the fastest growing mode at $q = q_0 = 1/\sqrt{2}$ and the system has stationary spatially periodic solutions (steady convection). For $\eta < \eta_c \simeq 0.080$, the system exhibits spatiotemporal intermittency regimes characterized by the coexistence of regular ("laminar") and disordered ("turbulent") patches slowly evolving in space and time (see fig. 1 for a summary of these bifurcations). Thanks to a binary reduction of the spatiotemporal information (to be generalized below for the full KSS model), the transition region was shown to possess most features of a continuous phase transition (Chaté and Manneville, 1987).

Ternary Reduction for the KSS Model

Figure 2 shows a typical instantaneous solution in an intermittent regime for a box of length L corresponding to 50 unstable modes ($L \simeq 444$). It is characterized by regions where the basic cellular solution is regular and by distorted cells, either of small or large local amplitude.

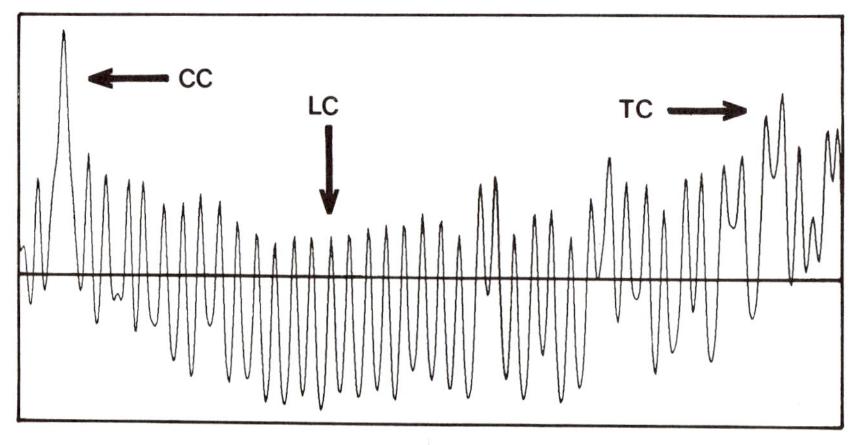

Figure 2. Typical instantaneous solution of the KSS model for a box of length $L = 444$ for $\eta = 0.05$ and $\delta = 0.046$. Laminar (LC), coherent (CC) and turbulent (TC) cells are indicated.

In order to unravel the essential features of the dynamics, we had to perform a reduction of the full field $\psi(x,t)$. The first step was to plot the extrema of the instantaneous solution along time (an equivalent choice would have been to show the zeros of the solution). Figure 3a shows such a spatiotemporal representation for a typical intermittent regime (same parameter values as in fig. 2). Important remarks are now in order: large regular regions differentiate themselves from a background of disordered patches characterized by spatiotemporal dislocations of cells. An overall propagative mode can also be noticed, which will appear much more clearly under the next step of the reduction. Finally, wide, non oscillating cells distinguish themselves from the normal width, sometimes oscillating ones forming the regular regions. These remarks are at the base of the reduction: normal wavelength, normal peak-to-peak amplitude cells, which can be seen as portions of the regular stationary solutions for larger values of η and/or δ, are represented by a particular color according to a criterion on the local amplitude and called *laminar cells*. More precisely, a cell delimited by three consecutive extrema of ψ, say x_{i-1}, x_i and x_{i+1}, is "laminar" if

$$| \psi(x_{i-1}) - 2\psi(x_i) + \psi(x_{i+1}) | \in \left[2(1-\Delta)\bar{A}\ ,\ 2(1+\Delta)\bar{A}\right]$$

where $\bar{A}$ is the mean peak-to-peak amplitude of the regular solutions and Δ is a "tolerance" factor. The physical relevance of such a criterion is insured by the independence of the results on the precise value of the tolerance factor Δ provided it remains a small but non vanishing number (e.g. 0.1). Other criteria can also be used, using for example the local wavelength, or even a procedure based on the time series at each point of the simulation. All give the same results, stressing the deep physical relevance of the reduction.

The large wavelength, large amplitude cells appearing in fig. 2 and 3a are characterized by a similar criterion using the fact the local peak-to-peak amplitude is roughly twice the normal amplitude $\bar{A}$ in this case (see fig. 2). Given their apparent dynamical robustness (see fig. 3a and next subsection) we call them *coherent cells.*

Finally, all cells not fulfilling one of the above criteria are called *turbulent cells* since they represent strongly distorted and unsteady regions.

This ternary reduction, summarized in fig. 2, produces three color spatiotemporal representations of the dynamics of the KSS model. Figure 4a shows the result of such a reduction on the run of fig. 3a. The propagating structures are now clearly shown. They are due to *deformation waves*: the local cellular structure supports a propagating deformation but the cells themselves are not propagating. Locally, the passage of waves is seen as the regular oscillations of the cellular structure (fig. 3a).

Again, the chosen criteria may have been different (and would probably be different for another model) but would have produced the same reduction, provided they are based on the elementary dynamical features of the KSS model.

Spatiotemporally Intermittent Regimes

The three basic ingredients of the dynamics have varying relative importance depending on the position in the (η, δ) parameter plane. There exists a large domain of this plane where the system exhibits spatiotemporal intermittency regimes with stationary statistical properties. They are characterized here by the coexistence of clusters of laminar and coherent cells with turbulent patches.

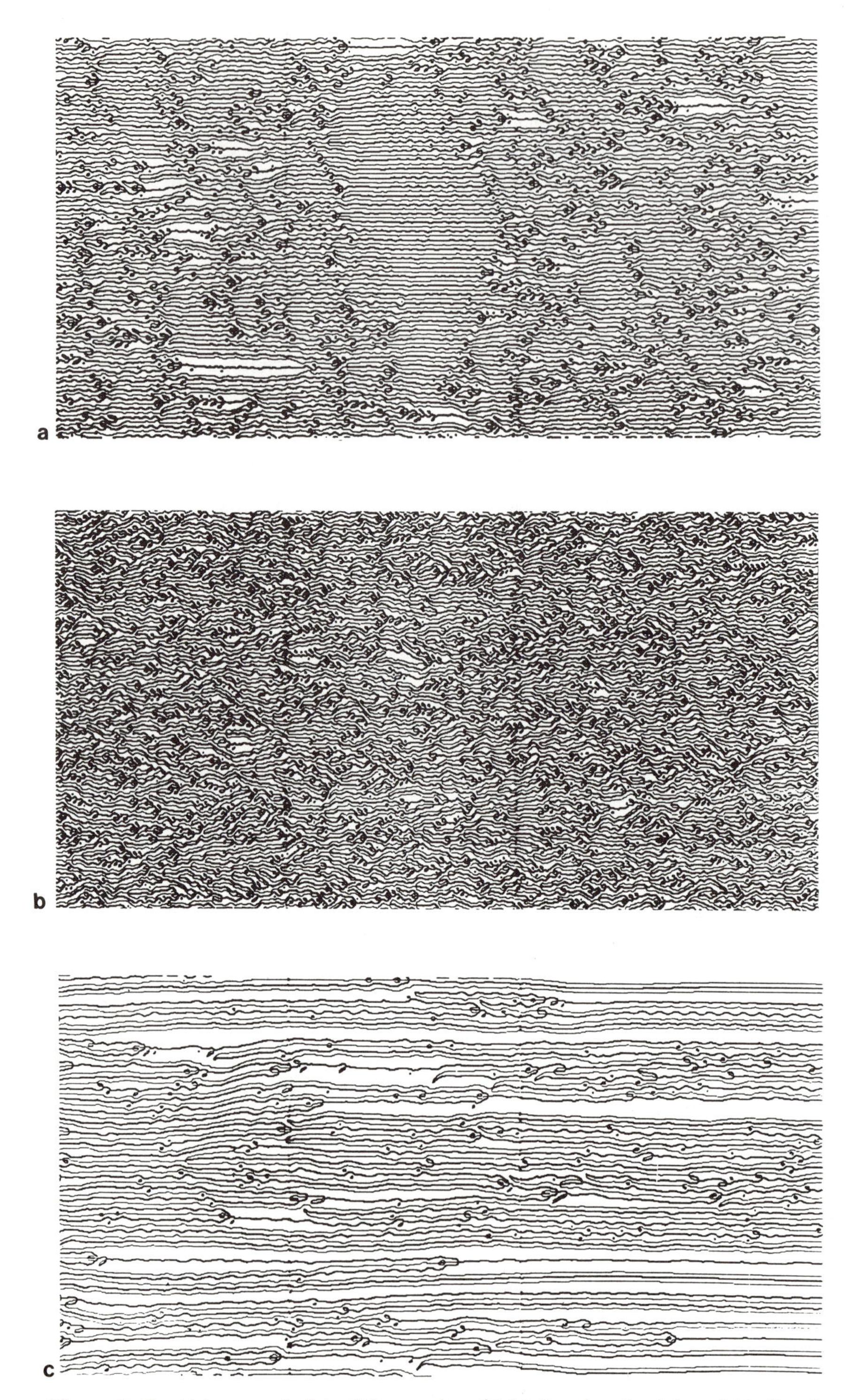

Figure 3. *Spatiotemporal plot of the maxima (thick lines) and minima (thin lines) of the solution. Time is running from left to right during $T = 2000$, the length of the box is $L = 444$. (a): $\eta = 0.05$ and $\delta = 0.046$. (b): $\eta = 0.025$ and $\delta = 0$. (c): $\eta = 0$ and $\delta = 0.25$.*

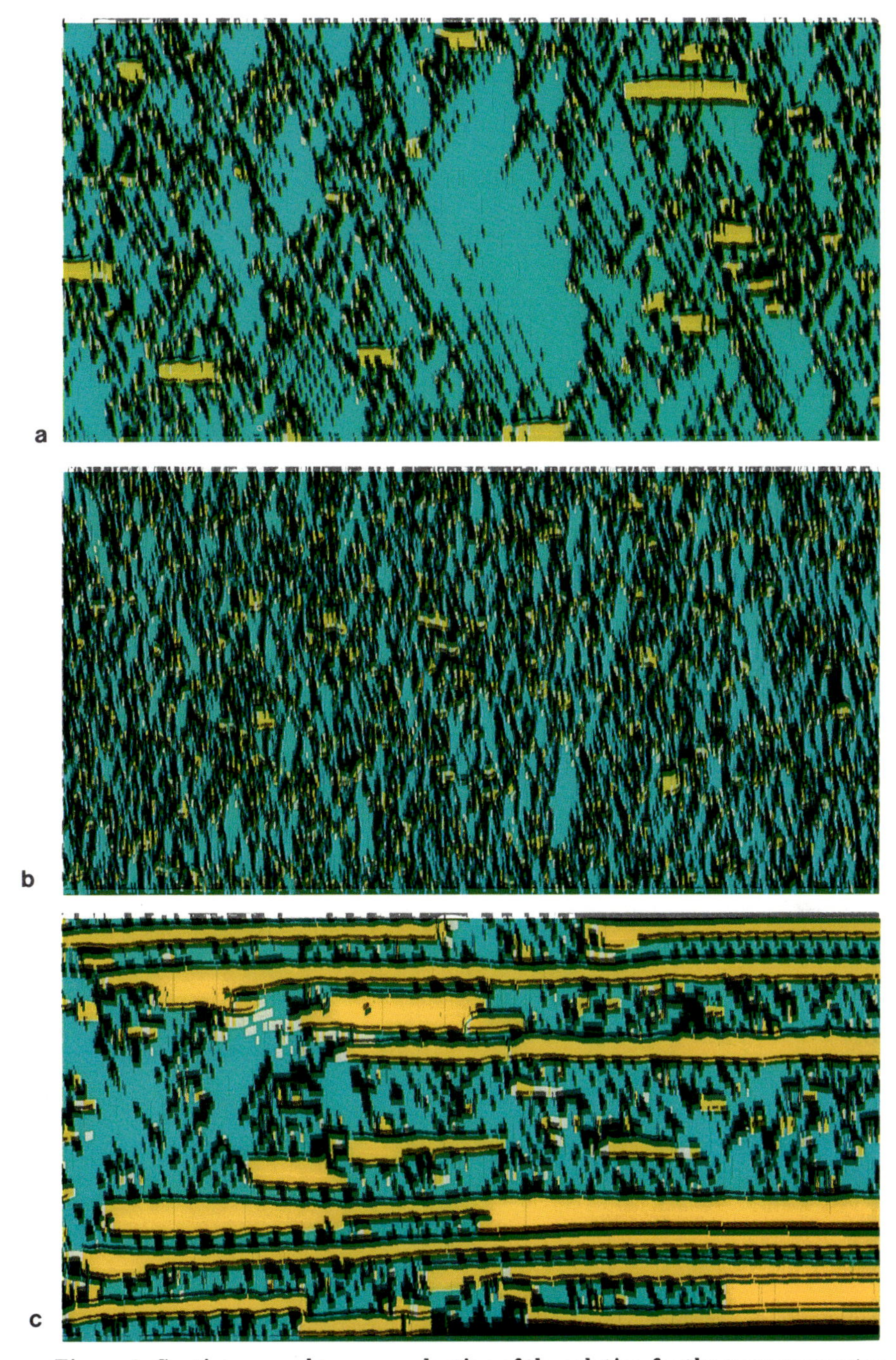

Figure 4. Spatiotemporal ternary reduction of the solution for the same parameter values as figure 3. Laminar cells are in blue, coherent cells in yellow and turbulent cells in black.

Along the $\delta = 0$ axis, coherent cells are so short lived that they do not appear at all in the reduction, except very near the pure KS model ($\eta \simeq 0$) where there is a rather high density of such transient structures (fig. 3b and 4b). In this respect, it is noteworthy to remind that such structures were evidenced for the pure KS model in small boxes (Aimar and Penel, 1983; Hyman *et al.*, 1986). Moreover, the absence of long lived coherent cells agrees with the binary reduction performed by Chaté and Manneville.

As soon as $\delta \neq 0$, the coherent cells can be very long lived, all the more near the transition region where they eventually never disappear. They may also occupy a large fraction of space, especially for small values of η and in particular on the $\eta = 0$ axis (see figure 3c and 4c).

Thus, the spatiotemporally intermittent regimes differ from the $\delta = 0$ case, where only laminar and turbulent cells are present, by the superposition of the coherent cells, which have a very particular dynamics. Indeed, inside a coherent cell, the local fluctuations of amplitude are orders of magnitude smaller than the oscillations of laminar cells submitted to deformation waves. This reminds us of the coherent structures in turbulence whose presence or absence change drastically the transport properties of the flow.

Along the $\eta = 0$ Axis: Dynamical Origin of Coherent Structures

The transition between stationary solutions and the maximum spatiotemporal disorder of the pure KS equation when varying δ and keeping $\eta = 0$ is characterized by the importance of coherent cells. Increasing δ, these cells become more and more frequent and agglomerate together so as to form complex multi-level structures (fig. 3c and 4c). These structures eventually become part of stationary solutions when δ is large enough and one may have many stationary or quasi-stationary solutions with different spatial structures coexisting for the same parameter values (fig. 5). A closer look at a stationary coherent cell shows that it can be obtained simply by flipping one arch of a laminar cell and compressing locally since the wavelength of a coherent cell (CC) is smaller than twice the wavelength of a laminar cell (LC): $\lambda_{CC} \simeq 1.75\ \lambda_{LC}$. Repeating this "micro-surgery" process several times, one can build a countable infinity (for $L \to \infty$) of complex coherent structures.

Rotating waves were also observed for which, thanks to the periodic boundary conditions, a complex frozen spatial structure moves with a finite, constant and very small angular velocity (we checked by energy considerations that this is not a numerical artifact). When δ is further increased, less and less complex stationary solutions are observed until only regular cellular solutions subsist.

Decreasing δ, and starting from regular cellular solutions, one observes (at fixed L) structures with more and more cells with long transients in between, until the actually spatiotemporally intermittent regimes appear. This happens for values of δ seemingly smaller than those at which the stationary solutions appear when increasing δ, although it is difficult to conclude to the existence of hysteresis loops due the multiplicity and the sometimes partially stationary nature of the solutions in this parameter domain.

The stationary solutions with complex spatial structures formed of coherent cells shed light on the dynamical origin of the coherent structures appearing in the spa-

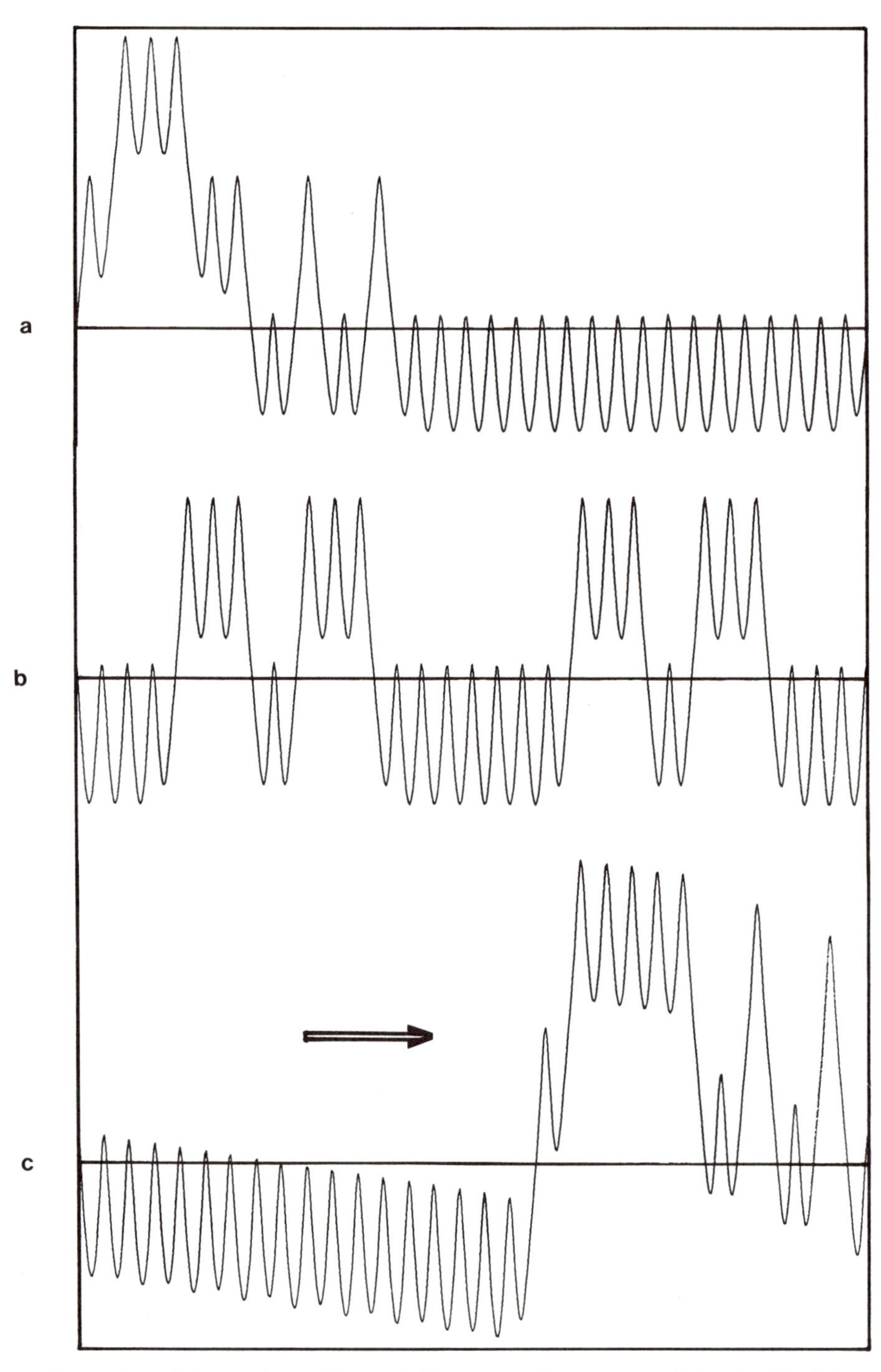

Figure 5. Solutions of the KSS model for $L = 444$, $\eta = 0$ and $\delta = 0.625$. (a): stationary solution with multi-level coherent structures. (b): symmetric stationary solution obtained from symmetric initial condition. (c): rotating wave with quasi-frozen spatial structure (the left part is slightly oscillating). The arrow shows the direction of rotation.

tiotemporally intermittent regimes (fig. 3c and 4c). Although maybe a little far-fetched, the relevance of this simple idea to "real" turbulence may not be negligible (the work of Moffatt on the fixed points of the Euler equations is based on similar ideas (Moffatt, 1987)). Coming back to the Kolmogorov flow, the coherent structures formed in the KSS model may be traced back to the large eddies appearing in the (local) inverse cascade phenomenon (for example by using a reconstruction procedure of the two dimensional flow (She, 1987b)).

TOWARDS QUANTITATIVE APPROACHES

Spatiotemporal Statistics

The transition at $\delta = 0$ and η varying was characterized by studying the distribution of the sizes of the clusters of laminar cells. Away from the threshold, this distribution is exponential with a characteristic length $\xi(\eta)$ of the order of one cell for the pure KS model. This length diverges at the threshold for spatiotemporal intermittency $\eta_c \simeq 0.080$, where there is a power-law distribution of laminar cluster sizes. The same phenomenon, characteristic of continuous phase transitions, was also observed when looking at the statistics of the time lapses during which a point remains in a laminar cluster.

Preliminary results on the full KSS model ($\delta \neq 0$) indicate that this picture is broken by the seemingly random occurrences of coherent cells. Indeed, if statistically stationary regimes are still observed, no power-law behavior is recovered in the threshold region. This puts forward the need for a different definition of a precise threshold. Other statistics are under current investigation, based on other dynamical objects: the statistics of the lifetimes of the coherent cells (their divergence then defines the threshold) and a study of the density of the various types of space-time dislocations (which must go to zero at threshold). First results do not show (yet) a smooth variation of these quantities, ruling out continuous transitions.

Towards a Cellular Automaton Description

In building the ternary reduction at the base of the spatiotemporal representations, we have made two steps towards a fully discrete analysis of the KSS model: thanks to the underlying basic cellular solution, space has been discretized into cells and the local phase space has been reduced to three possible states (laminar, coherent and turbulent). An important dynamical object was also put into light: the deformation waves which appear as the vectors of disorder, creating turbulent cells in a propagative manner. Moreover, they are at the origin of local oscillations of the cellular structure which could be used as a natural clock to discretize time.

At a qualitative level, the interaction rules between these objects translate themselves as *space-time dislocations* of cells (their importance has also been stressed by Shraiman (1986,1988)). A detailed study of the various interactions is out of the scope of this paper but it is important to stress already the dissymmetry between annihilation and creation of (laminar) cells. The latter are merely a relaxation process of the local structure submitted to stretching while the annihilations are the basic energetic events which emit the deformation waves when the structure resolves abruptly a local compression (Chaté *et al.*, 1988).

If the output of the interaction rules between the elementary objects described above were known as functions of the parameters η and δ, one would be left, at a formal level at least, with the problem of the phase transitions of a cellular automaton. This would yield a thermodynamical formulation of the original hydrodynamical problem, thanks to the large size limit and the stationary statistical properties of the spatiotemporally intermittent regimes. This approach in terms of critical phenomena of the transition to turbulence was proven fruitful by the study of the $\delta = 0$ case by Chaté and Manneville (1987) which originated in a conjecture of Pomeau (1986) saying that in some cases this transition could be of the directed percolation type. This justifies a direct cellular-automata like construction and suggests a general procedure to handle the spatiotemporal complexity of other fully continuous models.

CONCLUSION

Ths KSS model is directly related to hydrodynamical situations. In spite of its simplicity, it retains many features of the flows it is derived from. Most spectacular is the occurrence of spatiotemporal intermittency regimes with coherent cells. These cells eventually form complex coherent structures, which can be seen as the local remnants or partial replicas of stationary solutions exhibited by the model for other parameter values.

The problem of the transition to the spatiotemporal intermittency regimes is much richer for the full KSS model than for the $\delta = 0$ case. At this point, the scenario of a continuous phase transition does not seem to hold when $\delta \neq 0$, but the statistical stationarity of the intermittent regimes still suggest that a description of the transition in terms of critical phenomena may be suited. Ongoing work is devoted to this aspect of the problem.

At a more general level, the study of the transitions to spatiotemporal disorder in the KSS model can be seen as the unfolding of the intricate phase turbulence of the pure Kuramoto-Sivashinsky model in large boxes. With the help of the two control parameters η and δ, the most turbulent point in parameter space can be reached by different routes which may separate and put into light different aspects of its dynamics. This procedure could serve as a general guide, stressing again the need for a detailed investigation of the transitions in order to get insights on the problem of the "nature of turbulence".

Acknowledgements

The authors have benefited of numerous and fruitful discussions with P.C. Hohenberg, J.M. Hyman, P. Manneville, Y. Pomeau, Z.S. She, B. Shraiman, G. Sivashinsky, E.A. Spiegel and S. Zaleski. This collaboration was made possible thanks to NATO grant # 0509/85. Also deeply acknowledged is the friendly hospitality of CNLS and its director, D. Campbell.

REFERENCES

Aimar, M.T., and Penel, P., 1983, Some Numerical Aspects of a Disturbed Premixed Flame Front Study, in Publ. Math. Appl. Marseilles-Toulon, 2.

Bondarenko, N.F., Gak, M.Z., and Dolzhansky, F.V., 1979, Atmospheric and Oceanic Physics 15:711.

Chaté, H., and Manneville, P., 1987, Transition to Turbulence *via* Spatiotemporal Intermittency, Phys. Rev. Lett. 58:112.

Chaté, H., Manneville, P., Nicolaenko, B., and She, Z.S., 1988, in preparation.

Depassier, M.C., and Spiegel, E.A., 1981, The Large Scale Structure of Compressible Convection, Astro. Jour. 86(3):496.

Hyman, J.M., Nicolaenko, B., and Zaleski, S., 1986, Order and Complexity in the Kuramoto-Sivashinsky Model of Weakly Turbulent Interfaces, Physica D 23:265.

Kolmogorov, A.N., 1960, in Seminar Notes edited by V.I. Arnold and L.D. Meshalkin, Uspekhi Mat. Naut. 15:247.

Manneville, P., 1988, The Kuramoto-Sivashinsky Equation: a Progress Report, in *Propagation in Systems Far from Equilibrium*, J.E. Wesfreid *et al.* ed., Springer.

Moffatt, H.K., 1987, Geophysical and Astrophysical Turbulence, in *Advances in Turbulence*, G. Comte-Bellot and J. Mathieu ed., Springer, and references therein.

Nicolaenko, B., 1987, Large Scale Spatial Structure in Two-Dimensional Turbulent Flows, Nucl. Phys. B (Proc. suppl.) 2:453.

Pomeau, Y., 1986, Front Motion, Metastability and Subcritical Bifurcations in Hydrodynamics, Physica D 23:3.

Pomeau, Y., and Manneville, P., 1980, Wavelength Selection in Cellular Flows, Phys.Lett. A75:296.

Poyé, J.P., 1983, The Rayleigh-Bénard Two-Dimensional Convection in a Fluid between Two Plates of Finite Conductivity, Thesis, Columbia University.

She, Z.S., 1987a, Instabilités et dynamique à grande échelle en turbulence, Thèse de l'Université Paris 7.

She, Z.S., 1987b, Metastability and Vortex Pairing in the Kolmogorov Flow, Phys.Lett.A 124:161.

Shraiman, B.S., 1986, Order Disorder and Phase Turbulence, Phys. Rev. Lett. 57:325.; Hohenberg, P.C. and Shraiman, B.S., 1988, Chaotic Behavior of an Extended System, preprint.

Sivashinsky, G.I., 1985, Weak Turbulence in Periodic Flows, Physica D 17:243.

Yakhot, V., and Sivashinsky, G.I., 1987, Negative Viscosity Effects in Three Dimensional Flows, Phys. Rev. A 35:815.

PROPERTIES OF QUASI ONE-DIMENSIONAL RAYLEIGH BENARD CONVECTION

M. Dubois, P. Bergé and A. Petrov

Service de Physique du Solide et de Résonance Magnétique
C.E.N.-Saclay
91191 Gif-sur-Yvette, Cedex, France

Rayleigh Bénard convection[1] is a well known phenomenon which develops striking spatial periodic structures in a fluid layer submitted to a destabilizing temperature gradient. In a rectangular container of horizontal extensions L_x and L_y large compared to the depth d, nice straight parallel rolls can be observed under some particular conditions[2] near the critical Rayleigh number Ra_c. When the convection is achieved with a high Prandtl number fluid, an increase of Ra beyond a well defined value RaII (about 10 Ra_c) generates a new set of rolls superimposed on the critical roll pattern. The axes of the two sets are mutually perpendicular. In both cases, the convection is stationary. The velocity field associated with the simple critical rolls (below RaII) is two-dimensional while, in the case of the two perpendicular sets of rolls a three-dimensional velocity field is excited. As far as the horizontal planeform is concerned (meaning that we disregard the vertical dependence of the velocity) the spatial properties only depend on *one* coordinate, say X, in the case of the rolls below Ra II; in this context we speak of one-dimensional convection, while above Ra II, where the two sets of rolls coexist, we speak of two-dimensional convection.

A further increase of Ra above Ra II produces complex spatio-temporal turbulent states where spatial and temporal evolutions are so deeply intricated that a direct understanding of the relevant mechanisms is very difficult.[3,4,5].

A well known approach, used to simplify this turbulent situation, consists in freezing the spatial structure so that only temporal degrees[6,7,8,9] of freedom are excited, namely convection in confined geometries. Another kind of simplification which keeps part of the spatial richness of the flow consists in studying the approach of the turbulent states in a one-dimensional (or quasi one-dimensional) convective state. The interest for these 1-D systems has been growing up in the last few years, at first theoretically, then experimentally, since they can provide intermediate situations between dynamical systems and complex spatio-temporal behaviours. The aim of this paper is to report results obtained in such a case.

New Trends in Nonlinear Dynamics and Pattern-Forming Phenomena
Fdited by P. Coullet and P. Huerre
Plenum Press, New York, 1990

The first step is to study under what conditions a one-dimensional chain of rolls can be preserved up to Ra numbers for which interesting time dependences occur. One trick is to reduce one of the two horizontal dimensions down to values for which, even at the highest Ra values of interest, the convection remains mainly one-dimensional. Let us emphasize again that "one-dimensional" means here that the properties of the structure, seen from above, is fully described by a single coordinate, namely X, axis of which is parallel to the longest extension of the container.

EXPERIMENTAL APPARATUS

To study the influence of the transverse aspect ratio on the Ra domain of existence of the one-dimensional pattern, a fourfold cell was constructed. Four channels of decreasing width were machined in a plexiglass plate (fig.1) inserted between a pair of 6mm thick thermostated glass plates. The larger aspect ratio $\Gamma_x = L_x/d$ is the same for all the cells ($\Gamma_x = 180mm/8.56mm \simeq 21$) and the transverse aspect ratios $\Gamma_y = L_y/d$ are $1.17, 0.93, 0.58$ and 0.35 respectively. Shadowgraphic visualization, associated with a time-lapse video recorder provides detailed qualitative information about the structural behaviour in the horizontal plane, allowing to evaluate, by direct comparison, the effect of the transverse aspect ratio. Quantitative measurements were also performed in a rectangular container inserted between thermalized copper plates with aspect ratios $L_x/d = 36$ and $L_y/d = 0.4$. The shadowgraphic image was also observed, the light beam crossing the cell along the direction parallel to the rolls' axes. A part of this image was focused on a photodiode array of 256 pixels. This system, associated with a computer allowed to record spatio-temporal characteristics.

Both kinds of cells were filled with silicon oil of viscosity $0.65\ 10^{-2}$ Stokes (Prandtl number 7.5).

Note that Ra_c depends strongly on the Γ_y value when $\Gamma_y < 1$. In the case of the studied channels, the ratio $Ra_c(\Gamma_y)/Ra_c(\infty)$ is respectively 1.28, 1.6, 2.8 and 5.55 for the channels with $\Gamma_y = 1.17, 0.93, 0.58$ and 0.35 respectively. (from[12] and laboratory measurements)

SPATIAL PROPERTIES

The spatial properties of the convective patterns have been studied particularly in the four channels cell for which the horizontal configuration is visualized through the glass plates. Near the threshold Ra_c, whatever Γ_y is, we observe a perfect structure with rolls' axis parallel to the short side of the cell. The wavelength is close to λ_c for each channel (λ_c is the critical wavelength = 2d).

When Ra is increased, a 3-D stationary motion develops in the channels with $\Gamma_y = 1.17$ and $\Gamma_y = 0.93$, while the time dependence sets-in at larger Ra (In the following, we will be interested no longer by the behaviour in these channels). On the contrary, in the other channels ($\Gamma_y = 0.58$ and $\Gamma_y = 0.35$) time dependence appears far before the evidence of a 3-D motion in the pattern and then depends only on the X space variable. So, from the qualitative shadowgraphic observations, we can state that the related dynamical behaviour will be that of a quasi-one dimensional pattern.

Experimentally, it is found that in these channels, with $\Gamma_y < 0.6$, short wavelengths are favoured when Ra is increased. Starting from a pattern in which the wavelength is near λ_c, new rolls appear, inducing a shortening of the local wavelength, as shown in

Figure 1

Shadowgraphic pictures obtained with the fourfold cell, and seen from above. Dark = hot uprising streams, bright = cold downgoing streams Depth $d = 8.56mm$. *Channel a)* $\Gamma_y = 0.35$, *b)* 0.58, *c)* 0.93, *d)* 1.17.

Just below the channel a) one can see a length scale made of 5mm spaced dashes. ΔT_T, *the temperature difference applied to the glass plates, is* $3.2^\circ C$. *In the widest channels c) and d) two-dimensional (time dependent) patterns are clearly present.*

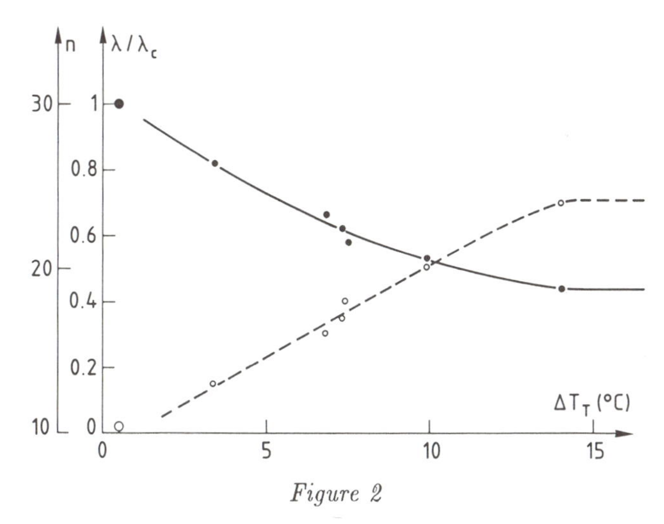

Figure 2

Evolution of the pattern in the channel with $\Gamma_y = 0.35$. *The variation of the number of wavelengths n is reported as a function of the temperature difference* ΔT_T *(open circles)* $\left(\mathrm{Ra} \simeq 10^6 \ for \ \Delta T_T = 10^\circ\right)$. *The full circles correspond to the correlative variation of the wavelength normalized to the critical one.*

fig.2. These short wavelengths can also be generated directly by a quick ΔT increase. This property is related to the confinement along the Y direction: the lower the Γ_y value, the shorter the observed wavelength by increasing Ra. Nevertheless, there is a lower limit λ_m which is such that the dimension of the roll $\lambda_m/2$ is very close to the width L_y of the channel for the smallest Γ_y values. It is the first time that so short and stable wavelengths are observed in R.B. convection, though a similar wavelength evolution was seen in Hele-shaw cells [10]. In this later case, this behaviour has been predicted theoretically, in the frame of a model involving a porous medium [11].

Another characteristics of these narrow channels is the appearance of a stable and stationary dispersion in the size of the different rolls, even under fixed external conditions. This striking phenomenon has to be studied in more details. The first observations indicate that it does not appear in the channels with $\Gamma_y > 1$, while it occurs reproducibly in the narrowest channels. This dispersion may appear at first as a broad distribution of the actual local wavelength[13]. Above a given *Ra* value which depends on Γ_y, multiples of λ_0 appear in the pattern, where λ_0 is the basic wavelength (Fig.4). This behaviour introduces a strong spatial discontinuity, revealing that in this situation, the diffusion of the phase of the rolls does not play an important role. This phenomenon is particularly important when the fluid is confined between glass plates.

COLLECTIVE OSCILLATIONS IN A RECTANGULAR CELL

In a certain range of (high) Ra numbers and (short) wavelengths, the first time dependence, observed with increasing Ra, manifests itself by a remarkable collective oscillation of the rolls. When the rolls size is not homogeneous, these oscillations concern the regions with the shortest wavelengths in which they are localized (cell with $\Gamma_y = 0.35$ between glass plates and annular geometry[13]) On the contrary, in the case of the rectangular cell, ($\Gamma_y = 0.4$ and between copper plates) the wavelengths were sufficiently homogeneous in the pattern so that the collective oscillation was very pure and synchronized all along the pattern.

As the detailed results are to be published elsewhere [14], let us just summarize the main features.

1) With a typical wavelength $\lambda = 0.38\lambda_c$ stable in a wide Ra domain, collective oscillations are present in the range $1.17\ 10^6 < \text{Ra} < 1.7\ 10^6$.

2) These oscillations have the following characteristics: the main vertical streams move periodically around their mean position with phase oposition between hot and cold streams and there exists a striking phase coherence of the oscillations all along the cell

3) The amplitude of these displacements varies periodically along the X direction. Some rolls do not oscillate, giving the equivalent of nodes of vibrations: then a stationary wave can be defined along the X axis of the cell.

These behaviours are illustrated in the figure 3.

Note that this collective oscillating behaviour has not been observed in the cells with $\Gamma_y > 0.5$.

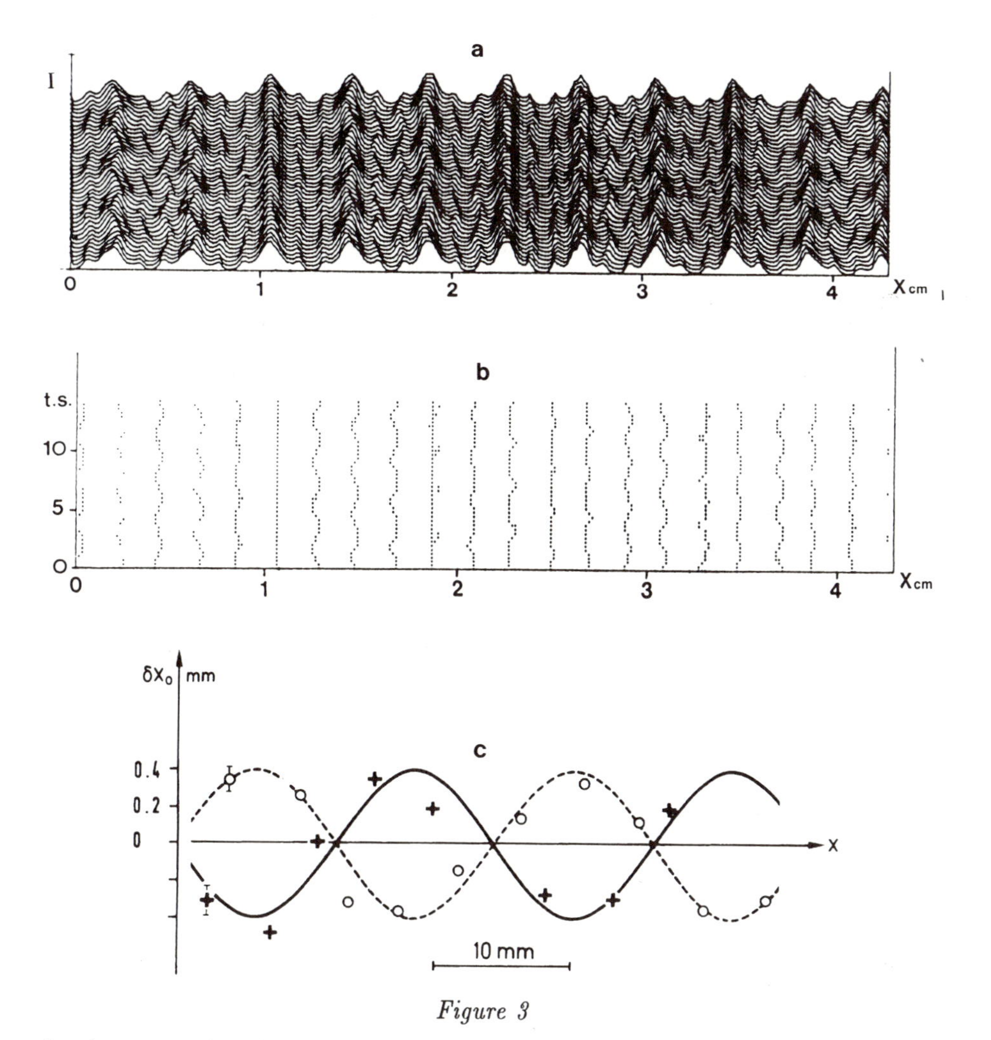

Figure 3

Spatio-temporal evolution of the pattern in the rectangular cell between copper plates (monoperiodic regime. Ra $= 1.27\ 10^6$).

a) Intensity versus X of a part of the shadowgraphic image taken at the midheight of the cell. Minima: uprising hot streams. Maxima: descending cold streams. The time spacing between each line record is 350ms. ($f_1 = 0.275H_z$).

b) Positions of the intensity extrema versus time. (Processed from the data of the fig.3.a)

c) Amplitude of the displacement $\delta X_0 = X_m - X_0$ *of the streams versus X for the data shown in fig.1b. The origin* X_0 *of the positions is taken arbitrarily at the instant* $t = 0$. X_m *is the maximum displacement for each stream. o : hot streams; + Cold streams. In spite of the poor spatial resolution (one pixel corresponds to* 0.17*mm, i.e.* $\sim \lambda_0/20$; δX_0 *max* $\simeq$ *3 pixels), the periodicity of the spatial modulation of* δX_0 *is clearly evidenced.*

SPATIAL CHAOS

The collective oscillations of the rolls around their mean position is the first manifestation of a time dependence of the pattern for the narrow channels. When the Rayleigh number is increased, this mechanism plays an important role in the appearance of turbulent regions with a typical behaviour characterized by spatio-temporal intermittencies.[15,16] In fact, this last situation has only been observed, up to now in our experiments, when the fluid is confined between copper or sapphire plates. Curiously, an other type of behaviour has been observed in the channels between glass plates, behaviour which probably does not exclude the existence of spatio-temporal turbulent states, but at higher *Ra* values. The appearance of wavelengths multiple of λ_0 where λ_0 is the basic wavelength of the pattern, introduces a specific chaotic evolution. It will be described, in the following, in the case of the channel $\Gamma_y = 0.35$.

At Ra $\simeq$ 1.6 10^6, the pattern is essentially formed with wavelengths equal to λ_m or slightly greater. When *Ra* reaches the value 2 10^6, local destabilizations initiate the formation of wavelengths which are close to $2\lambda_m$ and $3\lambda_m$ as it is shown in the fig.4. In a small *Ra* domain, the spatial organization remains stable, when Ra is fixed , at least during 2 or 3 days, but can evolve as Ra is increased. The channel then looks like a sequence of small boxes with different length, somewhat independant of each other. In some wavelengths, specific oscillators may appear, like the spatial oscillations described previously for the smallest wavelengths, or plumes in the widest ones, but the main part of the pattern is stationary. Then, if we look at the convective situation, locally there is a stable order but the global structure exhibits spatial disorder.

At Ra around 2, 15 10^6, the same type of organization is always present, but a given pattern is only stable during a lapse of time (Fig.4). This behaviour is illustrated in the figure 4, where we can follow the transformation undergone by the pattern from one configuration to one another: new rolls may re-emerge in the greatest rolls and larger wavelengths may appear, by a collective displacement of a part of the small rolls, this anywhere in the cell. The collective displacement of the small rolls is performed through the propagation of a phase defect, due to the presence, locally, of two (or three) wavelengths, larger than λ_m.The propagation is linear with time and reflexions can occur at the lateral boundaries of the cell (see fig.5). This motion looks like compression waves propagating along a linear spring. The time sequences during which the pattern is changing are short compared to the stability periods. Their lengths vary in an erratic way and are reminiscent of the characteristic behaviour of intermittencies[17], as they have been described in dynamical systems, although, we are actually dealing with dynamical spatial chaos.

CONCLUDING REMARKS

The observations on different geometries have shown that when $\Gamma_y < 0.6$, the convective pattern remains one-dimensional until high *Ra* values, at least in the first steps of time dependence. Then, convection in such narrow channels provides a good system to study the 1-D turbulence approach and to allow comparison with relevant theoretical models, like those derived from the Kuramoto-Sivashinsky equation [18] or similar models[19].

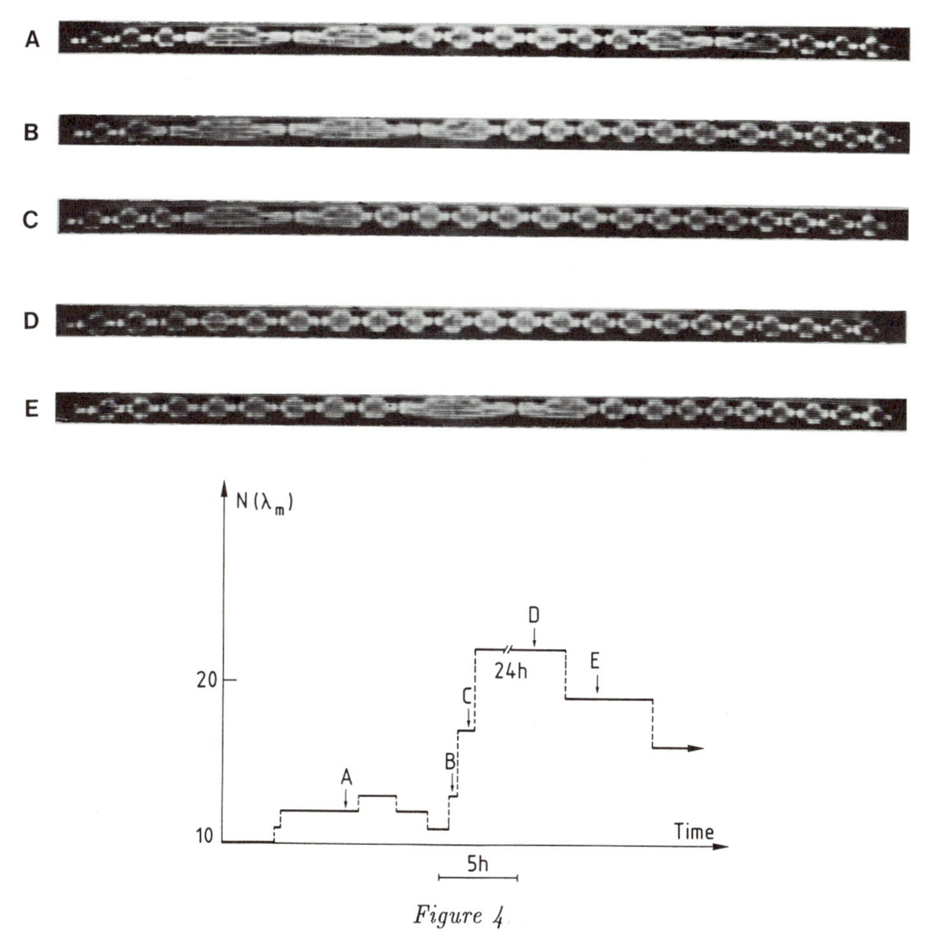

Figure 4

Shadowgraphic snapshots of the channel $\Gamma_y = 0.35$, Ra $\simeq$ 2.15 10^6. *Each pattern is stable during a long lapse of time which depends on the pattern. This is shown in the lower part of the figure, where the plateaus corresponding to the structures A, B, C, D, E, are marked as examples.* N (λ_m) *is the number of the shortest wavelengths* $\lambda = \lambda_m$ *present in each different pattern.*

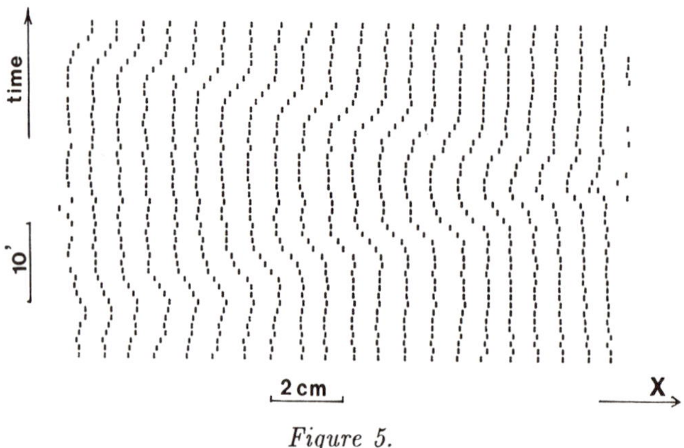

Figure 5.

Successive positions of the hot streams versus time, in the pattern D *of the fig.4.* (Ra $=$ 2.15 10^6) *. Each horizontal line consists in a succession of dashes indicating the position of the hot streams at a given instant; two consecutive lines are spaced by 1 minute. One can observe the propagation of a phase defect (three elongated wavelengths) from the right to the left of the chain of rolls, then from the left to the right after a reflection on the lateral boundary (this kind of motion may be observed during several hours).*

In these narrow channels, the fundamental variable is the spatial phase of the rolls. The first time dependence is given by its periodic variations. Then the evolution towards chaos by increasing Ra depends on the nature of the horizontal boundaries. When they are of good thermal conductivity, the time evolution at fixed Ra is characteristic of spatio-temporal intermittencies, with spatial exchange between regions which are ordered and those which are "fused". When the fluid is confined between glass plates, at a given instant, there are no "fused" zones and, locally, the pattern is always ordered. Nevertheless, the presence of different wavelengths makes the system globally chaotic in space. The time evolution is then given by an intermittent spatial variation of the different local arrangements.

To our knowledge, no theory or model directly related to this last situation, has been developed up to now. But similar behaviours can be found with the KSS model[19], though the relatively low aspect ratio Γ_x of our experiments introduces certainly qualitative differences in the chaotic dynamics.

Acknowledgement

The authors thank P. Hède for his contribution to data processing and F.Daviaud, H. Chaté and P. Manneville for constructive discussions. They thank too B. Ozenda and M. Labouise for their efficient technical assitance.

REFERENCES

/1/ P. Bergé, M. Dubois, Contemporary Physics, **25**, 535 (1984)

/2/ A. Pocheau, V. Croquette, J. de Phys., **45**, 35 (1984)

/3/ P. Bergé, "Chaos and Order in Nature", Elmau 1981, ed. By H. Haken (Springer-Verlag), P.14.

/4/ M.S. Heutmaker, P.N. Fraenkel, J.P. Gollub, Phys. Rev. Lett., **54**, 1369 (1985)
/5/ A. Pocheau, V. Croquette, P. le Gal, Phys. Rev. Letters, **55**, 1094 (1985)
/6/ Le Chaos. Théorie et experiences, ed. by P. Bergé. Eyrolles Paris (1988)
/7/ J.P. Gollub, S.V. Benson, J. of Fluid Mech., **100**, 449 (1980)
/8/ A. Libchaber, S. Fauve, C. Laroche, Physica 7D, 73 (1983)
/9/ M. Dubois, P. Bergé, Physica Scripta, **33**, 159 (1986)
/10/ J.N. Koster, U. Müller, J. Fluid Mech., **125**, 429 (1982)
/11/ O. Kvenvold, Int. J. Heat Mass Transfer, **22**, 395 (1979)
/12/ K. Stork, U. Müller, J. Fluid Mech., **71**, 231 (1975)
/13/ F. Daviaud, P. Bergé, M. Dubois, J. de Phys. Colloques "Non linear coherent structures in Physics, Mechanics and Biological Systems", Paris (1988) to appear.
/14/ M. Dubois, R. Da Silva, F. Daviaud, P. Bergé, A. Petrov, Europhysics Letters. To appear.
/15/ F. Daviaud et al, to be published
/16/ S. Ciliberto, P. Bigazzi, Phys. Rev. Lett., **60**, 286 (1988)
/17/ P. Bergé, M. Dubois, P. Manneville, Y. Pomeau, J. de Phys. Lettres, **41**, L341 (1980)
/18/ H. Chaté and Manneville, Phys. Rev. Lett., **58**, 112 (1987)
/19/ B. Nicolaenko, H. Chaté, these proceedings

PROBABILISTIC CELLULAR AUTOMATON MODELS FOR A FLUID EXPERIMENT

R. Livi and S. Ruffo

Istituto Nazionale di Fisica Nucleare Sez. di Firenze
Largo E. Fermi 2-50125 Firenze-Italy

Introduction

In the study of cellular automata (CA) there is often the problem of understanding if the observed spatio-temporal behaviour may be significant from a physical point of view. In this contribution we compare the behaviour of an experimental system - a fluid in an annular cell heated from below - with that of suitably chosen probabilistic CA rules. This has been made possible by the reduction of the space-time evolution of the experimental system to a symbolic dynamics.

Models and Experiment

In these proceedings S. Ciliberto has reported about the results obtained in an experiment on a fluid in an annular cell (see also Ciliberto and Bigazzi(1988) and Bagnoli et al.(1988)). The space-time behavior of the system has been analyzed by the reduction of the experimental signal to a binary sequence, where 0's and 1's represent laminar and turbulent flow regions respectively.

As a first attempt we have tried to model the observed patterns by an interacting probabilistic process. Restricting ourselves to nearest neighbour interactions, we have introduced eight transition probabilities (TP) from the time step t to $t+1$, $p(x_{i-1}(t), x_i(t), x_{i+1}(t)|x_i(t+1) = 1)$, where i is a space label and each $x(t)$ can be 0 or 1 (TP to $x_i(t+1) = 0$ are obtained by normalization). The TP have been measured from the experiment and, due to periodic boundary conditions (PBC), they show the symmetry properties : $p(001|1) = p(100|1)$, $p(011|1) = p(110|1)$. This reduces the probability space to six dimensions. We have considered a one dimensional Boolean probabilistic CA of length N with PBC, evolving in time according the probabilities extracted from the experiment. We have obtained a qualitative agreement with the experimentally observed patterns for values of the control parameter (the temperature difference between the plates) where turbulent regions diffuse; this agreement gets worse as the control parameter is lowered.

Chaté and Manneville (1987) have introduced the probability distribution $\delta_0(x)$ of laminar regions of size x to characterize the transition to spatio-temporal intermittency in coupled map lattices (see also Oppo and Kapral (1986) and Kaneko (1985)). This transition should reflect itself in a change from a power law to ex-

New Trends in Nonlinear Dynamics and Pattern-Forming Phenomena
Edited by P. Coullet and P. Huerre
Plenum Press, New York, 1990

ponential decay of $\delta_0(x)$ at large x. Besides this quantity we have also considered the probability distribution $\delta_1(x)$ of turbulent regions. For N$\sim$ 100 (recall that in the experiment N=128) $\delta_0(x)$ approximately reproduces this behavior, while $\delta_1(x)$ is always exponential. Reducing finite size efects, e. g. increasing N up to 10^4, we observe for both quantities an exponential decay at large x:

$$\delta_{0,1}(x) \sim exp(-m_{0,1}x) \tag{1}$$

Nevertheless, if m_0 and m_1 are plotted as a function of the control parameter they cross at the value of the control parameter where Ciliberto and Bigazzi (1987) locate the transition of $\delta_0(x)$ to a power law decay (see Fig.1). We conclude that the interacting probabilistic model indeed catches some features of the experiment, although it does not support the existence of the power law decay of $\delta_0(x)$ on larger systems.

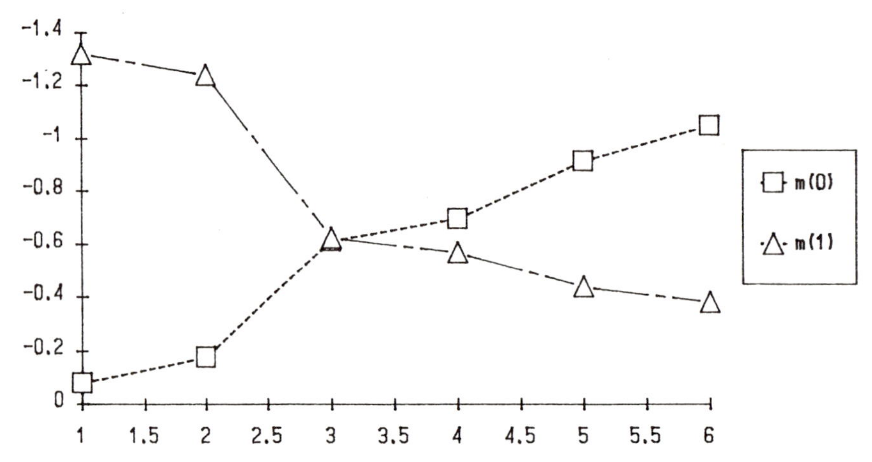

Figure 1 . m_0 and m_1 obtained from the experimental probabilities as a function of the control parameter on an appropriate scale

In order to simplify the model we have considered a reduced probability space (Bagnoli et al. (1988)), which we construct on the basis of some heuristic considerations. One easy choice is to mix a limited number of deterministic CA rules, with normalized probabilities. First of all we observe experimentally a region of the control parameter where turbulence is spatially localized. If a turbulent region appears it does not spread and eventually it dies. The simplest localization mechanism is produced by a majority rule. If we restrict to nearest-neighbor interactions this is rule 232 in the notation by Wolfram (1983). A random initial configuration evolves, after a short transient, to a stable pattern of stripes of 0's and 1's. In the average the initial densities are conserved. In another region of the control parameter, turbulence ceases to be localized and spreads over space. An interpretation of this *interpenetrating* phenomenon in terms of directed percolation has been proposed by Pomeau (1986). In our philosophy, rule 90 seems to be a good choice to represent a percolation process (see Kinzel (1985)). In fact, a strict correspondence between probabilistic growth processes in 1-D, and 2-D statistical models on disorder varieties has been derived by Georges (1987) and by Rujan (1987).

Our simplified model is, therefore, a probabilistic C.A., which is a mixture of rule 90 with probability P and rule 232 with probability Q. At each step the updating of each site is performed according to this probabilistic recipe (we are considering the "annealed" case (see Vichniac et al. (1986)).

The experimental data show that a turbulent region may originate from a laminar one. These fluctuation effects due to thermal noise can be introduced in our C.A. model by using "illegal rules" (Wolfram (1983)), which generate a site with value 1 from a triplet [000]. Instead of choosing an illegal rule randomly we have considered the rules which are complementary to 90 and 232, i.e. rule 165 and 23 respectively. We have assigned to them a much smaller probability, according to the relations:

$$p_{23} = \nu P \quad ; \qquad p_{165} = \nu Q \tag{2}$$

with $\nu = O(10^{-2})$. The normalization condition involves the following relation among P, Q and ν : $(P + Q)(1 + \nu) = 1$. With this choice the number of turbulent (laminar) outputs is constant for any input configuration. Moreover, the probability of a mostly turbulent region to become laminar is non-zero but $O(\nu)$. These two requests seem physically reasonable.

A few results are known for this model. A mean field analysis and numerical experiments show the occurrence of a phase transition at $P \sim .5$, where the asymptotic fraction of laminar sites changes abruptly from 1 to 1/2 (a result which agrees with the fluid experiment). However in the *laminar* phase m_0 does not reach a zero value, but it remains of the order of the noise ν (which is always present in the fluid experiment) in large systems (N $\sim 10^4$).

Therefore, although a transition is present in the density of laminar (turbulent) regions, we observe no transition in the functional form of the probability distribution of laminar regions of length x, at variance with what happens in directed percolation (for "wet" sites) . We should conclude that an *apparent* transition to a power law behavior of the probability distribution of laminar regions may be due both to experimental uncertainty and to finiteness of the lattice.

References

Bagnoli F. et al., 1988, in Proceedings of the Workshop *CHAOS AND COMPLEXITY*, Torino , October 1987, R.Livi, S. Ruffo, S. Ciliberto and M. Buiatti eds., World Publishing, Singapore.

Ciliberto S. and BigazziP., 1988, Phys. Rev. Lett., 60 : 286.

Chaté H. and Manneville P., 1987, Phys. Rev. Lett., 58 : 112 and Saclay - Preprint.

Georges A. and Le Doussal P., 1987, Ecole Normale preprint LPTENS 87/21 Oct.

Kaneko K., 1985, Prog. Theor. Phys., 74 : 1033.

Kinzel W., 1985, Z. Phys., 58 : 229.

Oppo G.L.and Kapral R., 1986, Phys. Rev., A33 : 4219.

Pomeau Y., 1986, Physica, D23 : 3.

Rujan P., 1987, J. Stat. Phys., 49 : 139.

Vichniac G., Tamayo P. and Hartmann H., 1986, J. Stat. Phys., 45 : 875.

Wolfram S., 1983, Rev. Mod. Phys., 55 : 601.

SPACE-TIME CHAOS AND COHERENT STRUCTURES IN THE COUPLED MAP LATTICES

L.A. Bunimovich

Centre de Physique Théorique
CNRS - Luminy, Case 907
F-13288, Marseille Cedex 09 (France)

Coupled map lattices (CML) are a useful tool to study properties of nonlinear spatially extended systems such as spatial patterns, space-time chaos, space-time intermittence, etc. The purpose of this paper is to present the first rigorous results on the subject. Our approach is based on the representation of the correspondent infinite-dimensional dynamical systems as lattice models of statistical mechanics. This allows to prove (Bunimovich and Sinai, 1988) that in CML generated by expanding one-dimensional maps with diffusion-like coupling for small enough coupling there exists a unique Gibbs distribution that is invariant with respect to dynamics and to space translations and mixing. It means that under these conditions the system exhibits turbulent (chaotic in space and time) behaviour.

From the other side one of the most appealing problem in the theory of nonlinear spatially distributed dynamical systems is an explanation of a widely observed (especially in hydrodynamics) phenomenon of the appearance of coherent structures (CS) from chaos. A general opinion which has now emerged considers CS as a new type of bifurcations. The representation of CML as lattice models of statistical mechanics allows to connect (at least in some cases) these bifurcations with the phase transitions (e.g. with the appearance of new ground states). In these models one dimension corresponds to the time direction generated by the dynamics while the others correspond to the space translations. The result of Bunimovich and Sinaï (1988) on the existence of space-time chaos is analogous to high-temperature results in statistical mechanics. The low-temperature domain of parameters corresponds to the large values of the space coupling.

New Trends in Nonlinear Dynamics and Pattern-Forming Phenomena
Edited by P. Coullet and P. Huerre
Plenum Press, New York, 1990

For some CML generated by expanding and quadratic maps of the unit interval with diffusion coupling it was shown in (Bunimovich et al., 1988) that if the coupling is increased then the bifurcations take place that give rise to CS. These CS correspond to some standing or moving waves in the space. If the coupling is increased further then the new ground states appear that correspond to the coexistence of the large scale standing or moving waves in space with the small scale chaotic (turbulent) motion. This approach based on the analogy with statistical mechanics allows to introduce the new general notion of space intermittence. Usually this phenomenon is considered as the coexistence of domains with chaotic and regular motion.

Let us assume now that in the corresponding system of statistical mechanics there are several ground states. If we take then random initial condition (initial configuration of the lattice), generally speaking it could be divided into parts that (locally) are typical for different ground states. So we get the regime of motion where there is intermittence not only between domains with chaotic and regular motion but also between domains with chaotic motion of different types. Many examples of hydrodynamical flows with such type of behaviour could be found in (Davis and Lumley, 1985) and in many papers of the volume. We want to mention especially in this respect experiments of S. Ciliberto. This type of intermittence was obtained also in the numerical experiments with lattices of one-dimensional quadratic maps (Bunimovich et al, 1988).

REFERENCES

[1] Bunimovich, L.A., Lambert, A. and Lima, R., The Emergency of Coherent Structures in Coupled Map Lattices, 1988, Phys. Rev. Lett. to appear.

[2] Bunimovich, L.A. and Sinaï, Ya.G., Space-Time Chaos in the Coupled Map Lattices, 1988, Nonlinearity, to appear.

[3] Davis, S.H. and Lumley, J.L., ed., 1985, Frontiers of Fluid Mechanics", Springer, New York.

TOPOLOGICAL DEFECTS IN VORTEX STREETS BEHIND TAPERED CIRCULAR CYLINDERS AT LOW REYNOLDS NUMBERS

C.W. Van Atta* and P. Piccirillo

Applied Mechanics and Engineering Sciences
*Also Scripps Institution of Oceanography
University of California, San Diego
La Jolla, California 92093 U.S.A.

ABSTRACT

Exploratory hot-wire measurements show that vortex shedding behind uniformly tapered cylinders with taper ratios in the range 32:1-13:1 is two-frequency quasiperiodic or chaotic, depending on the value of the local spanwise Reynolds number. For quasiperiodic shedding the modulation frequency of the vortex shedding is found to be equal to the frequency difference between equally spaced multiple-shedding-frequency spectral peaks. The relative energies, but not the frequencies, of these discrete peaks evolve in a continuous fashion with spanwise location, implying a long range spanwise correlation over most of the cylinder. Spanwise interactions are thus much stronger than in the cellular shedding observed by Gaster for mildly tapered cylinders, for which adjacent cells had their own individually different, but cellwise-constant shedding frequencies. Comparison with flow visualization pictures showed that chaotic regimes may be associated with topological defects in the vortex pattern.

INTRODUCTION

Vortex shedding from a two-dimensional circular cylinder at low Reynolds numbers (40-200) produces a periodic "vortex street" array of parallel vortex filaments whose velocity signature at a fixed point is characterized by a spectrum consisting of a single sharp peak and its harmonics. When, as in the present case, the cylinder is tapered, the vortex shedding frequency f varies with spanwise coordinate z so that the vortex lines cannot be continuous along the full span as in the two-dimensional case, and dislocations in the vortex structure can be anticipated. This leads to complex three-dimensional vortex patterns in which curved vortex lines may be shed from distinct cellular spanwise regions.

The only previous studies of this flow are those of Gaster (1969, 1971). From Gaster's (1969) water tunnel experiments on two cones of taper ratio 36:1 and 18:1 the local Roshko number fd^2/ν, where $d(z)$ is the local diameter and ν the fluid kinematic viscosity, appeared to be a universal function of the local Reynolds number Ud/ν, where U is the free stream velocity. Gaster concluded that the frequency of shedding was controlled by the local diameter and had a slightly lower value than that for a nontapered cylinder of the same diameter. Gaster found that the shedding process was modulated by a low-frequency oscillation, which he concluded was dependent only on U^2/ν and independent of any physical dimension of the cylinder. For a slightly tapered (120:1) cylinder Gaster (1971) found that at the lowest Reynolds numbers the velocity signals were singly periodic in distinct cells along the span except at cell boundaries, where the signals had low frequency

New Trends in Nonlinear Dynamics and Pattern-Forming Phenomena
Edited by P. Coullet and P. Huerre
Plenum Press, New York, 1990

modulations characteristic of the more highly tapered cylinders, with a modulation frequency precisely equal that which would be produced from a summation of the two velocity signals of differing frequency in neighboring cells. At higher speeds his hot-wire signals were similar to those observed on steeper cones, with modulated signals occurring over the entire span.

Recent experiments by Van Atta and Gharib (1987) and by Van Atta et al (1988) showed that mode competition between natural and forced vortex shedding frequencies in vibrating cylinder wakes could produce quasiperiodic and chaotic vortex shedding. The inherent mode competition present in the tapered cylinder case is therefore of intrinsic interest in this respect, as is the possibility of transition from two-to three-frequency quasiperiodicity or chaotic shedding in the context of Ruelle and Takens (1971) or other scenarios for transition to chaos in dynamical systems. The present experiments were undertaken to investigate the origin of the quasiperiodic behavior associated with the low frequency modulation observed by Gaster and possible connections with dynamical systems theory for open fluid mechanical systems.

EXPERIMENTAL ARRANGEMENT

The present experiments were carried out with tapered cylinders placed in the exit plane of a small open circuit wind tunnel, as shown in figure 1.

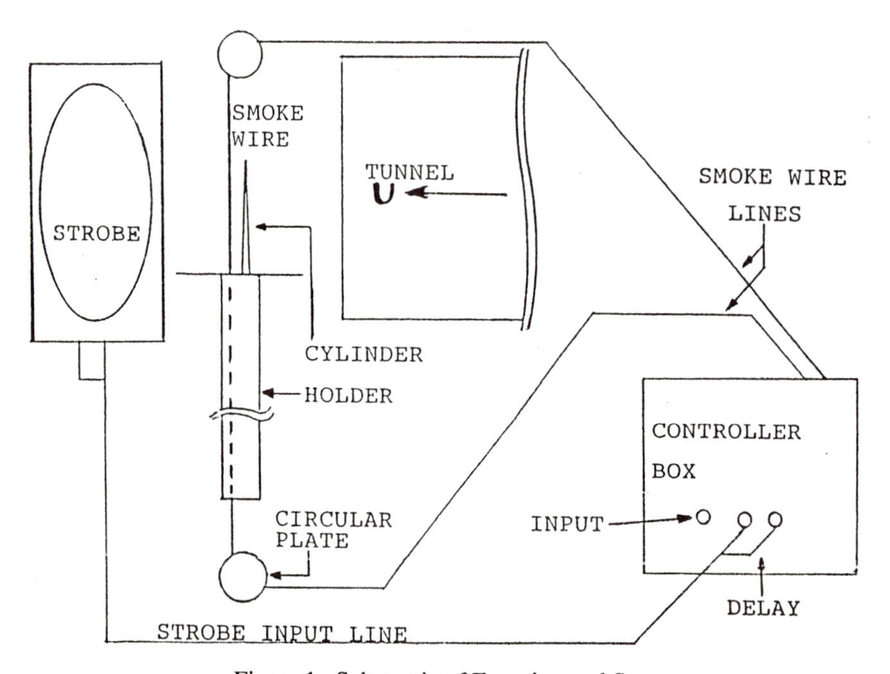

Figure 1. Schematic of Experimental Setup

The tapered cylinders used consisted of three different uniformly tapered 2.5 cm long stainless steel cones, with taper ratios of 32:1, 22:1, and 13:1. The geometrical parameters of the cylinders were measured both with micrometer tools and with the aid of a microscope and ruled steel scale. The tip and base diameters ranged from 0.015 to 0.03 cm, and 0.07 to 0.18 cm, respectively. Each of these slender cones was mounted on a hollow airfoil cross-sectioned strut supported outside the tunnel, with a 1 cm dia. circular endplate at the base of the cone separating it from the supporting strut. A hot-wire probe positioned by a micromanipulator traverse entirely outside the tunnel was located between 5 and 10 diameters downstream of the cone and in the center of the row of vortices shed from one side of the cone. The hot-wire was operated with a DISA 55M10 anemometer circuit. The freestream velocity U was measured with a small pitot-static tube and, as a check, from the vortex shedding frequency of a second nontapered cylinder also mounted in the test section exit plane. Power spectra of the velocity signals were measured on-line with a Spectral Dynamics SD380 Signal Analyzer.

Van Atta and Gharib (1987) found that even very small cylinder vibrations can have a profound effect on the vortex wake structure. A vibration detector of the type used in their study was used in some preliminary experiments to check for such effects in the present work. No vibrations of the tapered cylinders were detected under any of the conditions employed.

For flow visualization, a smoke wire located three base diameters downstream of the cone passed through the strut and cone endplate and was held tautly in place by two circular end plates. A controller sent a pulse to the wire and a delayed pulse to a strobe backlighting the flowing smoke, which was then photographed with a camera.

RESULTS

Quasiperiodic Velocity Spectra

The visual appearance of the hot-wire signals was generally very similar to that reported by Gaster for cones with similar taper ratios, but the present spectral analysis revealed a number of essential features not observed by him. For a fixed freestream velocity, the nature of the signals evolved with z, the spanwise coordinate with origin at the tip of the cone. At and near the tip, no vortex shedding was observed. With increasing z, the appearance of the hot wire signal indicated low-frequency-modulated vortex shedding as reported by Gaster. Spectral analysis of the velocity signals revealed some unexpected features. The spectra typically did not simply contain a single vortex shedding peak, as would be inferred from Gaster's description of his results, but instead the spectra contained multiple high frequency peaks near the expected shedding frequency. An example in which the two central "shedding peaks" have equal amplitudes is shown in figure.2. Typical evolution of the velocity power spectra with spanwise location is illustrated by the example in figure 3. The spectra at all locations contain a discrete series of strong peaks at a group of discrete frequencies which are independent of spanwise location. The most energetic peaks are a single dominant or two neighboring dominant high frequency vortex shedding peaks, and a less energetic but nevertheless dominant peak at a much lower (modulation) frequency. The frequencies of the vortex shedding peaks are all simply related to one another by adding or subtracting an integer multiple of the modulation frequency. Moving from the tip of the cone toward the base, as z and therefore d(z) increases each previously dominant shedding peak decreases in amplitude and is replaced in dominance by the adjacent higher frequency shedding peak, as its amplitude increases. The vortex shedding frequency itself is thus not uniquely spectrally defined and in a spectral sense is not a continuous function of spanwise position z. Gaster, who made very limited spectral measurements, did not report observing more than one dominant vortex shedding peak. However, the peak-counting method he used to measure shedding frequencies for most of his data is obviously incapable of detecting the presence of multiple spectral peaks. The presence of more than one high frequency peak allows one to interpret the low frequency modulation frequency as the difference frequency between two high frequency components, which could be associated with dominant frequencies in two adjacent spanwise cells. It is remarkable that the vortex shedding peaks are evenly spaced at multiples of the modulation frequency. The constant frequency between adjacent peaks suggests that the modulation frequency is a function of only the taper ratio and not the local scale or Reynolds number. The frequency that one would measure by counting peaks as Gaster did is thus composed of contributions from several discrete frequencies which do not change with z, but whose relative contributions to the observed frequency do change with z. This behavior suggests a kind of cell-like spanwise structure in which each cell is dominated by a single shedding frequency, but the continuous evolution of the vortex shedding peaks precludes the presence of the very simple independent cell structure observed by Gaster (1971) for much

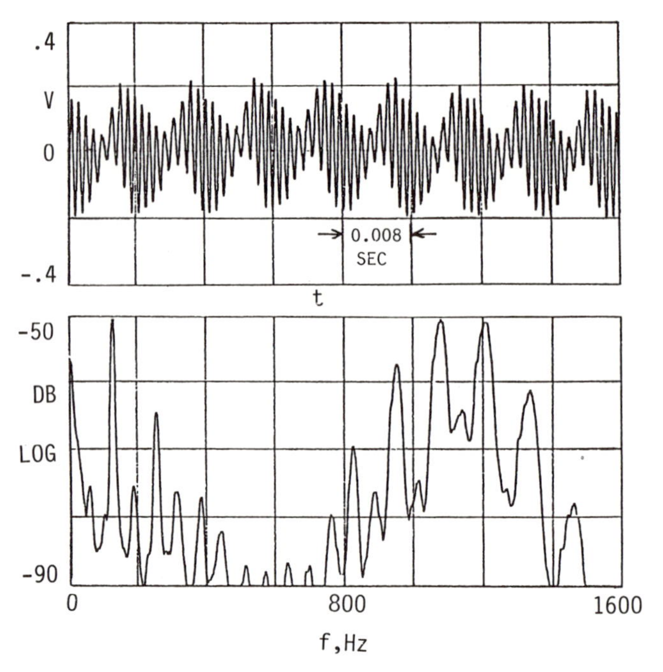

Figure 2. Example of quasiperiodic hot-wire velocity signal and corresponding spectrum. U = 3.64 m/s, Re=118, d(z)=0.03+0.031z cm

larger taper ratios. The frequency of the dominant low frequency peak is equal to the difference in frequency between adjacent dominant vortex shedding peaks and is independent of z, and the magnitude of the low frequency peak varies relatively little with z. Spectral peaks at higher order harmonics of the low frequency and at high frequencies separated from the dominant peaks by integer multiples of the modulation frequency are also present. The signals thus contain two basic incommensurate frequencies, a high frequency associated with vortex shedding and a much lower frequency modulation frequency. At a given z location the velocity signal consists of a weighted mixture of contributions from different spanwise locations with different vortex shedding frequencies, with a weighting factor which decreases with spanwise distance from the measurement point. This suggests a "leaky" cell structure of vortex shedding, unlike the independent cell structures observed by Gaster (1971) for much smaller taper ratios.

Because of the continuous evolution with z of neighboring vortex shedding peaks, at certain spanwise locations the velocity spectra contain two dominant peaks of equal amplitude. If the shedding frequency is taken as equal to the higher of these two frequencies, the corresponding values of the Roshko parameter and local Reynolds number (fd^2/ν, $Ud(z)/\nu$) fall right on the Roshko curve for a nontapered cylinder. Thus, whenever the competition between two dominant shedding frequencies of two neighboring cells is "equal", the upper frequency is the same as would be obtained from the Roshko relation for an untapered cylinder. The average frequency in this case is equal to the frequency obtained by simply counting peaks. When the largest shedding peaks are of substantially unequal magnitude and the shedding frequency is taken as that of the dominant peak, the fd^2/ν vs. Re points all lie below the nontapered cylinder curve, as noted by Gaster (1969). The departure from the Roshko curve increases with increasing taper angle α. The spanwise distance between the locations at which equal dual peaks occur can be used to define a "cell" wavelength. For the case shown in figure 3, the spanwise wavelength is 3.0 mm.

The presence of multiple vortex shedding peaks raises the awkward question of what frequency should be considered as the "true" shedding frequency and how to determine it. The spectrum appears to be operationally unsuitable for this purpose, and the peak counting method appears preferable.

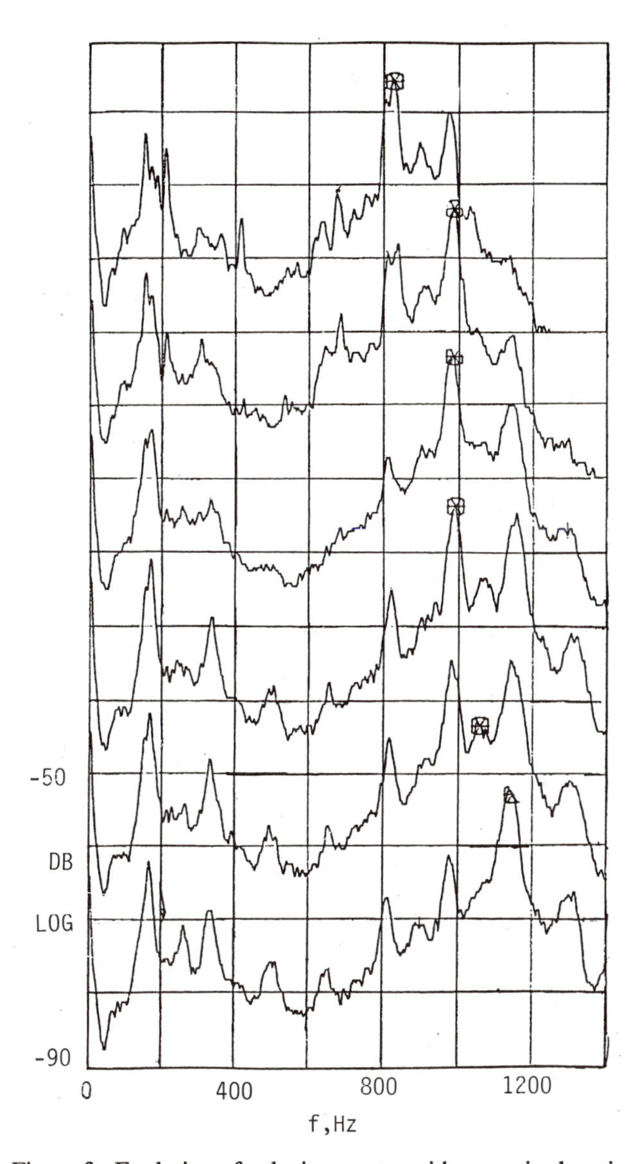

Figure 3. Evolution of velocity spectra with spanwise location
U=3.44 m/sec. d(z)=0.03+0.0^{31}z cm.
From bottom to top, spectra are at z=0.66, 0.74, 0.76, 0.81, 1.02, 1.12 cm. respectively

Chaotic Velocity Spectra

Broadened "chaotic" velocity spectra were often observed as the Reynolds number was increased, either by increasing U or at sufficiently large d(z). The spectra broadened gradually as the Reynolds number increased with the peaks becoming less pronounced until they were finally overcome by the increasing continuous background. Some examples of chaotic spectra are shown in figure 4. The broad peak at high frequency and the low frequency rise of these velocity spectra look similar to those of the chaotic spectra reported by Van Atta and Gharib (1987) and Van Atta et al (1988). In those cases the chaotic spectra were produced by high-order harmonic cylinder vibrations. Spanwise differences in local shedding frequency associated with differences between natural and forced shedding frequencies produced dislocations in the vortex patterns which in turn produced chaotic velocity spectra as they convected past the hot-wire. In the present case the mechanism for producing chaos is not aeroelastic, but purely fluid mechanical. Preliminary comparison with flow visualization pictures suggests that chaotic regions are associated with visual disruptions of regular vortex line patterns.

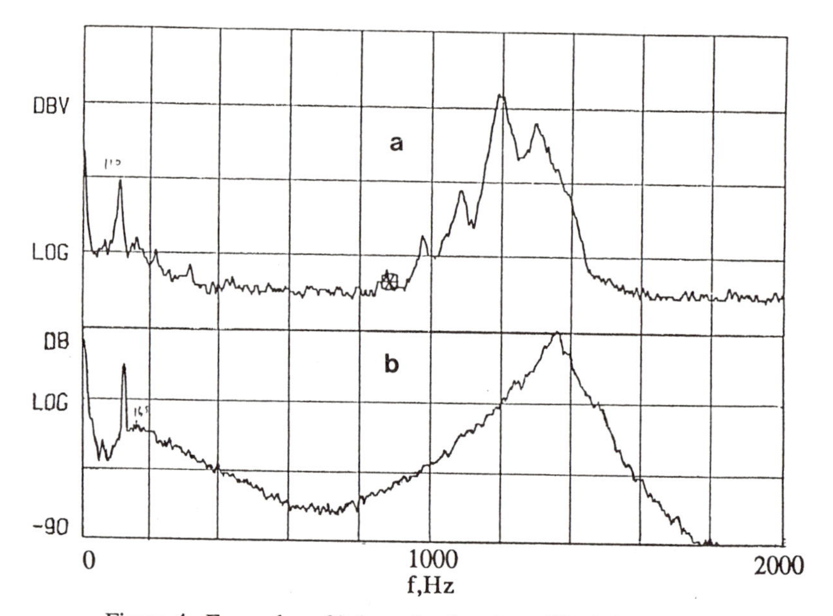

Figure 4. Examples of [a] nearly chaotic and[b] fully chaotic hot-wire velocity vortex shedding spectra

A typical flow visualization photograph is shown in figure 5. The vortex lines appear to be segmented into discrete spanwise segments or cells separated by dislocations along which the vortices become much less distinct. As each vortex line moves downstream its orientation changes so that it becomes more parallel to the axis of the cone. This results in a fan-like instantaneous vortex array for which the spanwise vortex spacing wavelength l_x increases with z, the distance from the tip of the cone. The spanwise wavelength $l_x(z)$ and its dependence on z were estimated from the photos by counting the number of vortices encountered along a given streamwise distance (3 cm) at fixed z. The shedding frequency measured at a single z location and the corresponding l_x were then used to estimate a convection velocity for the vortex lines, and the corresponding dimensionless frequencies were calculated for all values of z using this convection velocity. The result of one such calculation is shown in figure 5. Each individual curve corresponds to a separate cell of flow as defined from the photograph, each with its own shedding frequency. This plot is in general agreement with the one given in Gaster's paper, although the dimensionless frequency lines lie much lower in relation to the Roshko line than Gaster's data for much more gradually tapered cylinders. The difference is partially due to the fact that the deviations from Roshko's curve increase with increasing taper angle, and partly

Figure 5. Example of smoke-wire flow visualization photograph, U=3.12 m/sec, cylinder length = 1.9 cm

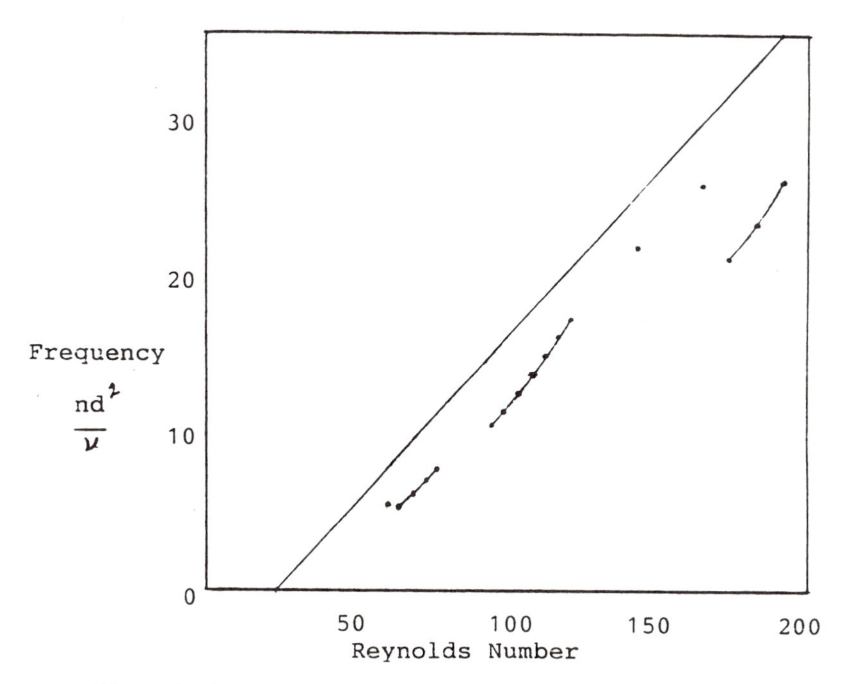

Figure 6. Dimensionless frequencies vs. local Reynolds number inferred from photograph of figure 5 assuming constant convection speed

because inserting the smoke wire in the flow decreased the observed frequencies by about 6%, while preserving the general character of the velocity power spectrum.

CONCLUSIONS

Vortex shedding at low Reynolds numbers from tapered circular cylinders with taper ratios in the range 32:1 to 13:1 is strongly modulated by a low frequency which is equal to the frequency difference between adjacent multiple shedding frequency peaks. The systematic continuous spanwise evolution of the magnitude of the vortex shedding peaks suggests a "leaky" cell structure with a wavelength defined as the spanwise distance between adjacent equal dual peak locations. Chaotic vortex shedding occurs at sufficiently large local Reynolds numbers, and is observed where significant large dislocations in the vortex pattern are seen in the flow visualization patterns.

ACKNOWLEDGMENTS

This work was supported by DARPA grant no. N00014-86-K-0758 and ONR Contracts N-00014-86-K-0690 and N-0014-88-K-0522.

REFERENCES

Gaster, M., 1969, Vortex Shedding from Slender Cones at Low Reynolds Numbers, J. Fluid Mech. 38:565-576.

Gaster, M., 1971, Vortex Shedding from Circular Cylinders at Low Reynolds Numbers, J. Fluid Mech. 46:749-756

Ruelle, D., and Takens, F., 1971, On the Nature of Turbulence, Commun. Math. Phys.

Van Atta, C.W., and Gharib, M., (1987), Ordered and Chaotic Vortex Streets behind Circular Cylinders at Low Reynolds Numbers, J. Fluid Mech. 174:113-133.

Van Atta, C.W., Gharib, M., and Hammache, M., (1988), Three-dimensional Structure of Order and Chaotic Vortex Streets Behind Circular Cylinders at Low Reynolds Numbers, Fluid Dynamics Research 3:127-132.

LARGE-EDDY SIMULATION OF A THREE-DIMENSIONAL MIXING LAYER

Pierre Comte, Marcel Lesieur and Yves Fouillet

Institut de Mécanique de Grenoble *
B.P. 53 X, 38041 Grenoble Cedex, France

Abstract

With the aid of an appropriate parameterization, we simulate numerically a three-dimensional turbulent mixing layer at high Reynolds number. Results are compared with corresponding direct simulations performed with the same pseudo-spectral numerical code. Periodicity is assumed in streamwise and spanwise directions. The evolution of a passive temperature is calculated simultaneously.

Both two- and three-dimensional instabilities grow naturally from a small amplitude three-dimensional random perturbation superimposed upon the basic hyperbolic tangent velocity profile.

Kelvin-Helmholtz billows, greatly distorted three-dimensionally, are found in both cases. We discuss the comparative growth of three-dimensional spanwise modes and two-dimensional instabilities.

Spatially-organized longitudinal structures appear, carrying streamwise vorticity which grows rapidly and levels off at about 2.5 times the initial vorticity brought by the inflexional shear. The spanwise spacing between these streamwise vortices is found and is in good agreement with recent high Reynolds number experiments.

Both in large-eddy and direct simulations, the growth rate of the vorticity thickness and the variance of the velocity fluctuations, obtained after streamwise and spanwise averaging, are also in good agreement with their experimental counterparts.

Spectra of kinetic energy and temperature fluctuations variance, obtained from large-eddy simulation, are both found to cascade with a slope of about -2. Calculations with a larger spatial resolution and a different spectral eddy-viscosity will soon be made.

Introduction

Quasi two-dimensional spatially-organized structures can be found in many high Reynolds number free shear flows, such as mixing layers [1]. They play an important part in momentum and temperature transport processes. In many cases, purely two-dimensional mechanisms allow one to understand most of the dynamics of these structures. For instance, reasonably intricate stability calculations account for formation [2] and pairing [3] of Kelvin-Helmholtz vortices in the case of incompressible mixing layer; two-dimensional numerical simulations reported in [4] have shown such a formation, from a small-amplitude white noise superimposed onto the basic hyperbolic tangent velocity profile, followed by three successive pairings. Evidence of the turbulent character of these vortices has been given there (unpredictability, broad-band kinetic energy and passive temperature spatial spectra). On the other hand, the afore-mentioned predictability study has been interpreted in terms of progressive spanwise decorrelation of initially two-dimensional Kelvin-Helmholtz billows.

---

* Unité Mixte de Recherche du CNRS

New Trends in Nonlinear Dynamics and Pattern-Forming Phenomena
Edited by P. Coullet and P. Huerre
Plenum Press, New York, 1990

In this paper, we simulate numerically a three-dimensional unforced mixing layer at high Reynolds number, by means of a pseudo-spectral method using a subgrid-scale parameterization. Results are compared with correponding direct simulations. Both two- three-dimensional instabilities are triggered by a broad-band three-dimensional noise of small variance, superposed initially upon the basic flow. Evolution of the flow is looked at with the aid of three-dimensional interactive visualization techniques; this enables us to evaluate the spanwise spacing between the spatially-organized structures we thus find, and to compare it with its experimental counterpart found in [5].

Numerical method

Incompressible Navier-Stokes equations written in Fourier space are solved using pseudo-spectral techniques (collocation methods) in the form

$$\frac{\partial \hat{\mathbf{u}}}{\partial t} = F\left[F^{-1}(\hat{\mathbf{u}}) \times F^{-1}(\hat{\omega})\right] - i\mathbf{k}\left[\frac{\hat{p}}{\rho} + \frac{1}{2}|\hat{\mathbf{u}}|^2\right] - \left[\nu + \nu_t(k|k_c)\right] \mathbf{k}^2\, \hat{\mathbf{u}} \qquad (1-a)$$

$$\mathbf{k}.\hat{\mathbf{u}} = 0 \quad , \qquad (1-b)$$

where $\hat{\omega} = i\ \mathbf{k} \times \hat{\mathbf{u}}$ is the vorticity in Fourier space ($i^2 = -1$), F is the direct fast Fourier transform operator, and $k = \sqrt{\mathbf{k}^2}$.

Large-eddy simulations are calculated with an eddy-viscosity $\nu_t(k|k_c)$, defined below, k_c standing for the cut-off wavenumber between the explicitely calculated scales and the modelled scales. We also perform direct numerical simulations, where all the scales are explicitely calculated. In this case, ν_t should be set to zero.

Equation (1-a) is solved first. A solution $\hat{\mathbf{u}}^*$ is found, which does not satisfy (1-b) a priori. $\hat{\mathbf{u}}^*$ is then projected on the plane perpendicular to the wave vector $\mathbf{k}$, providing the incompressible solution $\hat{\mathbf{u}}$ of (1). In practice, since the pressure head gradient is alignated with $\mathbf{k}$, it has been eliminated from (1-a). This modifies the value of $\hat{\mathbf{u}}^*$, but has no effect on its incompressible component $\hat{\mathbf{u}}$.

We solve simultaneously the following equation, verified by a passive temperature θ:

$$\frac{\partial \hat{\theta}}{\partial t} = -i\mathbf{k}.F\left[F^{-1}(\hat{\theta}).F^{-1}(\hat{\mathbf{u}})\right] - \left[\kappa + \kappa_t(k|k_c)\right] \mathbf{k}^2\, \hat{\theta} \quad . \qquad (1-c)$$

The spectral eddy-viscosity

$$\nu_t(k|k_c) = -T(k|k_c)/2k^2 E(k) \qquad (2)$$

and conductivity

$$\kappa_t(k|k_c) = P_r^{-1} \nu_t(k|k_c) \quad , \qquad (3)$$

introduced by Kraichnan [6], are calculated according to Chollet and Lesieur's formulation [7]. The kinetic energy transfer $T(k|k_c)$ through the cut-off wavenumber k_c is evaluated with the aid of isotropic E.D.Q.N.M two-point closure. The use of isotropic methods in the present case is justified by the experimental statement that turbulence at small scales does not depart much from isotropy, even in the case of shear flows like the mixing layer. Consequently, in all cases, provided we assume that turbulence is isotropic with a kinetic energy spectrum proportional to $k^{-5/3}$ for any $k \geq k_c$, one obtains the following form

$$\nu_t(k|k_c) = \left[0.267 + 9.21 \exp(-3k_c/k)\right] \sqrt{\frac{E(k_c)}{k_c}} \quad , \qquad (4)$$

$E(k_c)$ being the kinetic energy spectrum at the cut-off wavenumber k_c. This spectrum is evaluated by integration over a spheric shell S of radius k_c and of thickness $\delta_k = 2\pi/L$, which corresponds to the smallest non-zero Fourier mode

$$E(k_c, t) = \frac{1}{2} \int_S |\hat{\mathbf{u}}'(k_x, k_y, k_z, t)|^2 dk_x\ dk_y\ dk_z \quad , \qquad (5)$$

with

$$S = \left\{ \mathbf{k} : \sqrt{k_x^2 + k_y^2 + k_z^2} \in [k_c - \delta_k/2, k_c + \delta_k/2] \right\} \quad .$$

This spectral eddy-viscosity is then, in the spectral code, merely added to the molecular viscosity. It provides, in calculations of freely-decaying isotropic turbulence, inertial spectra decaying in a self-similar way, with a slope close to $-5/3$ in the vicinity of the cut-off scale. In these calculations, kinetic energy decreases with time according to a $t^{-1.4}$ law, in good agreement with grid-turbulence experiments.

In the following large-eddy simulations, spectral eddy-viscosity ν_t defined by (2) is, at any mode, considerably higher than molecular viscosity ν, which can be set to 0 without any noticeable change. For this reason, we cannot help considering those calculations as "infinite Reynolds number simulations". Corresponding direct numerical simulations are characterized by an initial Reynolds number $U\delta_i/\nu$ based upon half the velocity difference between the two streams, the initial vorticity thickness of the layer and the molecular viscosity. Values up to 150 will be reached, with a spatial resolution of $64 \times 32 \times 64$ Fourier modes.

Turbulent Prandtl number $\nu_t(k|k_c)/\kappa_t(k|k_c)$, where $\kappa_t(k|k_c)$ stands for spectral eddy-conductivity calculated with the aid of temperature spectrum and transfers, is chosen at a constant value of 0.6. In fact, recent works, reported in [8], show that it grows with k from 0.2 to 0.8. However, such a variation has no effect upon the dynamics of our simulations in which temperature is passive and acts only as a numerical dye.

The initial basic flow

$$\bar{u}(y) = U \tanh\left(2y/\delta_i\right) \tag{6}$$

corresponds approximately to the mean velocity profile of a turbulent mixing layer between two quasi-parallel co-flowing streams of respective velocity U_1 and U_2, viewed in a frame translating downstream at the velocity $(U_1 + U_2)/2$. Thus we have $U = (U_1 - U_2)/2$ (cf. Fig. 1). δ_i is the initial vorticity thickness.

Onto this basic flow is superimposed a non-divergent isotropic pertubation of small amplitude, defined in Fourier space in such a way that its kinetic energy spectrum decreases exponentially for any wavenumber larger than an adjustable value k_i, and decreases following a given power law for $k < k_i$. Here, k_i is chosen as the most amplified wavenumber predicted by linear instability theory. This perturbation is then modulated by a gaussian function defined in physical space by $f(y) = \exp\left[-0.5\,(y/\delta_i)^2\right]$, which limits its effects to the rotational zone of the basic flow. The filtered perturbation is eventually rendered non-divergent by projection, in Fourier space, on the plane perpendicular to $\mathbf{k}$. It models, in a somewhat primitive manner, residual turbulence brought about by the splitter plate and responsible for the mixing-layer's transition to turbulence.

Since the temperature θ is passive, the choice of its initial value is purely arbitrary. In order to visualize the mixing of the two flows, we have set $\theta(k_x, k_y, k_z, t = 0)$ equal to the streamwise velocity component $u(k_x, k_y, k_z, t = 0)$, of profile defined by (6), with a different realization of the perturbation having the same statistical properties.

The computational domain is, in physical space, a rectangular parallelepiped of sides L, $L/2$ and L in respectively streamwise x, transverse y and spanwise z directions. Periodic boundary conditions are applied in the streamwise and spanwise directions, while the flow is assumed quiescent for $y = \pm L/4$. The value of L is set to $m_i\lambda_a = 2\pi/k_a$, so that m_i Kelvin-Helmholtz billows of streamwise period λ_a are expected, $k_a \approx 0.4446\,(\delta_i/2)^{-1}$ being the most amplified mode predicted by linear stability theory [2].

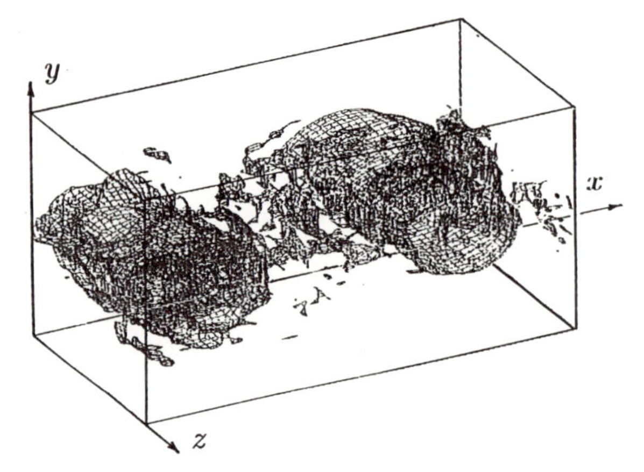

Figure 1. iso-surface $0.5 \max(\omega_z)$ at $t = 26\ \delta_i/U$ (large-eddy simulation, with $L = 2\lambda_a$).

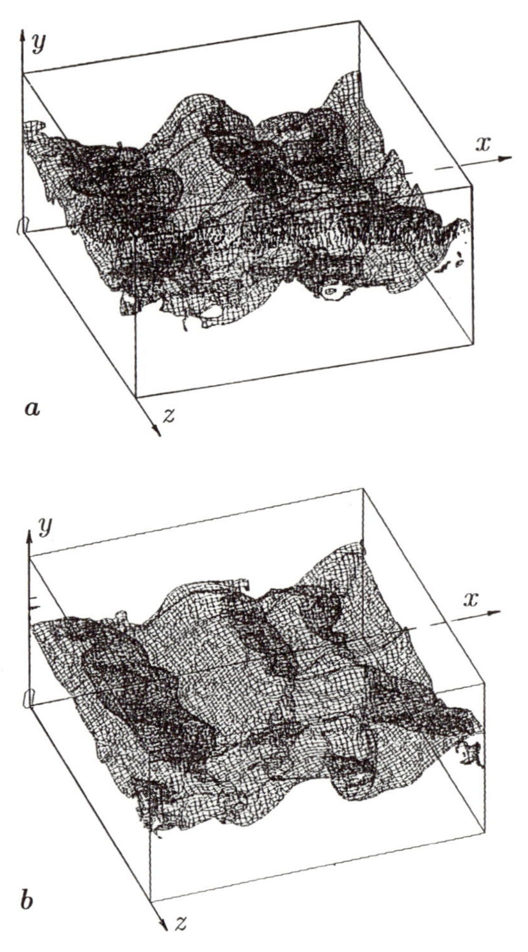

Figures 2-a and 2-b. iso-surface $\theta \approx 0$ at $t = 18\ \delta_i/U$, obtained respectively from large-eddy and direct simulations with $L = 4\lambda_a$.

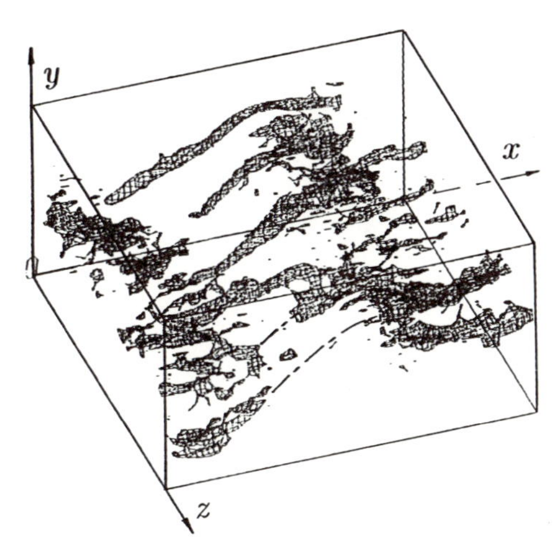

Figure 3. iso-surface $0.5\max(\omega_x)$ at $t = 26\ \delta_i/U$, showing anticlockwise streamwise vortices in large-eddy simulation. The dashed lines suggest possible vortex lines.

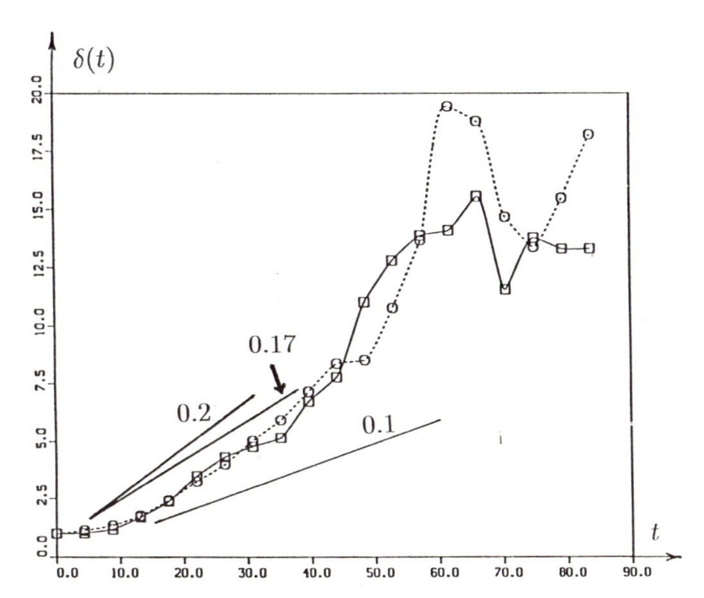

Figure 4. temporal evolution of the vorticity thickness $\delta(t)$ respectively in large-eddy (solid line) and direct (dashed line) simulation.

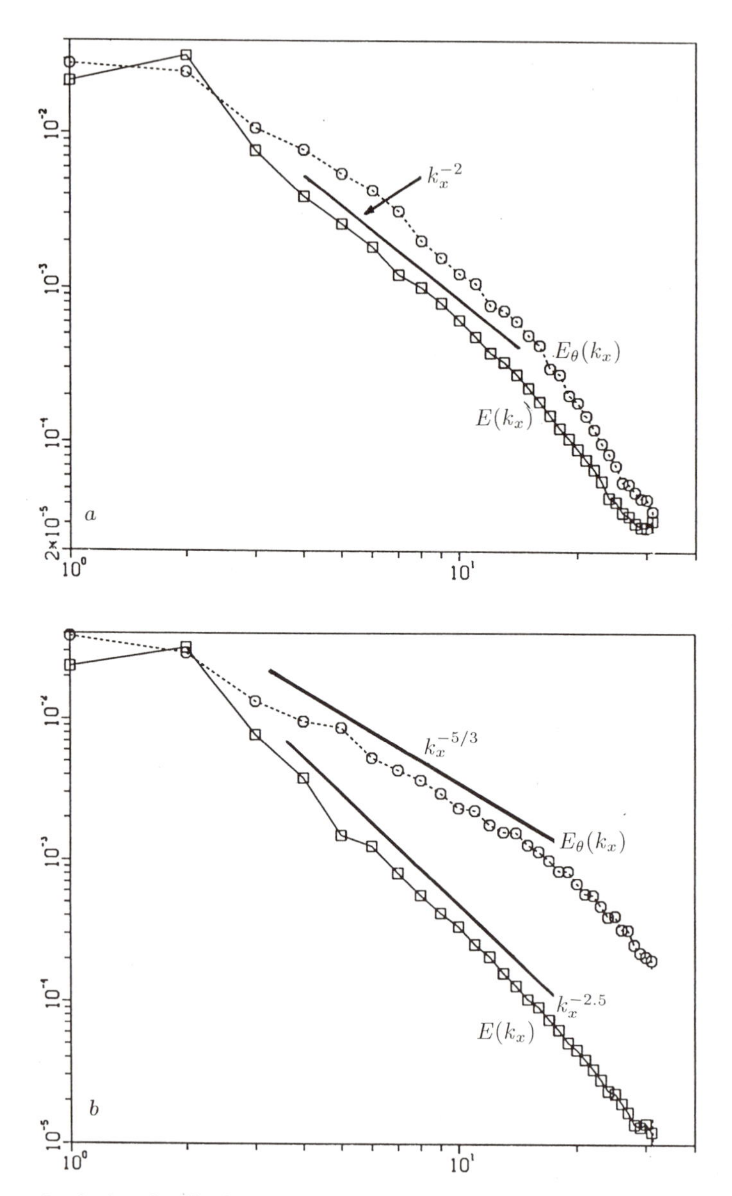

Figure 5-a and 5-b. longitudinal spectra of kinetic energy (solid line) and temperature variance (dashed line) at $t = 26\ \delta_i/U$, respectively in large-eddy and spanwise simulations.

Coherent structure dynamics

We first performed a large-eddy simulation in a domain of length $L = 2\ \lambda_a$, with an initial perturbation variance $< \hat{\mathbf{u}}'^2 >= 10^{-4}\ U^2$. The two expected Kelvin-Helmholtz billows formed at $t \approx 20\ \delta_i/U$ and paired at $t \approx 35\ \delta_i/U$, values in good agreement with two-dimensional simulations reported in [4]. Figure 1 shows the iso-surface corresponding to half the maximal value of the spanwise vorticity ω_z at $t = 26\ \delta_i/U$.

This large-eddy simulation is then repeated with $L = 4\ \lambda_a$ and compared with a corresponding direct simulation at an initial Reynolds number of 150. In both cases, it appears that we no longer have control over the growth of the fundamental mode, due to lack of spatial resolution: in fact, only three billows form, instead of the four expected (Fig. 2-a and 2-b). However, at about $20\ \delta_i/U$, two of the three billows pair, and we arrive back at the previous configuration shown in Fig. 1.

Three-dimensionality comes out mainly in the form of streamwise vortices, spatially organized along the braids joining the primary billows, and which strongly resemble those obtained experimentally in [5] and numerically in [9]. Figure 3 shows the anticlockwise streamwise vorticity, corresponding to half the maximal value of ω_x, obtained at $t = 26\ \delta_i/U$, from the large-eddy simulation. Analogous rib-like structures also appear when plotting the corresponding clockwise vorticity. In Fig. 3, it is possible to make out six interrupted streamwise vortices, which suggests a spanwise spacing s of about $L/6$, to be related, as in [5], to the local vorticity thickness δ_ω. The latter can be measured from contours of spanwise vorticity ω_z averaged along z. On measuring the streamwise distance λ between the centres of the two vortical cores, it is found that $\lambda \approx 0.56\ L$, a value which does not depend on contouring thresholds. Let us now consider that these two vortices issue from the first roll-up of a mixing layer of thickness δ. Hence we have, as shown in [4], $\delta_\omega = 2\ \delta$ and $\lambda = 7\ \delta$, which leads to $\delta_\omega = 2/7\ \lambda \approx 0.16\ L$. So we eventually find $s \approx \delta_\omega$, as in [5].

Evolution with time of the extreme values of streamwise vorticity actually encountered in the cores of these structures has been plotted, showing that positive and negative contributions evolve in the same way, levelling off after $15\ \delta_i/U$, at about 2.5 times the amplitude of the initial spanwise vorticity $2U/\delta_i$.

Figure 4 shows that the vorticity thickness grows almost linearly with time, at a rate close to the experimental value 0.17 found in [1], whereas a value of 0.1 was obtained in [4] from two-dimensional calculations.

Figures 5-a and 5-b show streamwise kinetic energy and temperature variance spectra at $t = 26\ \delta_i/U$. Starting from an exponential decay for $k > k_a$ at $t = 0$, these spectra point out the emergence of a small scale three-dimensional turbulence sharing some characteristics with isotropic turbulence. Unfortunately, the slope close to $-5/3$, followed by $E_\theta(k_x)$ in the case of the direct simulation, will decrease later on, until a peak forms at small scale, probably because the value of the molecular conductivity κ is too small. On the other hand, both energy and temperature spectra obtained from the large-eddy simulation follow approximately the same slope, which is of good omen. However, this slope's value of -2 and its slight steepening at small scale, instead of the expected cut-off at $k_x = 32$ suggests that the parameterization of subgrid scale could be improved upon by a better choice of coefficients of ν_t in (4).

We also calculated profiles of velocity fluctuation variances at the same time-step, and found, in both cases, $< u'^2 >\approx 0.11\ U^2$, $< v'^2 >\approx 0.07\ U^2$ and $< w'^2 >\approx 0.09\ U^2$, values in very good agreement with experiments, which are [10]

$$< u'^2 > = 0.109\ U^2$$

$$< v'^2 > = 0.068\ U^2$$

$$< w'^2 > = 0.083\ U^2 \quad .$$

References

[1]Brown, G.L. and Roshko, A., 1974, J. Fluid Mech., **64**, pp. 775-816.

[2]Michalke, A., 1964, J. Fluid Mech., **19**, pp. 543-556.

[3]Kelly, R. E., 1967, J. Fluid Mech., **27**, pp. 657-689.

[4]Lesieur, M. , Staquet, C. , Le Roy, P. and Comte, P. , 1988, J. Fluid Mech., **192**, pp. 511-534.

[5]Bernal, L.P. et Roshko, A., 1986, J. Fluid Mech., **170**, pp 499-525.

[6]Kraichnan, R.H., 1976, J. Atmos. Sci., **33**, pp 1521-1536.

[7]Chollet, J.P. and Lesieur, M., 1981, J. Atmos. Sci, **38**, pp. 2747-2757.

[8]Lesieur, M. et Rogallo, R., 1988, "Large-Eddy simulation of passive scalar diffusion in isotropic turbulence", submitted to Phys. Fluids.

[9]Metcalfe, R.W., Orszag, S.A., Brachet, M.E., Menon, S. et Riley, J., 1987, J. Fluid Mech., **184**, pp 207-243.

[10]Browand, F.K and Latigo B.O., 1979, Phys. Fluids **22** (6), pp. 1011-1019

THE EFFECT OF NONLINEARITY AND FORCING ON GLOBAL MODES

J. M. Chomaz

Meteorologie Nationale CNRM
42 Avenue G Coriolis
31057 Toulouse Cedex, France

P. Huerre and L. G. Redekopp

Department of Aerospace Engineering
University of Southern California
Los Angeles, California, USA 90089-1191

1 REVIEW OF RELEVANT CONCEPTS

Representation of the linear disturbance field in terms of local modes is firmly established for wave guides where the propagation space is homogeneous. Local modes, although discretized in the transverse or wave-guide dimension(s), are continuous in the streamwise or propagation-space dimension(s) by virtue of the translational invariance of the system. Specifically, considering a two-space-dimension problem with x being a coordinate in the propagation direction and y being a transverse, wave-guide coordinate, an arbitrary disturbance field is expressible as the superposition of eigenmodes of the form $\phi(y)\exp\{i(kx-\omega t)\}$, where the complex frequency ω and wave number k are related through a dispersion relation

$$D(\omega, k; \mu) = 0. \tag{1.1}$$

The symbol μ is used to represent the control parameter(s) of the problem. The dispersion relation specifies such stability characteristics of the local modes as the temporal growth rate $\omega_i = Im\omega$, the spatial growth rate $-k_i = -Imk$, the complex group velocity $\frac{\partial\omega}{\partial k}$, etc. In the context of temporal theory, for example, the wave- guide state is unstable provided $\omega_i > 0$ for any real k and the most unstable local mode has growth rate ω_i^{max}, where the superscript denotes the maximum over all real wave numbers at fixed control parameter.

New Trends in Nonlinear Dynamics and Pattern-Forming Phenomena
Edited by P. Coullet and P. Huerre
Plenum Press, New York, 1990

The foregoing criterion defines the temporal stability of a system, but it provides no information concerning the propagative character of the local modes emanating from and excited by a localized injection of energy. To describe the space-time response to any such excitation one must evaluate the group velocity and one can, therewith, specify whether any unstable modes grow *in situ* or whether all unstable modes propagate away from the source region. Clearly, a very important characteristic of an unstable system is, from this point of view, the growth rate of any mode having a vanishing group velocity (i.e., $\frac{\partial \omega}{\partial k}|_{k_0} = 0$). The growth rate of such a mode is termed the absolute growth rate and is denoted herein by ω_{0i}. The associated real frequency and complex wave number of this mode is denoted by ω_{0r} and $k_0 = k_{0r} + ik_{0i}$. If the absolute growth rate is negative, all unstable waves propagate away from the source and the system is termed convectively unstable. Alternatively, if the absolute growth rate is positive, at least one unstable mode grows in place and the system is termed absolutely unstable. For development of the technical aspects of these linear concepts the reader is referred to Sturrock (1961), Briggs (1964), Bers (1975,1983) and Lifshitz & Pitaevskii (1981).

The onset of instability as a control parameter μ exceeds its critical value for marginal stability in a wave guide possessing reflectional symmetry ($x \rightarrow -x$) is usually of absolute type. The instability fronts emanating from any localized impulse must, in such systems, spread symmetrically relative to the source and there will usually, although not necessarily, be an unstable mode with vanishing group speed. For more general wave guides without reflectional symmetry, one expects the linear dynamics to exhibit successive transitions from a stable state to a convectively unstable state to (possibly) an absolutely unstable state as the control parameter is continuously increased. This sequence with $\omega_i^{max} \geq \omega_{0i}$ for all x is expected to be generic because there is no *a priori* reason why the value of the control parameter for onset of instability μ_c should coincide with that for incipient absolute instability μ_t. The order of transitions just described can be readily verified by reference to the linear Ginzburg-Landau model

$$\left\{ \frac{\partial}{\partial t} + U \frac{\partial}{\partial x} - \mu - (1 + ic_d) \frac{\partial^2}{\partial x^2} \right\} A = 0. \tag{1.2}$$

The condition for marginal stability is $\mu_c = 0$, the parameter range for convective instability is $\mu_c < \mu < \mu_t$, and absolute instability exists for $\mu > \mu_t = U^2/[4(1 + c_d^2)]$ (cf., Chomaz, Huerre & Redekopp - hereinafter denoted as CHR, 1987,1988). When $U = 0$ the system possesses reflectional symmetry and the values of μ_c and μ_t coincide, the instability at onset being of absolute type.

The foregoing local mode description based on the assumption of translational invariance is no longer strictly valid when the basic state varies spatially along the wave guide and the propagation space is inhomogeneous. Nevertheless, as long as the inhomogeneity of the propagation space is weak, those concepts can be applied locally and one can determine the stability properties at any position along the wave guide based solely on the state at that position. The specific requirement on the strength of the inhomogeneity in order for a "local" analysis to be valid is that the ratio of the local mode wavelength to the scale of the inhomogeneity (δ, say) must be small compared to unity, a condition which is satisfied in many spatially-developing, hydrodynamic flows.

In such cases, the local state can be defined as being stable or unstable and, if it is unstable, whether it is absolutely unstable or convectively unstable. That is, one can define local values $\omega_i^{max}(\delta x), \omega_{0i}(\delta x)$, etc. which may vary slowly along the wave guide.

Different scenarios for the spatial variation of ω_i and ω_{0i} were considered in our earlier work (CHR, 1987, 1988), both for infinite and semi-infinite wave guides. In particular, it was revealed in either case that a synchronized mode exhibiting long-range spatial coherence is destabilized whenever an interval of sufficient spatial extent having local absolute instability (i.e., $\omega_{0i}(x) > 0$) exists. Such a mode is termed a global mode because it may extend over a domain encompassing the entire region of local instability, even extensive contiguous regions where the local instability is convective (i.e., where $\omega_i(x) > 0$, but $\omega_{0i}(x) < 0$). Where the inhomogeneous character of the wave guide satisfies the necessary conditions (to be prescribed later), the propagation-space dimension admits a denumerable set of eigenmodes, the gravest of which undergoes a Hopf bifurcation to a limit cycle. As the domain of absolute instability increases, a successive number of these modes are destabilized on a linear basis. The mode with the highest growth rate presumably prevails, although nonlinear effects may dramatically alter the observed dynamics, especially at high supercritical values of the control parameter. Some features of the nonlinear behavior are discussed later. The important point, however, is that global modes possess the characteristics of oscillators. This aspect is developed further with particular emphasis given to the forced response of a global mode excited by a spatially-compact source.

Criteria for specifying the onset or destabilization of global modes and their associated frequency ω_g have been derived by CHR (1989) based on a WKB analysis of a variable-coefficient, linear, Ginzburg-Landau equation (2) with arbitrary spatial variations of $\omega_0(x)$ and $k_0(x)$. The frequency of the first unstable global mode is determined, to first order in the WKB expansion parameter, by the local absolute frequency at the saddle point x_s nearest to the real axis of the function $\omega_0(x)$ in the complex x-plane. In this way we obtain that the complex frequency of a global mode satisfies the conditions

$$\omega_{gr} \sim \omega_r(x_s) + O(\delta), \tag{1.3a}$$

$$\omega_{gi} \leq \omega_{0i}(x_s) \leq \omega_{0i}^{max} \leq \omega_{i,max}^{max}, \tag{1.3b}$$

where $\omega_{i,max}^{max}$ denotes the maximum over both x and k, and where x_s is determined from the condition

$$\left.\frac{\partial \omega_0}{\partial x}\right|_{x_s} = 0. \tag{1.3c}$$

It is clear from these conditions that the destabilization of a global mode (i.e., $\omega_{gi} > 0$) requires a finite region with $\omega_{0i}(x) > 0$; that is, a finite interval of absolute instability in the wave guide. Consistent with the specific examples considered earlier (i.e., CHR 1987), we also observe that the condition

$$\int_{x_{t_1}}^{x_{t_2}} \sqrt{\omega_{0i}(x)}dx \geq O(1) \tag{1.4}$$

is satisfied for the existence of amplified global modes, where $x_{t_1} < x < x_{t_2}$ defines the interval on the real axis where the local state is absolutely unstable.

The global instability concepts described above have been tested in specific, hydrodynamic flows. The most extensive studies have been performed for the two-dimensional wake behind a bluff body. Local instability characteristics were calculated by Monkewitz (1988) using a two-parameter family of velocity profiles to model the spatially-developing flow. The onset of the discrete frequency, vortex-shedding mode was shown to correlate with the existence of a pocket of absolute instability and the destabilization of the global mode occurred for a Reynolds number significantly above that for local instability. Experiments by Mathis et al. (1984), Provansal et al. (1987), and Sreenivasan et al. (1987, 1989) demonstrate clearly that the onset of vortex shedding corresponds to a super-critical Hopf bifurcation to a global mode. Numerical simulations of bluff-body flows by Jackson (1987), Hannemann & Oertel (1988), and Yang & Zebib (1989) show clearly the existence of a global mode which appears at Reynolds numbers greater than that required for local instability and local absolute instability. For a comprehensive discussion of these concepts and their application to a number of spatially-developing systems, the reader is referred to a recent review by Huerre & Monkewitz (1990).

2 THEORETICAL FOUNDATIONS

Since a countable set of global modes exist which can be ordered with respect to their growth rates, only a single mode is marginal at the threshold value of the control parameter for global instability. Also, linear global modes have a discrete frequency of oscillation which is uniform over the entire extent of spatially-varying, local stability characteristics. For this reason the study of weak nonlinearity and weak forcing of the marginal mode by means of a multiple-scale analysis can be carried out in the same way as the classical results for oscillators. A study of these effects is pursued here because of the practically important issues pertaining to the control of the local and global dynamics of spatially-varying wave-guide states. The linear and nonlinear response to a localized external forcing, the relation between the response and the position of the forcing *vis-a-vis* the location of a pocket of local absolute instability, the gain or efficiency of the forcing, etc., are some important elements of global modes deserving clarification.

For the sake of brevity, the effect of weak forcing and nonlinearity will be illustrated using the model equation

$$\frac{\partial A}{\partial t} + \mathcal{L}(\frac{\partial}{\partial x}, x; \mu)A + c(x;\mu)|A|^2 A = f(x,t), \tag{2.1}$$

for the complex amplitude function $A(x,t)$. We assume that the unforced wave-guide dynamics admits $A \rightarrow Ae^{i\theta}$ symmetry and suppose that $A(x,t)$ vanishes at the boundaries of the infinite or semi-infinite domain. $\mathcal{L}$ is a linear differential operator, μ is the control parameter, and $c(x;\mu)$ is a complex coefficient of the nonlinear term. The linear homogeneous equation admits a solution of the form

$$A_g(x,t) = \phi_g(x)e^{-i\omega_g t} \tag{2.2}$$

for the gravest mode. The frequency ω_g and modal function $\phi_g(x)$ depend on μ and there exists a critical value μ_g such that the system is globally stable for $\mu < \mu_g$. At $\mu = \mu_g$ the system is neutral and $Im\omega_g(\mu_g) = 0$. The amplitude of the linearized marginal mode is unconstrained and may evolve slowly with respect to the time scale ω_g^{-1}, which we suppose is finite.

A multiple-scale analysis is performed using the small parameter ϵ ($0 < \epsilon \ll 1$) which measures the strength of the forcing and the departure from criticality:

$$f(x,t) = \epsilon^2 F \delta(x - x_f) e^{-i\omega_f t}, \tag{2.3a}$$

$$\omega_f = \omega_g + \epsilon^2 \Omega, \tag{2.3b}$$

$$\mu = \mu_g + \epsilon^2 \Delta_\mu. \tag{2.3c}$$

A slow time scale $T = \epsilon^2 t$ is introduced in order to avoid the appearance of secular terms in the perturbation expansion

$$A(x,t) = \sum_{n=1}^{\infty} \epsilon^n A_n(x,t,T). \tag{2.4}$$

Assuming that the operator $\mathcal{L}$ is analytic in the parameter μ we may write

$$\mathcal{L}\left(\frac{\partial}{\partial x}, x; \mu\right) = \mathcal{L}_g\left(\frac{\partial}{\partial x}, x; \mu_g\right) + \epsilon^2 \Delta_\mu \mathcal{L}_\mu \left(\frac{\partial}{\partial x}, x; \mu_g\right) + O(\epsilon^4). \tag{2.5}$$

Thus, the leading order term in (2.4) is described by the equation

$$\frac{\partial A_1}{\partial t} + \mathcal{L}_g\left(\frac{\partial}{\partial x}, x; \mu_g\right) A_1 = 0, \tag{2.6}$$

with the solution

$$A_1 = \mathcal{A}(T) e^{-i\omega_g t} \phi_g(x) \tag{2.7}$$

defining the neutral global mode with arbitrary amplitude $\mathcal{A}(T)$. The next order term A_2 satisfies the same homogeneous equation as (2.6) and provides no essential information or constraint concerning $\mathcal{A}(T)$. A compatibility condition for the avoidance of secular terms at third order leads to the following evolution equation for the global mode amplitude:

$$\frac{d\mathcal{A}}{dT} = \Delta_\mu \frac{< \psi | \mathcal{L}_\mu \phi_g >}{< \psi | \phi_g >} \mathcal{A} - \frac{< \psi | c(x;\mu) |\phi_g|^2 \phi_g >}{< \psi | \phi_g >} |\mathcal{A}|^2 \mathcal{A} - F \frac{\psi^*(x_f)}{< \psi | \phi_g >} e^{-i\Omega T}. \tag{2.8}$$

The quantity $< f | g >$ denotes the scalar product $\int f^* g \, dx$, superscript $*$ denotes the complex conjugate, and $\psi(x)$ is the solution of the equation

$$\left\{ i\omega_g^* + \mathcal{L}_g^A \left(\frac{\partial}{\partial x}, x; \mu_g\right)\right\} \psi = 0, \tag{2.9}$$

where $\mathcal{L}_g^A$ is the adjoint of $\mathcal{L}_g$.

The result (2.8) could be anticipated from the well-established results for weakly nonlinear oscillators. We emphasize here, however, that the dynamical state of the entire wave guide behaves like a single oscillator with a coherent spatial structure. When the forcing amplitude F vanishes, $\mathcal{A}(T)$ evolves according to the familiar Landau equation. When $F = O(1)$ an imperfect bifurcation occurs with the familiar result that an $O(\epsilon^2)$ forcing can generate an $O(\epsilon)$ response, even when the basic state is linearly damped (i.e., $\Delta_\mu < 0$). The forcing efficiency and phase shift of the response are determined by $\psi^*(x_f)$.

In order to exhibit explicit results for the forcing and to determine the sub- or super-critical nature of the bifurcation, we take a specific form for the operator $\mathcal{L}$. The form chosen is that of the Ginzburg-Landau equation with

$$\mathcal{L}\left(\frac{\partial}{\partial x}, x; \mu\right) = -\left\{p(x;\mu) - U\frac{\partial}{\partial x} + b\frac{\partial^2}{\partial x^2}\right\}. \tag{2.10}$$

where U is a constant, real advection velocity and b is a complex constant related to the curvature of the local dispersion relation for the wave guide. The adjoint operator is

$$\mathcal{L}^A\left(\frac{\partial}{\partial x}, x; \mu\right) = -\left\{p^*(x;\mu) + U\frac{\partial}{\partial x} + b^*\frac{\partial^2}{\partial x^2}\right\}. \tag{2.11}$$

The important difference between $\mathcal{L}$ and its adjoint is in the sign of the advection term which has been reversed in the two operators. A convenient choice for the term $p(x;\mu)$ for an infinite wave guide is

$$p(x;\mu) = \mu_0 + \frac{1}{2}\mu_2(x - z_0)^2 + \epsilon^2\Delta_\mu, \tag{2.12}$$

in which μ_0 is real and μ_2 and z_0 are complex constants. With this choice the solution for the global modes has

$$\omega_{g_n} = i\mu_0 - \frac{i}{4}\frac{U^2}{|b|}e^{-i\theta_b} - i(2n+1)\left|\frac{\mu_2 b}{2}\right|^{1/2} e^{i(\theta_{-\mu_2}+\theta_b)/2} \tag{2.13a}$$

$$\phi_{g_n}(x) = \exp\left\{\frac{1}{2}\frac{U}{b}x - \frac{1}{2}\left|\frac{\mu_2}{2b}\right|^{1/2}(x - z_0)^2 e^{i(\theta_{-\mu_2}+\theta_b)/2}\right\}\cdot H_n\left(\left|\frac{\mu_2}{2b}\right|^{1/4}(x - z_0)e^{i(\theta_{-\mu_2}+\theta_b)/4}\right), \tag{2.13b}$$

where $H_n(x)$ are the Hermite polynomials of order n, $\theta_{-\mu_2} = \arg(-\mu_2)$, and $\theta_b = \arg(b)$. The adjoint function is given by

$$\psi_n(x) = \exp\{-Ux/b^*\}\phi^*_{g_n}(x). \tag{2.14}$$

The reversed effect of advection in the adjoint is clearly evident and, by reference to (2.8), its role in regard to forcing efficiency is to shift the optimal location of forcing considerably upstream of the maximum signature of the global mode. This is demonstrated clearly in numerical simulations presented in the next section.

In order for the global mode to be marginal (i.e., $Im\omega_{g_n} = 0$), the control parameter μ_0 must take the value

$$\mu_0 \equiv \mu_{g_c} = \frac{1}{4}\frac{U^2}{|b|}\cos\theta_b + (2n+1)\left|\frac{\mu_2 b}{2}\right|\cos\left(\frac{\theta_{-\mu_2}+\theta_b}{2}\right). \tag{2.15}$$

Restricting attention to the first mode to be destabilized (i.e., $n = 0$), the Landau constant (i.e., the coefficient of the nonlinear term in (2.8)) can be evaluated analytically for specific choices of the coefficient $c(x;\mu_g)$ multiplying the nonlinear term in (2.1). We adopt the simple choice $c(x;\mu_g) = 1 + ic_n$ where c_n is a real constant in all that follows. For these conditions we obtain the following results:

$$< \psi_0 | \phi_{g0} > = \int_{-\infty}^{\infty} \psi_0^* \phi_g dx = \left| \frac{2b\pi^2}{\mu_2} \right|^{1/4} e^{-i\Theta/2}; \tag{2.16}$$

$$\begin{aligned} \ell &= \frac{< \psi_0 | c(x;\mu_g) | \phi_{g0} |^2 \phi_{g0} >}{< \psi_0 | \phi_{g0} >} \\ &= \frac{1 + ic_n}{[3\cos^2\Theta + 1]^{1/4}} \exp \left\{ \frac{\frac{U^2}{|b|^2}\cos^2\theta_b + 6\left|\frac{\mu_2 z_0^2}{b}\right| \sin^2\theta_{z0}}{4\left|\frac{\mu_2}{2b}\right|^{1/2}(3\cos^2\Theta + 1)} (2\cos\Theta - i\sin\Theta) \right. \\ &\quad + \frac{U|z_0/b|\cos\theta_b}{3\cos^2\Theta + 1}\left[\frac{5}{2}\cos\theta_{z_0} + \frac{3}{2}\cos(2\Theta + \theta_{z_0}) + 2i\sin\theta_{z_0}\right] \\ &\quad \left. - \frac{i}{2}\tan^{-1}\left(\frac{\sin\Theta}{2\cos\Theta}\right) + \frac{i\Theta}{2} \right\}, \end{aligned} \tag{2.17}$$

where

$$\Theta = \frac{1}{2}(\theta_{-\mu_2} - \theta_b), \quad \theta_{z_0} = \arg(z_0). \tag{2.18}$$

There is no apparent information to be gleaned from this expression aside from the fact that the Landau constant for the local bifurcation bears no specific relation to that for the global bifurcation. Furthermore, even though the local bifurcation in this instance is supercritical, the global bifurcation may be either subcritical or supercritical.

3 NUMERICAL EXPERIMENTS

3.1 Nonlinear effects in free global modes

Numerical simulations of the nonlinear equation (2.1) with $\mathcal{L}$ defined by (2.10), $b = 1 + ic_d$, and $c(x;\mu) = 1 + ic_n$, where c_d and c_n are real constants, have been performed for two different wave-guide configurations. One configuration had the quadratic variation of $p(x;\mu)$ given in (2.12) (with μ_0, μ_2 real and $z_0 = \Delta_\mu = 0$) on an infinite domain. In this case the local wave-guide state is stable at $x = \pm\infty$ and a single region of absolute instability, bordered by symmetric regions of convective instability, exists in the vicinity of the origin. The spatial extent of the region of local absolute instability is controlled by the parameter μ_0 for fixed μ_2. The other configuration had a semi-infinite domain $0 < x < \infty$ with a linear variation of $p(x;\mu)$ having negative slope

$$p(x;\mu) = \mu_0 + \mu_1 x, \qquad Re\mu_1 < 0,\ Im\mu_1 = 0. \tag{3.1}$$

A homogeneous boundary condition was imposed at $x = 0$ where the local state is absolutely unstable for sufficiently large values of μ_0 (i.e., $\mu_0 > \mu_t = U^2/[4(1 + c_d^2)]$). This region is followed by an interval of convective instability and then a region which is stable as x tends to infinity. These two configurations are simple examples of spatially-inhomogeneous wave guides possessing a single interval of absolute instability whose length is related to the control parameter μ_0.

The finite-difference code used in the present simulations was identical to that employed in our earlier studies (CHR 1987, 1988). It was demonstrated in that work that the analytically-derived linear eigenfunction was indeed the first mode to be destabilized. The predicted values of the threshold control parameter and the eigenfrequency were also verified in the limit as the spatial and temporal discretization were reduced (a one percent error for $\Delta x = 0.5$, $\Delta t = 7.5 \times 10^{-3}$ and a 0.3 percent error for $\Delta x = 0.25$, $\Delta t = 2.0 \times 10^{-3}$). The new results to be presented here were obtained using the following set of parameters: $U = 6$, $c_d = -10$, $\mu_1 = -1.19 \times 10^{-4}, \mu_2 = -2.534 \times 10^{-6}$, $\Delta x = 0.5$, $\Delta t = 7.5 \times 10^{-3}$. The value of c_n is taken to be zero except in cases where noted otherwise. The choice of parameters, particularly for U and c_d, was made in order to compare our results with those of Deissler (1985,1987) for a constant-coefficient, convectively unstable case.

The structure of the bifurcation to the lowest-order global mode is revealed in Figures 1 and 2. Figure 1 pertains to the quadratic form for $p(x;\mu)$ on the infinite domain and Figure 2 pertains to the linear form for $p(x;\mu)$ on the semi-infinite domain. Both figures exhibit the same feature; namely, a linear variation with slope unity of the initial growth rate with respect to μ_0/μ_t, where μ_t defines the value of the control parameter for the onset of local absolute instability at $x = 0$. The figures also show the equilibrium, finite amplitude $\bar{A}$ of the global mode by plotting the maximum of the saturated state as a function of the supercriticality. It is difficult to obtain results for very small supercriticality because the time to approach equilibrium tends to infinity as $\mu_0 - \mu_{g_c}$ tends to zero. For example, 1.2×10^6 time steps were required to arrive at a good

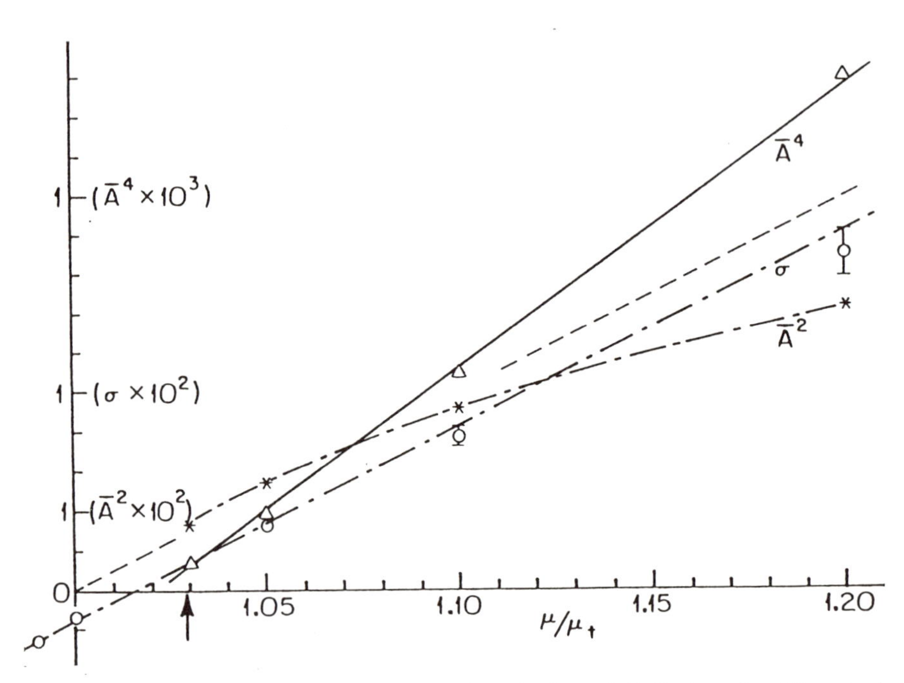

Figure 1 The bifurcation diagram for global modes in an infinite domain with a single interval of quadratically-varying absolute growth rate: – o – measured growth rate σ; – $*$ – variation of the square $\bar{A}^2$ of the peak saturation amplitude; – $\triangle$ – variation of the quartic $\bar{A}^4$ of the peak saturation amplitude. The dashed line has unit slope for σ vs. μ/μ_t. The vertical arrow denotes the theoretical bifurcation value μ_{g_c}/μ_t.

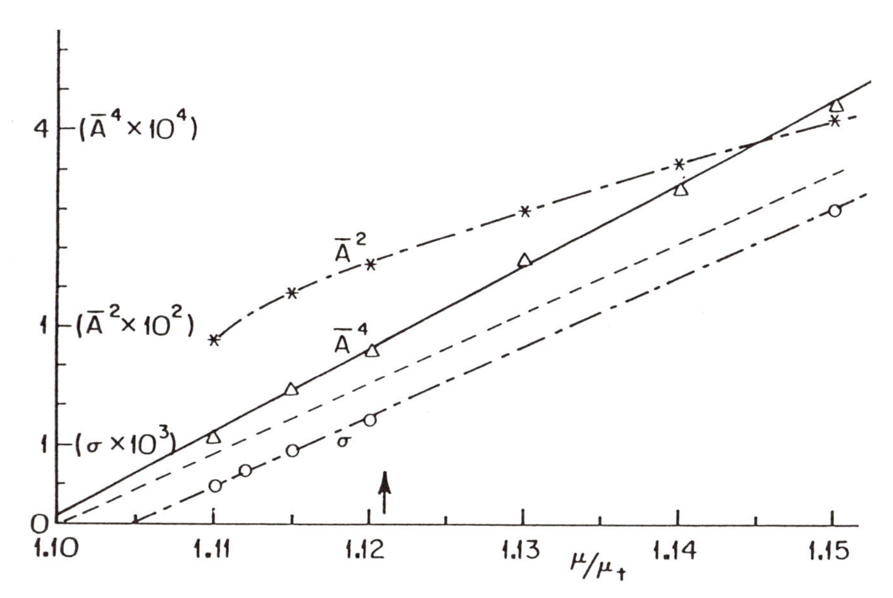

Figure 2 The same diagram as Fig. 1 for global modes in a semi-infinite domain with a single interval of linearly-varying absolute growth rate adjacent of the left boundary of the spatial domain.

estimate of the equilibrium amplitude at the smallest condition shown in Figure 2. On each figure we show variations of both quadratic and quartic values of the amplitude as well as the theoretically-derived value for the bifurcation parameter in an attempt to clarify the scaling law for the bifurcation. Using the specified numerical values of the parameters in equation (2.17), the Landau constant is $\ell = 1.1245 - 0.0526i$, suggesting that the bifurcation is supercritical. It is clear that either of three possibilities must exist: i) the values of $\mu_0 - \mu_{g_c}$ are too large to reveal the true scaling near the bifurcation point; ii) the next nonlinear term (i.e., $|\mathcal{A}|^4\mathcal{A}$) in the amplitude expansion dominates over the $|\mathcal{A}|^2\mathcal{A}$ term ; or iii), the true nonlinear evolution exhibits a subcritical bifurcation. For both wave-guide configurations, the data for $\bar{A}^4$ lie on a straight line which crosses the abscissa close to the point where the computed growth rate vanishes. This is strongly suggestive that the higher order nonlinear term dominates the equilibrium state. Of course, more simulations are needed to clarify the issue.

The theoretical development in the previous section deals exclusively with the global mode dynamics in the immediate vicinity of critical. It predicts the shape of the global mode very well for $|\mu_0 - \mu_{g_c}|/\mu_{g_c} \ll 1$, but the theory gives no indication as to how the mode characteristics change for finite values of the supercriticality. This issue was investigated and sample results are shown in Figure 3 for the infinite wave-guide configuration. It is evident that the maximum of the global mode moves upstream and the mode shape broadens as the supercriticality increases. The left column depicts the envelope of the mode with the advective factor $\exp(Ux/2b)$ subtracted (i.e., the nonlinear extension of the function $\phi_g(x)$ in (2.13b) with $U = 0$). One observes that this "transformed" mode is virtually unchanged in shape, but its maximum is shifted upstream as the nonlinearity increases. One interpretation of this effect is that it is due to the "mean flow" contribution to the growth rate (i.e., considering the combined effect of the terms $p(x;\mu) - (1 + ic_n)|A|^2$ as an effective growth rate). Based on this

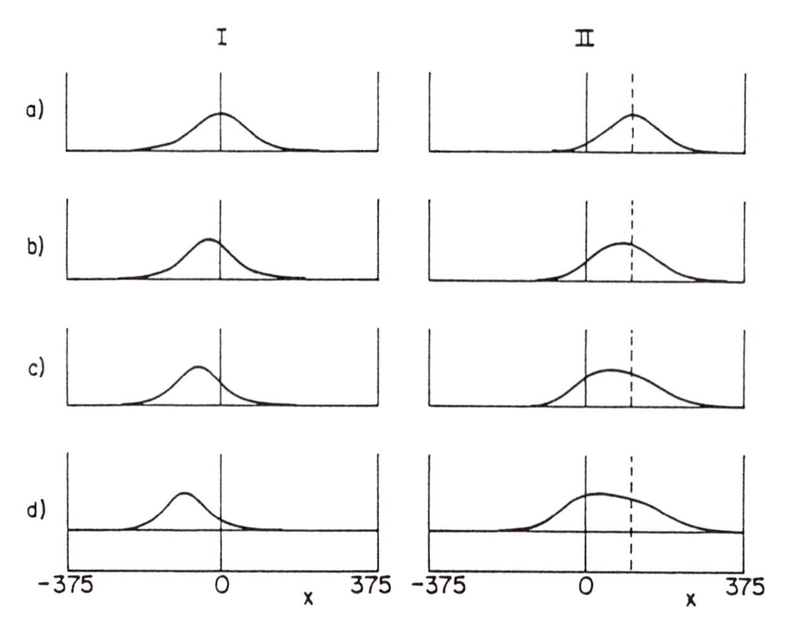

Figure 3 Nonlinear states for global modes in an infinite domain. The left column shows the modulus of the "transformed" global mode $|\phi_g(x)\exp(-Ux/2b)|$ and the right column shows the modulus of the global mode $|\phi_g(x)|$. a) The linear global mode. b) The nonlinear global mode for a supercriticality $(\mu_0 - \mu_{g_c})/\mu_{g_c} = 0.024$. c) Same as b) for a supercriticality of 0.073. d) Same as b) for a supercriticality of 0.171. The vertical line marks the location of the peak absolute growth rate and the dashed line marks the peak of the linear global mode.

point of view, the nonlinear effect shifts the position of the absolute instability and, therewith, the position of the global mode.

As indicated above, most of our numerical experiments were performed with $c_n = 0$ and, consequently, $1 + c_d c_n > 0$. In this case the uniform Stokes wave-train solutions of the constant-coefficient, Ginzburg-Landau equation are linearly stable with respect to the Benjamin-Feir mechanism (cf., Newell, 1974). The global modes are stable even for strong nonlinearity under these conditions. However, the criterion $1 + c_d c_n > 0$ is only a necessary condition for the stability of global modes. This is why a stable global mode was found for a numerical simulation with $c_n = 1$ $(1 + c_d c_n = -9)$ and a supercriticality $(\mu_0 - \mu_{g_c})/\mu_{g_c}$ of 43% (see Figure 4a). On the contrary, when $c_n = 10$ and all other parameters are unchanged, a very irregular state is observed (cf., Figure 4b). Furthermore, the irregular state persists even when the supercriticality is small. Figure 5 presents, for $(\mu_0 - \mu_{g_c})/\mu_{g_c} = 0.07$, the saturated regular mode for $c_n = 0$ and the corresponding irregular mode when $c_n = 10$. In both sets of figures the global mode for $c_n = 10$ exhibits a regular spatial structure and a complex time behavior over the initial regions in x. This early region is followed by a pulse-like structure with a fairly regular spatial periodicity.

The chaotic-type global mode for large negative values of $1 + c_d c_n$ merits a much more comprehensive study. The present exploratory simulations reveal a little of the

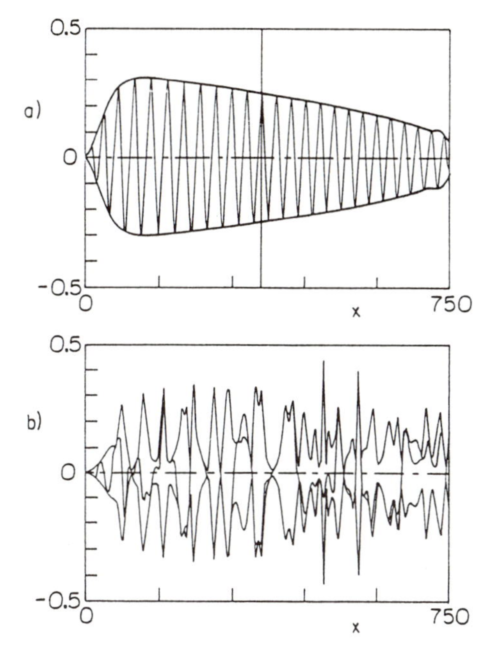

Figure 4 The envelope and the real part of the nonlinear global mode in a semi-infinite domain obtained for a supercriticality $(\mu_0 - \mu_{g_c})/\mu_{g_c} = 0.43$. a) $c_n = 1$. b) $c_n = 10$. The vertical line in a) marks the end of the absolutely unstable region and both structures are obtained at the same time $(3 \times 10^5 \Delta t)$.

variety of free, dynamical states possible in spatially-varying wave guides and their connection to regions of local absolute instability.

3.2 Forced response of global modes

The parameter space to be considered in a study on the effect of forcing is quite large since the strength, frequency, and position of forcing are added to those characterizing the homogeneous problem. For this reason the present results are limited to exploring the influence of the forcing location x_f (*vis-a-vis* the absolutely unstable region) and the shape of the response, which is predicted to be close to the free, linear global mode when the forcing is weak and the wave-guide state is close to critical.

The important elements in the forced response of the lowest (i.e., $n = 0$) linear, marginal mode for the infinite-domain configuration (see Eqn. 2.13 - 14) are shown in Figure 6 for the same set of parameters used in the earlier simulations; $U = 6$, $c_d = -10$, and $\mu_2 = -2.534 \times 10^{-6}$. The free mode, whose envelope is a simple Gaussian when $n = 0$, is shown in Figure 6a in its position relative to the maximum absolute growth rate denoted by the vertical line in the figure. The strongest action of the global mode is found downstream of the absolutely unstable region. When the factor $\exp(Ux/2b)$ is

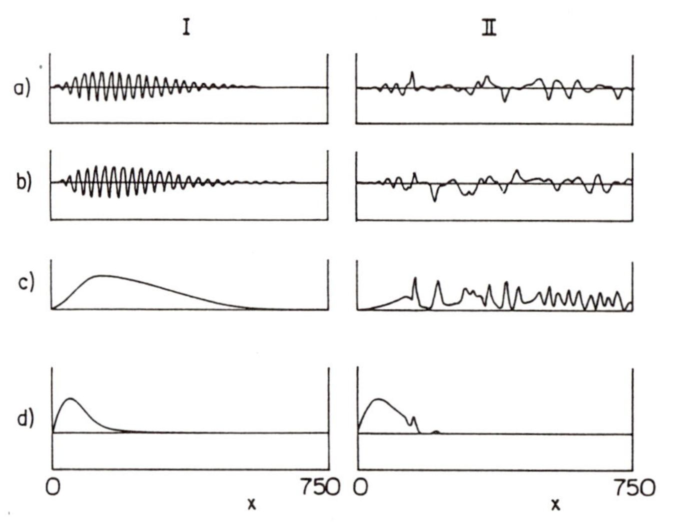

Figure 5 The nonlinear global mode in a semi-infinite domain obtained for a supercriticality of 0.07. The left column is for $c_n = 0$ and the right column is for $c_n = 10$. a) The real part of $\phi_g(x)$. b) The imaginary part of $\phi_g(x)$. c) The modulus of $\phi_g(x)$. d) The modulus of $\phi_g(x)\exp(-Ux/2b)$.

removed from the eigenfunction (cf., Eqn. 2.13b), the remainder is a Gaussian centered over the region of absolute instability as seen in Figure 6b. The relationship to the forcing efficiency is revealed in Figure 6c where the envelope of the adjoint function (2.14) is included. The adjoint is shifted upstream from the peak absolute growth rate by an amount ($\delta x = 112.5$) equal to the downstream shift of the global mode.

The important result just described regarding the forcing efficiency was tested in numerical simulations of the nonlinear equation with $c_n = 0$ and with $\epsilon^2\Delta_\mu = -0.027\mu_t$, $\epsilon^2 F = 10^{-5}$, $\epsilon^2\Omega = 0.006$. The neutral values for the free global mode are $\mu_0 = 1.017\mu_t$ and $\omega_g = 0.884$. The forced response is shown in Figure 7 as the forcing location is varied from upstream to downstream of the peak of the adjoint function which is located at $x = -112.5$. Figure 7a shows the free mode shape for this (weakly) nonlinear realization and it corresponds very closely to the linear mode. Since the simulation is performed for a slightly subcritical setting of the control parameter, this mode would ultimately decay. In the remaining panels of the figure, however, a finite equilibrium amplitude is sustained by virtue of the forcing which is most effective when it is positioned at or slightly downstream of the peak of the adjoint function. The forced response is observed to move upstream as the forcing is displaced downstream of its optimum location, but the response does not follow the theoretical prediction in that the equilibrium amplitude does not scale directly with $\psi^*(x_f)$ which is symmetric about $x = -112.5$. It is worth noting that the response reaches an amplitude which is several orders of

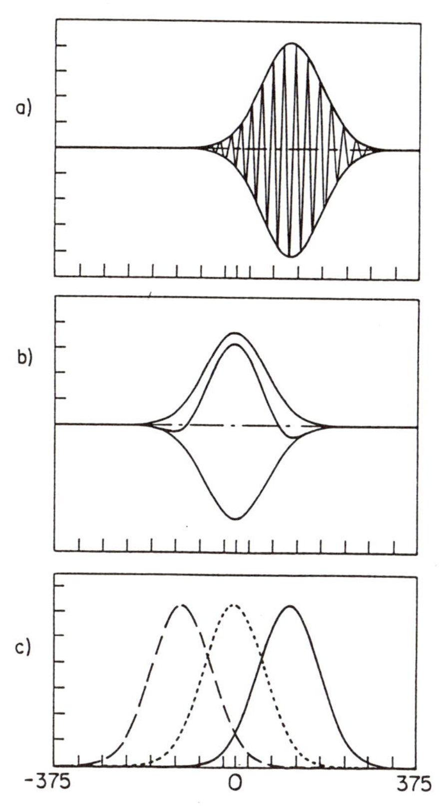

Figure 6 The linear, free global mode in the infinite domain. a) The modulus and real part of $\phi_g(x)$. b) The modulus and real part of the "transformed" mode $\phi_g(x)\exp(-Ux/2b)$. c) — $|\phi_g(x)|$; - - - - $|\phi_g(x)\exp(-Ux/2b)|$; - - - $|\psi_0(x)|$.

magnitude larger than the forcing, a result in agreement with predictions of weakly nonlinear theory. The numerical simulations for the last two cases shown in Figure 7 revealed an important nonlinear effect in that there was a distinct temporal oscillation of the forced response. This dynamical state having temporal quasi-periodicity with a coherent spatial structure is interesting and deserves further investigation. It may explain the observed departures from predictions based on weakly nonlinear theory as noted above and exhibited in more detail in Figure 8. Figure 8 shows two different measures of the forcing efficiency for the same set of experiments. One is simply the global maximum of the response and the other is the square-root of the total energy of the response. Both measures show that the peak efficiency occurs for forcing locations upstream of the maximum absolute growth rate, but downstream of the position suggested by linear theory. The discrepancies must derive from nonlinear effects.

4 CONCLUDING REMARKS

The results presented here, reveal important characteristics of global modes. The necessary condition of an interval of local absolute instability for the existence of a global

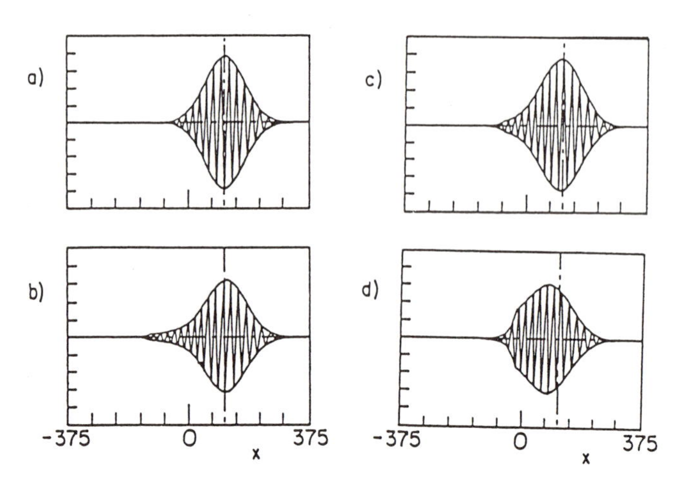

Figure 7 The forced, nonlinear response for a weakly-damped global mode in the infinite domain. a) The linear eigenfunction. b) The response with forcing at $x_f = -125$; the peak response amplitude is $\bar{A} = 2.5 \times 10^{-3}$. c) Same as b) with $x_f = -75$, $\bar{A} = (4.4 \pm 0.5) \times 10^{-3}$. d) Same as b) with $x_f = -25$, $\bar{A} = 2.5 \times 10^{-3}$.

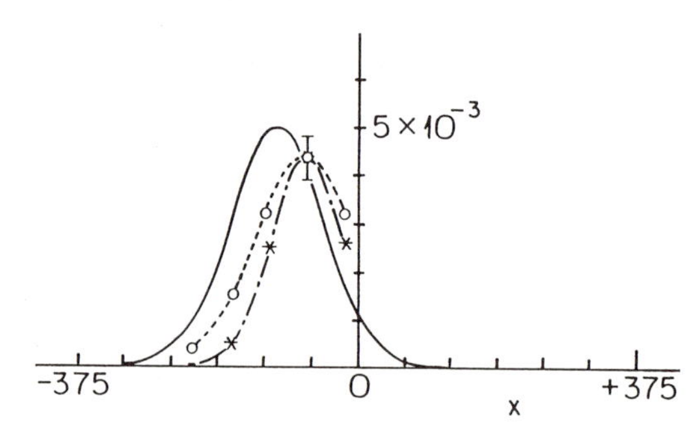

Figure 8 The forcing efficiency for the same parameters used in the simulations shown in Figure 7. — the linear, theorectical prediction; - - o - - the maximum response amplitude; – * – the square root of the total response energy.

mode, and the significance of these modes in the dynamics of spatially-inhomogeneous wave guides, is firmly established. However, there exists a rich and diverse spectrum of nonlinear effects which await careful study, even in the context of the simplified models discussed herein.

It is clear from the present work that the specific details and location of the absolutely unstable region do not necessarily determine the location of the "action" of global modes. Their "action" is felt in regions well beyond that where the local state is absolutely unstable and the most efficient location to force such modes also lies outside (and considerably upstream) of the region of absolute instability. In parameter domains where these modes are stable, they impose a long-range order to the wave guide and are responsible for the existence of discrete frequencies which are quite unrelated to those frequencies associated with the local state. The present work also shows that the effect of nonlinearity and forcing on these modes open the possibility of frequency and mode competition which can lead to a host of complex spatio-temporal dynamical states. Studies based on models and of spatially-developing wave guides in applications remain a fruitful area of study.

ACKNOWLEDGEMENTS

This work was supported by the U.S. Air Force Office of Scientific Research under Contract No. F49620-85-C-0080.

REFERENCES

1. Bers, A., 1975. Linear waves and instabilities. *Physique de Plasmas*, eds. C. DeWitt and J. Peyraud, pp. 117-215. Gordon and Breach, New York.

2. Bers, A., 1983. Space-time evolution of plasma instabilities-absolute and convective. *Handbook of Plasma Physics*, eds. M.N. Rosenbluth and R.Z. Sagdeev, pp. 1: 451-517. North-Holland, Amsterdam.

3. Briggs, R. J., 1964. *Electron-Stream Interaction with Plasmas.* MIT Press, Cambridge.

4. Chomaz, J. M., Huerre, P., and Redekopp, L. G., 1987. Models of hydrodynamic resonances in separated shear flows. Proc. 6th Symp. Turb. Shear Flows, pp. 3.2.1-6.

5. Chomaz, J. M., Huerre, P., and Redekopp, L. G., 1988. Bifurcations to local and global modes in spatially-developing flows. *Phys. Rev. Lett.*, 60: 25-28.

6. Chomaz, J. M., Huerre, P., and Redekopp, L. G., 1989. A frequency selection criterion in spatially-developing flows. *Stud. Appl. Math.*, (submitted for publication).

7. Deissler, R. J., 1985. Noise-sustained structure, intermittency, and the Ginzburg-Landau equation. *J. Stat. Phys.*, 40: 371-95

8. Deissler, R. J., 1987. Spatially-growing waves, intermittency and convective chaos in an open-flow system. *Physica D*, 25: 233-260

9. Hannemann, K., and Oertel, Jr. H., 1988. Numerical simulation of the absolutely and convectively unstable wake. *J. Fluid Mech.*, 199: 55-88.

10. Huerre, P., and Monkewitz, P. A., 1990. Local and global instabilities in spatially-developing flows. *Ann. Rev. Fluid Mech.*, Vol. 22 (in press).

11. Jackson, C. P., 1987. A finite-element study of the onset of vortex shedding in flow past variously shaped bodies. *J. Fluid Mech.*, 182: 23-45.

12. Lifshitz, E. M., and Pitaevskii, L. P., 1981. *Physical Kinetics.* Chapter VI. Pergamon, London.

13. Mathis, C., Provansal, M., and Boyer, L., 1984. The Bénard-von Karman instability: an experimental study near the threshold. *J. Phys. Lett.* (Paris), 45: 483-491.

14. Monkewitz, P. A., 1988. The absolute and convective nature of instability in two-dimensional wakes at low Reynolds numbers. *Phys. Fluids*, 31: 999-1006.

15. Newell, A. C., 1974. Envelope equations. *Lect. Appl. Math.*, ed A.C. Newell, Vol. 15, pp. 157-163. *Am. Math. Soc.*, Providence, RI.

16. Provansal, M., Mathis, C., and Boyer, L., 1987. Bénard-von Karman instability: transient and forced regimes. *J. Fluid Mech.*, 182: 1-22.

17. Sreenivasan, K. R., Strykowski, P. J., and Olinger, D. J., 1987. Hopf bifurcation, Landaw equation and vortex shedding behind circular cylinders. *Proc. Forum on Unsteady Flow Separation*, ed. K. N. Ghia. ASME FED. Vol. 52.

18. Sreenivasan, K. R., Strykowski, P. J., and Olinger, D. J., 1989. On the Hopf bifurcation and Landau-Stuart constants associated with vortex 'shedding' behind circular cylinders. *J. Fluid Mech.*, (submitted for publication).

19. Sturrock, P. A., 1961. Amplifying and evanescent waves, convective and non-convective instabilities. *Plasma Physics*, ed. J.E. Drummond, pp. 124-42. McGraw-Hill, New York.

20. Yang, X. and Zebib, A., 1989. Absolute and convective instability of a cylinder wake. *Phys. Fluids*, A1: 689-696.

MATHEMATICAL JUSTIFICATION OF STEADY GINZBURG-LANDAU EQUATION STARTING FROM NAVIER-STOKES

Gérard Iooss[†], Alexander Mielke[¥], and Yves Demay[†]

[†] Laboratoire de Mathématiques, U.A.CNRS 168, Université de Nice, Parc Valrose, F-06034 Nice

[¥] Mathematisches Institut A, Universität Stuttgart, Pfaffenwaldring 57, D-7000 Stuttgart 80

I. Introduction

Many classical hydrodynamical stability problems deal with flows in a very long domain. This is often theoretically modelized by an infinite domain, which simplifies the linear analysis. Here we consider cases of cylindrical domains of one or two dimensional bounded cross-section Ω. Examples of such a situation are i) the Taylor-Couette problem of the flow between two concentric rotating cylinders, where the section is a 2-dimensional annulus, ii) the Bénard convection problem of a liquid heated from below in a long box, and where the section is a rectangle. In both of these problems there are two very important symmetries. First, the problem is invariant under translations parallel to the generatrices of the cylinder, second the problem is invariant under the reflection symmetry through any cross-sectional plane.

In many mathematical treatments of nonlinear hydrodynamic stability problems, a given spatial periodicity is assumed. This then leads to bifurcated solutions actually spatially periodic (!). The aim of our analysis is to study the existence of bounded steady solutions other than these ones, bifurcating from the basic maximally symmetric one. In the present work, we derive the basic system of ordinary differential equations to be studied for obtaining all new solutions.

By not assuming a spatial periodicity of the solutions, we fall on the difficulty of dealing with a continuous spectrum for the linearized problem. Since about 20 years, this difficulty was solved by physicists in considering slow modulations in space of the amplitudes of critical modes. The envelope equation they obtain is usually called the Ginzburg-Landau (G-L) equation, see Newell-Whitehead [N-W], Segel [Se] . Among the few mathematical studies in this field, Collet and Eckmann [C-E 86,87] start with an equation simpler than Navier-Stokes, looking like (G-L) equation and give all bifurcating steady solutions. They also obtain propagating fronts and are able to study their stability.

Now, for steady solutions of Navier-Stokes equations, we consider the unbounded space variable x as an evolution variable varying from $-\infty$ to $+\infty$. This way for studying mathematically elliptic problems was initiated by Kirchgässner [Ki] and is now extensively used for water waves problems [Mi 86b][A-K], and elasticity problems (long beams) [Mi 88b]. Following this idea, the usual techniques: center manifold and normal form theories apply near criticality. The center manifold theorem is used here for finding solutions which are bounded at both infinities in x, and which are close to the basic fully symmetric solution. Using results of Mielke [Mi 88a], this method is shown to be applicable on steady Navier-Stokes equations in a cylindrical domain, once they are written as an evolution problem in the x-variable. We then obtain a reversible 4-dimensional system which is written in normal form, and whose relationship with the steady (G-L) equation is emphazised. The idea of using x as

New Trends in Nonlinear Dynamics and Pattern-Forming Phenomena
Edited by P. Coullet and P. Huerre
Plenum Press, New York, 1990

an evolution variable for obtaining steady bifurcating solutions in hydrodynamical nonlinear stability problems is due to Coullet and Repaux in [C-R], where they give heuristic arguments leading at first orders to the steady (G-L) equation. In fact, we give a way to compute the coefficients of our new system and we give the relationship between these coefficients and those of (G-L) equation.The study of our normal form allows to recover all known solutions of the (G-L) equation truncated at lowest orders [N-W][K-Z]. In addition we are able to answer the following questions: What is the meaning of (for instance) the third order spatial derivative in (G-L) equation ? Is a solution, such that amplitude vanishes while its gradient does not, justified for (G-L) equation ?

II. The classical frame

Let us denote by $Q = \Omega\times\mathbb{R}$ the domain of the flow where Ω is a bounded regular domain of $\mathbb{R}$ or $\mathbb{R}^2$. The classical mathematical studies are looking for solutions which are $2\pi/h$-periodic in $x\in\mathbb{R}$, where h is a wave number to be specified later. Without loss of generality, we simplify the exposition in assuming that the only equations of the problem are here the Navier-Stokes equations for incompressible fluids. One of the important points is that boundary conditions on $\partial\Omega\times\mathbb{R}$ are steady and independent of x. An example of such a problem is the Taylor-Couette flow between concentric rotating infinite cylinders, where Ω is the annulus $R_1< r <R_2$ in polar coordinates. For the Bénard convection problem in a long box, one has to add the (coupled) equation for energy conservation, and Ω is a rectangle $[-a,a]\times[-b,b]$. It is known that, once substracted a fully symmetric divergence free vector field satisfying the boundary conditions, the equations for the perturbation can be put into the form of a differential equation lying in a suitable function space:

$$\frac{dU}{dt} = L_\mu U + N(\mu,U). \tag{1}$$

In (1) U is in most of cases the velocity vector field in Q belonging to a function space $\mathcal{D}_h$, and $\mu \in \mathbb{R}$ represents a distinguished parameter among the set of parameters of the problem. For instance, for the Couette-Taylor problem and the Rayleigh-Bénard convection problem we can respectively take the Reynolds number built with the rotating rate of the inner cylinder, and the Rayleigh number. The operators L_μ and $N(\mu,.)$ are respectively linear and quadratic in U.

Now, there are very important symmetry properties of the system: the translational invariance ($x\to x+a$) and reflectional invariance ($x \to -x$). They are expressed by the commutativity of L_μ and $N(\mu,.)$ with a one parameter group of linear operators τ_a, $a \in \mathbb{R}$, and with a symmetry operator S (S^2=Id). Moreover we have the property

$$\tau_a S = S\tau_{-a}. \tag{2}$$

In fact, in the classical mathematical formulation, the assumed $2\pi/h$-periodicity leads to an O(2) invariant problem.

II.1. Linear stability problem

We start with (1) and study the stability of the maximally symmetric solution U = 0. Usual linear theory of hydrodynamical stability looks for perturbations of the form $\hat{U}_k e^{ikx}$, where $\hat{U}_k$ is a function of variables lying in Ω. The corresponding eigenvalues of L_μ are denoted by $\sigma(\mu,ik)$:

$$L_\mu(\hat{U}_k e^{ikx}) = \sigma(\mu,ik)\hat{U}_k e^{ikx}. \tag{3}$$

For each k, there is an infinite set of eigenvalues $\{\sigma_m;\ m\in\mathbb{N}\}$ and if the only restriction on the behavior in x is the boundedness of vector fields, it is clear that the set of all eigenvalues $\{\sigma_m(\mu,ik)\}$ is not discrete (since k can vary continuously). Once $2\pi/h$ periodicity is assumed, one only allows k to take values multiple of h. It is a classical result that the full spectrum of L_μ is then discrete.

Another important point here is the effect of the symmetry $x \to -x$. In fact $(S\hat{U}_k)e^{-ikx}$ is an eigenvector belonging to the same eigenvalue σ, hence we have

$$(4) \qquad \sigma(\mu,-ik) = \sigma(\mu,ik), \qquad \hat{U}_{-k} = S\hat{U}_k.$$

<u>Assume that the eigenvalue of largest real part</u> σ_0 <u>is real</u>, then classical theory deals with a neutral stability curve $\mu = \mu_c(k)$ (even function) defined by

$$(5) \qquad \sigma_0(\mu,ik) = 0,$$

and which passes through a minimum $\mu = 0$ at $k = k_c$. We arrange notations in such a way that

i) for $\mu < 0$, $\sigma_0 < 0$ for any k, and the 0 solution is exponentially stable, while

ii) for $\mu > 0$, $\sigma_0 > 0$ for some k, and 0 is linearly unstable.

We have a family of curves σ_0 as a function of k, for fixed values of μ, and we see that for $\mu > 0$, there are two symmetric intervals (see figure 1a) where the wave number k gives an exponential growth of the perturbation. In the (k,μ) plane, if we look at a fixed value of $\mu > 0$, then the values of $|k|$ giving points inside the parabolic region $\mu > \mu_c(k)$ lead to instability (see figure 1b), while outside of this region perturbations $\hat{U}_k e^{ikx}$ are damped.

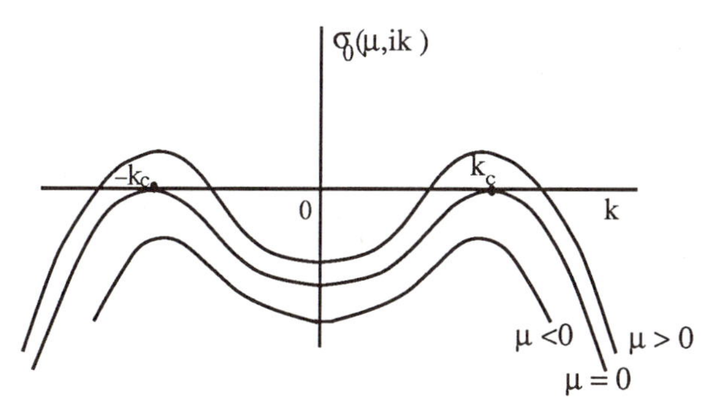

Figure 1 a.

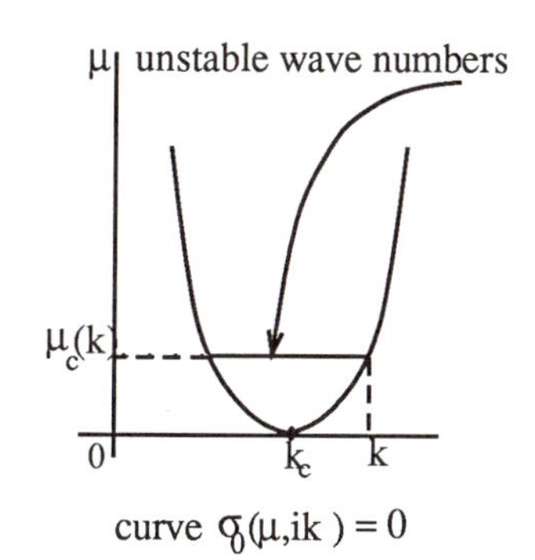

Figure 1 b.

If we choose a spatial period $2\pi/h$, then criticality is given by the multiple of h giving <u>the lowest point</u> on the neutral curve $\mu = \mu_c(k)$. This point is denoted by k (close to k_c if h is small). It is clear that the larger is the period, the smaller is the domain of validity of this analysis, since even for $\mu > 0$ close to 0, there will be many positive eigenvalues giving excited modes and the nonlinear interaction between modes will then be very strong. This is precisely the aim of Ginzburg-Landau equation to take account of all these excited wave numbers near k_c when period is infinite.

Let us go on the classical analysis assuming $2\pi/h$ -periodicity. It is then clear that the problem is O(2) <u>invariant</u>. Hence, for $\mu = \mu_c(k)$ we have <u>a double eigenvalue</u> 0 of the operator L_{μ_c} and eigenvectors:

$$(6) \qquad \hat{U}_k e^{ikx} \; ; \; \overline{\hat{U}_k} e^{-ikx}$$

can be chosen such that

$$(7) \qquad S\hat{U}_k = \overline{\hat{U}_k} \; (= \hat{U}_{-k}).$$

By construction, the remaining part of the spectrum of L_{μ_c} is of strictly negative real part and situated in a sector centered on the real axis as it results from a perturbation analysis from the Stokes operator [Ka][Iu][Io71].

II.2. Landau equation

The structure of (1), i.e. the properties of L_{μ_c} and of $N(\mu,.)$ allow us to use the Center Manifold theorem for μ near $\mu_c(k)$ (also called the slaving principle by physicists). See [He] for a proof on evolution problems satisfying partial differential equations like Navier-Stokes. See also [Va] for a modern proof on vector fields, easily adaptable. Let us denote by P_μ the projection operator (of rank 2) associated with the isolated eigenvalue $\sigma_0(\mu,ik)$ which commutes with L_μ, then we may assert the following:

There exists a neighborhood of $(\mu_c,0)$ *in* $\mathbb{R}\times\mathcal{D}_h$, *on which is defined a regular map* Φ, *such that* $U = X + \Phi(\mu,X)$, $X \in P_{\mu_c}\mathcal{D}_h$, *represents a manifold* $\mathcal{M}_\mu$ *with the following properties*:

i) $\mathcal{M}_{\mu_c}$ *is tangent to the space* $P_{\mu_c}\mathcal{D}_h$ *at the origin* ($\Phi(\mu_c,0) = 0$, *and* $D_X\Phi(\mu_c,0) = 0$),

ii) $\mathcal{M}_\mu$ *is locally invariant under equation* (1),

iii) $\mathcal{M}_\mu$ *is locally attracting under equation* (1): *if* U_0 *is an initial data such that the solution* $U(t)$ *stays in the neighborhood of 0 for all* t, *then* dist $[U(t),\mathcal{M}_\mu] \to 0$ *as* $t \to \infty$,

iv) *we can choose the manifold such that* $\Phi(\mu,.)$ *commutes with the group actions* τ_a *and* S.

It follows that the manifold $\mathcal{M}_\mu$ is two dimensional and that the asymptotic dynamics lies on it. Moreover the trace of equation (1) on $\mathcal{M}_\mu$ can be parametrized by $\mathcal{A} \in \mathbb{C}$ defined by:

$$(8) \qquad X = \mathcal{A}\hat{U}_k e^{ikx} + \bar{\mathcal{A}}\bar{\hat{U}}_k e^{-ikx}, \quad X \in P_{\mu_c}\mathcal{D}_h,$$

and the dynamics on the Center Manifold is now represented by an ODE which commutes with τ_a and S. It results easily that it can be written as

$$(9) \qquad \frac{d\mathcal{A}}{dt} = \mathcal{A} f_k(\mu,|\mathcal{A}|)$$

where f_k is even in its second argument, and real. An additional property of representation (8) is that one should have an equation invariant under the transformation: $(\mathcal{A},k) \to (\bar{\mathcal{A}},-k)$. This leads to coefficients which are even functions of k. The principal part of (9) becomes *(Landau equation):*

$$(10) \qquad \frac{d\mathcal{A}}{dt} = a_k[\mu-\mu_c(k)]\mathcal{A} + b_k\mathcal{A}|\mathcal{A}|^2 + \ldots$$

where a_k and b_k are real and even functions of k, and where we have

$$(11) \qquad \sigma_0(\mu,ik) = [\mu-\mu_c(k)](a_k + \text{h.o.t.}).$$

In what follows, we assume that :

$$(12) \qquad \text{H.1} \quad a_{k_c} > 0, \quad b_{k_c} < 0,$$

hence, for k near k_c (i.e. for a given period $2\pi/h$ large enough), we obtain a classical supercritical pitchfork bifurcation to a solution of (9) :

$$(13) \qquad |\mathcal{A}|^2 = \frac{-a_k}{b_k}[\mu-\mu_c(k)][1 + \text{h.o.t.}].$$

The solution U of (1) is now given by $U = X + \Phi(\mu,X)$ where X is given by (8) and (13). We observe that $\tau_a U$ is a shifted solution corresponding to a change of $\mathcal{A}$ into $\mathcal{A}e^{ika}$, so there is a group orbit of steady bifurcated solutions of (1) of spatial period $2\pi/k$.

III. Steady Navier-Stokes as a spatial dynamical system

III.1 Transformation into a dynamical system

The aim of this part is to write the steady Navier-Stokes equations in the infinite cylinder $Q = \Omega \times \mathbb{R}$, into the form of a differential equation in the space variable x.

The steady Navier-Stokes equations are as follows:

$$(14) \qquad \begin{cases} (V.\nabla)V + \nabla p = \nu\Delta V + f, \\ \nabla.V = 0 \qquad \text{in } Q, \\ V = g(\mu,.) \text{ on } \partial Q = \partial\Omega \times \mathbb{R}, \end{cases}$$

where μ represents a distinguished parameter, and f,g are functions of the cross-sectional variable $y \in \Omega$ (resp.$\partial\Omega$) only. The velocity vector field V will be decomposed into a longitudinal component V_x and a transversal component $V_\perp$.We assume the existence of a family of fully symmetric x-independent solutions $V = V^{(0)}(\mu,.)$. System (14) is reversible; because for every solution V of (14), the reversed flow $\hat{V} = SV$, defined by

(15) $\hat{V}(x,y) = [\hat{V}_x(x,y), \hat{V}_\perp(x,y)]$, with $\hat{V}_x(x,y) = -V_x(-x,y)$, and $\hat{V}_\perp(x,y) = V_\perp(-x,y)$,

[where $(x,y) \in \mathbb{R}\times\Omega$] is also a solution of the problem. For instance, the Taylor-Couette problem (Ω being the annulus $R_1 < |y| < R_2$, $f \equiv 0$ and $g \equiv \Omega_i x y|_{|y|=R_i}$) is reversible in this sense.

To do the bifurcation analysis we introduce the notations $U = V - V^{(0)}$, $W = \nu\frac{\partial U}{\partial x} - pe_x$ where $e_x = (1,0)$ in $\mathbb{R}\times\mathbb{R}^2$. We notice that W has the same number of components as U. Using the incompressibility written in the form

$$\frac{\partial U_x}{\partial x} + \nabla_\perp.U_\perp = 0,$$

we find for the pressure

$$(16) \qquad p = -\nu\nabla_\perp.U_\perp - W_x.$$

Moreover, setting $\mathcal{V} = (U,W)$, equation (14) takes now the form

$$(17) \qquad \frac{d\mathcal{V}}{dx} = \mathcal{A}_\mu\mathcal{V} + \mathcal{B}_\mu(\mathcal{V},\mathcal{V}),$$

where there are no longer differentiations in x on the right hand side and $\mathcal{A}_\mu$ and $\mathcal{B}_\mu$ are respectively linear and quadratic operators on $\mathcal{V}$. The reversibility of the problem is now expressed through the reflection $\hat{S}$ defined by

$$(18) \qquad \hat{S}\mathcal{V} = \hat{S}(U,W) = (SU,-SW),$$

and we then have

$$(19) \qquad \hat{S}\mathcal{A}_\mu = -\mathcal{A}_\mu\hat{S}, \qquad \mathcal{B}_\mu \circ \hat{S} = -\hat{S}\mathcal{B}_\mu.$$

Our aim in this section is to characterize all solutions V of (14) (resp.(17)) which exist on the whole unbounded region Q and are close to the trivial solution $V^{(0)}$ uniformly in x. Following the methods of Kirchgässner [Ki] further developed in [Mi 86a,88a] this can be achieved by constructing a center manifold for (17). It should be noted that (17) is not an evolutionary problem as it is derived from the elliptic problem (14). The spectrum of $\mathcal{A}_\mu$ is infinite on both sides of the imaginary axis; however there are only finitely many eigenvalues on the imaginary axis. Exactly these give rise to a locally invariant manifold for the flow of (17): the center manifold.

To apply the result of [Mi 88a] we have to show that the resolvent of $\mathcal{A}_\mu$ satisfies

$$(20) \qquad \|(\mathcal{A}_\mu - ik\text{Id})^{-1}\|_{\mathcal{L}(\mathcal{X})} = O(|k|^{-1})$$

for $k \in \mathbb{R}$ and $|k| \to \infty$. This estimate is established in [I-M-D] and it is shown that the spectrum of $\mathcal{A}_\mu$ is only composed with eigenvalues of finite multiplicities, not accumulating at a finite distance, and situated in a sector of the complex plane centered on the real axis (see figure 2). Moreover, we shall see below that for $\mu = 0$ there are only two eigenvalues on the imaginary axis. Hence the remaining part of the spectrum is at a finite distance of this axis. All this ensures the possibility of using the result of [Mi 88a].

To characterize the center manifold we have to construct the spectral part $\mathcal{X}_0$ corresponding to the spectrum lying on the imaginary axis. This amounts in solving the eigenvalue problem

$$\mathcal{A}_\mu \mathbb{V} = \lambda \mathbb{V} \tag{21}$$

for $\lambda = ik$, $k \in \mathbb{R}$. Note that $-\lambda$ is an eigenvalue whenever λ is one, as (21) is equivalent to $\mathcal{A}_\mu \hat{S}\mathbb{V} = -\lambda \hat{S}\mathbb{V}$.

At this point, it is important to remember that the linear part $\frac{d}{dx} - \mathcal{A}_\mu$ of (17) is only a reformulation of the linear operator L_μ in equation (1). Hence, the eigenvalue problem (21) with $\lambda = ik$ is exactly equivalent to the linear stability problem (3) with $\sigma(\mu,ik) = 0$. Now, we observed that for $\mu = \mu_c(k)$ we have $\sigma_0 = 0$, and

$$L_{\mu_c(k)}(\hat{U}_k e^{ikx}) = 0. \tag{22}$$

This is equivalent to the existence of an eigenvector $\mathbb{V}(ik)$ of $\mathcal{A}_{\mu_c(k)}$ such that

$$(\mathcal{A}_{\mu_c(k)} - ik)\mathbb{V}(ik) = 0. \tag{23}$$

Moreover, if we define a projection Π by $\Pi\mathbb{V} = \Pi(U,W) = U$, we have $\Pi\mathbb{V}(ik) = \hat{U}_k$.

For $\mu = 0$ the only eigenvalues on the imaginary axis are $\pm ik_c$; for $\mu > 0$ there are two pairs of eigenvalues on the imaginary axis, and for $\mu < 0$ they disappear from this axis. Because of reversibility, we know that the generic situation is that for $\mu = 0$ these eigenvalues are double, non semi-simple (geometric multiplicity one). In fact, starting with the relation (23) we successively obtain, using the property that $\frac{d\mu_c}{dk}(k_c) = 0$,

$$(\mathcal{A}_0 - ik_c)\mathbb{V}(ik_c) = 0, \tag{24}$$

$$(\mathcal{A}_0 - ik_c)\frac{d\mathbb{V}(ik_c)}{d\lambda} = \mathbb{V}(ik_c). \tag{25}$$

Denoting by $\mathbb{V}_0 = \mathbb{V}(ik_c)$, and $\mathbb{V}_1 = \frac{d\mathbb{V}(ik_c)}{d\lambda}$, we then have a Jordan basis for the generalized eigenspace belonging to the eigenvalue ik_c of $\mathcal{A}_0$, and we see that because of $\hat{S}\mathbb{V}(ik) = \mathbb{V}(-ik) = \overline{\mathbb{V}}(ik)$, we have the following representation of $\hat{S}$ on the generalized eigenspace:

$$\hat{S}\mathbb{V}_0 = \overline{\mathbb{V}}_0, \quad \hat{S}\mathbb{V}_1 = -\overline{\mathbb{V}}_1. \tag{26}$$

For $\mu = 0$ wc know by construction the position of all pure imaginary eigenvalues of $\mathcal{A}_0$, and because of the properties of the spectrum described above, all other eigenvalues λ of $\mathcal{A}_\mu$ are bounded away from the imaginary axis as long as $|\mu|$ is small (shaded regions on figure 2).

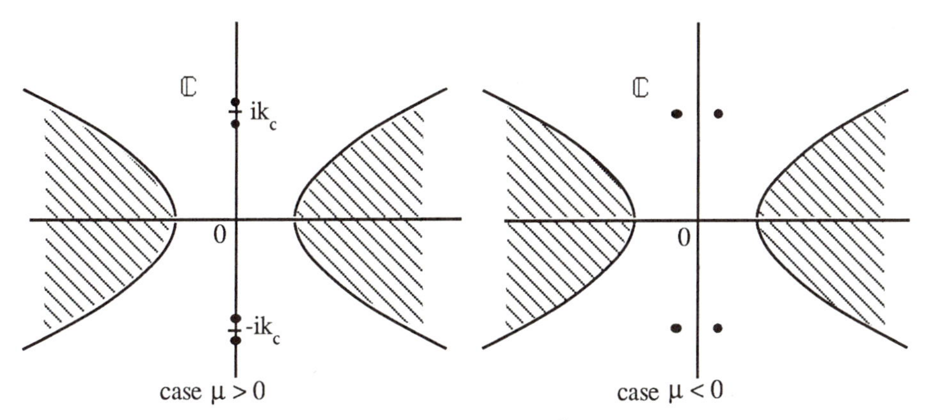

case $\mu > 0$ case $\mu < 0$

Figure 2. Location of the spectrum of $\mathcal{A}_\mu$ for $|\mu|$ close to 0.

We can then use the center manifold theorem in [Mi 88a], to arrive to the following result:

All solutions V of (14), *with* $|\mu|$ *small and being sufficiently close to* $V = V^{(0)}$ *for all* x *in* $\mathbb{R}$, *satisfy a relation*

$$\mathbb{V} = A\mathbb{V}_0 + B\mathbb{V}_1 + \bar{A}\bar{\mathbb{V}}_0 + \bar{B}\bar{\mathbb{V}}_1 + \mathbb{Q}(\mu,A,\bar{A},B,\bar{B}), \tag{27}$$

where $\mathbb{Q}$ *is a smooth function with* $\mathbb{Q} = O[|\mu|(|A|+|B|)+|A|^2+|B|^2]$, *and* $V-V^{(0)}=\Pi\mathbb{V}$. *Moreover* (A,B) *is solution of a reduced equation*

$$\begin{cases} \dfrac{dA}{dx} = ik_cA + B + f(\mu,A,\bar{A},B,\bar{B}) \\ \dfrac{dB}{dx} = ik_cB + g(\mu,A,\bar{A},B,\bar{B}) \end{cases} \tag{28}$$

where f and g do not contain any constant and pure (with no μ*) linear term in* A *and* B. *Moreover, due to reversibility, we have (see* [Mi 86a]*)*

$$\begin{cases} f(\mu,\bar{A},A,-\bar{B},-B) = -\overline{f(\mu,A,\bar{A},B,\bar{B})} \\ g(\mu,\bar{A},A,-\bar{B},-B) = \overline{g(\mu,A,\bar{A},B,\bar{B})} \end{cases}, \tag{29}$$

$$\mathbb{Q}(\mu,\bar{A},A,-\bar{B},-B) = \hat{S}\mathbb{Q}(\mu,A,\bar{A},B,\bar{B}). \tag{30}$$

III.2. Solutions of the normal form

Now, to simplify the form of (28), we can put it into normal form. This, of course, can only arrange coefficients up to a given order, but this simplifies a lot the further analysis. It is shown (see [E-T-B-C-I]) that a good choice of normal form associated with a critical linear operator such that expressed in (28) and satisfying the reversibility symmetry is as follows up to any order $O(|A| + |B|)^N$:

$$\begin{cases} \dfrac{dA}{dx} = ik_cA + B + iAP[\mu;\ |A|^2;\ \tfrac{i}{2}(A\bar{B}-\bar{A}B)] \\ \dfrac{dB}{dx} = ik_cB + iBP[\mu;\ |A|^2;\ \tfrac{i}{2}(A\bar{B}-\bar{A}B)] + AQ[\mu;\ |A|^2;\ \tfrac{i}{2}(A\bar{B}-\bar{A}B)] \end{cases} \tag{31}$$

Here P and Q are real polynomials in their two last arguments, with μ dependent coefficients, and are such that $P(0,0,0) = Q(0,0,0) = 0$.

In this case, we have two arbitrary functions P and Q, hence (31) is not hamiltonian in general. Nevertheless our system (31) is integrable (like in the hamiltonian case). To easily solve this problem, let us change variables:

(32) $$A = r_0 e^{i(k_c x+\psi_0)}, \qquad B = r_1 e^{i(k_c x+\psi_1)},$$

then two first integrals are :

(33) $$\begin{cases} r_0 r_1 \sin(\psi_1 - \psi_0) = K \\ r_1^2 - G(\mu, r_0^2, K) = H \end{cases},$$

with $G(\mu,u,v) = \int_0^u Q(\mu,s,v)ds$. To study more precisely the behavior of solutions, let us define the principal part of P and Q:

(34) $$\begin{cases} P(\mu,u,v) = p_1\mu + p_2 u + p_3 v + O(|\mu|+|u|+|v|)^2 \\ Q(\mu,u,v) = -q_1\mu + q_2 u + q_3 v + O(|\mu|+|u|+|v|)^2 \end{cases}.$$

We can specify the meaning of coefficients p_j and q_j by taking account of what we assumed on the eigenvalues of the linear operator $\mathcal{L}_\mu$, and also on what we know on the steady spatially periodic solutions obtained in part II, with assumption H.1 [see(12)].

For the linear operator occuring in (31), eigenvalues are

(35) $$i[k_c+P(\mu,0,0)] \pm \sqrt{Q(\mu,0,0)}$$, and the complex conjugate.

This shows that $Q(\mu,0,0)$ is < 0 for $\mu > 0$. The generic situation is then when

(36) $$q_1 > 0.$$

The neutral stability curve, given at figure 1b, is obtained by solving with respect to μ the equation

(37) $$[(k-k_c) - P(\mu,0,0)]^2 + Q(\mu,0,0) = 0.$$

This leads to the following expansion:

(38) $$\mu_c(k) = \frac{1}{q_1}(k-k_c)^2 - \frac{2p_1}{q_1^2}(k-k_c)^3 + O(k-k_c)^4,$$

which will be used for the comparison with the Ginzburg-Landau equation .

The steady spatially periodic solutions obtained in §II.2 correspond to stationary solutions in r_0, r_1: they are given, up to a phase shift, by:

$$\psi_1 - \psi_0 = \pi/2 + \ell\pi, \ \psi_0 = \alpha x, \ \ \alpha = k-k_c, \ \ \ell = 0 \text{ or } 1, \text{ with}$$

(39) $$\begin{cases} [\alpha - P(\mu,r_0^2,(-1)^\ell r_0 r_1)]r_0 = (-1)^\ell r_1 \\ Q(\mu,r_0^2,(-1)^\ell r_0 r_1) + [\alpha - P(\mu,r_0^2,(-1)^\ell r_0 r_1)]^2 = 0. \end{cases}$$

Solving $(39)_1$ with respect to $r_o r_1$, and replacing into $(39)_2$ we can solve with respect to μ:

(40) $$\mu - \mu_c(k) = \frac{q_2}{q_1} r_0^2 + O[r_0^2(|\alpha|+r_0^2)].$$

With the assumption H.1 made at (12), the bifurcation is <u>supercritical</u> and <u>non degenerate</u>, hence

(41) $$q_2 > 0.$$

Let us now consider the level curves Γ_{HK} in the (r_0,r_1) plane given by

(42) $$r_1^2 = G(\mu,r_0^2,K) + H,$$

these curves are sketched at figure 3.

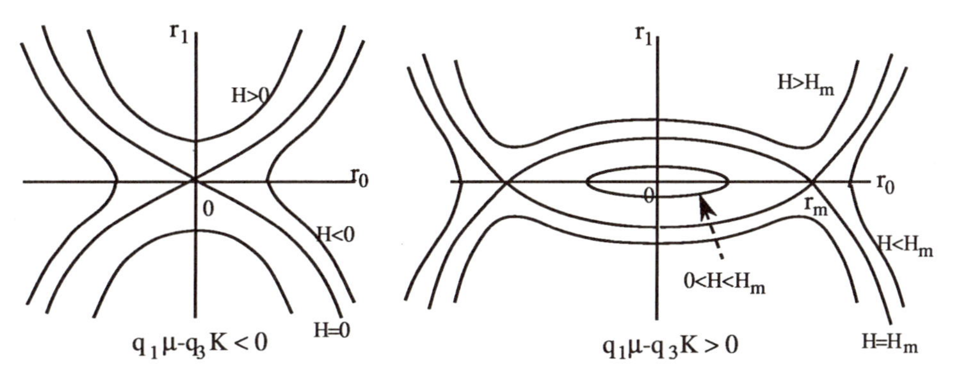

Figure 3. Level curves Γ_{HK}

When $q_1\mu - q_3K > 0$, Γ_{HK} is in 3 connected parts if $0 < H < H_m$. For $H = H_m$, all parts connect at two double points on the r_0 axis of abscissas $\pm r_m$ .Now, the integral $(33)_1$ leads to the property that

$$(43) \qquad r_0 r_1 > |K|.$$

For $K \neq 0$, this condition corresponds to a region limited by a set $\mathcal{H}_K$ of symmetric hyperbolas in the (r_0,r_1) plane. It is shown in [I-M-D] that the relevant solutions (bounded solutions) correspond to parts of the curves Γ_{HK} indicated on figure 4, bounded by intersections of Γ_{HK} with $\mathcal{H}_K$. All relevant solutions are obtained as follows: i) if the bounds are simple intersections this gives a spatially quasi-periodic solution, with wave numbers $k=k_c+\alpha$ and $\beta = O(\mu^{1/2})$; ii) if Γ_{HK} is tangent in $(r_0^{(1)},r_1^{(1)})$ to $\mathcal{H}_K$ we have a periodic solution given by (39); iii) if Γ_{HK} is tangent in $(r_0^{(2)},r_1^{(2)})$ to $\mathcal{H}_K$ we have a "pulse like" solution (homoclinic solution of (31)), corresponding to a solution spatially periodic in most of the space except in a little region where the amplitude becomes small. This amplitude cancels in the case K=0 and gives a "front like" solution (heteroclinic solution).

Remark 1. The problem of knowing whether these quasi-periodic solutions of the normal form (31) *persist* when we add the previously neglected terms of order $O(|A|+|B|)^N$, and then give quasi-periodic solutions for the original equation (1) is delicate, and needs further technical analysis.

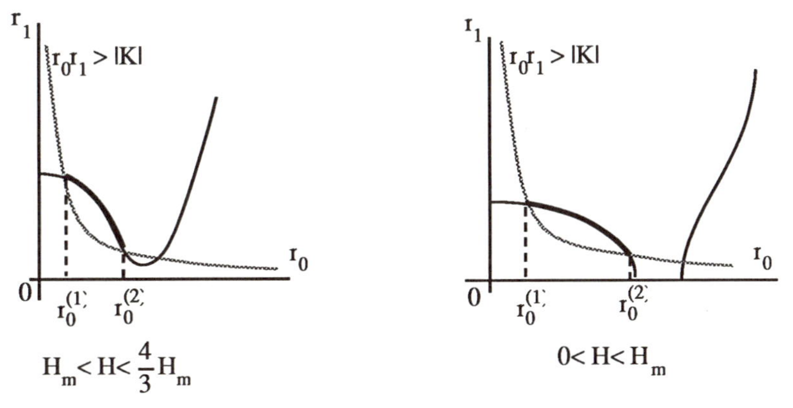

Figure 4. Relevant part of curves Γ_{HK}

Remark 2. Physically, the quasi-periodic flows we obtained correspond to large scale modulations in amplitude and phase on periodic solutions (39).

Remark 3. The delicate problem on what happens to the "pulse like" and "front like" solutions when we add the neglected terms of order $O(|A|+|B|)^N$ into (31), is open. We might conjecture that some of them persist, while some other give spatially chaotic solutions (analogue to the chaotic behavior generated by a transverse homoclinic intersection).

III.3. Eckhaus instability limit

A special limit case is obtained when hyperbola $\mathcal{H}_K$ and the level curve Γ_{HK} have an order 3 contact. This happens when $H=(4/3)H_m$, then K is fixed. Points $r_0^{(1)}$ and $r_0^{(2)}$ are the same and this situation corresponds to the limit when quasi-periodic solutions become periodic with a cancelling wave number β. It can be shown (see[I-M-D]) that this special bifurcation from periodic solutions is just associated with the Eckhaus instability limit. This limit is in fact obtained by starting with the Ginzburg-Landau equation, and studying the linear stability of the classical spatially periodic solutions. It was shown by Eckhaus [Ec] that these solutions are stable (temporally) when (k,μ) lies inside a parabolic region bounded by the graph of a function $\mu = \mu_E(k) \approx 3\mu_c(k)$.

IV. Comparison with Ginzburg-Landau equation. Second order O.D.E.

We now want to rewrite the fourth order system (31) into the form of a second order complex equation, for helping the comparison with the steady (G-L) equation . Let us set

$$(44)\qquad \begin{cases} A' = Ae^{-ik_cx} \\ B' = \{B+iAP[\mu,|A|^2,\tfrac{i}{2}(A\bar{B}-\bar{A}B)]\}e^{-ik_cx}, \end{cases}$$

then

$$u' = |A'|^2 = u, \qquad\qquad v' = \tfrac{i}{2}(A'\bar{B}'-\bar{A}'B') = v+uP(\mu,u,v),$$

Now the system (31) becomes

$$(45)\qquad \frac{d^2A'}{dx^2} + A'Q'(\mu,u',v') + i\frac{dA'}{dx}P'(\mu,u',v') = 0,$$

where $u' = |A'|^2$, $v' = \frac{i}{2}(A'\frac{d\bar{A}'}{dx} - \bar{A}'\frac{dA'}{dx})$, and the principal part of (45) reads:

$$(46)\qquad \frac{d^2A'}{dx^2} + q_1\mu A' - 2ip_1\mu\frac{dA'}{dx} - q_2A'|A'|^2 + \frac{i}{2}(q_3-6p_2)|A'|^2\frac{dA'}{dx} +$$

$$-\frac{i}{2}(2p_2+q_3)A'^2\frac{d\bar{A}'}{dx} + p_3\left\{A'\left|\frac{dA'}{dx}\right|^2 - \bar{A}'\left(\frac{dA'}{dx}\right)^2\right\} = 0.$$

Now, if we come back to the definition of A in (27) and consider the projection $\Pi\mathcal{V} = U$ (see §III.1), we have in fact a decomposition of the velocity field of the form

$$(47)\qquad U(x) = A'(x)\hat{U}_{k_c}e^{ik_cx} + \frac{dA'}{dx}(x)\hat{U}_1e^{ik_cx} + \text{c.c.} + \Psi(\mu,A',\bar{A}',\frac{dA'}{dx},\frac{d\bar{A}'}{dx}),$$

where "c.c." means "complex conjugate", $\hat{U}_1 = \frac{d}{d(ik)}\hat{U}|_{k_c}$, and where we took account of the fact that $B-\frac{dA'}{dx}e^{ik_cx}$ is of higher order and can be incorporated into Ψ. This decomposition looks very much like the following decomposition used for obtaining (G-L) equation (see [N-W],[Se],[C-R]):

$$(48)\qquad U = \mathcal{A}(x,t)\hat{U}_{k_c}e^{ik_cx} + \bar{\mathcal{A}}(x,t)\bar{\hat{U}}_{k_c}e^{-ik_cx} + \Phi(\mu,\mathcal{A},\bar{\mathcal{A}},\partial_x)$$

where ∂_x is considered as a small parameter since the amplitude is slowly varying in x and t. Since in deriving (48) we consider the terms in $\partial_x\mathcal{A}$ of order higher than $\mathcal{A}$, these terms belong to Φ. Note that in the study of solutions of (45), completely made in part III.2, we obtained in particular solutions where $\left|\frac{dA'}{dx}\right| >> |A'|$ in the neighborhood of some front or pulse and even for a family of quasi-periodic solutions, which then violate the imposed condition on $\partial_x\mathcal{A}$ and $\mathcal{A}$ for the (G-L) equation.

Finally, equation (46) associated with decomposition (47) and the steady (G-L) equation associated with decomposition (48) have to give the same solutions in U. The (G-L) equation reads ([N-W],[Se]):

$$(49)\ \frac{\partial\mathcal{A}}{\partial t} = c_0\mu\mathcal{A} + ie_1\mu\partial_x\mathcal{A} + e_2\partial_x^2\mathcal{A} + ie_3\partial_x^3\mathcal{A} + d_0\mathcal{A}|\mathcal{A}|^2 + id_1|\mathcal{A}|^2\partial_x\mathcal{A} + id_2\mathcal{A}^2\partial_x\bar{\mathcal{A}} + \ldots = 0.$$

We see that the steady equation (49-S)(without $\frac{\partial\mathcal{A}}{\partial t}$) is (unfortunately) of <u>infinite</u> order, hence there is a problem for the identification of the two equations (49-S) and (46). However, we already obtained (38) and (40) which lead, after computing the eigenvalue $\sigma_0(\mu,i(k_c+\alpha))$ with (49) and deducing $\mu_c(k)$ from it, and computing the bifurcated solution of the form Const.$e^{i\alpha x}$ given by (8),(13), to the following correspondances:

$$(50)\qquad q_1 = \frac{c_0}{e_2},\qquad -q_2 = \frac{d_0}{e_2},\qquad -2p_1 = \frac{e_1}{e_2} - \frac{e_3c_0}{e_2^2}.$$

Hence we check the terms in $\mu A'$, $\frac{d^2A'}{dx^2}$, $A'|A'|^2$, and we now have a rule for the coefficient of $\mu\frac{dA'}{dx}$ which takes into account coefficients of $\mu\partial_x\mathcal{A}$ and of $\partial_x^3\mathcal{A}$ in (49-S). In fact, it may be obtained by replacing $\partial_x^2\mathcal{A}$ by $-\frac{c_0}{e_2}\mu\mathcal{A}$ in (49-S). This leads to an idea for obtaining (45) from the steady (G-L) equation (49-S): i) write (G-L) equation (49-S) on the form $\partial_x^2\mathcal{A}$ = r.h.s. and ii) replace on the right hand side all $\partial_x^p\mathcal{A}$, $p\geq 2$ by the derivative of order p-2 of the full right hand side, then iii) iterate indefinitely the process. The resulting equation, truncated at some arbitrary order, is a second order complex ODE, of the form

$$\partial_x^2\mathcal{A} = h(\mu,\mathcal{A},\bar{\mathcal{A}},\partial_x\mathcal{A},\partial_x\bar{\mathcal{A}})$$

which satisfies the following invariances

$$\begin{cases} h(\mu,e^{i\phi}\mathcal{A},e^{-i\phi}\bar{\mathcal{A}},e^{i\phi}\partial_x\mathcal{A},e^{-i\phi}\partial_x\bar{\mathcal{A}}) = e^{i\phi}h(\mu,\mathcal{A},\bar{\mathcal{A}},\partial_x\mathcal{A},\partial_x\bar{\mathcal{A}}) \\ h(\mu,\bar{\mathcal{A}},\mathcal{A},-\partial_x\bar{\mathcal{A}},-\partial_x\mathcal{A}) = \overline{h(\mu,\mathcal{A},\bar{\mathcal{A}},\partial_x\mathcal{A},\partial_x\bar{\mathcal{A}})}. \end{cases}$$

Hence, it is clear that this equation contains more terms in the power expansion than (45). For instance, there are two additionnal types of monomials in $(\mathcal{A},\bar{\mathcal{A}},\partial_x\mathcal{A},\partial_x\bar{\mathcal{A}})$ of degree 3, and 6 more for degree 5. Our conjecture is the following:

Conjecture

Assume $e_2\neq 0$, *then the ODE* (45) *plays again the role of a normal form for the steady (G-L) equation (49-S) on a 4-dimensional center manifold.*

However, we may observe that for terms such that the total order of derivation is ≤ 1, we have not suppressed any monomial in this normalization, hence we might identify corresponding coefficients. Finally, if the conjecture is correct, we obtain a new form for the principal part of equation (49-S):

$$\partial_x^2 \mathcal{A} + \frac{c_0}{e_2}\mu\mathcal{A} + \frac{d_0}{e_2}\mathcal{A}|\mathcal{A}|^2 + i\left[\frac{e_1}{e_2} - \frac{c_0 e_3}{e_2^2}\right]\mu\partial_x\mathcal{A} + i\left[\frac{d_1}{e_2} - \frac{2d_0 e_3}{e_2^2}\right]|\mathcal{A}|^2\partial_x\mathcal{A} + \tag{51}$$

$$+ i\left[\frac{d_2}{e_2} - \frac{d_0 e_3}{e_2^2}\right]\mathcal{A}^2\partial_x\bar{\mathcal{A}} + \ldots = 0,$$

which leads, in addition to (50), by identification with (46), to the correspondances:

$$q_3 = \frac{d_1 - 3d_2}{2e_2} + \frac{d_0 e_3}{2e_2^2}, \qquad p_2 = -\frac{d_1 + d_2}{4e_2} + \frac{3d_0 e_3}{4e_2^2}. \tag{52}$$

Since it is possible to compute analytic formulas for coefficients p_j, q_j, c_0, e_j, d_j, ..as indicated in [C-S], we were able to check the validity of (50),(52) (see [I-M-D] for explicit formulas). This, of course, is a good presage for the rightness of the conjecture.

Bibliography

[A-K] C.J.Amick, K.Kirchgässner. Arch.Rat.Mech.Anal. (to appear).
[C-E 86] P.Collet, J.P.Eckmann. Comm.Math.Phys. 107, 39-92, 1986.
[C-E 87] P.Collet, J.P.Eckmann. Helvetica Phys.Acta 60, 969, 1987.
[C-S] P.Coullet, E.A.Spiegel. SIAM J.Applied Math. 43,774-819, 1983.
[C-R] P.Coullet, D.Repaux. Instabilities and Nonequilibrium Structures. E.Tirapegui, D.Villaroel ed.,179-195,Reidel, 1987.
[Ec] W.Eckhaus. Studies in nonlinear stability theory. Springer tracts in Nat. Philo. Vol.6,1965.
[E-T-B-C-I] C.Elphick, E.Tirapegui, M.E.Brachet, P.Coullet, G.Iooss. Physica 29D, 95-127, 1987.
[He] D.Henry. Springer Lecture Notes in Math.840,1981.
[Io 71] G.Iooss. Arch.Rat.Mech.Anal. 40,3,166-208,1971.
[I-M-D] G.Iooss, A.Mielke, Y.Demay. Eur.J.Mech.B,1989 (to appear).
[Iu] V.I.Iudovich. Dokl.Akad.Nauk. SSSR, 161,5,1037-1040, 1965.
[Ka] T.Kato. Perturbation theory for linear operators. Springer Verlag, Berlin, 1966.
[Ki] K.Kirchgässner. J.Diff.Equ. 45,113-127, 1982.
[K-Z] L.Kramer, W.Zimmermann. Physica 16D, 221-232,1985.
[Mi 86a] A.Mielke. J.Diff.Equ. 65,68-88, 1986.
[Mi 86b] A.Mielke. J.Diff.Equ. 65,89-116, 1986.
[Mi 88a] A.Mielke. Math.Meth.Appl.Sci.10,51-66, 1988.
[Mi 88b] A.Mielke. Arch.Rat.Mech.Anal.102,205-229, 1988.
[N-W] A.Newell, J.Whitehead. J.Fluid Mech. 38,2, 279-303, 1969.
[Se] L.A.Segel. J.Fluid Mech. 38,1, 203-224, 1969.
[Va] A.Vanderbauwhede. Dynamics Reported 2, 1989 (to appear).

NON LINEAR SPATIAL ANALYSIS: APPLICATION TO THE STUDY OF FLOWS SUBJECTED TO THERMOCAPILLARY EFFECTS

P. Laure†, H. Ben Hadid and B. Roux

† Lab. de Mathématiques, Parc Valrose, F-06034 Nice (France)
Institut de Mécanique des Fluides, 1 Rue Honnorat F-13003 Marseille (France)

INTRODUCTION : THE PHYSICAL MODEL

In the long term the motivation of this study is to look at the convective motions occurring during the growth of metals and semiconductor crystals in open horizontal boats (e.g. Bridgman technique).

In your study, we consider low–Prandtl number fluids, Pr, in long open parallelepipedic cavity (A=height/length). The vertical endwalls of this cavity are maintained at different temperatures and a linear temperature distribution is assumed along the horizontal walls ("perfectly conducting" type condition). The upper horizontal surface is subjected to a constant shear stress resulting from the variation of the surface tension with the temperature (thermocapillary effect), and it is assumed to remain flat. Flow motions and instabilities in the melts come from these surface (or thermocapillary) forces. This type of convection is dominant for shallow layers (small height) or for microgravity conditions (small g). The stress strength is characterized by the Reynold-Marangoni number Re = Ma/Pr .

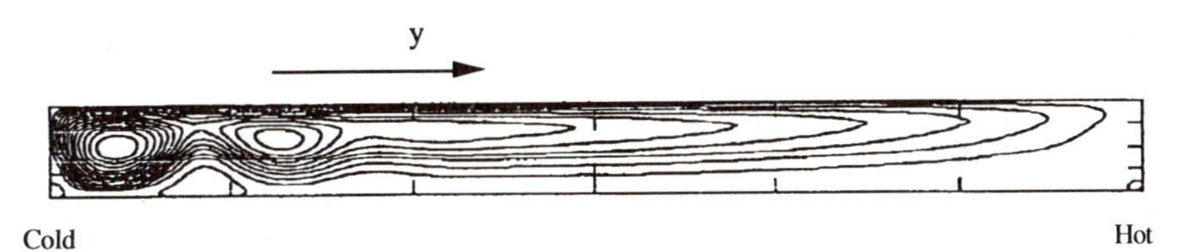

Fig. 1. Streamline Pattern for A = 12.5 , Pr = 0.015 and Re = 2000

The present work has been guided by numerical results (see Ben Hadid and Roux,1987, Strani and al. ,1983,) which exhibits for large cavity three regions in the horizontal direction, y (see Fig. 1): a central one which correspond well to a basic flow (which is observed for small Marangoni number) , a region in which the melt is accelerated , and a region in which the melt is decelerated (in this case we can observe a spatial periodicity along the y axis). Then, we propose to explain the appearance of this flow pattern by using a spatial analysis of the exact solution obtained for infinite cavity (see Birikh, 1966). A review of temporal thermocapillary instabilities has also been given by Davis, 1987, but these perturbations are essentially three dimensional. In the following, we describe only the main steps of the non linear analysis, more details can be found in Ben Hadid and al., 1988.

New Trends in Nonlinear Dynamics and Pattern-Forming Phenomena
Edited by P. Coullet and P. Huerre
Plenum Press, New York, 1990

FORMULATION

In order to look for spatial steady state perturbation of the basic flow along the y-axis (see also Bye, 1966), we write the Navier-Stockes equation as an evolution equation in y. It takes the symbolic form (the variable y replacing the usual variable t) :

$$F(dZ,dy) = F(Z) = L(Z) + M(Z,Z) \qquad (1)$$

where $Z = (\varphi, \frac{\partial\varphi}{\partial y}, \frac{\partial^2\varphi}{\partial y}, \frac{\partial^3\varphi}{\partial y})$ represents the state of the system, φ the perturbation of the basic flow, L and M are linear and bilinear operators. As we are in the approximation of infinite cavity, the boundary conditions at the vertical endwalls are replaced with the condition that the solution of (1) is bounded for infinite y.

The study of the differential system (1) starts with the computation of the spectrum of L which will give the spatial stability of the core flow. We obtained two families of eigenvalues, with positive and negative real part, μ. The smallest absolute values of μ are complex for low Re, but for Re > 15.45 the positive eigenvalue becomes real. We can explain the appearance of three regions with different flow motions (for large aspect ratio A), by the interaction of one real and two complex eigenvalues (α and $\beta \pm i\,\omega$).

Though we are in a hyperbolic case (we have both eigenvalues with positive and real negative part), we can suppose that we are in the neighborhood of a critical situation where eigenvalues of linear operator belong to imaginary axis. We propose now to make local study of bifurcated solutions by using center manifold theorem (see Iooss,1987), because all bounded perturbations belong to this manifold (see Vanderbauwhede, 1988). Moreover the dynamic of perturbations is given by amplitude equation which correspond to the interaction between Hopf and saddle node bifurcation (it is a codimension 2 bifurcation).

RESULTS

In the nondegenerate case, we have three couples of solutions which are called solution 1 (basic flow), 2 (stationary flow) and 3 (oscillatory in y flow) respectively (see Guckenheimer and Holmes, 1983)). In the following, we note R and C the amplitudes of perturbations 3 and 2 respectively.

Now, it is necessary to compute the coefficients of the amplitude equation at first order in order to draw the phase portraits in the plane (R,C) and show the different paths between these bounded solutions.The general method to compute these coefficients are described in Demay and Iooss,1987, or Laure and Demay, 1988. But in our case, we don't known the true form of the operator occurring in the model problem, then we can only compute approximative values of coefficients .

Numerical results give two possible phase portraits in the plane (R, C), which are plotted on Fig. 2 (the arrows show the evolution when y increases from $-\infty$ to $+\infty$):

for $15.45 < Re < 400$, the solution 3 does not exist . In this case, we can only see the transition of the basic flow (1) to the stationary solution (2) (diagram (a))

for $400 \leq Re \leq 1400$, the phase portrait (diagram (b)) shows two types of transition as y increases, between oscillatory solution 3 and solution 1 (basic flow), and between solution 1 and solution 2.

In the cavity with two vertical endwalls, the two types of transitions from the basic flow are observed. In fact, the vertical endwalls perturb the basic flow and induce spatial loss of the stability in the two y–directions. But in other hand, the boundary conditions at vertical endwalls prevent the total appearance of the final solutions 2 and 3. This means that the amplitudes of the disturbances do not reach the values obtained by the non linear stability theory. We only observe what it happens in the neighborhood of solution 1 (Basic flow) as shown in Fig. 2 (thick lines).

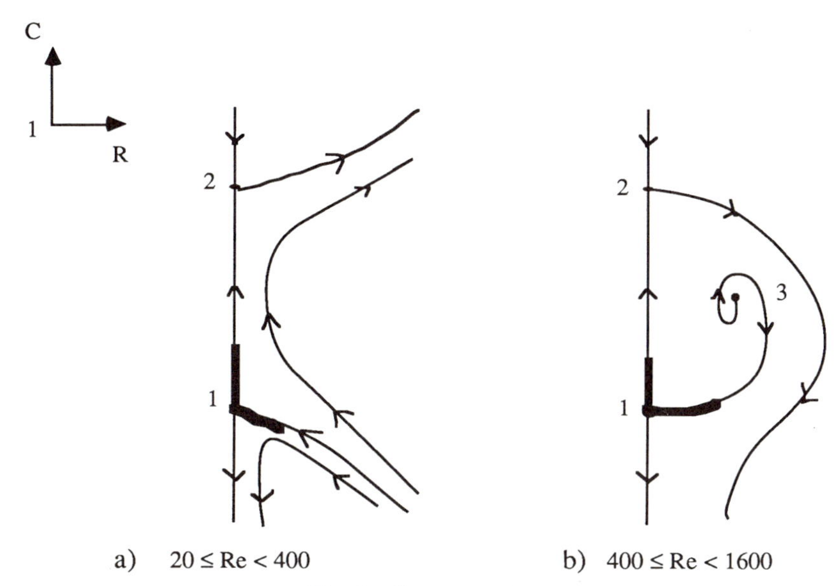

Fig. 2. Phase portraits.

Acknowledgements

This research was financially supported by the Centre National d'Etudes Spatiales (Division Microgravité Fondamentale et Appliquée) and the Direction Recherches et Etudes Techniques (Groupe Hydrodynamique).

REFERENCES

Ben Hadid, H., and Roux, B., 1987, Thermocapillary convection in long horizontal layers of low–Prandtl number melts subject to an horizontal temperature gradient, J. Fluid Mech. (to appear).

Ben Hadid, H., Laure, P., and Roux, B., 1988, Non linear study of the flow in a long rectangular cavity subjected to thermocapillary effect, Physics of Fluid (submitted).

Birikh, R. V., 1966, Thermocapillary Convection in a Horizontal Layer of Liquid", J. of Applied Mechanics and Technical Physics, 3, 69-72.

Bye, J. A. T., 1966, Numerical solutions of the steady-state vorticity equation in rectangular basins, J. Fluid Mech., 26, part. 3, 577-598 .

Davis, S. H., 1987, Thermocapillary instabilities , Ann. Rev. Fluid Mech., 19 , 403-435.

Demay, Y., and Iooss, G., 1984, Calcul des solutions bifurquées pour le problème de Couette-Taylor avec deux cylindres en rotation, J. de Meca. Théor. et Appliq., numéro spécial, 196–216 .

Guckenheimer, J., and Holmes, P., 1983, Nonlinear oscillations, Dynamical Systems, and bifurcations of vector fields, Appl. Math. Sciences, A2, Springer .

Iooss, G., 1987, Reduction of the dynamics of a bifurcation problem using normal forms and symmetries, E. Tirapegui and D. Villaroel (eds), Instabililities and Nonequilibrium Structures, 3-40. D. Reidel Publishing Company .

Laure, P., and Demay, Y.,1988, Symbolic computation and equation on center manifold : application to the Couette–Taylor problem, Computers & Fluids, Vol. 6, N° 3, 229-238 (1988).

Strani, M., Piva, R., and Graziani, 1983, G., Thermocapillary convection in a rectangular cavity: asymptotic theory and numerical simulation, J. Fluid Mech., 130, 347–376 .

Vanderbauwhede, A., 1988, Center Manifolds , normal forms and elementary bifurcations, Dynamics Reported, 2, J. Wiley & B. G. Teulner (To appear).

DISCRETIZATIONS AND JACOBIAN

IN FUNCTIONAL INTEGRALS

Enrique Tirapegui

Facultad de Ciencias Físicas y Matemáticas
Universidad de Chile, Santiago

The evolution of macroscopic systems is often modelized by differential equations. The technique of functional integration is especially interesting here when gaussian noise sources or parameters with gaussian distributions appear linearly in the equations. We examine briefly a technical point which arises in this method, namely the appeareance of a Jacobian due to change of variables [1,2]. This problem also appears in the theory of spin glasses [3]. Consider the equation (for a set of variables $q_\mu(t)$ the discussion is the same)

$$\dot{q}(t) = A(q(t)) + f(t) \tag{1}$$

where $f(t)$ is a gaussian white noise with zero mean and correlations $\langle f(t)\, f(t')\rangle = \delta(t-t')$.
Consider the interval $|t_0, T|$, $q(t_0) = Q_0$, $t_j = t_0 + j\varepsilon$, $t_{N+1} = T$, $q(t_j) = q_j$, $f(t_j) = f_j$, $q_{j-1}^{(\alpha)} = q_{j-1} + \alpha\Delta q_j$, $\Delta q_j = q_j - q_{j-1}$. A discretized version of (1) is $G(q_j) \equiv \Delta q_j - \varepsilon(A(q_{j-1}^{(\alpha)}) + f_j) = 0$. Correlation functions $\langle q(\tau_1)\, q(\tau_2) \ldots q(\tau_n)\rangle$ of the Markov process defined by (1) are obtained by functional differentiation from the generating functional [4] $(q(t_0) = Q_0)$

$$Z|J,J^+| = \int_{\gamma(\alpha)} DqDp \exp i\int_{t_0}^{T} dt\, |p(\dot{q}-A(q)) + \frac{i}{2}\, p^2 + i\alpha A'(q) + Jq + J^*p| \tag{2}$$

Here $A'(q) = \partial A/\partial q$ and the symbol $\gamma(\alpha)$ in the functional integral means that in the discretized version of which (2) is the limit $A(q)$ is discretized as $A(q_{j-1}^{(\alpha)})$. Separating the linear part of $A(q) = -\mu q + \overline{A}(q)$, $\mu>0$, we can calculate Z by a perturbation expansion in the nonlinearities

$$Z|J,J^*| = \exp i\int dt\, |-p(t)\overline{A}(q) + i\alpha A'(q)|\, \Big|_{ip = \frac{\delta}{\delta J^*},\ iq = \frac{\delta}{\delta J}} .\ Z_0|J,J^*| \tag{3}$$

New Trends in Nonlinear Dynamics and Pattern-Forming Phenomena
Edited by P. Coullet and P. Huerre
Plenum Press, New York, 1990

where the gaussian functional integral

$$Z_0|J,J^*| = \exp\left|iQ_0 \int_{t_0}^{T} dt'D(t'-t_0)\, J(t') - \right.$$

$$\left. - \int_{t_0}^{T} dt'\, dt''\, J(t') \left(\frac{1}{2}\Delta(t',t'')\, J(t'') + S(t'-t'')\, J^*(t'')\right)\right|. \tag{4}$$

Here $\Delta(t + \eta, t) = \Delta(t, t + \eta)$, but $S(\eta) \neq S(-\eta) = 0$, $\eta \to + 0$, since $S(t) = i\theta(t)D(t)$, $D(t) = \exp(-\mu t)$. One arrives to (2) as follows. Let $Q^f(t;Q_0,t_0)$ be the solution of (1) with $Q^f(t_0)=Q_0$ and put $Q^f(t_j)=Q_j^f$. Let $F|q|$ be any functional of $q(t)$. Then $F|Q^f|$ has the discretized form

$$F\,|Q^f_{N+1}, \ldots Q^f_0| = \int \prod_{i=1}^{N+1} dq_i\, F|\{q_K\}| \prod_{j=1}^{N+1} \delta(G(q_j)).\, J|\{q_k\}|,\; q_0 = Q_0,$$

where the Jacobian $J = |\det M_{kj}|$, with $M_{kj} = \partial G(q_j)/\partial q_k$, is $\exp(-\, \varepsilon\alpha \sum_{j=1}^{N+1} A'(q_{j-1}^{(\alpha)})) + 0(\varepsilon^2)$. The limit $N \to \infty$ ($\varepsilon \to 0$) gives $F|Q^f|$.

Using $2\pi\delta(G(q)) = \int dp \exp ip\, G(q)$ one has

$$F|Q^f| = \int \prod_{j=1}^{N+1} \left(dq_j \frac{dp_j}{2\pi}\right) \prod_{j=1}^{N+1} \exp(ip_j G(q_j) - \alpha A'(q_{j-1}^{(\alpha)})).\, F|\{q_k\}|.$$

Here in each term of the product $\prod_j$ we need only to keep terms up to $0(\varepsilon)$ due to the limit $N \to \infty$. We do the average over $\{f_j\}$ with the weight $\prod_{j=1}^{N+1} df_j \sqrt{\frac{\varepsilon}{2\pi}} \exp(-\frac{\varepsilon}{2} f_j^2)$. The result of the gaussian integral in the limit $N \to \infty$ is written $\langle F|Q^f|\rangle \int_{\gamma(\alpha)} Dq\, Dp\, F|q| \exp i \int_{t_0}^{T} dt\, |p(\dot{q}-A(q)) + \frac{i}{2} p^2 + i\alpha A'(q)|$.

Correlation functions correspond to $F|q| = \prod_{j=1}^{n} q(\tau_i)$ and $Z|J,J^*|$ to $F|q| = \exp i\int dt(J(t)q(t) + J^*(t)p(t))$ which gives (2). In (3) the action of the exponential is not defined since $p(t)\bar{A}(q(t))$ produces $S(0)$ which is not defined. However the discretization $\gamma(\alpha)$ in (2) solves this problem since it states that in (3) one has to interpret [4] $p(t)\bar{A}(q(t))$ as $p(t)\bar{A}((1-\alpha)q(t-\eta) + \alpha q(t+\eta))$, $\eta \to +0$. One can then check that Z is independent of α. In particular for $\alpha=0$ the term $i\alpha\bar{A}'$ coming from the Jacobian vanishes and at the same time $p(t)\, A(q(t-\eta))$ says that $S(0) = S(-\eta) = 0$, i.e. all tadpoles $\langle pq\rangle$ vahish. Thus the value of the Jacobian is a typical discretization problem (for rigorous results concerning discretizations see [5]). In this simple case (additive noise) the prescription concerning the role of the Jacobian is : eliminate the Jacobian and put all tadpoles $\langle pq\rangle$ equal to zero ($S(0) = 0$). However even here if we are calculating the transition probability $P(Q, T|Q_0, t_0)$ which is given by (2) with $q(T) = Q$, $J = J^* = 0$, the previous prescription fails. The reason is that although the perturbation expansion is again given by (3) now $Z_0|J,J^*|$ is different and in particular the function which replaces

$S(t'-t'')$ will not vanish for $t' < t''$, and consequently in the prepoint discretization $\gamma(0)$ the tadpoles $<pq>$ survive. If one has multiplicative noise, i.e. if (1) is replaced by $\dot{q} = A(q) + (1+D(q))f(t)$, things are again different. If we discretize as $\Delta q_j = A(q_{j-1}^{(\alpha)})+(1+D(q_{j-1}^{(\beta)}))f_j$ and proceed as before then in equation (3) the exponential acting on Z_0 (Z_0 does not change) is now $\exp i\int dt\ |-p\bar{A}(q) + i\ p^2\ D(q) + \frac{i}{2}\ p^2\ D(q)^2 +$ $\beta\ p\ D'(q)\_(1+D(q))+ i\ \alpha\ \bar{A}'(q)|$. In this last expression the q - dependence in $\bar{A}(q)$ is to be interpreted as in the previous case and the q - dependence in $D(q)$ as $D((1-\beta)q(t-\eta) + \beta\ q(t+\eta))$, $\eta \to +0$. Once again if we take $\alpha = \beta = 0$ the tadpoles $<pq>$ vanish and at the same time the terms coming from the Jacobian disappear. But now the important point is that the stochastic process depends on β, i.e. $Z|J,J^*|$ depends explicitly on the value of β. The case $\beta = 0$ corresponds to the Ito interpretation of the stochastic differential equation and $\beta = \frac{1}{2}$ to Stratonovic, other values correspond to intermediate cases. The choice of β is dictated by physical reasons, for example if the white noise $f(t)$ is limit of a colored noise $\beta = \frac{1}{2}$. We see then that in the case of multiplicative noise the effect of the Jacobian remains and has to be properly taken into account.

REFERENCES

1. C. De Dominicis, L. Peliti, Phys. Rev. B18, 353 (1978).

2. H.K. Janssen, Z. Phys. B23, 377 (1976); R. Bausch, H.K. Janssen, H. Wagner, Z. Phys. B24, 113 (1976).

3. H. Sompolinski, A. Zippelius, Phys. Rev. B25, 6860 (1982); A. Crisanti, H. Sompolinsky, Phys. Rev. A36, 4922 (1987) and A37, 4865 (1987).

4. F. Langouche, D. Roekaerts, E. Tirapegui Functional integration and semiclassical expansions, Reidel (1982).

5. R. Alicki, D. Makowiec, J. Phys. A18, 3319 (1985).

HOMOCLINIC BIFURCATIONS IN ORDINARY AND PARTIAL DIFFERENTIAL EQUATIONS

A.C. Fowler

Mathematical Institute, Oxford University
24-29 St. Giles', Oxford, OX1 3LB
England

ABSTRACT

The formal analysis of bifurcations from homoclinic orbits in low-dimensional ordinary differential equations is here extended to deal with ordinary differential equations in n dimensions, and to certain partial differential equations in one space variable on the infinite real axis. For ordinary differential equations, results are equivalent to various cases treated by Shil'nikov: depending on the eigenvalues at the fixed point, an infinite number of periodic orbits can bifurcate at the critical parameter value. By contrast, homoclinic bifurcations for partial differential equations can produce an infinite number of quasi-periodic (modulated travelling wave) solutions.

1. INTRODUCTION

One of the few direct methods of analysis which connects chaotic behaviour with trajectories of differential equations is the analysis of homoclinic bifurcations. These occur in the neighbourhood of parameter values for which homoclinic orbits (orbits which are bi-asymptotic to a fixed point as $t \to \pm \infty$) exist. Many examples of ordinary differential equations exist, for which the existence of chaotic behaviour is closely associated with the existence and nature of homoclinic bifurcations.

The most striking example is perhaps that of the Lorenz equations (Lorenz 1963), whose qualitative behaviour can be largely understood (Sparrow 1982) by an analysis of the associated homoclinic structures. The pioneer for these kinds of study was Shil'nikov (1965, 1967, 1970), and his methodology underlies the treatment of later authors (e.g. Arneodo et al. 1982, Tresser 1984, Glendinning and Sparrow 1984, Gaspard et al. 1984). A recent review is by Glendinning (1988).

If homoclinic bifurcations are as important for our understanding of chaotic behaviour in ordinary differential equations as these papers suggest, then it is clear that one should aim to understand their effect in partial differential equations. This is particularly true

New Trends in Nonlinear Dynamics and Pattern-Forming Phenomena
Edited by P. Coullet and P. Huerre
Plenum Press, New York, 1990

for (flow) systems whose behaviour seems entirely unrelated to finite-dimensional systems, of which the outstanding example is the transition to turbulent flow in parallel wall-bounded shear flows: Poiseuille flow, boundary layer flow. Emmons' turbulent spot is an observed feature of the breakdown of laminar shear flow, which is conceptually at odds with the descriptions afforded by classical linear and nonlinear stability theories (Widnall 1984, Stuart 1960).

Our aim in this brief paper is therefore to examine the possible bifurcations which may arise from the (postulated) existence of homoclinic orbits in partial differential equations, with a view to establishing what may be different to the case of ordinary differential equations. We proceed heuristically; in certain cases, the results may be proved, but there are substantial technical difficulties involved. Furthermore, brevity compels us to be a little inexact.

2. ORDINARY DIFFERENTIAL EQUATIONS

We first illustrate the procedure in $\mathbb{R}^n$, in order to draw a connection between the classical Shil'nikov theory, and the more convoluted exercise for partial differential equations.

Suppose the system

$$\dot{x} = f(x,\mu) \ , \tag{2.1}$$

$x \in \mathbb{R}^n$, $\mu \in \mathbb{R}$, with f smooth, has a fixed point $x = 0$ for all μ, and that when $\mu = 0$, there is a homoclinic orbit Γ:

$$x = x^*(t), \ x^* \to 0 \text{ as } t \to \pm \infty \ . \tag{2.2}$$

It is convenient to choose the phase of x^* so that $x^* = O(1)$ when $t = 0$, whence

$$x^* \sim \exp(tD)\alpha^* \ , \quad t \to -\infty \ ,$$

$$x^* \sim \exp(tD)\beta^* \ , \quad t \to +\infty \ ,, \tag{2.3}$$

where $\alpha^*, \beta^* = O(1)$, and D is the Jacobian of f at 0, which we may assume to be diagonal. Let W_S and W_U be the stable and unstable manifolds of 0, and let e^S and e^U be eigenvectors in W_S and W_U,, with corresponding eigenvalues whose real part is closest to zero. We define surfaces Σ and Σ' by

$$\Sigma \ : \ |\langle x, e^S\rangle| = \nu \ ,$$
$$\Sigma' \ : \ |\langle x, e^U\rangle| = \nu \ , \tag{2.4}$$

where $\nu \ll 1$, as shown in Fig. 1. Σ is our Poincaré surface and we aim to construct an approximate map from Σ to Σ', and thence back to Σ.

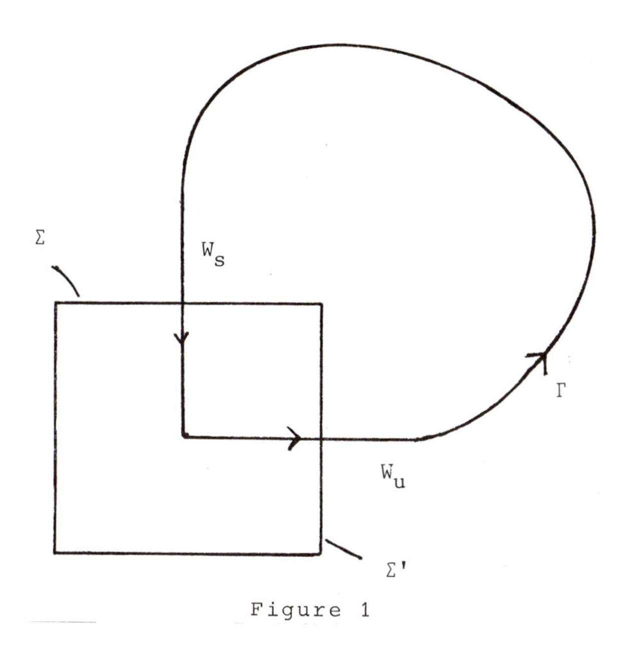

Figure 1

In fact, it is algebraically convenient to use the rescaled coordinates α and β on Σ' and Σ, which are defined by analogy with (2.3): if $x^* \in \Sigma'$ at $t = -t_U$, $x^* \in \Sigma$ at $t = t_S$, then we define

$$x = \exp[-t_U D]\alpha \quad \text{on} \quad \Sigma' ,$$
$$x = \exp[t_S D]\beta \quad \text{on} \quad \Sigma . \tag{2.5}$$

The arcane choice of α and β ensures that the Poincaré map we shall derive for β is properly scaled: in this way, it is transparent to ascertain which terms are in practice small (this is not normally done).

The approximation to the flow consists of two parts. Near 0, from Σ to Σ', we have

$$\dot{x} \approx Dx , \tag{2.6}$$

whence

$$x \approx \exp[(t-t_0)D]x_0 , \tag{2.7}$$

and use of (2.5) leads to

$$\alpha \approx e^{PD}\beta , \tag{2.8}$$

where $P = t_U + t_S + \tilde{t}$, $\tilde{t}$ is the transit time from Σ to Σ' (which depends on β): P is essentially the interval between passages through Σ. On the other hand, between Σ' and Σ, we linearise about x^*, so that

$$x = x^* + y , \quad \dot{y} \approx Df[x^*(t),0]y + \mu \frac{\partial f}{\partial \mu} , \tag{2.9}$$

whence

$$y \approx \Psi(t)\Psi^{-1}(t_0)y_0 + \mu\Psi(t)\int_{t_0}^{t}\Psi^{-1}(s)\,\frac{\partial f}{\partial\mu}\,ds\ , \tag{2.10}$$

where Ψ is a fundamental matrix for Df. We define

$$\Psi = \exp(tD)H(t)\ , \tag{2.11}$$

and assume $H(-\infty) = I$, $H(\infty) = M$. (Strictly, this requires the eigenvalues of D to be independent of μ.) In this case, we find the image β' in Σ is given by

$$\beta' - \beta^* \approx M(\alpha - \alpha^*) + \mu c\ , \tag{2.12}$$

where we may take $c = O(1)$ and $M = O(1)$.

The composition of (2.8) and (2.12) now provides an approximate (n-1)-dimensional map (since P is unknown, and $\beta \in \Sigma$, essentially $|\langle\beta, e^S\rangle| = 1$). The further vital observation is that P is large, so that (2.8) and (2.12) are (differentiably) close to a one-dimensional map. To make this explicit, we label elements of W_S and W_U with suffices S and U respectively. With an obvious notation, we write

$$\alpha_U = \alpha_U^* + a_U\ ,\quad \beta_S = \beta_S^* + b_S\ , \tag{2.13}$$

so that

$$\begin{aligned}
\beta_U &= \exp[-PD_U](\alpha_U^* + a_U)\ ,\\
\alpha_S &= \exp[PD_S](\beta_S^* + b_S)\ ,\\
b_S' &= M_{SU}a_U + M_{SS}\alpha_S + \mu c_S\ ,\\
\beta_U' &= M_{UU}a_U + M_{US}\alpha_S + \mu c_U\ .
\end{aligned} \tag{2.14}$$

Notice $\alpha_U^*, \beta_S^* \sim O(1)$, $a_U, b_S, b_S' \ll 1$; if we restrict attention to only those points in Σ and Σ' which will be in the invariant set of trajectories, i.e. which remain close to Γ for all past and future iterations, then clearly we must have

$$\beta_U \approx \exp[-PD_U]\alpha_U^*\ ,$$

$$\alpha_S \approx \exp[PD_S]\beta_S^*\ , \tag{2.15}$$

(note $\mathrm{Re}\ \sigma(D_U) > 0$, $\mathrm{Re}\ \sigma(D_S) < 0$), and so β_U' must be approximately given by

$$\beta_U' \approx \exp[-P'D_U]\alpha_U^* \tag{2.16}$$

for some P' (otherwise subsequent iterates would not be close to Γ). Hence $(2.14)_4$ is an approximate equation for a_U; however, one can show that rank $(M_{UU}) = k-1$ (if $\dim W_U = k$, so that M_{UU} is $k \times k$) (this is because of the autonomy, i.e. time-translation invariance, of (2.1)). If η is the unique eigenvector of the Hermitian adjoint of M_{UU}, then P' is explicitly determined by the orthogonality criterion

$$\langle\eta, \exp(-P'D_U)\alpha_U^*\rangle = \langle\eta, M_{US}\exp(PD_S)\beta_S^*\rangle + \mu\langle\eta, c_U\rangle\ . \tag{2.17}$$

The exact map is differentiably close to this one.

It is straightforward to read from this map the usual Shil'nikov (1965,1967) results for fixed points: the proof uses the implicit function theorem in a straightforward manner. In particular, the results extend to n dimensions in a seemingly straightforward manner. In order to make statements about aperiodic orbits (that there can be uncountably many which are homeomorphic to a shift on N symbols, where $N \to \infty$ as $\mu \to 0$), one has to be more careful. One cannot use the one-dimensional map on its own. In fact, the proof involves the full (n-1)-dimensional map, but uses the fact that the action of the map is naturally partitioned into the effect on P, together with a hyperbolic structure on the remainder of W_U and W_S.

3. PARTIAL DIFFERENTIAL EQUATIONS

Here we cursorily sketch the analogous ideas to those of the preceding section. We consider the partial differential equation

$$A_t = N[\partial_x](A;\mu) \quad , \quad -\infty < x < \infty \; , \tag{3.1}$$

where N is autonomous in both space and time. We assume $N(0;\mu) = 0$, and that when $\mu = 0$, there is a homoclinic orbit Γ:

$$A = A^*(x,t) \quad , \quad A^* \to 0 \;\; \underline{\text{uniformly}} \text{ as } \;\; t \to \pm\infty \tag{3.2}$$

(i.e. not a soliton). Moreover, we assume A^* is localised in both space and time. If the dispersion relation at the origin for modes $\exp[ikx + \sigma(k)t]$ is $\sigma = \sigma(k)$, then we choose the phase of A^* so that

$$A^* \sim \int_{-\infty}^{\infty} \alpha^*(k)\exp[ikx + \sigma(k)t]dk \quad \text{as} \quad t \to -\infty \; ,$$

$$A^* \sim \int_{-\infty}^{\infty} \beta^*(k)\exp[ikx + \sigma(k)t]dk \quad \text{as} \quad t \to +\infty \; ; \tag{3.3}$$

the definitions of Σ and Σ' are entirely analogous to the finite-dimensional case, as are those of α and β, and we quickly derive

$$\alpha(k) \approx \beta(k)e^{\sigma(k)P} \; , \tag{3.4}$$

in analogy to (2.8). Going round the outside is a little more cumbersome, since for example if (3.1) is a semi-flow (e.g. if parabolic) then the analogy to Ψ^{-1} in (2.10) does not exist (in the same sense). However, with appropriate care, one can indeed derive the analogue of (2.12). Specifically, put

$$A = A^* + v \quad , \quad v_t \approx N'(A^*)v + \mu \frac{\partial N}{\partial \mu} \; , \tag{3.5}$$

where N' is the derivative of N. Let $T(t,\tau)$ be the (semi) flow corresponding to (3.5), with $\mu = 0$, and let $\psi(x,k;t,\tau)$ be a family of orthonormal eigenfunctions of T^*T satisfying

$$\int_{-\infty}^{\infty} \psi(x,k) \; \overline{\psi}(x,\ell)dx = \delta(k-\ell) \; ; \tag{3.6}$$

then we $\underline{\text{assume}}$

$$\psi(s,k;t,0) \to \psi_{\pm}(s,k) \quad \text{as} \quad t \to \pm\infty \tag{3.7}$$

(the definition of ψ for $t < 0$ follows from the formal inverse). In this case, we find

$$\beta'(k) - \beta^*(k) = \int_{-\infty}^{\infty} M(k,\ell)\{\alpha(\ell) - \alpha^*(\ell)\}d\ell + \mu c(k) \tag{3.8}$$

as an analogue to (2.12).

One continues by observing that $P >> 1$. The procedure is essentially the same as before except that the subsequent homoclinic pulse will in general be phase shifted by a length L, say; thus (with $\beta_U = \beta(k)|_{k\epsilon U}$, $U = \{k \text{ s.t. } Re\sigma > 0\}$)

$$\beta_U \approx \exp[-\sigma_U(k)P]\alpha_U^* \quad \text{in} \quad \Sigma , \tag{3.9}$$

and

$$\beta_U' \approx \exp[-\sigma_U(k)P - ikL]\alpha_U^* \quad \text{in} \quad \Sigma , \tag{3.10}$$

for some P' and L. The analogue to the matrix M_{UU} is the operator $\int_U M_U(k,\ell)a_U(\ell)d\ell$ ($M_U = M(k,\ell)|_{k\epsilon U}$), which has <u>two</u> null vectors, corresponding to time translation invariance <u>and</u> spatial translation invariance, and hence we get two orthogonality conditions which serve to define P'and L. The general form is

$$\int_U \exp[-\sigma_U(k)P' - ikL]w_j(k)dk = \int_S \exp[\sigma_S(k)P]y_j(k)dk + \mu \tag{3.11}$$

with $j = 1,2$ and w_j, y_j known functions.

Yet further simplification uses the fact that $P' >> 1$ $(= O(1/\mu))$ to approximate the integrals in (3.11) using the method of steepest descents. The result now depends on the dispersion relation. A typical case is $\sigma = \alpha - \gamma k^2$, corresponding to

$$A_t = \alpha A + \gamma A_{xx} + f(A) , \tag{3.12}$$

where f is a nonlinear operator. Then one gets

$$\sum_m c_{jm}\exp[-i\omega_m P' - ik_m L]/(P' + i\lambda_m L) = \sum_m d_{jm}\exp[i\omega_m P]/P + \mu , \tag{3.13}$$

where ω_m, k_m, λ_m depend on $\sigma(k)$. A simple example is where N is real and symmetric, and k_m, λ_m, ω_m occur in complex conjugate pairs. Then (3.13) can be simplified to

$$\frac{e^{-ikL}}{P' + i\lambda L} = \frac{B}{P} + \mu C , \tag{3.14}$$

where we assume one positive zero of Re σ at k, and B and C are generally complex.

Periodic orbits in the flow would correspond to $P = P'$ with $L = 0$, and in general do not exist. Multiple fixed points with $P = P'$, $L \neq 0$, do exist, and have $P \sim 1/\mu$ as $\mu \to 0$. (This contrasts to the logarithmic dependence, $P \sim \ell n(1/\mu)$, in $\mathbb{R}^n$.) For example, if B and C are real, then

$$L \sim n\pi/k , \quad \mu \sim P^{-1} \tag{3.15}$$

as $P \to \infty$, $n \in \mathbb{Z}^+$. There are countably many of these solutions which have the form of travelling modulated 'solitary' waves. The period of modulation is $P \sim \mu^{-1}$ and the wave speed is $L/P \sim n\pi/\mu k$.

If N(A) is real but not symmetric, then $\omega_m \neq 0$ in (3.13); if $k_m = \pm k$ as before, then $\lambda_m = \pm \lambda$, and the equivalent of (3.14) is

$$\exp[-i\omega P' - ikL]/(P' + i\lambda L) = (B/P)\cos(\omega P + \theta) + \mu C . \qquad (3.16)$$

Fixed points $P = P'$, L = constant exist, with

$$\mu \sim \frac{1}{P}\cos(\omega P + \tilde{\theta}) ,$$

$$\omega P + kL \sim n\pi , \quad n \in \mathbb{Z}^+ , \qquad (3.17)$$

corresponding to Shil'nikov's results (Glendinning and Sparrow 1984). There are countably many such modulated travelling waves, with periods P and $2\pi/\omega$, and the wave speed is $L/P \sim -\omega/k$. Figures 2 and 3 illustrate the general dependence of P on μ and L.

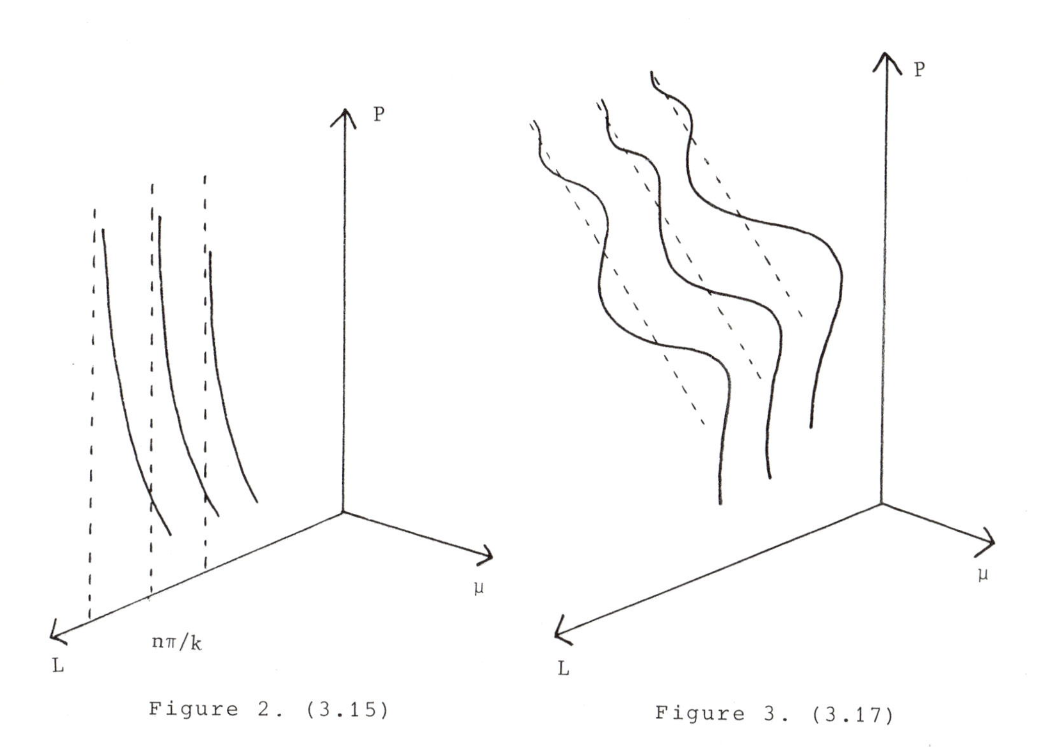

Figure 2. (3.15)　　Figure 3. (3.17)

4. DISCUSSION

The resumé here is obviously very truncated, and will be developed elsewhere. However, it is clearly formally possible to develop results for partial differential equations which are analogous to those which can be obtained for ordinary differential equations. The principal conclusion is that periodic solutions are not generated, but rather modulated travelling waves with one very long period. There is plenty of suggestive evidence of such types of solutions in results reported at this conference. However, the details obviously depend closely on the form of the dispersion relation $\sigma(k)$, amongst other things.

As a final comment, let us note that if, in (3.12), α, γ or f(A) is complex, then in general, we obtain two complex equations for the two real unknowns P' and L. (When α,γ,f(A) are real, these are complex conjugates of each other.) This is inconsistent, and we pose as a preliminary conjecture the idea that in this case, pair production occurs: two homoclinic pulses are produced by one parent, with values P_i', L_i determined by the two complex equations. This conjecture is preliminary, since it would be destroyed by a further two orthogonality relations, which one might suppose could arise from taking the complex conjugte of the equation. At the moment, this question is unresolved. However, if homoclinic orbits induce pair production, then the consequent trajectories have no counterpart in finite systems, but might provide a possible framework for the interpretation of turbulent spot generation in shear flows.

ACKNOWLEDGEMENTS

I thank Andy Bernoff, Paul Glendinning and Jeff DeWynne for discussion. I am very grateful to NATO, and to Oxford University and the Lockey Bequest for travel support to attend this conference.

REFERENCES

Arneodo, A., P. Coullet and C. Tresser 1982 Oscillations with chaotic behaviour: an illustration of a theorem by Shil'nikov. J. Stat. Phys. 27, 171-182.

Gaspard, P., R, Kapral and G. Nicolis 1984 Bifurcation phenomena near homoclinic systems: a two-parameter analysis. J. Stat. Phys. 35, 697-727.

Glendinning, P. 1988 Global bifurcations in flows. In: New directions in dynamical systems, ed. T. Bedford and J. Swift. LMS Lect. Note Ser. 127, pp.120-149.

Glendinning, P. and C. Sparrow 1984 Local and global behaviour near homoclinic orbits. J. Stat. Phys. 35, 645-696.

Lorenz, E.N. 1963 Deterministic non-periodic flow. J. Atmos. Sci. 20, 131-141.

Shil'nikov, L.P. 1965 A case of the existence of a countable number of periodic motions. Sov. Math. Dokl. 6, 163-166.

Shil'nikov, L.P. 1967 The existence of a denumerable set of periodic motions in four-dimensional space in an extended neighbourhood of a saddle-focus. Sov. Math. Dokl. 8, 54-58.

Shil'nikov, L.P. 1970 A contribution to the problem of the structure of an extended neighbourhood of a rough equilibrium state of saddle-focus type. Mat. Sb. 10, 91-102.

Sparrow, C.T. 1982 The Lorenz equations: bifurcations, chaos, and strange attractors. Springer-Verlag, Berlin.

Stuart, J.T. 1960 On the non-linear mechanics of wave disturbances in stable and unstable parallel flows. J. Fluid Mech. 9, 353-370.

Tresser, C. 1984 About some theorems by L.P. Shil'nikov. Ann. de l'Inst. Henri Poinc. 40, 441-461.

Widnall, S.E. 1984 Growth of turbulent spots in plane Poiseuille flow. In: Turbulence and Chaotic Phenomena in Fluids, ed. T. Tatsumi, pp.93-98. North-Holland, Amsterdam.

LARGE-SCALE VORTEX INSTABILITY IN HELICAL CONVECTIVE TURBULENCE

A.V. Tur, S.S. Moiseev, P.B. Rutkevich, and
V.V. Yanovsky

Space Research Institute, Academy of Sciences
Moscow, USSR

Various structures that can be observed in convective flows have recently become a subject of active research (see, e.g., Westfried and Zaleski, 1984). Convection has turned into a sort of laboratory for investigation of structures and structural transitions. Here we discuss a new principal possibility typical of the turbulent convection with non-vanishing mean helicity, i.e. with non-vanishing correlation $\langle \mathbf{v}\cdot\nabla\times\mathbf{v}\rangle \neq 0$. We show that in turbulent convection there exists a new type of instability that leads to generation of large-scale vortex structures with non-trivial topology of streamlines (see also Moiseev et al., 1988). Such structures can be called topological solitons. Below we consider a simplified version of the problem which, however, preserves all principal physical characteristics of this phenomenon. We consider the helical turbulence created by a random external force and weak deviations from linearity in equations of motion. From physical point of view, non-zero helicity means that turbulence is non-invariant with respect to reflection, i.e. the number of vortices with right-handed screwedness is not equal to the number of vortices with the left-handed one. This is the case when the random external force F_i is an axial vector.

The system of equations describing convection under the action of a random force is as follows (with standard notations):

$$\frac{\partial v_i}{\partial t} - \nu\Delta v_i + v_k\nabla_k v_i + \frac{\nabla_i p}{\rho} + ge_i = F_i, \tag{1}$$

$$\frac{\partial T}{\partial t} - \chi\Delta T + v_k\nabla_k T = 0, \tag{2}$$

$$\nabla\mathbf{v} = 0, \tag{3}$$

$$\rho = \rho_0(1-\beta T), \ \mathbf{e} \equiv (0,0,1), \ \beta = \frac{1}{\rho}\left[\frac{\partial\rho}{\partial T}\right]_p. \tag{4}$$

The temperature gradient of the basic state is considered constant:

New Trends in Nonlinear Dynamics and Pattern-Forming Phenomena
Edited by P. Coullet and P. Huerre
Plenum Press, New York, 1990

$$\nabla T_0(z) = -Ae,\ A > 0.$$

The random force F_i ($\langle F_i \rangle = 0$) determines the small-scale helical turbulence which we consider to be homogeneous, isotropic and stationary. Fourier transform of the correlator of such turbulence is well known to be (Monin and Yaglom, 1975)

$$\hat{Q}_{ij}^{t}(t_1 - t_2, k) = B(t_1 - t_2, k)\left[\delta_{ij} - \frac{k_i k_j}{k^2}\right] + iG(t_1 - t_2, k)\epsilon_{ijk}k_k, \qquad (5)$$

where

$$\int k^2 G(t_1 - t_2, k)\,dk = 4\pi^3 \langle \mathbf{v}\cdot\nabla\times\mathbf{v}\rangle.$$

Our aim here is to obtain the Reynolds equation corresponding to to the system (1)-(4) with the Reynolds stresses obtained by the means of ensemble averaging procedure with the use of the properties (5) and (6) of the given turbulent velocity field at small Reynolds numbers,

$$Re = \frac{u\lambda}{\nu} \ll 1 \qquad (6)$$

To accomplish the averaging we present the velocity field v_i as a sum of the averaged, $\langle v_i \rangle$, and fluctuative, v'_i ($\langle v'_i \rangle = 0$) parts:

$$v_i = v'_i + \langle v_i \rangle. \qquad (7)$$

If $\langle v_i \rangle$ vanishes the random velocity field is determined by the external force F_i. Let us denote this component of the velocity field as v_i^t. It is a homogeneous, isotropic and helical random field. Now we can consider the evolution of a weak averaged field $\langle v_i \rangle$, assuming that

$$\langle v \rangle \ll v^t \sim \langle (v^t)^2 \rangle^{1/2} \qquad (8)$$

In this case the random velocity component acquires a weak inhomogeneous correction $\tilde{v}_i$, $\tilde{v} \ll v^t$ and can be expressed as:

$$v'_i = v_i^t + \tilde{v}_i\ .$$

As a result the total velocity is given by

$$v_i = \langle v_i \rangle + v_i^t + \tilde{v}_i\ . \qquad (9)$$

It should be noted that $\tilde{\mathbf{v}}$ is a functional of $\mathbf{v}^t$ and $\langle \mathbf{v} \rangle$:

$$\tilde{\mathbf{v}} = \tilde{\mathbf{v}}\{\mathbf{v}^t, \langle \mathbf{v} \rangle\}.$$

Performing the averaging of the system (1)-(4) we obtain the following equations for $\langle v_i \rangle$ and $\tilde{v}_i$:

$$L_{ij}\langle v_j\rangle = -D_\chi P_{im}\nabla_k(\langle v_k^t\tilde{v}_m\rangle + \langle\tilde{v}_k v_m^t\rangle)$$
$$-\beta AgP_{im}\nabla_k e_m e_j(\langle\tilde{v}_k D_\chi^{-1} v_j^t\rangle + \langle v_k^t D_\chi^{-1}\tilde{v}_j\rangle), \quad (10)$$

$$L_{ij}\tilde{v}_j = -D_\chi P_{im}\nabla_k(\langle v_k\rangle v_m^t + v_k^t\langle v_m\rangle)$$
$$-\beta AgP_{in}e_n e_m(v_k^t D_\chi^{-1}\langle v_m\rangle + \langle v_k\rangle D_\chi^{-1} v_m^t). \quad (11)$$

Here L_{ij} is the following linear operator:

$$L_{ij} = D_\nu D_\chi \delta_{ij} - \beta AgP_{im}e_m e_j, \quad (12)$$

where

$$D_\nu = \frac{\partial}{\partial t} - \nu\Delta,\ D_\chi = \frac{\partial}{\partial t} - \chi\Delta,\ P_{im} = \delta_{im} - \frac{\nabla_i\nabla_m}{\Delta}.$$

Differential operators that appear in the denominators are understood as integral operators with corresponding Green's functions.

Equation(10) for the average velocity includes averages of quadric combinations of velocities (Reynolds' stresses). They can be expressed in terms of the average field $\langle v_i\rangle$ and turbulence correlator using the functional dependence of the field $\tilde{v}_i$ on the turbulent field v_i^t with the help of the Furutsu-Novikov formula:

$$\langle v_i^t(t,\mathrm{x})v_m(t,\mathrm{x})\rangle = \lim_{\substack{t_1\to t\\ \mathrm{x}_1\to \mathrm{x}}} \int ds\int dy \langle v_i^t(t,\mathrm{x})v_e^t(s,\mathrm{y})\rangle \left\langle \frac{\delta\tilde{v}_m(t_1\mathrm{x}_1)}{\delta v_e^t(s,\mathrm{y})}\right\rangle$$

In order to apply this formula we have to assume that the turbulent noise is gaussian. In problems of interaction of a large-scale field and small-scale turbulence this assumption is quite admissible even when Reynolds stresses are large, since these are the energy-range vortices that make a principal contribution to the interaction of large-scale motions and turbulent pulsations.

The effect of the large-scale, mean velocity field generation is associated with the first derivatives of averaged fields with respect to coordinates. The terms with second derivatives determine the turbulent viscosity, where the non-helical character of turbulence is not important and corrections to the turbulent viscosity associated with helical turbulence are negligible. Since the right-hand side of the averaged field equation (10) contains the first derivatives with respect to coordinates, in equation (11) we should take into account only the terms without the averaged field gradients.

Performing calculations of the Reynolds stresses we obtain the equation for the averaged velocity field. Consider a plane layer of thickness h and introduce the dimensionless coordinates x/h and time $t\nu/h^2$. Large-scale convection in this layer is then described by the equation:

$$(\frac{\partial}{\partial t} - \Delta)(\Pr\frac{\partial}{\partial t} - \Delta)\langle v_i\rangle - \mathrm{Ra} P_{im} e_m e_j \langle v_j\rangle$$
$$= \mathrm{Ra}\ s\ P_{im} \nabla_k \Big[\mu_1 (Pr\frac{\partial}{\partial t} - \Delta)(e_m \epsilon_{klj} + e_k \epsilon_{mlj}) - \mu_2 e_m \epsilon_{klj}\Big] e_l \langle v_j\rangle. \quad (13)$$

Here $\mathrm{Ra} = \frac{\beta A g h^4}{\nu \chi}$ is the Rayleigh number, $\Pr = \frac{\nu}{\chi}$ is the Prandtl number, $s = (30\pi)^{-1} G_0 \mathrm{Re}^2 (\lambda/h)$ is the helical turbulence parameter.

Dependence of the helical coefficient on typical velocity U, correlation length λ and correlation time τ can be taken in as

$$G(t_1 - t_2, \mathrm{k}) = G_0 U^2 \lambda^2 (1 + \lambda^2 \mathrm{k}^2)^{-2}. \quad (14)$$

Then the coefficients μ_1 and μ_2 can be expressed as

$$\mu_1 = \frac{\lambda^2}{h^2} \frac{\Pr^{-1}}{1+\Pr^{-\frac{1}{2}}} \frac{3\frac{\lambda}{\sqrt{\tau\nu}} + 3(1+\Pr^{-\frac{1}{2}}) + \frac{\sqrt{\tau\nu}}{\lambda}(1+\Pr^{-\frac{1}{2}}+\Pr^{-1})}{(1+\frac{\lambda}{\sqrt{\tau\nu}})^3 (\Pr^{-\frac{1}{2}} + \frac{\lambda}{\sqrt{\tau\nu}})^3};$$

$$\mu_2 = \frac{5}{2} \frac{\frac{\lambda^2}{\tau\nu}}{(1+\Pr^{-\frac{1}{2}})(1+\Pr^{-1})} \frac{1+\Pr^{-\frac{1}{2}} + 2\frac{\sqrt{\tau\nu}}{\lambda}\Pr^{-\frac{1}{2}}}{(1+\frac{\lambda}{\sqrt{\tau\nu}})^2 (\Pr^{-\frac{1}{2}} + \frac{\lambda}{\sqrt{\tau\nu}})^2}.$$

In the laminar limit equation (13) reduces to usual plane layer convection equation(Chandrasekhar, 1981; Gershuni and Zhukhovitsky,1976). The obtained averaged field equations, in contrast to their laminar counterparts, contain the terms with the tensor ϵ_{ira}. These terms result in a positive feedback loop between the toroidal and poloidal components of the velocity field and thus, lead to the large-scale instability.

To analyze these effects we represent the velocity field in the form:

$$\langle \mathrm{v}\rangle = \langle \mathrm{v}_T\rangle + \langle \mathrm{v}_P\rangle, \quad (15)$$

$$\langle \mathrm{v}_T\rangle = \nabla\times(\mathrm{e}\psi), \ \langle \mathrm{v}_P\rangle = \nabla\times\nabla\times(\mathrm{e}\varphi). \quad (16)$$

Here $\langle \mathrm{v}_T\rangle$ and $\langle \mathrm{v}_P\rangle$ are toroidal and poloidal components of the solenoidal field $\langle \mathrm{v}\rangle$ respectively, and $\psi(\mathrm{x}, t)$ and $\varphi(\mathrm{x}, t)$ are pseudoscalar and scalar functions, respectively.

Substituting representation (15) into equation (13) we obtain the system of equation describing the large-scale convection

$$(\frac{\partial}{\partial t} - \Delta)\psi = -\mathrm{Ra}\ s\mu_1 (\mathrm{e}\nabla)^2 \varphi, \quad (17)$$

$$(\frac{\partial}{\partial t}-\Delta)(\mathrm{Pr}\frac{\partial}{\partial t}-\Delta)\Delta\varphi - \mathrm{Ra}\Delta_{\perp}\varphi = - \mathrm{Ras}\left[\mu_1 (\mathrm{Pr}\frac{\partial}{\partial t}-\Delta)(\Delta_{\perp}-(\mathrm{e}\nabla)^2) - \mu_2\Delta_{\perp}\right]\psi,$$

where $\Delta_{\perp}$ is the Laplace operator in horizontal coordinates.

It is clear from the system (17) that connection of toroidal and poloidal fields is realized only through the helical parameter s.

For the sake of simplicity we consider the case Pr=1 and $\mu_2 << \mu_1$, which is achievable, for example, for sufficiently small λ: $\lambda^2/(\tau\nu) << \lambda/h$. In this case the system of equations for the scalar functions ψ, φ and θ reduces to:

$$(\frac{\partial}{\partial t} - \Delta)\psi = -\mathrm{Ra}\ s\ \mu_1\Delta_z\varphi,$$

$$(\frac{\partial}{\partial t} - \Delta)\Delta\varphi + \mathrm{Ra}\ \theta = \mathrm{Ra}\ s\mu_1(\Delta_z-\Delta_{\perp})\psi\ , \qquad (18)$$

$$(\frac{\partial}{\partial t} - \Delta)\theta = -\Delta_{\perp}\varphi\ .$$

Restricting ourselves to the so-called rigid boundary conditions on Γ,

$$\varphi_{\Gamma}=0,\ \varphi'_{\Gamma}= 0,\ \psi_{\Gamma} = 0,\ \theta_{\Gamma} = 0, \qquad (19)$$

let us seek the solution for each scalar in the following form :

$$\varphi(\mathrm{x},t) \sim \exp(\gamma t+\mathrm{ikr}_{\perp})\varphi(z).$$

If the helicity parameter is zero, $s = 0$, the properties of $\mathrm{Ra}_{cr}(\mathrm{k}^2,0)$ dependence are well known: $\mathrm{Ra}_{cr}(\mathrm{k}^2,0) \to \infty$, when $\mathrm{k} \to 0$, and min $\mathrm{Ra}_{cr}(\mathrm{k}^2,0) \approx 1708$, $k_{min} \approx 3.117$. In the case of helical turbulence, $s \neq 0$, one obtains the following long-wavelength asymptotics ($k^2 << 1$):

$$\mathrm{Ra}_{cr}(\mathrm{k}^2,s) \approx \frac{2\pi}{\mu_1 s} + \frac{2k^2}{\pi\mu_1 s}(1 - \frac{s_0}{s}) + \dots\ ,$$

$$\mu_1 s_0 = \frac{30+(2\pi)^2}{48(2\pi)^3} \approx 0{,}006$$

With increase of the parameter s the minimal value of the critical Rayleigh number decreases, and position of the minimum shifts towards lower values of the horizontal wave number, i.e. the horizontal size of the cells increases. When the helicity parameter reaches the value $s = s_0$, the minimum is at $k_{min} = 0$. Formally, this means that the horizontal scale of the instability turns out to be infinite. In its turn this also means the whole change of the convection due to which the system tends to organize a single large cell (vortex) instead of a great number of convective cells, and its size in this case is determined by the horizontal inhomogeneity of the system. Note that as can be seen from the system (17) or (18) the vortex that appears as the result of the instability, always has linked toroidal and poloidal velocity fields and this leads to topologically non-trivial configuration of streamlines.

To obtain explicitly the field configurations and the instability growth rate for $s > s_0$, consider the system (18) with small supercriticality δR = Ra $-\mathrm{Ra}_{cr}(0,s)$. It is also necessary to introduce into the system the horizontal inhomogeneity, for example, in the simplest form:

$$\delta R(\mathrm{r}_\perp) = \delta R_0(1-\mathrm{r}_\perp^2/\mathrm{r}_0^2), \tag{20}$$

If supercriticality is rather small and $r_0 \gg 1$, it is possible to simplify the system (18) by neglecting higher derivatives with respect to time and horizontal coordinates $(\frac{\partial}{\partial t} \to \gamma)$:

$$\gamma\varphi(\mathrm{r}_\perp) = \mu_1 s\pi\delta R(\mathrm{r}_\perp)\varphi(\mathrm{r}_\perp) + 2(1 - \frac{s_0}{s})\Delta_\perp\varphi(\mathrm{r}_\perp). \tag{21}$$

Solution of equation (21) can be easily found in terms of Laguerre polynomials. The characteristic horizontal size of the structure (in dimensional variables) is then:

$$L = h\left[\frac{r_0^2}{h^2}\frac{2(1-\frac{s_0}{s})}{\mu_1 s\pi\delta R_0}\right]^{1/4} \gg h \ , \tag{22}$$

The lowest mode is determined by the growth rate:

$$\gamma = \mu_1 s\pi\delta R_0\left\{1 - \frac{2h}{r_0}\left[\frac{2(1-\frac{s_0}{s})}{\mu_1 s\pi\delta R_0}\right]^{1/2}\right\}\frac{\nu}{h^2} \tag{23}$$

Thus, from the point of view of convection, helical turbulence can lead to a considerable change of instability character and complete reorganization of the convective structure.

Helical turbulence in a stably stratified fluid can also lead to a large-scale instability. Below we restrict ourselves to the case of a stable stratification associated with temperature gradient which is considered constant here, $A = -N^2/(g\beta) < 0$. The system (1)-(4) then describes dynamics of internal waves in homogeneously stratified fluid, the constant N being the Brunt-Väisälä frequency. Since the internal waves are usually considered through entropy conserving equations, the heat conduction should be neglected, i.e. $\chi \to 0$.

Deriving the Reynolds equation that describes the internal wave dynamics in the presence of helical turbulence, it is convenient to introduce the helicity parameter $\Lambda = (30\pi)^{-1}G_0\mathrm{Re}^2\lambda$. The Reynolds equation takes the form:

$$(\frac{\partial}{\partial t} - \nu\Delta)\frac{\partial}{\partial t}\langle v_i\rangle + N^2 P_{im} e_m e_j \langle v_j\rangle =$$

$$= -N^2 \Lambda \nabla_k [\tau m_1 \frac{\partial}{\partial t}(e_m \epsilon_{krj} + e_k \epsilon_{mrj}) - m_2 e_m \epsilon_{kra}] e_r \langle v_j\rangle,$$

where

$$m_1 = \frac{3+3\frac{\sqrt{\tau\nu}}{\lambda} + \frac{\tau\nu}{\lambda^2}}{(1 + \frac{\lambda}{\sqrt{\tau\nu}})^3} \quad \text{and} \quad m_2 = \frac{5}{2}\frac{1}{(1 + \frac{\lambda}{\sqrt{\tau\nu}})^2}.$$

In the laminar limit this equation describes viscous internal waves.

Similarly to analysis of the convective problem [see (5) and (6)], we now introduce poloidal Φ and toroidal Ψ fields and obtain the following equations coupled through the helicity parameter Λ:

$$(\frac{\partial}{\partial t} - \nu\Delta)\Psi = N^2 \tau\Lambda m_1 (\mathrm{e}\nabla)^2\Phi,$$

$$(\frac{\partial}{\partial t} - \nu\Delta)\frac{\partial}{\partial t}\Phi + N^2\Delta_\perp\Phi = N^2\Lambda\left[m_1 \tau\frac{\partial}{\partial t}(\Delta_\perp-(\mathrm{e}\nabla)^2) - m_2\Delta_\perp\right]\Psi. \tag{24}$$

For solutions for Φ and Ψ of the form $\exp(\gamma t+i\mathrm{k}_\perp \mathrm{r}_\perp+i\mathrm{k}_z z)$ we obtain, for sufficiently long horizontal waves,

$$\gamma = -\nu\mathrm{k}_z^2 \pm [(N^2\Lambda\tau\mathrm{k}_z m_1)^2 - \mathrm{k}_\perp^2\mathrm{k}_z^{-2}N^2]^{1/2}. \tag{25}$$

In the laminar limit expression (25) describes ordinary damping of viscous internal waves:

$$\gamma = -\nu\mathrm{k}_z^2 \pm iN\mathrm{k}_\perp/\mathrm{k}_z. \tag{26}$$

Helical turbulence ($\Lambda \neq 0$) leads to decrease of the internal wave frequency and the longer the horizontal wavelength of the wave is, the lower is the frequency. Very long internal waves ($\mathrm{k}_\perp/\mathrm{k}_z \leq N\Lambda\tau\mathrm{k}_z m_1$) cannot exist in such environment; unstable motions replace internal waves. Scale of the instability is controlled by the horizontal inhomogeneity of the background parameters, and can be evaluated in the same way as in the convective case.

To be specific, consider the case when the Brunt-Väisälä frequency has large horizontal scale r_0 and its dependence on the horizontal distance $r_\perp$ is the same that of the supercriticality of the Rayleigh number in (20). keeping in equations (24) only dependence on slow variables (horizontal coordinates) we obtain, in the laminar limit $r_0\mathrm{k}_z \gg 1$, equation of the type (21), which also can be easily solved in terms of the Laguerre polynomials.

Denoting by N_0 the amplitude of $r_\perp$-variations of the Brunt-Väisälä frequency, we obtain the following expression for the horizontal scale of the solution:

$$\mathrm{L} \approx [2^{-1/2} r_0 (N_0\tau\Lambda\mathrm{k}_z^2 m_1)^{-1}]^{1/2}.$$

The growth rate of the lowest unstable mode can be expressed as:

$$\gamma = -\nu k_z^2 + N_0^2 \tau \Lambda k_z m_1 \left(1 \pm 2^{1/2} (N_0 \tau \Lambda k_z^2 \Lambda m_1 r_0)^{-1}\right) \tag{27}$$

The maximal growth rate and the corresponding vertical wave number are given by

$$\gamma_{max} = (N_0^2 \tau \Lambda m_1)^2 / (4\nu) \text{ and } k_{z\ max} = N_0^2 \tau \Lambda m_1 / (2\nu).$$

Note that in the horizontally homogeneous limiting case equation (27) is similar to the corresponding expression for the magnetohydrodynamic α^2-dynamo. Thus, in a stably stratified fluid helical turbulence can lead to appearance, after the time $\sim \gamma_{max}^{-1}$, of a large-scale vortex with the vertical size $\sim k_{z\ max}^{-1}$ and with the horizontal size

$$L_{max} = [8^{1/2} \nu^2 r_0 N_0^{-5} (\tau \Lambda m_1)^{-3}]^{1/2}.$$

REFERENCES

Westfreid, J.E., and Zaleski, S. (eds.), 1984, "Cellular Structures in Instabilities. Lecture Notes in Physics, v. 210", Springer-Verlag, Berlin.

Moiseev, S.S., Rutkevich, P.B., Tur, A.V., and Yanovskii, V.V., 1988, Vortex Dynamo in a Convective medium with Helical Turbulence, Sov. Phys. JETP, 67(2):294.

Monin, A.S., and Yaglom, A.m., 1975, "Statistical Fluid mechanics", v. 2, Cambridge, mass., mIT Press.

Chandrasekhar, S., 1981, "Hydrodynamic and Hydromagnetic Stability", Dover.

Gershuni, G.Z., and Zhukhovitskii, E.m., 1976, "Convective Stability of Incompressible Fluids", keter Publishing House, Jerusalem.

GENERATION OF LARGE SCALE STRUCTURES IN THREE-DIMENSIONAL ANISOTROPIC INCOMPRESSIBLE FLOW LACKING PARITY-INVARIANCE

P.-L. Sulem[1,2], U. Frisch[1], H. Scholl[1,3] and Z.S. She[4,5]

[1] - *CNRS, Observatoire de Nice, BP 139, 06003 Nice Cedex, France*
[2] - *School of Mathematical Sciences, Tel Aviv University, Israel*
[3] - *Astron. Rechen-Inst. Heidelberg, FRG*
[4] - *Appl. Comput. Math., Princeton University, USA*
[5] - *Nanjing University, Nanjing, P.R. China*

In MHD, an instability, called the alpha-effect, is believed to be at the origin of the generation of large-scale magnetic fields in many cosmical objects (Moffatt 1978, Zeldovich, Ruzmaikin and Sokoloff 1983).

We have shown that there is also an analog of the alpha-effect for ordinary incompressible flows. The crucial ingredients are anisotropy (not necessary in MHD) and presence of a small-scale forcing lacking parity-invariance. This means that the flow has no center of symmetry. Helicity in a flow implies lack of parity-invariance, but the latter is a broader concept. The new instability, which we call the Anisotropic Kinetic Alpha (AKA) effect, arises from the change in the average Reynolds stresses of a small-scale flow which are induced by the presence of a large-scale flow. The effect has been obtained from a multi-scale analysis and demonstrated by full three-dimensional simulations of the Navier-Stokes equations (Frisch, She and Sulem 1987).

The nonlinear studies (theory and simulations) have been developed in detail for a simple forcing function (Sulem, She, Scholl and Frisch 1989). When there is a limited range of unstable scales, nonlinear saturation of the instability arises by a feed-back mechanism. When the unstable range becomes wide, numerical simulations on a CRAY-2 have revealed the existence of an inverse cascade (Fig.1). At first, the linearly most unstable modes are dominant. The maximum excitation then grows and migrates to larger and larger scales. Eventually, a very energetic steady state is obtained with dominance of the largest available scale.

This inverse cascade is amenable to a two-stage asymptotic analysis. The first stage uses the separation of scale between the forcing and the unstable modes; it leads, for the large-scale field, to a set of two coupled PDE's in one space and one time variable having a rather unusal (denominator) nonlinearity. The second stage uses the existence of a wide range of unstable modes to obtain an asymptotic description of the steady-state solutions of the PDE's and study their stability.

Acknowledgements

Computations were done on the CRAY-2 of CCVR (Palaiseau). This work was supported by CNRS (ATP Dynamique des Fluides Geophysiques et Astrophysiques) and by EEC grant ST-2J-0029-1-F.

New Trends in Nonlinear Dynamics and Pattern-Forming Phenomena
Edited by P. Coullet and P. Huerre
Plenum Press, New York, 1990

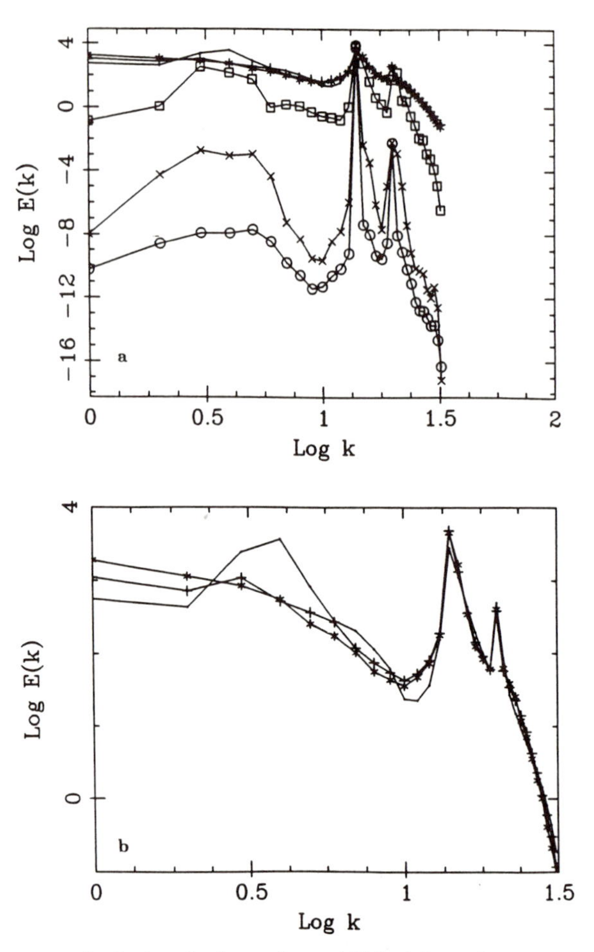

Fig 1. Direct numerical simulation of an AKA-driven inverse cascade, using 64^3 modes. The range of linearly unstable modes extends from $k = 1$ to $k = 8$. Forcing is at wavenumber 14. The figure shows the energy spectrum $E(t, k)$ for various times in log-log coordinates. The circles are for $t = 0.05$, the x-signs for $t = 0.1$, the squares for $t = 0.15$, the dots for $t = 0.2$, the crosses for $t = 0.3$, and the stars for $t = 0.4$. For clarity, late times are shown enlarged in figure 1b.

References

Frisch, U., She, Z.S. and Sulem, P.L. 1987 Physica **28D**, 382.

Moffatt, H.K. 1978 *Magnetic Field Generation in Electrically Conducting Fluids.* Cambridge Univ. Press.

Sulem, P.L., She, Z.S., Scholl, H., Frisch, U. 1989 J. Fluid Mech. **205**, 341.

Zeldovich, Ya.B., Ruzmaikin A.A. and Sokolov D.D. 1983 *Magnetic fields in Astrophysics*, Gordon and Breach.

A NEW UNIVERSAL SCALING FOR FULLY DEVELOPED TURBULENCE:

THE DISTRIBUTION OF VELOCITY INCREMENTS

Yves Gagne[α], Emil J. Hopfinger[α] and Uriel Frisch[β]

[α] Institut de Mécanique de Grenoble *
B.P. 53X, 38041 Grenoble Cedex, France

[β] Observatoire de Nice **
BP 139; 06003 Nice Cedex, France

At the Cargèse meeting a brief presentation was given by U. Frisch of experimental results obtained recently in the Modane wind tunnel (O.N.E.R.A.) by Y.Gagne and E.Hopfinger. The present summary is only intented to highlight one of the main results. For detailed presentation, the reader is referred to Gagne (1987).

Abstract

It is well known that the probability density functions (p.d.f.) of two point velocity differences measured in fully developed turbulence are non gaussian, a signature of internal intermittency. Measurements of $\Delta u(r) = u(x) - u(x+r)$ were performed at high Reynolds number ($R_\lambda = 2720$). The novel results are that: (i) the functionnal behaviour of the tails of the p.d.f. can be represented by $P(\Delta u) \sim \exp(-b(r)|\Delta u/\sigma_{\Delta u}|)$ and (ii) the logarithmic decrement $b(r)$ scales as $b(r) \sim r^{0.15}$ when the separation r lies in the inertial range.

Experiments and Results

The measurements were conducted in the wind tunnel S1 of the O.N.E.R.A. in Modane which consists of a very large pipe of 24 m diameter and 150 m length. The velocity sensor, a standard DISA gold plated hot wire, was positioned on the axis of the pipe; a DISA 55M10 constant temperature system was used. The Taylor microscale Reynolds number was equal to $R_\lambda = 2720$, when the mean velocity was 20 m/s. We recall that experimentalists measure their Reynolds number R_λ, based on the Taylor microscale λ (roughly the geometric mean of the integral and dissipation scales). The integral scale Reynolds number scales approximatively as $1/15R_\lambda^2$. A value $R_\lambda = 2720$ is one of the highest Reynolds number case ever studied.

At this Reynolds number, the power spectrum of the longitudinal velocity fluctuations, shown in figure 1, has a power law over three decades with a slope close to $-5/3$. The fact that the slope, shown in figure 1, is slighty steeper than $-5/3$ will not be discussed here.Here, we only concentrate on the inertial range behaviour of moments of the

* Unité Mixte de Recherche du CNRS

** Unité Associée du CNRS

New Trends in Nonlinear Dynamics and Pattern-Forming Phenomena
Edited by P. Coullet and P. Huerre
Plenum Press, New York, 1990

velocity structure functions defined by $< (u(x) - u(x+r))^p >$.It is well known that for p=3, we have in the inertial range $< (u(x) - u(x+r))^3 >= -4/5 < \varepsilon > r$, where $< \varepsilon >$ is the energy transfer rate. This relation has been deduced by Kolmogorov from the Navier-Stokes equations without recourse to self-similarity. The proof may also be found in Landau and Lifschitz (1954). Consequently,we use this relation to define the extent of the inertial range. Figure 2 shows the "compensated" (premultiplied by suitable powers of separation to make them as flat as possible), structures functions of orders 2,3,6. The third order structure function $< (\Delta u(r))^3 > r^{-1}$ has a plateau indicating the existence of an inertial range over two decades, which is considerably narrower than the "$-5/3$" range. The second point seen in figure 2, is that the structure functions of order 2,3 and 6 exhibit oscillations (for separation r lying in the inertial range). These oscillations which are superposed on the power law scaling $< (\Delta u(r))^p >\sim r^{\zeta_p}$, are not due to insufficient convergence, but are a consequence of the lacunar aspect of the energy transfer in the inertial range (Smith et al. 1986).

The main purpose of this work was to study the tails of the p.d.f. of the velocity increments $\Delta u(r)$. For one thing, the tails give information on the intensity of the vortex stretching events. Furthermore, the convergence of moments of the structure functions depends on the behaviour of the tails of the p.d.f.. In figure 3, we show the p.d.f. of α, where $\alpha = \Delta u(r)/\sigma_{\Delta u}$, $\sigma_{\Delta u}$ is the r.m.s. value of the velocity increment $\Delta u(r)$.

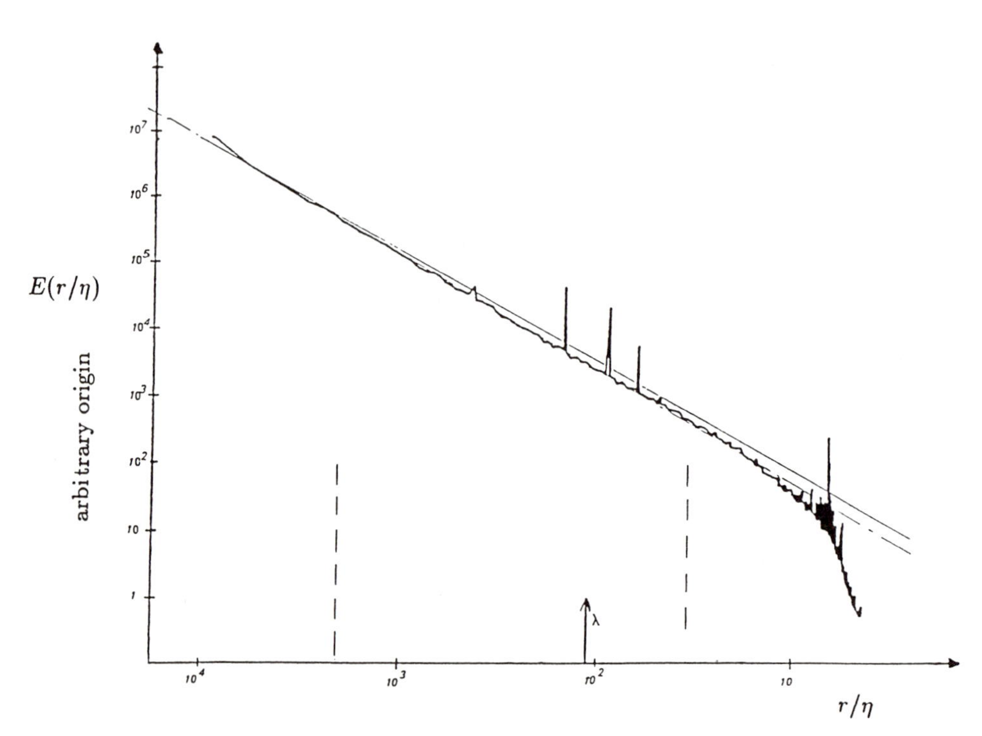

Fig.1 Power spectrum of the longitudinal velocity $u(x)$. λ : Taylor scale; η : Kolmogorov scale;
$R_\lambda = 2720$.
——— slope $-5/3$,
—— · —— best linear fit,
— — —extent of the inertial range defined with $< (\Delta u(r))^3 >$.

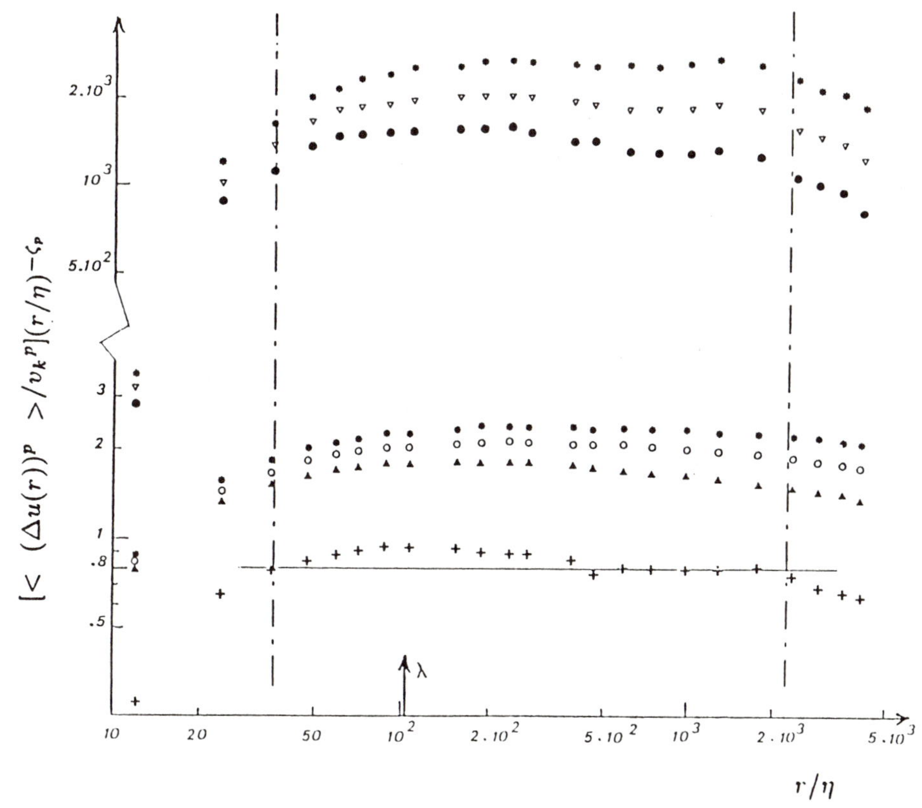

Fig.2 Compensated velocity structure functions $[< (\Delta u(r))^p >/v_k{}^p](r/\eta)^{-\zeta_p}$.

v_k is the Kolmogorov velocity defined by :$(\varepsilon\nu)^{1/4}$.
Suitable values of ζ_p :
p=2 (✱ 0.66, ○ 0.69, ▲ 0.73)
p=3 (+ 1)
p=6 (✱ 1.75, ▽ 1.80, ● 1.85)
—— - ——Inertial range limits.

In this case, the separation r corresponds to scales in the middle of the inertial range. The functional behaviour of the tails (related to large fluctuations with low probability) is clearly exponential; that is $P(\alpha) \sim \exp(-b|\alpha|)$, with $b > 0$. The value of the logarithmic decrement b^+ of the positive velocity fluctuations is larger than the logarithmic decrement b^- of the negative ones. This property leads to a negative skewness coefficient, a necessary condition for the statistical stretching of vortex lines (cf. Monin and Yaglom (1975)).

Figure 4 shows the scaling of the logarithmic decrements b^+ and b^-. The experimental data give the power law $b(r) \sim r^{0.15}$, for both b^+ and b^-, when r belongs to the inertial range. In figure 4, are also included, the experimental results of Van Atta and Park (1971), taken in an atmospheric boundary layer, and results obtained in an axisymmetric jet and in a rectangular duct by Anselmet et al. (1984). Despite the experimental error bars, the same scaling holds for all these very different kinds of flows. This suggests that the above power law is universal. Note that such a distribution of the velocity increments $\Delta u(r)$ implies, in the limit of the large order p, a linear dependence of the exponents ζ_p on the order p. In contrast, Kolmogorov's (1962) log-normal model has a parabolic dependence and the β-model (Frisch et al.(1978)) has a linear dependence for all p's.

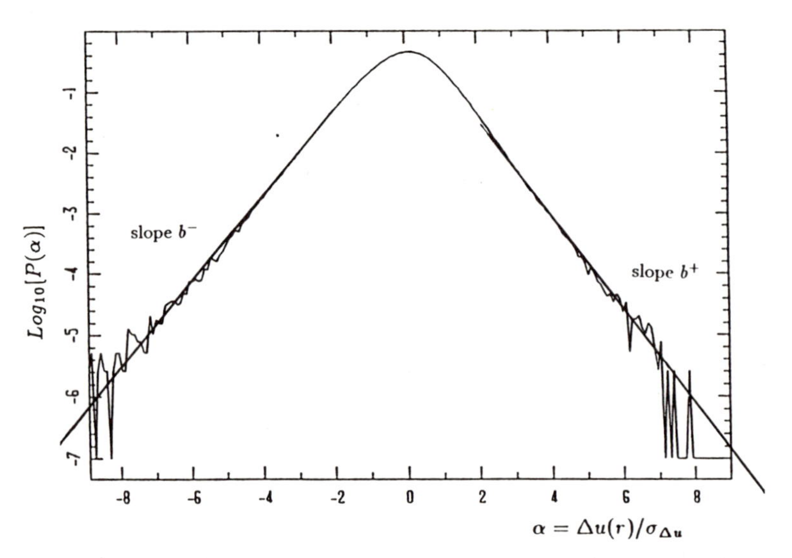

Fig.3 Probability density functions of $\Delta u(r)$ for the separation $r/\eta = 392$

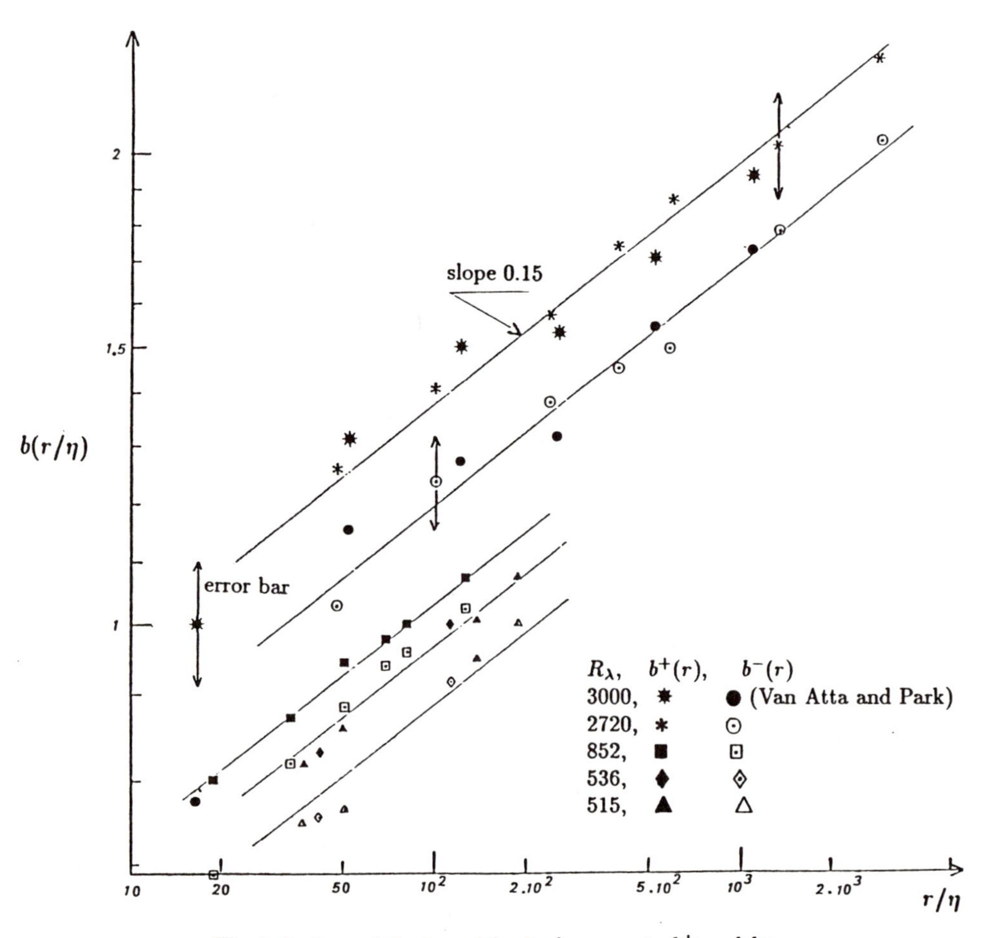

Fig.4 Scaling of the logarithmic decrements b^+ and b^-.

It is of interest to point out that the tails of the p.d.f. of temperature functions measured by Castaing et al. (1988) in high Rayleigh number thermal convection (hard turbulence) also have an exponential behaviour. Furthermore, recent numerical simulations of isotropic turbulence (Métais and Herring (1988), Kida and Murakami (1988)) show an exponential full off of the p.d.f. of the velocity gradient.

Acknowledgements

This work was supported by contract D.R.E.T. n 83/314.

References

Anselmet, F., Gagne, Y., Hopfinger, E. J., and Antonia, R.A., 1978, High order velocity structure functions in turbulent shear flows, J.Fluid.Mech., 140;63,89.

Castaing, B., Gunaratne, G., Heslot, F., Kadanoff, L., Libchaber, A., Thomae, S., Wu, X., Zaleski, S., and Zanetti, G., 1988, Scaling of hard thermal turbulence in Rayleigh Bénard convection, J.Fluid.Mech., to appear.

Frisch, U., Sulem, P.L., and Nelkin, M., 1978, A simple dynamical model of intermittent fully developed turbulence , J.Fluid.Mech., 87;719,736.

Gagne, Y., 1987, Etude expérimentale de l'intermittence et des singularités dans le plan complexe en turbulence développée, Thesis; Université de Grenoble;France.

Kida, S., and Murakami, Y., 1988, Fluid Dynamics Res., to appear.

Kolmogorov, A.N., 1941, Local structure in an incompressible fluid at very high Reynolds number, Dokl.Akad.Nauk., 26;115,118.

Kolmogorov, A.N., 1962, A refinement of previous hypotheses concerning the local structure of turbulence in a viscous incompressible fluid at high Reynolds number, J.Fluid.Mech., 13;82,84.

Landau, L.D., and Lifchitz, E.M., 1958, in " Fluid Mechanics", Addison Wesley.

Métais, O., and Herring, J.R., 1988, Numerical simulations of freely evolving turbulence in stably stratified fluids , J.Fluid.Mech., in press.

Monin, A.S., and Yaglom, A.M., 1975, in " Statistical Fluid Mechanics", Vol.2, M.I.T. Press.

Smith, L.A., Fournier, J.D., and Spiegel, E.A., 1986, Lacunarity and intermittency in turbulence, Phys. Lett., A114;465.

Van Atta, C.W., and Park, J., 1971, Statistical self- similarity and inertial range turbulence, in "Lectures Notes in Physics", 12;402, Springer Verlag.

VORTEX DYNAMICS AND SINGULARITIES
IN THE 3-D EULER EQUATIONS

Alain Pumir[1,2] and Eric D. Siggia[1,3]

[1] Laboratoire de Physique Statistique
Ecole Normale Supérieure
24, rue Lhomond, 75231 Paris Cedex
France

[2] SPT, CEN Saclay
91191 Gif sur Yvette, France

[3] LASSP, Cornell University
Ithaca, N.Y. 14853, USA

The serious study of singularities in the equations of three dimensional hydrodynamics goes back at least to 1934, and a proof still has not been given that solutions of the Navier-Stokes equations remain smooth.[1] Far less is known rigorously about solutions to the Euler equations. Although on physical grounds one might expect a singularity here, the numerical evidence is unpersuasive.
Other than being a challenging mathematical problem, singularities may furnish a paradigm for three dimensional flows. An analogy should be drawn with two dimensional shear flows, specifically the mixing layer, where the first visualization studies revealed large vortex blobs.[2] Their persistence came as a complete surprise since this same flow had been studied earlier with hot wires and generally found to evolve in accord with simple mixing length or dimensional scaling ideas. The coexistence of simple scaling and coherent structures has subsequently been verified in other 2-D flows.[3]

The analogy we wish to draw for 3-D is between coherent structures in space and those in space-time; or singularities. If we take a turbulent boundary layer as the prototypical 3-D flow, we are forced to consider the bursting phenomena which accounts for an appreciable fraction of the Reynolds stress.[4] Boundary layer bursts also came as a surprise since the mean velocity profile was known to obey the von Karmen scaling law. There is still debate about the proper interpretation of bursts but there is incontrovertible evidence of very large gradients on very small scales.[5] Furthermore the bursts contribute an appreciable fraction of the Reynolds' stress which is an important component of the drag.

Two further more technical arguments for studying singularities should be advanced. Some form of collapse is the only way in which a computer can simulate a large range of scales. Under favorable circumstances, the simulation time will grow logarithmically with scale size rather than algebraically. Analytic calculations may also prove feasible for a collapse and point the way towards more generally applicable approximations.

Our numerical approach to singularities should be contrasted with two previous efforts. Brachet et. al. [6] computed a power series in time for the enstrophy with Fourier modes, which was then Padé approximated. A subsequent theorem showed this analysis to be superfluous. Namely, Beale et. al. proved that the enstrophy can only diverge if the time integral of the supremum (over space) of the vorticity does.[7] In Ref. 6 this quantity grew by no more than a factor of 4-5, hence no singularity.

New Trends in Nonlinear Dynamics and Pattern-Forming Phenomena
Edited by P. Coullet and P. Huerre
Plenum Press, New York, 1990

One can also question whether Fourier modes are the optimal way of describing spacial singularities. The equations are local in real space except for the pressure which would tend to average out for a singularity with fractal support. Our working assumption is that the first singularity develops with a single well defined scale in the x,y,z directions, though more complicated objects could develop later.

Chorin applied a vortex algorithm to this problem but in our view its implementation was inadequate.[8]

Pumir and Siggia solved the Biot-Savart model for a single vortex filament with a locally variable core size chosen to preserve volume.[9] A singularity was found, which modulo logarithmic terms, would persist with viscosity present. This model is perhaps the simplest possible which contains vortex stretching and can be solved with reasonable confidence both numerically and analytically. It invalidates any casual arguments that viscosity will control a singularity in the Euler equations.

This model has one serious shortcoming which was explicitly noted in Ref. 9, namely that the vortex cores undergo secular distortion when they pair and stretch.[10] Vortex reconnection is not an issue since it clearly doesn't occur for Euler prior to the singularity, and for Navier-Stokes, large gradients diverging with the inverse viscosity are necessary for reconnection to occur in a finite time[11]. Granting that the collapse does preserve some core shape, then it is of no importance whether the shape is circular or not. It is necessary to adjust the core size locally since comparison with the correct equations shows that there is insufficient time for the core volume to redistribute[9]. Lack of manifest energy conservation is also not a problem since under these assumptions the energy in the Euler flow to which the filaments are asymptotic is finite at the singularity but has a square root cusp in time. The distortion however can be viewed as the consequence of an insufficient energy flux into the singular region.

To simulate a 3-D singularity without modeling in the absence of boundaries, we have mapped the line onto the interval with the tangent or similar function and then finite differenced the Euler equations in each direction separately. The code is second order accurate in space and time and explicitly respects incompressibility, and momentum and energy conservation. We initially adjusted scales continuously to preserve certain norms, but found this both to interfere with the conservation laws and to bias the result through the choice of norms. Currently, we periodically interpolate from the old mesh to a new one using splines under tension.

If we initialize this code with two antiparallel vortex tubes, designed to imitate the Biot-Savart model, the tubes first press together and flatten into sheets or ribbons with the direction of vorticity and its magnitude changing little. Vortex amplification then ensues at the leading edge of the pair of ribbons which assume a *V* shape. Vorticity is periodically shed from the tip and is folded back by the mean flow (fig.1). The shape of the developing singularity changes with time even after separate length scale adjustments are made in the x, y and z- directions. This intrinsic time dependence does not seem to be an initial transient and is associated with the "vortex shedding" noted above. The two eigenvalues of largest magnitude of the rate of strain tensor (one positive, one negative) are

associated with strain in the x,y plane. The third eigenvalue, which stretches ω, can be as small as 5% of the other two. The same relation between strain and vorticity has been noted in other simulations[12].

To give some impression of the ranges of scales we have simulated to date, if the initial value of the x-scale, y-scale, z-scale, maximum velocity, maximum vorticity, maximum rate of strain squared are respectively 1, 1, 1, 3, 3, 3, then, the current values are 10^{-3}, 10^{-1}-10^{-2}, 10^{-1}-10^{-2}, 3, 120, 6000. The sheets have lateral scales 10-100 times their spacing. It is still unclear whether the maximum vorticity diverges in a finite time or only grows exponentially.

The best qualitative account we can give of the fluid mechanical origins of the vortex self-stretching in this flow is by reference to the Biot-Savart simulations. In spite of qualitative differences, the relation of vorticity and strain is the same, as well as the resultant velocity.

The numerical calculation is basically not a difficult one since there is one dominant scale in each direction and the flow is smooth. The power of our method stems from the successive mesh refinements which are common enough in other contexts but have not yet been employed on the 3-D Euler equations. If a singularity exists for our initial conditions, we expect that our simulations will exhibit it and provide strong guidance for an analytic-perturbative treatment of the collapsing solution.

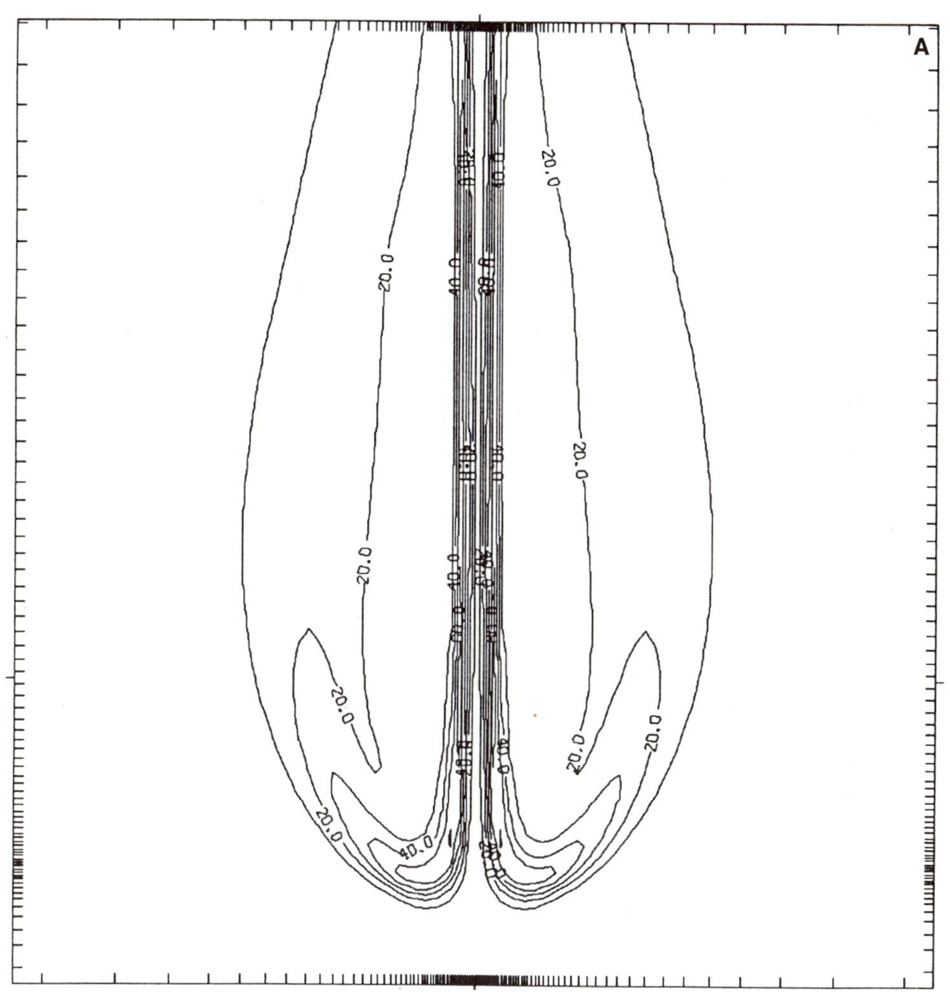

Fig.1 Contours of vorticity (entirely parallel to the z direction) in the x-y plane with physical scales (A) and with amplified and stretched scales (B), ($x = 10\ y$). The region around the point of maximum vorticity is shown and the direction of propagation is -y. The plane $x = 0$ is a plane of symmetry and the vorticity is negative for $x < 0$. (cont.)

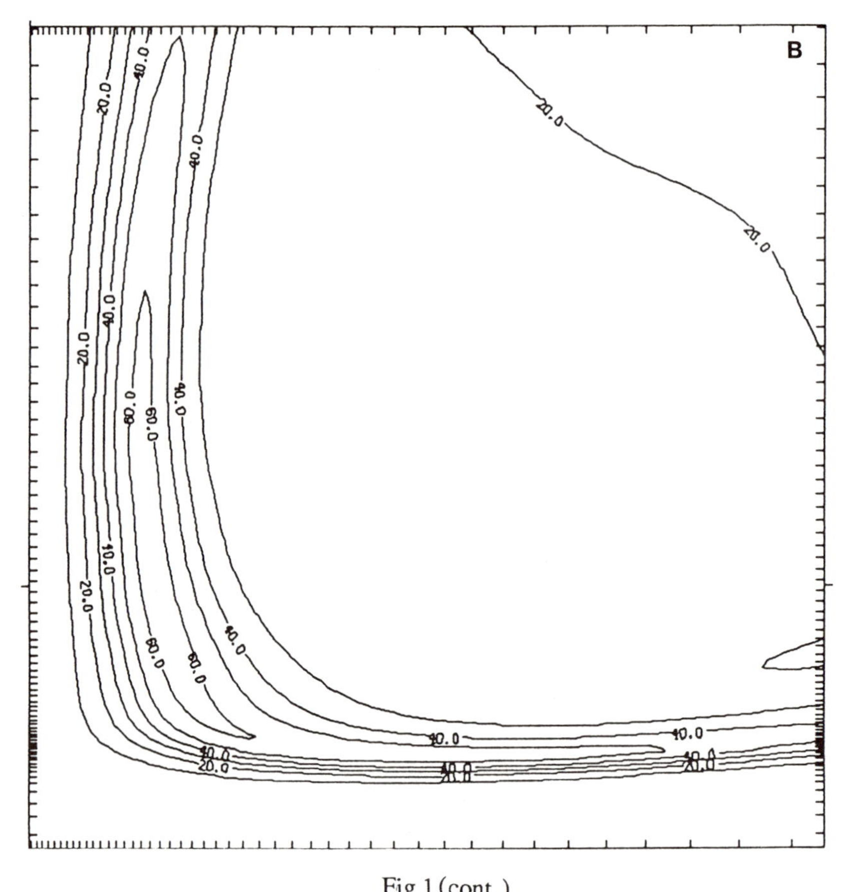

Fig.1 (cont.)

References

1) J. Leray, Acta Math, **63**, 193 (1934); L. Caffarelli, R. Kohn, L. Nirenberg, Commun. Math. Phys. **61**, 41 (1978).
2) G. L. Brown, A. Roshko, J. Fluid Mech. **64**, 775 (1974).
3) H. Aref, E. D. Siggia, J. Fluid. Mech. **100**, 705 (1980) and J. Fluid Mech. **109**, 435 (1981).
4) M. Head and P. Bandyopadhyay, J. Fluid Mech. **107**, 297 (1981).
5) W. Willmarth and T. Bogar, Phys. Fluids. **20**, S9 (1977); S. Kline, W. Reynolds, F. Schraub, and P. Runstadler, J. Fluid Mech. **30**, 741 (1967).
6) M. Brachet, D. Meiron, B. Nickel, S. Orszag, and U. Frisch, J. Fluid Mech. **130**, 411 (1983). 7) J. T. Beale, T. Kato and A. Majda, Commun. Math. Phys. **94**, 61 (1984).
8) A. J. Chorin, Commun. Math Phys. **83**, 517 (1982).
9) A. Pumir and E. D. Siggia, Phys. Fluids, **30**, 1606 (1987).
10) A. Pumir and R. Kerr, Phys. Rev. Lett. **58**, 1636 (1987).
11) W. Ashurst and D.Meiron, Phys. Rev. Lett. **58**, 1632 (1987)
12) W. Ashurst, A. Kerstein, R. Kerr and C. Gibson, Phys. Fluids **30**, 2343

DIRECTIONAL GROWTH IN VISCOUS FINGERING

V. Hakim+, M. Rabaud*, H. Thomé* and Y. Couder*

+ Laboratoire de Physique Statistique de l'Ecole Normale Supérieure
* Groupe de Physique des Solides de l'Ecole Normale Supérieure
24 rue Lhomond, 75231 Paris Cedex 05

I. INTRODUCTION

During the past few years there has been a renewed interest in the Saffman-Taylor instability (Saffman and Taylor (1958)), as this instability between strongly viscosity-contrasted fluids gives rise to shape selection processes. Its dynamics is very similar to that of another pattern forming instability, the Mullins-Sekerka instability, occurring during the free growth of a crystal in an undercooled liquid (Mullins and Sekerka (1964)). The theoretical analogy between the growth of viscous fingers in Hele-Shaw cells and the growth of crystalline dendrites is now well-known (e.g. Langer (1988) ; Kessler and Levine (1988 a) for a review) and the similarity between the two phenomena has been demonstrated experimentally (Rabaud *et al.* (1988)). Both processes are free growths in which the non linear evolution of the pattern defines the field in which it keeps growing.

The Mullins Sekerka instability has a different dynamics in directional solidification where a dilute binary mixture is drawn at constant velocity across a linear temperature gradient (Langer (1980) and references therein). In this set-up there is a return force towards a linear front fixed in the laboratory frame of reference, the instability has a threshold ; above a critical pulling speed, the interface between the liquid and solid phases forms a cellular pattern. Numerous experimental (Jackson and Hunt (1966) ; Trivedi (1984) ; de Cheveigné (1986) ; Billia (1987), Bechhoefer *et al.* (1988)) and theoretical works (Ungar and Brown (1985) ; Dombre and Hakim (1987) ; Muller Krumbhaar (1988) ; Kessler and Levine (1988 b)) have been devoted to the understanding of the fundamental parameters involved in the evolution of the instability. The remaining unsolved problems concern the origin of wavelength selection and the exact role of crystal anisotropy.

We sought an analog of this situation in viscous fingering. Ben Jacob *et al.* (1985) had already noticed that separating the glass plates of a Hele Shaw cell created a localized pressure gradient. However in this case the conditions are unsteady and the phenomenon transient. We were thus led to revisit experiments performed some time ago by Pearsons (1960), Pitts and Greiller (1961), Taylor (1963) and McEwan and Taylor (1966).

The first two papers were initially motivated by the needs of the photographic and painting industry, and gave a convincing reason for the origin of the lines left behind when a wall is painted with a smooth roller. Pitts and Greiller (1961) studied the flow emerging from the narrow space between two contra-rotating cylinders, half immersed at the surface of a large tank filled with viscous fluid. They observed ripples at the surface of cylinders when the velocity reached a threshold value. Pearsons (1960) tried a more general approach for the flow between any slowly divergent and moving walls. In the limit of large velocity, he presented results on the evolution of the observed wavelength versus minimum thickness and for a cylinder of constant radius rolling or sliding over a flat plate.

Taylor (1963) analyzed a large number of experimental situations where cavitation takes place. He pointed out the importance of the thickness of the coating fluid, and analyzed its

New Trends in Nonlinear Dynamics and Pattern-Forming Phenomena
Edited by P. Coullet and P. Huerre
Plenum Press, New York, 1990

evolution below the instability threshold. He also presented photographs of cavitation fingers in the case of eccentrics co-rotating cylinders. McEwan and Taylor (1966) showed that a comparable process takes place during the peeling of a flexible strip attached to a glass plate by a viscous adhesive.

In these experiments, moving boundaries force the viscous fluid to flow in a widening gap, thus creating a pressure gradient. The less viscous fluid (air) penetrating in this gap forms a front which can be stationary in the laboratory frame of reference. This front, moving in the frame of reference of the boundaries, tends to be unstable due to the Saffman-Taylor instability but is stabilized by the gradient. These are situations of the type we are looking for.

The most symmetrical situation would have been to investigate the stability of the menisci of an oil layer placed between two contra-rotating cylinders in the region where they come in near contact (an experiment close to that of Pitts and Greiller (1961)). However in this configuration the unstable region is difficult to observe visually. We briefly investigated a situation where a glass plate moved beneath a rotating cylinder so that the equality of the velocities of the two surfaces was maintained. But the finite length of the glass plates limited the duration of each experiment. So finally we chose the asymmetrical case described below which has a similar dynamical behaviour near the instability threshold which we will investigate in the present article. Differences in the large constraint regions will be described elsewhere.

Our set-up is presented in part II and the experimental results in part III. In part IV the experiment is analyzed in terms of the lubrication approximation. A linear analysis of the planar front is then presented in part IV 3. A short conclusion follows which emphasizes the analogy with directional solidification.

II. EXPERIMENTAL SET-UP

The cell was built so as to get a stationary interface in the laboratory frame of reference. A sketch of the apparatus is presented on figure 1a. It consisted of a horizontal cylinder of length L (L = 300 mm) and radius R (R = 50 mm) made of perspex and rectified so that its diameter was constant with a precision better than 0.01 mm. This cylinder was driven in rotation by a regulated motor. The tangential velocity, $V_0 = \Omega R$, was in the range [0, 1 m/s] with a precision of 0.1 mm/s. Underneath was placed a horizontal, motionless, glass plate

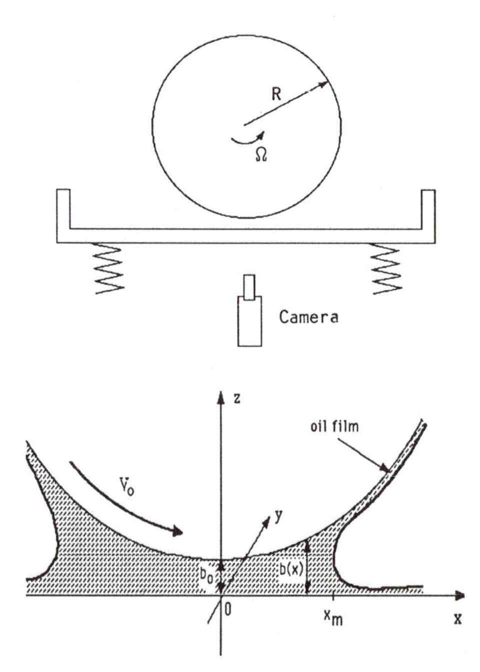

Fig. 1 . a/ Sketch of the cylinder rotating over a stationary glass plate.
b/ Sketch of the boundaries conditions used in the analysis of the flow.

held up against the cylinder by means of springs, the two extremities of the cylinder being separated from the glass plate by small spacers. Changing these spacers the minimum distance b_o of the cylinder to the plate could be chosen at will in the range 40 μm to 1 mm. The value of b_o was always found to exceed the spacers thickness because of the oil films lubricating the spacers. These films are approximately 20 μm thick, but introduce an uncertainty in thickness from one experiment to another.

Oil was initially spread on the glass plate. As we will see our results were independent of the initial thickness of this layer provided that it was larger than, but of the order of, b_o.

We used a silicon oil (Rhododendron 47 V 100) with dynamical viscosity $\mu = 96.5$ 10^{-3} Kg/m.s and surface tension $T = 20.9$ 10^{-3} N/m at 25°C. This oil insured a perfect wetting of the perspex and glass surfaces so that we avoided any influence of dynamical contact angles.

We observed the interface from underneath through the horizontal plate. The menisci appeared as dark lines on a white background and, as the cell was thin, our resolution on meniscus position and instability wavelength was of order of 0.5 mm. The dynamical evolution was recorded on video tapes.

III. EXPERIMENTAL RESULTS

We will be particularly interested in the dynamics of what we will call the back meniscus which is situated on the right of figure 1b where the upper boundary lifts away from the lower one. We will only present and discuss here two main results ; the position of the back meniscus when increasing the velocity and the instability threshold at which a cellular type of instability shows up. The results will be discussed with the conventions sketched on figure 1b. The cylinder is rotating anticlockwise. The axis Ox and Oy are in the plane of the glass plate, Oy being parallel to the cylinder axis, Oz is vertical.

1) Interface Position

At zero velocity, we observe two wetting menisci between oil and air, symmetric with respect to the vertical plane which contains the axis of the cylinder. The position of these interfaces results from a balance between gravity and surface tension forces and depends on the amount of oil spread on the horizontal glass plates.

At low angular velocity, both menisci are stable. The position of the back meniscus is seen to change as the velocity increases. We plotted its abscissa x_m as a function of velocity on figure 2. At first there is a small increase of x_m, then a rapid decrease. For velocities lower than a critical value V_c, the front is straight in the y direction. Above the threshold V_c, the front is no longer flat and we then plot on figure 2 both the minimum and the maximum values, x_{m1} and x_{m2}, of its position.

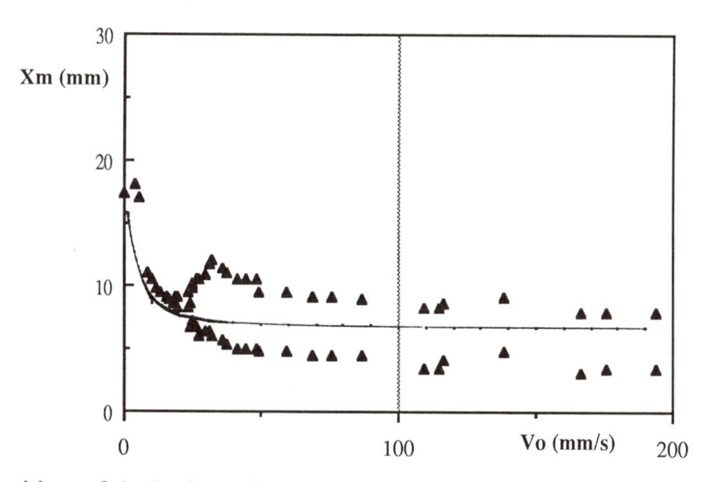

Fig. 2 . Position of the back meniscus, (increasing or decreasing tangential velocity V_o) in a cell where $b_o = 0.17$ mm. Above the critical velocity the nearest and furthest positions of the front (x_{m1} and x_{m2}) are shown. The line is the theoretically predicted position.

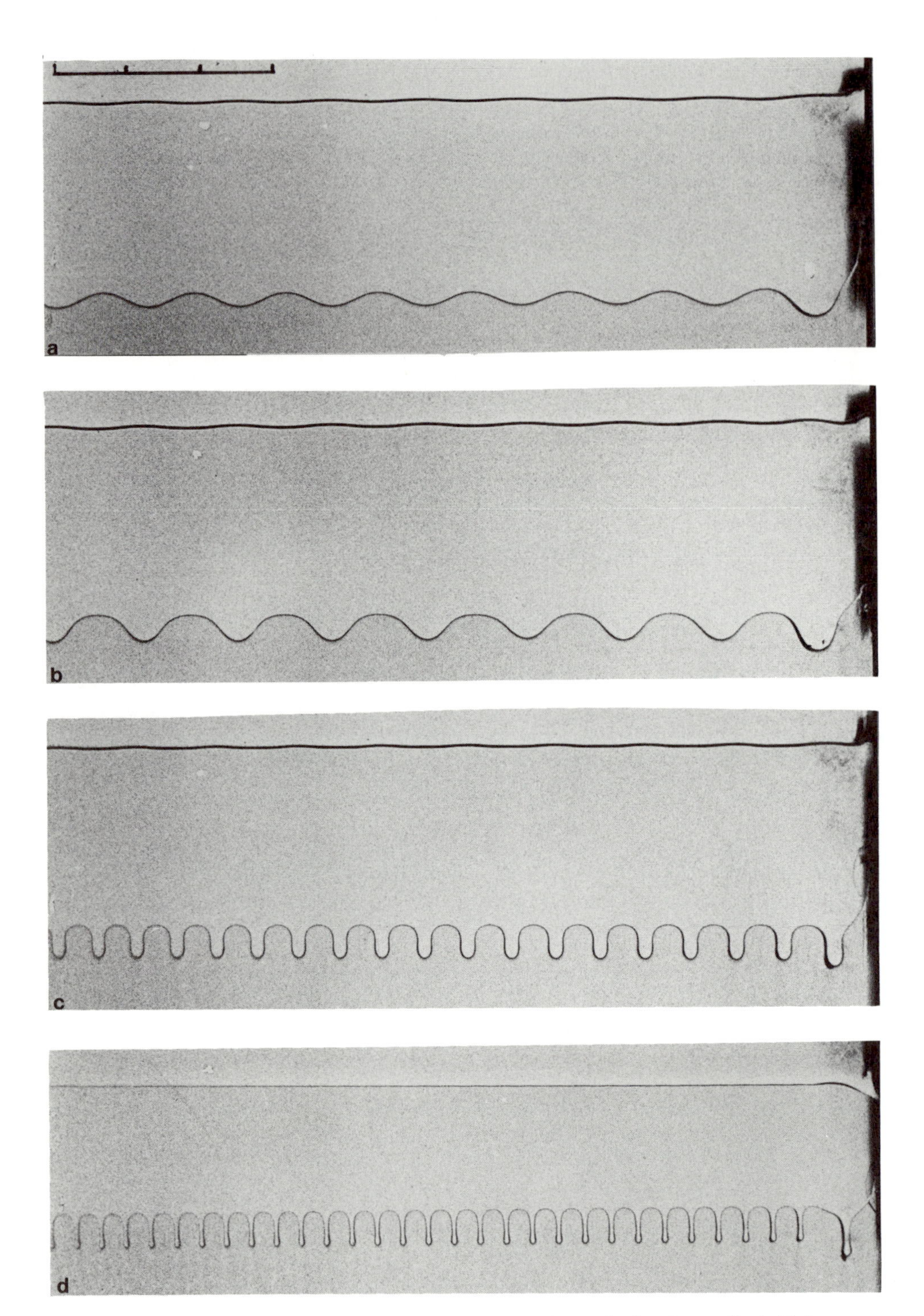

Fig. 3 . Sequence of pictures of the interface for increasing velocity, in a cell of thickness $b_o = 0.17$ mm. The line at the top of each picture is the front meniscus, at the bottom the back meniscus. (a) $V_o = 31.5$ mm/s, (b) $V_o = 35$ mm/s, (c) $V_o = 67$ mm/s, (d) $V_o = 185$ mm/s. The scale at the top of (a) is in centimeters.

We checked that in the range of velocity shown on figure 2, the position of the meniscus was insensitive to the amount of oil set on the horizontal plate. At smaller velocity, the effect of gravity cannot be completely neglected and the meniscus is then slightly affected by the oil layer's thickness.

2) Instability Threshold and Wavelength Selection

For velocities smaller than a critical value V_c the front is straight in the Oy direction except near the spacers where the boundary creates small dips. At the threshold the front destabilizes and becomes wavy. It takes a sinusoidal shape of wavelength λ_c but the process of setting up a regular shape is long ; it is of the order of the viscous diffusion time over the length of the front $\tau \sim L^2/\nu$. The amplitude of the sinusoid is very small at threshold (Fig. 3a) and increases with increasing velocity (Fig. 3b). In a given state the amplitude is the same independently of whether this state has been reached by increasing or decreasing the velocity. No discontinuity in the amplitude is ever observed near the threshold. The bifurcation is thus clearly supercritical. In the range $0.02 < b_0 < 1$ mm the critical velocity V_c is approximately proportional to the thickness (Fig. 4) ; we find V_c (mm/s) $\approx 180\ b_o$ (mm).

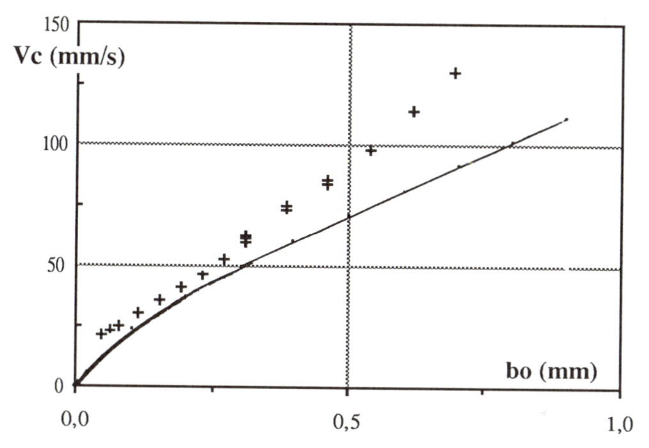

Fig. 4 . The critical velocity V_c as a function of the thickness b_o. (+) are the experimental points and the line represents the calculated ones.

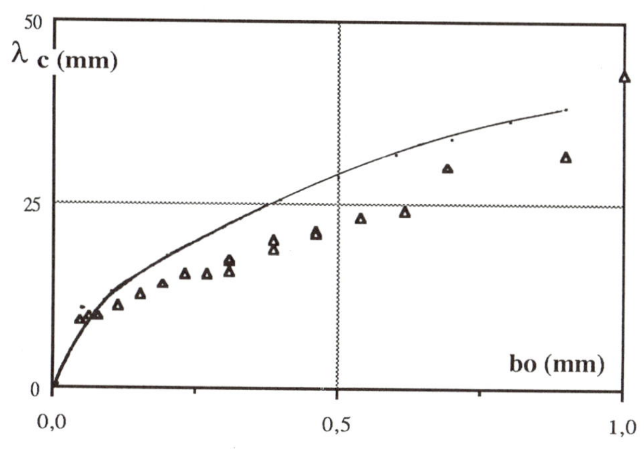

Fig. 5 . The critical wavelength λ_c as a function of the thickness b_o. (Δ) are the experimental points and the line represents the calculated ones.

This result can be expressed in dimensionless form as

$$\frac{\mu V_c}{T}\frac{R}{b_0} \approx 40 \qquad (1)$$

Pitts and Greiller (1961) found a similar expression but a value larger than ours. A possibility is that, as they could not observe the meniscus directly, they measured the threshold only when the wave had reached a larger amplitude. Figure 5 shows the dependence of the critical wavelength λ_c on b_0. For a cylinder of radius R = 50 mm, we found $\lambda_c \approx 40\ b_0$. Both these relations, giving V_c and λ_c are empirical and we will see in part IV that the physical relation linking V_c and λ_c with b_0 and db/dx is more complex. Figure 3 shows successive states of the front when the velocity is increased. As the amplitude of the instability grows larger, the front departs from its sinusoidal shape. It becomes formed of a series of parallel fingers separated by thin oil walls (connecting the oil film on the cylinder to the oil on the plate). The general shape (figure 3d) resembles strikingly that of the front observed in directional solidification. The oil walls observed here are the equivalent of the liquid grooves separating the cells in solidification.

It is worth noting that beyond the front the oil film left on the cylinder has a modulated thickness because the detachment of the walls creates periodic crests. This modulation is still present after one rotation and causes the stable meniscus (the upper one on figure 3) to be slightly wavy (e.g. figure 3c). We checked that this does not induce a feedback effect on the wavelength selection. For example we can collect or spread the coating film on the cylinder, affecting the position of the stable front meniscus. However the unstable back meniscus, which appears extremely resistant, remains unaffected.

The wavelength λ of the pattern decreases rapidly as the velocity increases (and as the meniscus abscissa decreases) and seems to saturate at large velocity (figure 6). At all the velocities that we investigated, the pattern is stable, regular and quite insensitive to vibrations. For small thicknesses, fronts with more than a hundred regular and stable cells have been observed. At any time the velocity can be decreased, and provided the structure has time to evolve, no hysteresis is observed. We also checked that the structure was not sensitive to the amount of oil spread on the stationary glass plate.

The evolution of the wavelength gives rise to a rich dynamics. For instance by rapidly increasing the velocity, it is possible to observe simultaneous tip-splittings of all the fingers resulting in a halving of the wavelength. The dynamics of the front will be described in details elsewhere.

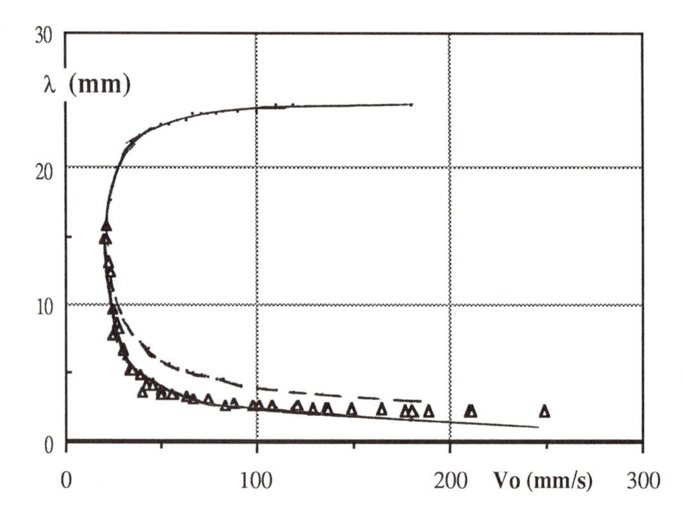

Fig. 6 . (Δ) : Experimentally observed wavelengths versus tangential velocity V_o (for either increasing or decreasing velocity) in a cell of thickness b_o = 0.1 mm. (———) : Calculated marginal stability curve. (------) : Calculated most unstable wavelengths.

IV. EQUATIONS OF MOTION

In the first part of this section the mechanisms which fix the position of the meniscus in the cell are explained. In a second part, a linear analysis is used to predict when the planar interface is unstable and becomes cellular.

1) Laminar Flow between Two Close Moving Boundaries

The notations are introduced on figure 1b ; V_o is the velocity of the cylinder boundary and b(x) is the vertical distance between the lower fixed plate and the rotating cylinder. We suppose an infinite and homogeneous cell in the y direction. All the analysis will be done in the laboratory frame of reference. We make use of the lubrication approximation which is applicable to the motion of a viscous fluid between two boundaries which are sufficiently close together and such that the thickness of the layer varies sufficiently slowly. (Experimentally the gradient $G(x) = db/dx \approx 0.2$ and the Reynolds number, $Re = V_o b_o/\nu$, is smaller than unity.) If the motion of the boundaries had been symmetrical (e.g. between two rollers) we would have had the Poiseuille flow of the Saffman Taylor instability. Here, owing to the asymmetry of the system, we have a local Poiseuille-Couette flow. That is, the fluid velocity **u** is assumed to be

$$\mathbf{u}(x,z) = z\left(\frac{V_o}{b(x)}\hat{\mathbf{x}} + \frac{z - b(x)}{2\mu}\vec{\nabla}_h P\right) \qquad (2)$$

Where ∇_h denotes the two-dimensional horizontal gradient. So the local mean fluid velocity across the stratum is :

$$\mathbf{u}(x) = \frac{V_o}{2}\hat{\mathbf{x}} - \frac{b^2(x)}{12\mu}\vec{\nabla}_h P \qquad (3)$$

The equation for the pressure field comes from the incompressibility of the oil

$$\mathrm{div}\,(b(x)\mathbf{u}(x)) = 0 \qquad (4)$$

Let us first recall the deduced classical expression (Martin (1916) ; Bank and Mill (1954)) for the distribution of pressure between an immersed rotating cylinder and a plate, that is neglecting any influence of a meniscus. Introducing the constant flux Q through the gap, equation 3 gives readily the variation of pressure

$$\frac{dP}{dx} = \frac{6\mu V_o}{b^2(x)} - \frac{12\mu Q}{b^3(x)} \qquad (5)$$

This can be easily integrated if one approximates the cylinder by its osculating parabola (i.e. $b(x) \approx x^2/2R + b_o$) and introduces θ such that $x = (2Rb_o)^{1/2}\tan\theta$:

$$P(x,Q) - P_0 = \frac{12\mu V_o R^2}{(2Rb_0)^{3/2}}\left[(\theta(x)+\frac{\pi}{2}) + \frac{1}{2}\sin(2\theta)\right] - \frac{48\mu Q R^3}{(2Rb_0)^{5/2}}\left[\frac{3}{4}(\theta(x)+\frac{\pi}{2}) + \frac{1}{2}\sin(2\theta) + \frac{1}{16}\sin(4\theta)\right] \qquad (6)$$

where P_o is the atmospheric pressure (at $x = -\infty$). In particular since the exit pressure is the atmospheric pressure (we neglect gravity) $P(\infty) = P_o$, and this determines the total flux Q to be :

$$Q = \frac{2}{3} b_0 V_0 \qquad (7)$$

and thus P(x) is given by : $\quad P(x) = -\mu V_o x\, b(x)^{-2}$.

2) What is the position of the planar interface ?

i) a simple answer. The simplest way of finding the meniscus position is to use a balance argument : the flux of oil Q_{en} that enters between the cylinder and the plate should equal the flux of oil Q_{ex} that exits coated on the cylinder. This last flux is simply given by the velocity of the rotating cylinder multiplied by the thickness of the wetting film that covers it (see figure 1b) and is given dimensionally by

$$Q_{ex} = V_0 \, b(x_m) \, F(\frac{\mu V_0}{T}) \qquad (8)$$

where $b(x_m)$ is the thickness at the tip of the meniscus and T the surface tension between oil and air. At very low capillary number ($Ca = \mu V_0/T$), the function F can be computed by a matching procedure (Landau and Levich (1942) ; Bretherton (1961) ; Park and Homsy (1984)) and otherwise should be taken from experiment. As we did not measure the thickness of the wetting film, we use the fit of the experimental data of Tabeling *et al.* (1987) obtained in a classical Saffman Taylor experiment.

$$F\,(Ca) = 0.12\,[\,1 - \exp\,(\,-\,8.6\,Ca^{2/3})\,] \qquad (9)$$

This would really correspond to the case where the lower plate is also moving at speed V_0 so that the flow in front of the meniscus is locally a Poiseuille flow instead of Poiseuille-Couette as in our case. We neglect the difference in order to avoid an ad-hoc curve fitting.

Now let us suppose that the meniscus has a negligible influence on the total flux between the cylinder and the plate. Then Q_{en} is simply equal to the flux in the absence of meniscus (equation 7) and the interface position is very simply determined by equating Q_{en} and Q_{ex} :

$$b(x_m) = \frac{2\,b_0}{3\,F(Ca)} \qquad (10)$$

It is also clear that x_m is a stable position for the planar interface. Suppose that the meniscus is behind x_m (x larger than x_m on figure 1b) then equation 8 shows that the flux of oil that goes out, Q_{ex}, is larger than the flux oil that goes in. Therefore the hatched region of figure 1b tends to be emptied, the interface moves toward the locus of minimum gap and therefore moves to the left toward x_m. If, on the contrary the meniscus is located at the left of x_m, more oil goes in than goes out, the hatched region grows and x moves to the right toward x_m.

This simple argument is based on the a priori supposition that the meniscus has a negligible influence on the flux between the cylinder and the plate. We present now a more refined discussion which takes into account the modification of the total flux due to the meniscus and show the approximate validity of the above argument.

ii) a refined answer. Q_{ex} is still given by equation 8 and is still equal to Q_{en}, but now we should determine the flux Q_{en} that enters between the cylinder and the plate. For low velocities ($Ca \ll 1$) the pressure drop due to surface tension at the interface is approximately equal to the static value i.e.

$$P(x_m) = P_0 - T(\frac{2}{b(x)} + \frac{1}{\rho}) \qquad (11)$$

where ρ is the radius of curvature of the meniscus in the horizontal plane (xoy) (for a planar interface $1/\rho = 0$). (Correction to this static condition are discussed by Park and Homsy (1985) ; Reinelt and Saffman (1985).)

This equation, together with equation 6 (which is valid for a planar interface) gives the supplementary condition between x_m and Q_{en} we were looking for. We have to solve simultaneously for x_m and Q_{en} :

$$\begin{cases} Q_{en} = V_0\, b(x_m)\, F(\frac{\mu V_0}{T}) \\ P(x_m, Q_{en}) = P_0 - \frac{2T}{b(x_m)} \end{cases} \qquad (12)$$

When Q is expressed as $\alpha b_o V_o$ and the angle θ introduced, the two equations becomes :

$$\begin{cases} \alpha = \frac{1}{\cos^2(\theta_m)}\, F(\frac{\mu V_0}{T}) \qquad (13) \\ \frac{T}{6\mu V_0}\sqrt{\frac{2b_0}{R}} = \frac{-1}{\cos^2(\theta_m)}\left[\left[(\theta_m+\frac{\pi}{2})+\frac{1}{2}\sin(2\theta_m)\right] - 2\alpha\left[\frac{3}{4}(\theta_m+\frac{\pi}{2})+\frac{1}{2}\sin(2\theta_m)+\frac{1}{16}\sin(4\theta_m)\right]\right] \end{cases}$$

In the experiment, the term

$$-\frac{T}{6\mu V_0}\sqrt{\frac{2b_0}{R}}$$

is small and this alone implies that α is close to 2/3 using the second equation (13). It means that the total flux is approximately the same as it would be without a meniscus (which was the a priori assumption of section i). The calculated values for the planar front are shown on figure 2 together with the experimental data. The agreement is satisfactory below the instability threshold, especially when one takes into account the fact that there is no free parameter and the approximate nature of function F. Above threshold, the calculated position of the planar interface is slightly behind the tips of the cells but well in front of the end of the grooves, as could be expected.

3) Linear Analysis of the Planar Front

The equations for the pressure field in front of the interface are (3) and (4) together with the boundary condition on the interface given by equation 11. This defines entirely the pressure field once the interface position is given. In order to complete the dynamical equations, we now only need to write the equation of motion for the interface. This is, as usual, given by the conservation of fluid :

$$V_n\, b(x_m)\left[1 - F(\frac{\mu V_0}{T})\right] = \left[\mathbf{n}.\mathbf{u}\ (\mathbf{r}_m) - V_0\, F(\frac{\mu V_0}{T})\right] b(x_m) \qquad (14)$$

where V_n is the normal velocity of the interface at the point $\mathbf{r}_m$.

Now a linear analysis of the planar interface can be carried out. The position $\xi(y)$ of the interface is supposed to be :

$$\xi(y) = x_m + \varepsilon(t)\sin(ky) \qquad (15)$$

The pressure field is of the form :

$$P(\mathbf{r}) = P^{(0)}(x) + \eta(t)\, q_k(x)\sin(ky) \qquad (16)$$

where $P^{(o)}(x)$ is the pressure field in front of a planar interface determined above. Equations (3) and (4) for the pressure field show that $q_k(x)$ is the solution of

$$\frac{d^2 q_k}{dx^2} + \frac{3}{b}\frac{db}{dx}\frac{dq_k}{dx} - k^2 q_k = 0 \qquad (17)$$

which goes to zero when $x \to -\infty$.

The boundary condition for the pressure on the interface gives

$$\eta(t)\, q_k(x_m) = \varepsilon(t)\left\{ -\left[\frac{d\,P^{(0)}}{dx}\right]_{x=x_m} + \frac{2T}{b^2}\left[\frac{db}{dx}\right]_{x=x_m} + Tk^2 \right\} \tag{18}$$

The conservation equation at the interface (equation 14) determines the time dependence of ε and η. Namely

$$\frac{d\varepsilon}{dt}\left(1 - F(Ca)\right) = -\,\varepsilon(t)\,\frac{Q}{b^2(x_m)}\left[\frac{db}{dx}\right]_{x_m} - \frac{b^2(x_m)}{12\mu}\, q'_k(x_m)\,\eta(t) \tag{19}$$

where we denote $dq_k/dx = q'_k$. So ε varies exponentially ($\varepsilon \sim \exp(\sigma t)$) with σ given by

$$\sigma = \frac{1}{1 - F(Ca)}\left\{ -\frac{Q}{b^2(x_m)}\left[\frac{db}{dx}\right]_{x_m} + \frac{b^2(x_m)}{12\mu}\,\frac{q'_k(x_m)}{q_k(x_m)}\left[+ \left[\frac{d\,P^{(0)}}{dx}\right]_{x_m} - \frac{2T}{b^2}\left[\frac{db}{dx}\right]_{x_m} - Tk^2 \right]\right\} \tag{20}$$

In order to obtain the instability parameters (critical velocity, wavelength at the threshold) in a given geometry one should compute $q_k(x)$ (equation 17). For simplicity, we make the approximation of computing $q_k(x)$ as if $(1/b)(db/dx)$ was constant (exponential profile). That is, we define

$$\frac{1}{h} = \frac{3}{2b}\left(\frac{db}{dx}\right)_{x=x_m} \tag{21}$$

and take

$$q_k \sim e^{m_k x} \quad \text{with} \quad m_k = \frac{1}{h}\left(\sqrt{1 + h^2 k^2} - 1\right) \tag{22}$$

This is only approximate as experimentally (hk) is of order one.

The computed critical velocities as a function of the gap between the cylinder and the plate are shown on figure 4. The agreement is fairly satisfactory for small gaps, but deteriorates for larger gaps. This may be due to the fact that the critical velocities are increasing with the gap so that the capillary number at threshold is increasing (from a value of order 0.1 for $b_o = 0.1$ mm to a value of order 1 for $b_o = 1$ mm) and the hypotheses of our computation are breaking down (especially equation 11). Similar comments apply also to the wavelengths at threshold shown on figure 5. We have also plotted the linear stability boundaries (equation 20) and the most unstable wavelength on figure 6, together with the experimentally observed wavelength of the non linear rippled interface. Surprisingly enough, the experimental points are right on the marginal stability boundary and would thus seem to be in the Eckaus unstable region. It remains to be seen whether this is due to the approximate nature of our calculation or to the fact that the boundary of the Eckaus instability tends to the linear stability boundary as the amplitude of the cells grows.

V. CONCLUSION

It may be appropriate to conclude by emphasizing the similarity between the physics and the equations that we have described, and those of directional solidification. In both cases, one has a *diffusion-controlled* instability restabilized by the variation of an external parameter. The role played there by impurity concentration is played here by the pressure. The relation between the velocity of oil and the pressure (equation 3) gives for the pressure a diffusion-like equation analogous to the one governing the diffusion of impurity. Here and there the gradient of the diffusing field at the interface gives it its motion through a conservation equation for the fluxes (flux of oil or impurity flux) and this generates the instability. For directional solidification large wavelengths are stabilized by the externally varying temperature. In the experiment described here they are stabilized by the variation of cell thickness. Surface tension is responsible for the stabilization of short wavelengths in both cases. Each experimental situation has its own advantages. In directional solidification the most usual experiments, performed with organic materials, show a sub critical threshold of the front instability (de Cheveigné *et al* 1986). A direct bifurcation to the cellular state, though it is believed to exist also in metals, has only been observed in the case of liquid crystals (Bechhoeffer *et al* 1988). Here, on the contrary, the experimental results lead us to conclude that it is indeed a normal bifurcation (note also, in support of this result, the good agreement of the experimental thresholds with the ones given by the linear analysis). A non linear analysis of the equation remains of course to be done. We hope that the viscous fingering experiment described here can help to shed some light on the behaviour of these cellular structures and in particular on the origin of wavelength selection.

ACKNOWLEDGMENT

We acknowledge stimulating discussions about this work with C. Caroli, B. Caroli and B. Roulet.

Note added in proofs

Since this work has been completed, we found out that M.D. Savage (J. Fluid Mech. **80**, 743 (1977)) had already made a linear analysis of this instability, with a different assumption for the position of the meniscus. The results of our calculation are in complete agreement with his. Further references to previous works on this problem can be found in K.J. Ruschak (Ann. Rev. Fluid Mech. **17**, 65 (1982)).

References

W. H. Bank and C. C. Mill (1954), Proc. Roy. Soc. London, A **223**, 414.
J. Bechhoefer, A. Simon and A. Libchaber (1988), in these proceedings.
E. Ben-Jacob, R. Godbey, N. D. Goldenfeld, J. Koplik, H. Levine, T. Mueller and L. M. Sander (1985), Phys.Rev. Lett. **55**, 1315.
B. Billia (1987), J. of Crystal Growth **82**, 747 and these proceedings.
F. P. Bretherton (1961), J. Fluid Mech. **10**, 166.
S. de Cheveigné, C. Guthmann and M. M. Lebrun (1986), J. Physique, **47**, 2095 and these proceedings.
T. Dombre and V. Hakim (1987), Phys.Rev. A, **36**, 2811.
D. A. Kessler and H. Levine (1988 a), Adv. in Physics **37**, 255.
D. A. Kessler and H. Levine (1988 b) Preprint.
K. A. Jackson and J. D. Hunt (1966) Trans. Met. Soc. of AIME **236**, 1129.
L. D. Landau and B. Levich (1942), Acta Phys. Chim. URSS, **17**, 42.
J. S. Langer (1980), Rev. Mod. Phys. **52**, 1.
J. S. Langer (1988), preprint.
H. Martin (1916), Engineering **102**, 119.
A. D. McEwan and G. I. Taylor (1966), J. Fluid Mech. **26**, 1.
H. Muller Krumbhaar, these proceedings.
W. W. Mullins and R. F. Sekerka (1964), J. Appl. Phys., **35**, 444.
C. W. Park and G. M. Homsy (1985), Phys. Fluids **28**, 1583.
J. R. A. Pearson (1960), J. Fluid Mech. **7**, 481.
E. Pitts and J. Greiller (1961), J. Fluid Mech. **11**, 33.
M. Rabaud, Y. Couder and N. Gerard (1988) Phys. Rev. A, **37**, 935.
D. A. Reinelt and P. G. Saffman (1985), Journal Sci. Stat. Comp. **6**, 542.
P. G. Saffman and G.I. Taylor (1958), Proc. Roy. Soc. London, A **245**, 312.
P. Tabeling, G. Zocchi and A. Libchaber (1987), J. Fluid Mech. **177**, 67.
G. I. Taylor (1963), J. Fluid Mech. **16**, 595.
R. Trivedi (1984), Metal. Trans. A **15**, 977 and Part I, p. 967.
L. H. Ungar and R. A. Brown (1985), Phys. Rev. B **31**, 5931.

NEW RESULTS IN DENDRITIC CRYSTAL GROWTH

J. Maurer, P. Bouissou, B. Perrin and P. Tabeling

Groupe de Physique des Solides de l'ENS
24, rue Lhomond
75231, Paris (France)

1 INTRODUCTION

Crystal growth experiments have led, for a few years now, to clarify the characteristics of the dendrites from both the geometrical and dynamical points of view[1]. In the present study, we test the classical results concerning crystal shape selection in new physical situations : (i) when there are facets along the flanks of the dendrite, and (ii) in the presence of a controlled hydrodynamic flow around the crystal.

2 EXPERIMENTAL SYSTEM

The liquid phases from which crystals grow are either aqueous solutions of ammonium bromide[2] or alcohol solutions of pivalic acid[3]. In all the experiments, the operating temperatures range from 10 to 50°C ; this corresponds to situations well below the roughening transition in the case of NH_4Br and well above the roughening transition in the case of PVA.
The experimental system is composed of a cell (either circular or parallepipedic), which is entirely immersed in a circulation of water, regulated thermally within 20 mK. For PVA experiments, we allow for a circulation of the fluid in the direction parallel to the larger side of the rectangular cell. For NH_4Br experiments, the cell is sealed. The images of the growing crystals are sent from a microscope to a video camera, and recorded on a video tape. They are further digitized at a resolution of 512 by 512 pixels, leading to an overall magnification ranging from .2 to .5 μm/pixel.

3 RESULTS ON AMMONIUM BROMIDE SYSTEM : FACETED DENDRITES

At large velocities (a few microns per seconds), the shape of crystals growing along (100) direction looks like an usual dendrite : the tip is parabolic and side branching develops, at a few radii of curvature behind. At intermediate velocities, the crystal has an overall parabolic shape but facets, developing in (110) direction are perceptible ; they become clearly visible at smaller velocities. Because of the coexistence of a parabolic shape and facets, we call them "faceted dendrites". At small velocities, the crystal is entirely faceted.
The sizes of the facets appearing on the surface of the crystal, close to the tip, are estimated by using the following method : we first determine a parabola which fits the interface over distances appreciable compared to its own tip radius of curvature. Further, we estimate the deviation λ between the parabola and the interface at the tip. A simple geometric model, assuming that the facets lie along (110) plane and match tangentially the interface gives $l_f = 2\sqrt{2}\,\lambda$, where l_f is the facet size. Therefore, measuring λ allows for determining facet sizes. The dependance of λ with growth velocity V is then studied. We find that λ continuously decreases with increasing V,

New Trends in Nonlinear Dynamics and Pattern-Forming Phenomena
Edited by P. Coullet and P. Huerre
Plenum Press, New York, 1990

following approximately a power law. The exponent, calculated by using least mean square method leads to a law of the form :

$$\lambda \sim V^{-.5 \pm .1},$$

Thus, there seems to be no sharp transition leading to a disappearance of the facets as velocity is raised up, but rather a continuous decrease of their size. The so-called "dynamical roughening" phenomenon amounts, in this experiment, to a gradual reduction of the extension of the facets on the crystal interface.
Another interesting result - concerning now the dynamics and not the geometry - is the dependance of product ρ^2V with velocity. We usually find two regimes : one for the range of large velocities, for which product ρ^2V is a constant. This regime corresponds to the case where the crystal interface looks rough, i.e there are no facets along the flanks of the dendrite.Another regime, at small velocities, where the product ρ^2V is also a constant, but the constant value is different from the preceding one. In this case, facets are clearly visible on the flanks of the dendrite. Between these two limits, product ρ^2V is velocity dependant.

4 RESULTS ON PIVALIC ACID SYSTEM : INFLUENCE OF AN EXTERNAL FLOW

Experiments on pivalic acid are performed in the presence of an external controlled flow. All the experiments are performed in a parapepipedic cell with a fluid thickness equal to 1 mm. A typical dendritic shape observed in the presence of a forced flow looks like that shown in Picture 1. In this case, the external hydrodynamic velocity U is 15 µm/s and the crystal growth velocity V is 0.8 µm/s. One observes that, even in the presence of the external flow, the shape of the tip looks parabolic. On the other hand, side branching develops dissymetricaly along the sides : we thus observe - in agreement with earlier studies - that external flows favour side-branching development.

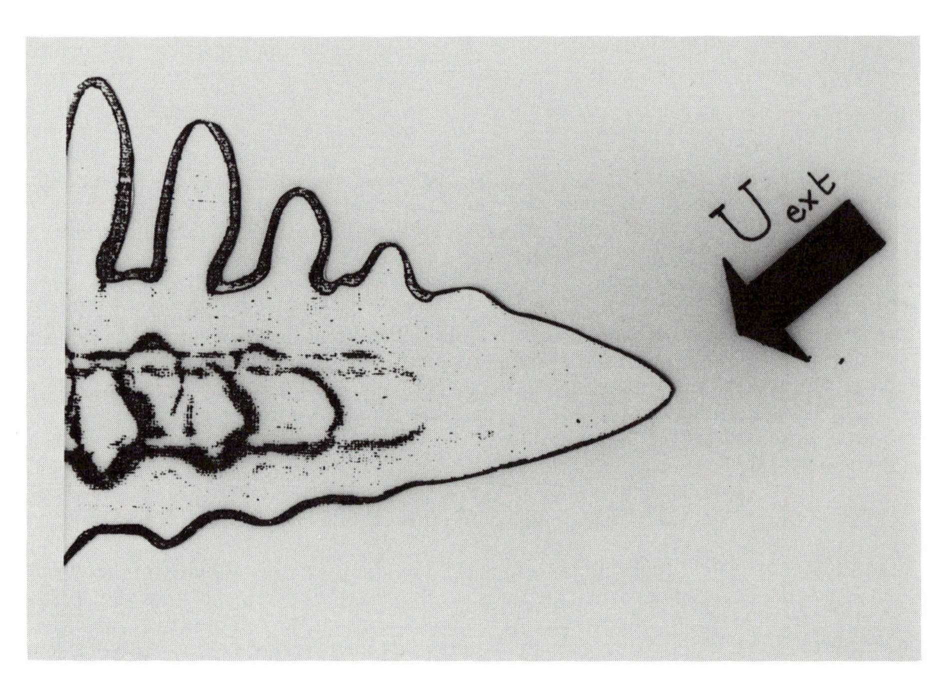

Picture 1 . Pivalic acid dendrite growing along (100) direction at 0.8 µm/s, in the presence of an external flow with a velocity of 15 µm/s

We further have studied the quantity ρ^2V as a function of V in two cases : with and without external flow. In the no flow case, a plateau is observed. The remarkable fact is that when there is a flow ρ^2V remains constant - but at a higher level. Thus it appears that external flows do not modify the form of the power law relating ρ to V but only changes the numerical prefactor involved in the relation. Moreover, we have studied the evolution of the plateau levels as a function of the transversal (i.e normal to the growth axis) and longitudinal (i.e parallel to the growth axis) components $U_\perp$, $U_{//}$ of the external flow. We first find that $\rho^2 V$ can be considered as independant of $U_\perp$. On the other hand, the influence of $U_{//}$ is studied ; results have been obtained for two different states in the phase diagram, characterized by two distinct equilibrium concentrations of impurities . In both cases, the experimental data can be reasonably approximated by linear laws in the form :

$$\rho^2 V = 2\, D d_0/\sigma\, (1 + \beta\, U_{//} d_0/D),$$

in which D is the chemical diffusion coefficient, d_0 the capillary lenght, and σ and β numerical constant, found respectively equal to 0.020 and 5300 (such estimates have been deduced by using $D = 2\ 10^{-6}\ cm^2/s$ for the chemical diffusion coefficient). The estimate for σ is in reasonable agreement with other experimental results. There is no available data for constant β so far.

5 CONCLUSION

Two new physical situations concerning dendritic growth are presented : dendrites with facets along the flanks and dendrites in the presence of a controlled external flow. Theoretical attempts have been done to interpret some of the results which we have obtained[4]. Actually we are still far from a complete understanding of most of the striking phenomena that we have observed.

ACKNOWLEDGMENTS

We acknowledge Y. Pomeau, S. Balibar, M. Benamar, P. Pelce, C. Caroli for many stimulating discussions during this experimental work.

REFERENCES

1. See for instance S.C. Huang, M.E. Glicksman, Acta Metall. 29, 717 (1981) ; A. Dougherty, J. P. Gollub Phys. RevA (to be published).
2. J. Maurer, P. Bouissou, B. Perrin, P. Tabeling, Europhysics Letters (to appear, 1988).
3. P. Bouissou, B. Perrin, P. Tabeling, preprint 1988
4. M. Ben Amar, P. Bouissou, P. Pelcé, to appear en J. Crystal Growth (1988); M. Ben Amar, Y. Pomeau, to appear in Europhysics Letter (1988).

MICROSTRUCTURAL TRANSITIONS DURING DIRECTIONAL SOLIDIFICATION OF A BINARY ALLOY

Bernard Billia*, Haïk Jamgotchian* and Rohit Trivedi**

*Laboratoire de Physique Cristalline (U.A.797), Faculté de St Jérôme, Case 151, 13397 Marseille Cedex 13, France
**Ames Laboratory USDOE and Department of Materials Science and Engineering, Iowa State University, Ames, IA 50011, USA

INTRODUCTION

Controlled solidification is one of the major techniques to produce engineering components as a close control of the microstructure, and thus of the properties of the material, is allowed. Therefore, it is important to establish the precise correlation between the microstructure and the processing conditions.

In directional solidification, the growth rate V, thermal gradient G and solute concentration C_∞ are the control parameters.For given G and C_∞, the interface profiles change from planar to cellular to dendritic as the velocity is increased from a small value. It is appropriate to refer to the variation of the primary spacing λ to define these morphological transitions. Indeed, such criteria become very clear on the schematic plot of λ versus the distance ν from the onset of morphological instability (Fig. 1). The transition from small amplitude cells (C) to deep cells (DC), also called dendritic cells, occurs at the minimum in the curve, wheras the transition from DC to dendrites (D) occurs at the maximum in λ.

The aim of the paper is to examine critical results on succinonitrile-acetone alloys and develop appropriate criteria for these morphological transitions. All the physical properties have been measured for this system and the experiments have been carried out in Hele-Shaw cells where convection is negligible (Esaka and Kurz, 1985; Esaka, 1986; Somboonsuk et al., 1984; Eshelman et al., 1985).

3D-REPRESENTATION OF THE STRUCTURE PARAMETERS

The morphology of the nonplanar interface is determined by thermal diffusion, solute diffusion and capillarity which are respectively characterized by the thermal length $\mathbf{l_t}$, solutal length $\mathbf{l_s}$ and capillary length $\mathbf{d_o}$ (Trivedi, 1980; Billia et al., 1988). For a given solute partition coefficient k, the dimensionless shape parameters (primary spacing λ^*, tip radius R^* ...) depend only on the relative magnitude of these three length scales, i.e. on the distance $\nu = \mathbf{l_t} / \mathbf{l_s}$ and on the Sekerka number $A = k\,\mathbf{d_o} / \mathbf{l_s}$, so that 3D-representations are possible.

For the sake of brievity, only a nondimensional spacing linked to classical capillary parameters (Kessler et al., 1986; Ben Amar and Moussalam, 1988)

$$\lambda^* = P_\lambda\,[k(\nu-1)/\nu A]^{1/2}\ , \qquad (1)$$

New Trends in Nonlinear Dynamics and Pattern-Forming Phenomena
Edited by P. Coullet and P. Huerre
Plenum Press, New York, 1990

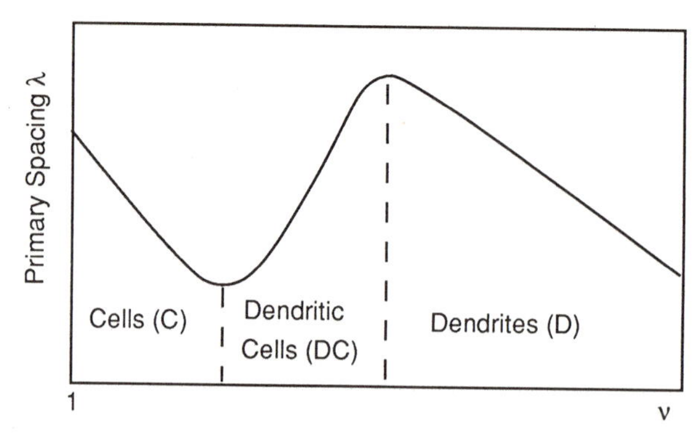

Fig. 1. Diagram showing the general variation in the primary spacing λ with the distance ν from the onset of morphological instability, for given G and C_∞.

is considered in the following, where P_λ is the Peclet number based on the spacing. By tabulating and then displaying the data points on a computer, rather flat sheets are evidenced for both the celular and dendritic regimes which, after proper rotations can be seen from the side, perpendicular to the plane of the figure (Fig. 2). Under a planar approximation, the third dimension can then be ignored and lines fitted to cells and dendrites. Going back to ν and A, two relationships are obtained, which are labelled by subscripts c and d

$$\lambda_c^* = 8.07\ A^{-0.04}\ \nu^{1.09} \tag{2.a}$$

$$\lambda_d^* = 5.24\ A^{-0.22}\ \nu^{0.38} \tag{2.b}$$

The cell to dendrite transition corresponds to $\lambda_c^* = \lambda_d^*$, i.e. to

$$\nu = 0.54\ A^{-0.25} \tag{3.a}$$

or

$$l_s = d_o^{1/5}\ l_t^{4/5} \tag{3.b}$$

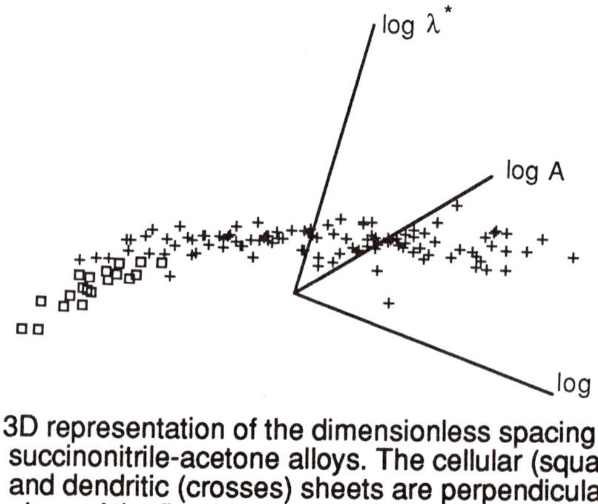

Fig. 2. 3D representation of the dimensionless spacing λ^* for succinonitrile-acetone alloys. The cellular (squares) and dendritic (crosses) sheets are perpendicular to the plane of the figure.

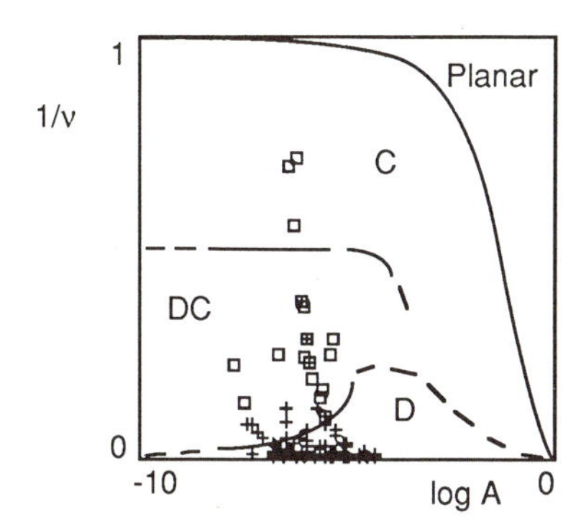

Fig. 3. Microstructural stability diagram for succinonitrile-acetone alloys (k=0.1). □ : cells, +: dendrites.

Relation 3a can be mapped to give a microstructural stability diagram (Fig.3). For constant values of G and C_∞, the variation of λ with V can be deduced from eq. 2a. The coefficient $(\nu -1)/\nu$ in the definition of λ^* leads to a minimum for $\nu = 2$, which fits well with the few available data for low amplitude cells so that it is shown on Fig. 3 as a tentative demarcation between the C and DC regimes. Dashed lines correspond to the limits which are most likely from the present knowledge about cellular and dendritic domains, but out of the range of experimental measurements on the succinonitrile-acetone system.

A complete development of this analysis of the cell to dendrite transition will be presented in a forthcoming paper (Billia and Trivedi, 1988).

REFERENCES

Ben Amar, M., and Moussalam, B., 1988, Absence of selection in directional solidification, Phys. Rev. Letters, 60:317.

Billia, B., Jamgotchian, H., and Capella, L., 1988, Unifying representations for cells and dendrites, submitted.

Billia, B., and Trivedi, R., 1988, Cellular and dendritic regimes in directional solidification, submitted.

Esaka, H., and Kurz, W., 1985, Columnar dendrite growth: experiments on tip growth, J. Crystal Growth, 72:578.

Esaka, H., 1986, "Dendrite growth and spacing in succinonitrile-acetone alloys", PhD Thesis, Ecole Polytechnique Fédérale de Lausanne, Lausanne.

Eshelman, M. A., Seetharaman, V., and Trivedi, R., 1988, Cellular spacings - I. Steady-state growth, Acta Met., 36:1165.

Kessler, D. A., Koplik, J., and Levine, H., 1986, Dendritic growth in a channel, Phys. Rev., 34A:4980.

Somboonsuk, K., Mason, J. T., and Trivedi, R., 1984, Interdendritic spacing: Part I. Experimental studies, Met. Trans., 15A:967.

Trivedi, R.,1980, Theory of dendritic growth during the directional solidification of a binary alloy, J. Crystal Growth, 49:219.

NONSTATIONARY CELL SHAPES IN DIRECTIONAL SOLIDIFICATION

S. de Cheveigné, C. Guthmann and P. Kurowski

Groupe de Physique des Solides de l'E.N.S.
Université Paris VII
Tour 23, 2 Pl. Jussieu, 75251 Paris Cedex 05

In directional solidification, a sample, initially liquid, is pulled at a given velocity in a temperature gradient in such a manner as to solidify it progressively. If the material is not pure - in the present case we consider dilute binary alloys - the solid liquid interface presents a cellular instability above a certain critical pulling speed (Langer 1970). This deformation is associated with the segregation of solute in the cusps behind the cells. The mechanism behind the instability was explained by Mullins and Sekerka (1964): it is the competition between the destabilizing effect of solute diffusion and the stabilizing effects of the temperature gradient and the interfacial tension. To allow direct observation of the phenomenon, we use thin (50 μ) samples of a transparent material, tetrabromomethane, containing 0.12 % excess brome. The experimental set-up is described elsewhere (de Cheveigné et al, 1986).

The observations described here concern a secondary instability: the periodic formation of liquid droplets in the cusp behind the cells (Fig 1). These droplets are strongly enriched in solute and therefore remain liquid as they are advected back into the solid. They do not in fact remain immobile in the solid but migrate back towards the solid liquid interface by temperature gradient zone-melting (de Cheveigné et al, 1988).

The details of the shape of the solid-liquid interface are not accessible to analytic calculations, especially in a strongly non-linear regime in which the interface is clearly non-sinusoidal. Numerical calculations on the other hand are very promising (Ungar and Brown, 1985, Müller-Krumbhaar, 1988, Levine, 1988). One interesting feature that appears in various models is a droplet at the bottom of deep cusps. The study of an eventual instability of this droplet is nevertheless beyond the present scope of such calculations.

The formation of solute-rich inclusions in directionally solidified metals has already been observed (they should be distinguished from the liquid trapped between dendrite branches) but little is known about the exact conditions of production. We have determined that, generally speaking, their apparition is related to variation or modulation of the pulling speed V, probably via the necessary adjustment of the cellular wavelengths. (The wavelength varies as $V^{-1/2}$ (de Cheveigné et al, 1986)).

New Trends in Nonlinear Dynamics and Pattern-Forming Phenomena
Edited by P. Coullet and P. Huerre
Plenum Press, New York, 1990

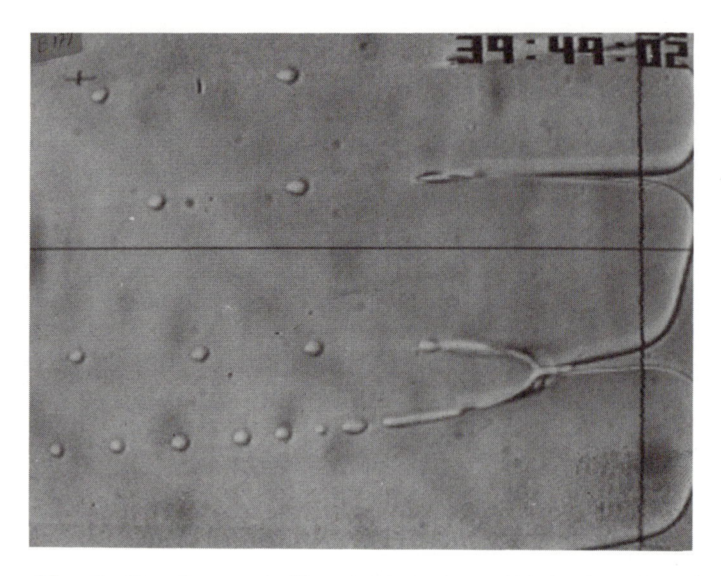

Fig. 1. Droplet production behind the solidification front.

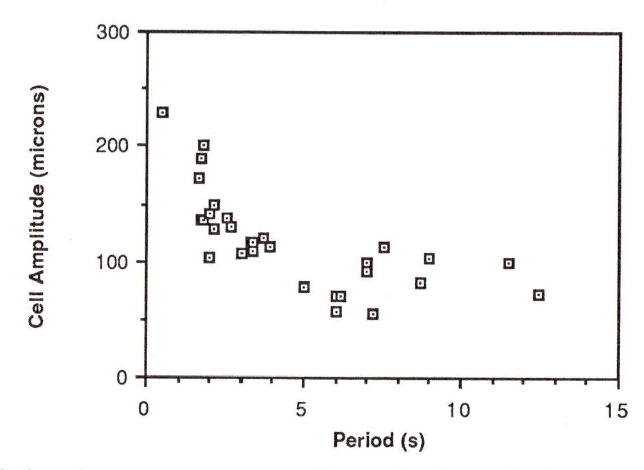

Fig. 2. Relation between average cell amplitude and time interval between droplets

To be more precise, one must distinguish different cell amplitudes.

a) low amplitude cells: (≈ 1-3 λ). These cells do not spontaneously emit droplets. Emission in a few cells can be provoked by modulation of the pulling speed (of the order of 10% at frequencies of about 0.02 Hz). The frequency of emission (0.1 to 2 Hz) is much higher than that of the the modulation and is related to the cell amplitude (Fig 2). The phenomenon is not stationary: cells start or stop emitting, apparently spontaneously. The droplets appear in a manner very similar to drops appearing one by one at the edge of a faucet.

b) Large amplitude cells: (≈ 10 λ). The emission is much easier to provoke, either by modulation or by a sudden variation of the pulling speed. The cells respond within a few seconds, but are very slow in stopping emitting (1/2 hour). This emission is perhaps due to a perturbation initiating at the cell tip and advecting back to the cusp. Nearly all the cells emit again at a frequency much higher than the external modulation. In this case the droplets are produced by what looks like a Rayleigh instability of the liquid trail behind the cusp (Goedde and Yven, 1970). In such a mechanism, the periodicity of the droplets l_D is related to the radius r of the liquid tube by the relation $2\pi r/l_D = 0.697$. We find the ratio to be in the range of 0.5 to 0.8. It is interesting to observe that such an instability normally due to air-liquid interfacial tension effects takes place in a similar manner under the effect of the solid-liquid interfacial tension. It should also be noted that these observations imply that the cusp is not a two dimensional curtain but rather is like a funnel pointing back into the solid.

As mentioned above, the emission of droplets seems related to an adjustment of the wavelength (Fig 1). This adjustment takes place by the splitting or the disappearing of a cell. Before this takes place, the system can be pushed fairly far from its equilibrium wavelength, perhaps because there is a much longer response time to wavelength adjustment than for example to amplitude adjustment. For example, for a modulation at 0.02 Hz of 80% amplitude, the wavelength remains roughly constant.

In conclusion, the emission of droplets seems related to the adjustment of cell shape and wavelength, and illustrates the great sensitivity of the cell cusp to external perturbations.

References

de Cheveigné, S., Guthmann, C. and Lebrun M.M.,1986 Cellular instabilities in directional solidification,J. Physique, 47 : 2095.

de Cheveigné, S., Guthmann, C. and Kurowski, P.,1988, Liquid droplet production and migration in directional solidification , in preparation.

Goedde E.F., and Yven, M..C (1970), J. Fluid Mech. 40 : 495

Langer,J.S., 1980, Instabilities and pattern formation in crystal growth, Rev. Mod. Phys. 52 : 1

Levine,H., 1988, preprint.

Müller-Krumbhaar, H.,1988 private communication.

Mullins, W.W. and Sekerka, R.F. , 1964, Stability of a planar interface during solidification of a dilute binary alloy , J. Appl. Phys. 35 : 444.

Ungar,L. and Brown R.A., 1985, Cellular interface morphologies in directional solidification. IV. The formation of deep cells, Phys. Rev. B, 31 : 5931.

INDEX

ABC-flow, 92
Absolute instability, absolute growth rate and, 260, 261
Absorbing state, 211-212
Acoustic forcing, 166-168
Active chemical medium, defined, 1
Advection, 148-149, 152
Alloy solidification, 343-345
Alpha-effect, 313
Amplitude equations, 27-31, 137-138
  anisotropic systems, 29-31
  degenerate, 69-71
  interfacial instabilities in presence of transverse electric current, 101-102
  isotropic systems, 27-29
  localized structures generated by subcritical instabilities,85
  transition to periodic spatially uniform state, 185
Amplitude turbulence, 141
Anisotropic Kinetic Alpha (AKA) effect, 313-314
Anisotropic systems, amplitude equations and pattern selection, 29-31
Annulus, 237
  collective oscillations, 230
  confined state formation, 77-82
  spatiotemporal intermittency, 193-198
Antidefects, 183
Antikinking solutions, 31
Archimedean spirals, 25
Aspect ratio, 78, 199-200
Asymptotically stable waveforms, 3
Asymptotic regimes, 200, 313-314
Asymptotic velocity, chemical wave, 4
Attractors, 200
Axisymmetric convection, 73-75
Axisymmetric jet, 317

Bardeen-Herring sources, 33
Basic flow, 288
Belousov-Zhabotinsky (BZ) media, 1-8, 11-18, 25
Beltrami property, 92
Benjamin-Feir-Eckhaus instability, 127, 134, 176
Benjamin-Feir instabilities, 127
Benjamin-Feir limit, 120
Benjamin-Feir mechanism, 268
Benjamin-Feir-Newell (BFN) instability, 80, 81
Benjamin-Feir resonance, 113
Benjamin-Feir stable range, 118
Bernouilli shift, 210
Beta model, 317
Binary alloy solidification, 343-345
Binary convection
  confined states, 77-82
  degenerate amplitude equation near codimension-2 point in, 9-71
  football states, 81
  noise sustained structures, 79-82
  slugs, 80
  traveling waves, 55-60, 77
Biot-Savart model, 322
Bistable media, chemical, 1
Blob oscillation, 38, 40
Boundary conditions
  liquid crystal patterns and defects, 120
  rigid, 201
  single vortex, 190
  torque free, 120
  traveling waves in binary convection, 58-60
  vortex dynamics in coupled map lattice, 186
Boundary layer bursts, three-dimensional flows, 321
Boussinesq equations, 56, 147
Bridgman technique, 287
Burgers nonlinearity, 216
Bursting phenomena, 80, 321
Busse balloon, 146, 149
Busse instability, 82, 133

Cascade, inverse, 223, 313-314
Cascade collapse efficiency, 28
Cell-like regions, mixing layers, 163
Cell shapes; *see also* Annulus
  in directional solidification, 347-349
  laminar, 201, 218, 219, 223
  KSS model, 218-219
  rectangular, 78, 81
    collective oscillation in, 230-231
    straight roll patterns, 146
Cellular automata, 202-205
  deterministic, 203, 204, 205, 210-211
  KSS model, 223-224
  minimal CML as, 210-211
  probabilistic, 205, 237-239
Cellular pattern formation, anisotropic systems, 30
Cellular solution, KSS model, 218-219
Center manifold dynamics, 199, 278, 281
Channel flow, 77

Chaos, 199
to quasiregular patterns, 89-92
spatial, dynamical, 232
transient, conversion to global sustained turbulence, 204
Chaotically oscillating front, chemical system, 22
Chaotic collective behavior, 211
Chaotic velocity spectra, 248-250
Chemical systems, one-dimensional reaction-diffusion, 21-23
Chemical waves
flow induced by, 14-15
hydrodynamic flow interaction, 11-18
in oscillatory media, 1-8
convection, 4-6
definition, 1-2
dispersion relation, 6-7
trigger wave origin, 7-8
Chomaz, Huerre, and Redekopp instability, 260
Codimension-2 point, 69-71, 77
Coherent structures, 216
in coupled map lattices, 241-242
dynamics
large-eddy simulation in 3-D mixing layer, 257
origin of, in phase transition, 221-223
KSS model, 218, 219
Collocation methods, 252
Complex envelope, 112, 199
Complex Ginzburg-Landau equation: *see* Ginzburg-Landau equation
Conductivity, spectral eddy, 253
Confined states, 77-82
Confinement effects, 199, 200, 202
Constant phase gradients, 86
Continuous phase transition, 208
Continuous spectrum for linearized problem, 275
Convection
axisymmetric, 73-75
binary fluid, 55-60, 69-71
chemical waves, 4-6, 18
electrohydrodynamic, 65-67
knot, 40-43
large-scale vortex instability in, 305
one-dimensional, 200-201
onset of time-dependence, mean flow and, 145-156
Rayleigh-Benard: *see* Rayleigh-Benard convection
solar, 216
thermal, 37-45
Convection diffusion equation, in free-free-permeable case, 57
Convection rolls, 37-42
Convective instability, 82
absolute growth rate and, 260, 261
thermal convection, 37-42
Couette-Taylor flow, 21, 276
Counterpropagating waves, 85-87, 130
Counterpropagating pulse-like solutions, 86
Coupled map lattices (CML), 202-205
mean field approach, 211-212
minimal, 206-208
Coupled map lattices (CML) (*cont'd*)
minimal (*cont'd*)
transition in higher dimensions, 211-212
transition in one dimension, 206-208
transition in two dimensions, 208-210
space-time chaos and coherent structures in, 241
transition to spatio-temporal intermittency, 237
vortex dynamics in, 185-192
Coupling, diffusion, 242
Critical Rayleigh number, binary convection, 56, 57
Cross-coupling, 80
Cross-Newell equation, 146, 154
Crystal growth, dendritic, 339-341

Damping, 80
Damping term, 216
Defects; *see also* Topological defects
complex Ginzburg-Landau equation, 181-183
nucleation, 127-128, 129, 130, 131, 176
and transition to disorder, 125-134, 137-143
symmetry, 27
vortex, natural development of, 160-166
Deformation waves, 202-205, 219
Deformed materials, defect structures, 27
Degenerate amplitude equation, 69-71
Degrees of freedom, 200
Dendritic crystal growth, 339-341
Deterministic cellular automata, 203, 204, 205, 210-211
Deterministic cellular automata rules, 238
Deterministic chaos, 215
Differential equations, homoclinic bifurcations, 295-302
Diffusion
of dislocations, 33
reaction-diffusion chemical system, 21-23
Diffusion coefficient, tunable, 21
Diffusion coupling, and tolerant structures, 242
Diffusion-free limit, 2
Diffusion-induced chaos, 23
Dimensionless cross-coupling between temperature and concentration variations, 78
Dirichlet boundary conditions, 21, 23
Discontinuous phase transition, 208, 210
Discretizations in functional integrals, 291-293
Dislocation patterns, 26-27
Dislocations
nucleation of, 176-178
slip band formation simulation, 33-36
Dispersion relation, 5, 6-7
Distortion field, 122
Dynamical instabilities, 26-27
Dynamical spatial chaos, 232
Dynamical systems theory, 215

Eckhaus-Benjamin-Feir instability, 127, 134, 176
Eckhaus instability, 38, 49, 50, 116-117, 174
Eckhaus instability limit, 284
Eckhaus mode, time-dependent, 133
Eckhaus processes, 113
Eckhaus-unstable wavenumber state, 117
Eddies, 216, 223, 251-257

Electric current, transverse, 95-102
- amplitude evolution equation, 101-102
- and interfacial instabilities, 95-102
- linear stability analysis, 96-101

Electroconvective rolls, 78
Electrohydrodynamic convection, 65-67
Electrohydrodynamic instability, 77-82
Emmons' turbulent spot, 296
Enstrophy, 321
Entanglement, 188
Euler equations, 321-324
Excitable media, chemical, 1

Faceted dendrites, 339-341
Ferroin-catalyzed BZ system, 3
Fingering, viscous, 327-337
Five-roll state, 73, 74, 75
Flame propagation, 107-109
Floquet's theorem, 38
Focusing effect, nonlinear, 80
Football states, 78, 80, 81
Forcing
- advection, 148-149, 152
- flame propagation, 108
- and global modes, 259-265, 269-271
- small-scale, anisotropic kinetic alpha effect, 313-314
- vortex defect production, 166-168

Fourier spectrum, 122
Four-roll state, 75
Frank-Read sources, 33
Free-free-permeable case, traveling waves, 57-58
Freeing, of dislocations, 33
Free shear flow, 251-257
Frequency shift, nonlinear, 67
Friction, in Kolmogorov flow, 216
Front patterns, chemical system, 21-23
Functional integrals, discretization and Jacobian in, 291-293

Galerkin coefficients, 50
Geometry, cell: *see* Cell geometry
Ginzburg-Landau equation
- complex, 79, 80, 181-183
  - defects in, 118-120
  - statistical properties of defects in, 181-183
  - vortex dynamics in coupled map lattice, 185-192
- liquid crystal patterns and defects, 118, 120
- onset of destabilization of global modes, 261
- stable solutions of, 28
- steady, mathematical justification, 275-286
- time-dependent, 133-134

Ginzburg-Landau model, linear, 260
Ginzburg-Landau type equations, liquid crystals, 112
Global instability, mean flow effects, 149
Global models, nonlinearity and forcing effects, 259-273
- concepts, 259-262
- experiment, 265-273
- theory, 262-265

Goertler instability, 77
Grain-boundaries space-time, 128
Group velocity, 260
Growth
- dendritic, 339-341
- probabilistic, 238
- in viscous fingering, 327-337

Growth rate, global modes, 260, 261

Hard turbulence, 319
Hele-Shaw cell, 147, 327
Helical convective turbulence, 305-312
Homoclinic bifurcations, 295-302
Hopf bifurcation: *see* specific systems
Hydrodynamic flow, chemical wave interactions, 11-18
Hysteresis loop, 210

Ideal pattern, 199
Incompressible flows, 313-314
Inhomogeneous propagation space, 260
Initial conditions, 186, 199
Interfacial instabilities, 95-102
Intermittency: *see* Spatiotemporal intermittency
Interpenetrating phenomenon, 238
Invariant manifold-stochastic web, 92
Invariant problem, 276
Inverse cascade phenomenon, 223, 313-314
Isola, 75
Isolated right moving pulse, 85-86
Isotropic systems, 27-29, 319

Jacobian, 291-293

Kelvin-Helmholtz flow, 101, 172, 251
Kicked oscillator, 89-90
Kinetic energy spectrum, large eddy simulation of 3-D mixing layer, 252-253
Kinetic equation, for defect concentrations, 27-28
Kink-antikink solutions, 31
Knot convection, 40-43, 45
Knot oscillations, 38
Kolmogorov log-normal model, 317
Kolmogorov-Spiegel-Sivashinsky (KSS) model, 215, 216-217, 234
- coherent structure origin, 221-223
- dynamics at large L, 219-220
- ternary reduction for, 218-219

Kosterlitz-Thouless phase transition, 185
Kuramoto-Sivashinsky (KS) model, 79, 80, 138, 200-202, 217-218, 232

Labyrinth structures, 26, 29, 30
Lagrangian turbulence, 92
Laminar cells: *see* Cell shapes
Laminar cluster, 223
Laminar shear flow, 216, 296
Laminar state, 239
- absorbing, 203
- transition to turbulence, 202, 210

Landau-Ginzburg equation: *see* Ginsburg-Landau equation

Large aspect-ratio systems, 78, 200
Large eddy simulation, 251-257
Large scale flow, 120, 148
Large-scale phase modulations, 133
Large scale structures, 305-314
Lattice models of statistical mechanics, 241-242
Lattices, 27, 119; *see also* Coupled map lattices
Leslie-Erickson equation, 65
Lewis number, 55, 71
Lifshitz point, 112, 113, 114-116
Limit cycle, 2
Linear analysis, 121
Linear friction, 216
Linear instability threshold, vortex dynamics in coupled map lattice, 186
Linear stability analysis, 28, 96-101, 137-138, 218
  interfacial instabilities in presence of transverse electric current, 96-101
  liquid crystal patterns and defects, 111
  traveling wave, 66-67
  of uniform amplitude steady state, 30-31
Linear stability problem, 276-277
Liquid crystals
  patterns and defects in, 111-123
    complex Ginzburg-Landau equations, 118-120
    Lifschitz point vicinity, 114-116
    mean flow effects, 120-122
    wavelength selection far from Lifschitz point, 116-118
  transition to disorder in, 125-134
Localized phase modulations, 127
Localized structures, subcritical instabilities and, 85-87
Local mixing, transition to turbulence, 210
Local mode description, 259, 260
Low dimensional dynamical systems theory, 199
Low temperature mixtures, 78
Lyapunov exponent, 210

Magnetic field, liquid crystal patterns, 120
Magnetic Reynolds number, 95, 98
Map: *see* Coupled map lattices
Map iteration, 199
Martenistic transformation, 133
Maxwell's equations, 95
Mean field analysis, 211-212, 239
Mean flow
  liquid crystal patterns and defects, 120-122, 123
  and onset of time-dependence in convection, 145-156
    mean flows as non-local phenomenon, 147-149
    pattern distortion and time dependence, 145-147
Mean flow field, 154
Mean velocity profile, 321
Metal solidification, 343-345, 347, 349
Metastable media, chemical, 1
Microstructural transitions, 343-345
Mixing layers, 159, 251-257
Mobile dislocations, 27
Mullins Sekerka instability, 327

Navier-Stokes equations, 95, 215, 252
  steady solution, 275-276, 279-281
  three-dimensional flow, 321, 322
Negative viscosity, 216
Newell-Whitehead-Segel equation, 117
Noise, 77-82, 119
Non-central rolls, 73, 74, 75
Nonlinear analysis
  binary convection, traveling waves in, 55-60
  thermocapillary effects, 287-289
Nonlinear focusing effects, 80
Nonlinearity, and global modes, 259-265
Nonlinear roll solution, 121
Nonlocal coupling, 152-153
Nonlocality, mean flows, 147-149
Non-potential pattern evolution model, 178
Non-universality, 210-211
Nucleation
  by dislocation, 177-178
  by phase perturbation, 176-177
  space-time, 127-128, 129, 130, 131
Nusselt number, 52, 53, 78

Oberbeck-Boussinesq equations, 47
Oblique rolls, 115, 130
Ohm's law, 95
One-dimensional lattice, 204
One-dimensional models
  chemical system, reaction-diffusion, 21-23
  convection model variant, 200
  probabilistic growth processes, 238
  Rayleigh-Benard convection, 227-234
  transition to turbulence in, 206-208
Open Couette-Flow Reactor (OCFR), 21, 23
Order parameter, 185
Oscillations
  collective, in rectangular cell, 230-231
  periodic, 21-22
  uniform, 25
Oscillator, kicked, 89-90
Oscillatory flow, in y, 288
Oscillatory instability, 38, 133
Oscillatory knot convection, 42, 45
Oscillatory media, chemical, 1

Parabolic flames, 108-109
Parity invariance, flow lacking, 313-314
Partial differential equations, 200-202, 215, 295-302
Pattern distortion, and time dependence, 145-147
Pattern evolution in 2-D mixing layers, 159-169, 171-179
Pattern selection, 27-31
  anisotropic systems, 29-31
  isotropic systems, 27-29
Peclet number, 55
Percolation
  directed, 205, 208, 210, 238
  spatiotemporal intermittency, 193-198
Periodically oscillating front, chemical system, 21-22
Periodic boundary conditions, 183, 189, 191
Persistent slip bands (PSB), 26, 29, 30, 33-36
Phase advection within stationary distortion, 152

Phase-amplitude-mean drift equation, 47
Phase defect, Rayleigh-Benard convection and, 234
Phase diffusion, 2, 115
Phase dynamics, defect-mediated turbulence, 138-139
Phase equation, 199-213
Phase field, mean flows and onset of time dependence in convection, 153
Phase gradients, localized structures, 86
Phase modes, 200
Phase modulation, space-time patterns, 127-133
Phase perturbations, nucleation of dislocations by, 176-177
Phase portraits, 22, 89-92
Phase transitions; *see also* Spatiotemporal intermittency
  from chaos to quasiregular patterns, 89-92
  probabilistic cellular automaton models, 239
  Rayleigh-Benard convection in annulus, 193-198
Phase turbulence; *see also* Spatiotemporal properties; Turbulence
  KSS model, 215, 216-217
  qualitative dynamics at large L, 217-223
    dynamical origin of coherent structures, 221-223
    KSS model, ternary reduction for, 218-219
    spatiotemporally intermittent regimes, 219-221
  quantitative approaches, 223-224
Phenomenological model, pattern evolution, 172-174
Pinning, 31, 33, 192
Pleated slow manifold, 23
Poincaré map, 297
Poincaré surface, 199, 296
Point defects, 27, 122
Poiseuille flow, 77, 296
Pole decompositions, 107-108
Power law behavior, 239
Probabilistic cellular automata (PCA), 205, 237-239
Probabilistic growth process, 238
Probability density functions, 315-319
Probability distribution, 237, 239
Propagation-space dimension, 261
Propagation-space inhomogeneity, 260, 261
Propagation velocity, 2

Quasicoherent modes, space-time patterns, 133
Quasi-one-dimensional Rayleigh-Benard convection, 227-234
Quasiperiodic velocity spectra, 245-246
Quasiregular patterns, transition to, 89-92

Random pinning, kink, 31
Rayleigh-Benard convection, 276
  at finite Rayleigh number in large aspect ratio boxes, 47-50
  liquid crystal patterns and defects, 121-122
  quasi-one-dimensional, 227-234
  spatiotemporal intermittency in, 193-198
  transition between different symmetries, 51-53
Rayleigh number
  binary convection, 56, 57
Rayleigh number (*cont'd*)
  R-α-P parameter space, 38, 39, 40
Reaction-convection couplings, 18
Reaction-diffusion systems, 1-2
  one-dimensional, 21-23
  pattern-forming instabilities, 28
  persistent slip band formation, 36
  spatial characteristics of patterns, 12-13
Reaction-transport model, 29, 30
Rectangular cells, 78, 81, 146
Reflections, 79-80, 81
Reynolds stress, bursting phenomena, 321
R-α-P parameter space, 38, 39, 40
Right moving waves, 85-86
Rigid boundary conditions, 201
Rigid-rigid-impermeable boundary conditions, 58-60
Roll patterns, 120-121
  axisymmetric convection, 73-75
  liquid crystals, 123
  mean flow effects, 149, 150
  Rayleigh-Benard convection, 51-53
Rolls, 37-40
  distortions, mean flows generation, 148
  in football states, 80
  liquid crystal patterns, 114-116
Room temperature mixtures, 78
Rotating waves, 3

Saddle-node bifurcations, 62, 74
Saffman-Taylor instability, 327-328
Scaling, universal, 315-319
Separation ratio, 78
Shear flow, 216, 251-257, 296
Shear mode, liquid crystal patterns and defects, 121
Single vortex filament, 322
Singularities, 134, 321-324
Sinks, space-time grain boundaries, 128, 132
Skew-varicose instability, 38, 49, 50, 120, 149, 174, 176
Slip bands, 26, 29, 30, 33-36
Slip planes, anisotropy effects, 27
Slow dislocations, 34
Slow manifold, 23
Slugs, 77-82
Small-scale forcing, 313-314
Soft undulations, 116
Solar convection, 216
Solidification, binary alloy, 343-349
Soret effect, 55
Spatial analysis, nonlinear, 287-289
Spatial chaos, dynamical, 232
Spatial coupling, 204
Spatial distortion, mean flow effects, 149
Spatial dynamical system, Navier-Stokes as, 279-284
Spatiotemporal chaos, 241-242
Spatiotemporal dislocations, 223
Spatiotemporal grain-boundaries, 128, 132
Spatiotemporal intermittency
  KSS model dynamics at large L, 219-221
  Rayleigh-Benard convection, 193-198, 234
  transition to turbulence via, 199-213

Spatiotemporal intermittency (*cont'd*)
transition to turbulence via (*cont'd*)
coupled map lattice and cellular automata, 202-205, 210-211
critical phenomena, 205-212
higher dimension, 211-212
one dimensional, transition in, 206-208
in partial differential equation, 200-202
two dimensions, transition in, 208-210
Spatiotemporal properties
cellular automaton, 223
in coupled map lattices, 241-242
defect-mediated turbulence, 137-143
KSS model, 223
localized injection of energy and, 260
mean flows, 145-156
non-linear waves, defects and transition to disorder, 125-134
mean flows, 145-156
Spectral eddy-viscosity, 253
Spikes, 81
Spiral-antispiral lattice, 119
Spoke pattern convection, 40
Stability domain, transition to turbulence, 210
Standing blob oscillations, 40
Standing waves, 62, 67, 128
Stationary bifurcation, 69
Stationary flow, 288
Stationary states, 22, 120, 154-155
Statistical mechanics, lattice models, 241-242
Steady four-roll state, 75
Steady Ginzburg-Landau equation, 275-286
Steady solutions of Navier-Stokes equations, 275-276, 279-281
Steady state, uniform amplitude, 30-31
Step-function approximation, 203
Stochastic web, 89
Straight roll patterns, 146
Strange attractors, 23
Streamlines, chaotic, 92
Stresses, persistent slip band formation, 33-36
Stuck pair of vortices, 191
Subcritical bifurcations, 28, 61-63, 80
Subcritical bursting, 80
Subcritical instabilities, 85-87, 200
Supercritical bifurcation, 28
Supercritical conditions, 200
Surface tension inhomogeneties, 6
Sustained turbulence, 204
Sustained waves, 2
Swift-Hohenberg model, 47-50, 218
Symmetry
competition in materials instabilities, 25-31
convection rolls, 38
transition between convective patterns, 51-53
traveling wave convection, 43-44

Target patterns, 25
Taylor-Couette flow, 159, 275
Taylor vortex flow, 21, 77, 103-105
Temporal chaos, 199
Temporally evolving patterns, 172; *see also* Spatiotemporal intermittency; Spatiotemporal patterns; Time dependency
Temporal modulation of Hopf bifurcation, 65
Thermal convection, 319
in binary fluid mixtures, 78
transitions to more complex patterns in, 37-45
instabilities of convection rolls, 37-40
three-dimensional solutions in knot convection, 40-43
Thermocapillary effects, 287-289
Three-dimensional hydrodynamics
anisotropic incompressible flow, 313-314
knot convection, 40-43
large-eddy simulation, 251-257
vortex dynamics and singularities in 3D-Euler equations, 321-324
Tiling, 89-92
Time dependency
transition to, 145-156
time-dependent Landau-Ginzburg equation, 133-134
Topological defects, 185
and spatio-temporal patterns, 140
traveling wave convection, 66
vortex streets behind tapered circular cylinder at low Reynolds number, 243-250
Topological singularities, 134
Transient chaos, 204
Transition probabilities, 237
Transition to disorder, defects and, 125-134
Transition to time-dependency, 145-146
Transition to turbulence: *see* Phase transitions; Spatiotemporal intermittency; Spatiotemporal properties
Transitions to standing waves, 67
Transverse electric current: *see* Electric current, transverse
Transverse instabilities, interfacial, 97
Traveling waves, 3, 43-44, 125
in axisymmetric convection, 73-75
in binary convection, 55-60, 71
in electrohydrodynamic convection, 65-67
temporal modulation of subcritical bifurcation to, 61-63
transition to disorder
grain boundaries, space-time, 128-133
large scale phase modulations and quasi-coherent modes, 133
nucleation of space-time dislocation, 127-128
order structures, 126
space-time complexity, 133
Trigger waves, 2, 3, 7-8
Truncation parameter, 45
Tunable diffusion coefficient, 21
Turbulence, 202, 215-224
amplitude, 141
defect-mediated, 140-143
fully developed, 315-319
hard, 319
helical convective, 305-312
KSS model, 218, 219

Turbulence (*cont'd*)
Lagrangian, 92
local mixing, 210
one-dimensional, 206-208
phase: *see* Phase turbulence
sustained, 204
transition to: *see* Phase changes; Spatiotemporal intermittency
vortex dynamics in coupled map lattice, 186
Two-dimensional systems
chemical waves in oscillatory media, 3
convection, 37-40, 227
Kolmogorov flow, 216
pattern evolution, 159-169, 171-179
probabilistic growth processes, 238
transitions to turbulence in, 208-210
vortex dynamics in coupled map lattice, 188-192

Undulated states, 114-115, 116
Uniform amplitude steady state, 30-31
Universal scaling for fully developed turbulence, 315-319

Vacancy loop density, 27
Vanishing group velocity, 260
Variable-coefficient, linear, Ginzburg-Landau equation, 261
Velocity increments, fully developed turbulence, 315-319
Velocity spectra, vortex dynamics, 243-250
Viscosity
large-eddy simulations, 252, 253
negative, 216
Viscous fingering, 327-337
Viscous fingering (*cont'd*)
equations of motion, 333-336
instability threshold and wavelength selection, 331-332
interface position, 329-331
von Karmen scaling law, 321
Vortex-antivortex pairs, 186, 189
Vortex defects
artificial, 166-168
natural development of, 160-166
2-D circular cylinder at low Reynolds numbers, 243-245
chaotic velocity spectra, 248-250
quasiperiodical velocity spectra, 245-247
Vortex dynamics, 172
coupled map lattice, 185-192
in Euler equations, 3-D, 321-324
in helical convective turbulence, 305-312
Taylor flow, 103-105
Vortex states, 186

Wave front geometry, 16-17
Wave patterns, defect-mediated turbulence in, 141-143
Wave propagation in dispersive media, 6
Weak diffusion limit, 2
Weakly confined systems, 200
Weakly nonlinear analysis, 120
Weak turbulence, 113
Williams domains, 78
WKB analysis, 261

Zigzag instability, 38, 49, 50